ATOMIC ENERGY LEVELS
and
GROTRIAN DIAGRAMS

ATOMIC ENERGY LEVELS
and
GROTRIAN DIAGRAMS

Volume I. Hydrogen I - Phosphorus XV

Stanley Bashkin
and
John O. Stoner, Jr.

Department of Physics, University of Arizona, Tucson, Arizona 85721

1975

NORTH-HOLLAND PUBLISHING COMPANY - AMSTERDAM.OXFORD
AMERICAN ELSEVIER PUBLISHING COMPANY, INC. - NEW YORK

North-Holland ISBN: 0 7204 0322 7
American Elsevier ISBN: 0 444 10827 0

Published by:

North-Holland Publishing Company - Amsterdam
North-Holland Publishing Company, Ltd. - Oxford

Sole distributors for the U.S.A. and Canada:

American Elsevier Publishing Company, Inc.
52 Vanderbilt Avenue
New York, N.Y. 10017

PRINTED IN THE NETHERLANDS

INTRODUCTION

Energy-level and transition diagrams have been indispensable since their first appearance in the literature of atomic structure. The pioneering work by Grotrian[1] was so important that his name has since been associated with pictorial representations of electronic transitions. In the years following the publication of Grotrian's books, the growth of information was so rapid that subsequent compilations were restricted to special cases, such as transitions of interest to astrophysics, or transitions within a given range of wavelengths. However, our own experience suggested that a collation of all present information concerning electronic transitions in monatomic systems would be useful to the scientific community, and we have attempted to make such a collation.

Despite our concern with the entire area of atomic spectroscopy, we have found it necessary to exclude certain parts from our compilation. Thus we have largely neglected hyperfine effects, radio-frequency spectroscopy, and inner-shell transitions. We have used principally the information contained in the energy-level compilations by Moore[2] and Kelly,[3] the spectral data in the publications by Kelly and Palumbo[4] and by Striganov and Sventitskii,[5] and a small number of papers not summarized in the foregoing works.

DIAGRAMS

The diagrams are of two kinds. One shows the energy levels, the other the electronic transitions and associated wavelengths.

a) Energy-Level Diagrams
The general format of these diagrams is quite standard — a level is represented by a short, horizontal line which is located by two coordinates. The ordinate is the level's energy, always given in inverse cm, and the abscissa indicates some combination of quantum numbers. With the exception of some hydrogenic and helium-like systems, the levels' energies are derived strictly from experimental spectra.

Preparation of these diagrams was hampered by the fact that there are numerous cases where different authors give different quantum numbers for the same level. We have simply used the values compiled by Kelly[3] and kindly supplied by him from his unpublished work. In a few instances, Kelly himself lists two or more values for a given level. Here we have either taken the most recent data, or we have indicated by, "*see Kelly", that a unique value has not yet been determined.

For the abscissa, we have used familiar combinations of spin and orbital angular momenta, in most cases, taking the final designation given by Kelly[3] with a certain exception noted below. In a number of systems, there are discrepancies between the designations listed by Kelly[3] and those used by Striganov and Sventitskii.[5] For example, in Si I, Kelly describes many levels in terms of j,l or j,j coupling, whereas the other authors use LS coupling. We have followed Kelly, and have transposed the designations of Ref. 5 into those given by Kelly.

The exception mentioned above is that we have adopted the "primed" symbols in Moore's tables,[2] whereas Kelly omits primes entirely. Sometimes Ref. 5 uses primes which do not conform to the pattern followed by Moore; we have used Moore's notation.

It became apparent with our first diagrams that the variety of levels is so great that no single format can serve to represent all of them. It was also clear that some level systems are too densely packed to permit inclusion of all the data. We have tried to exhibit each system according to a format which seems best for it, sometimes avoiding any redundancy, and sometimes deliberately including duplicate information, so as to give a clear picture. When level densities are inconveniently high, we have indicated the topmost level and one or two below until the density is at such a value that all the lower levels can be shown. The level energy as derived from experiment is given to the full precision with which it is listed in the literature. However, some of the calculated levels are given with less precision than is stated in the original source.

In many ions, a number of different core configurations produce similar final configurations. Often a single diagram suffices for displaying all the levels, but there are also instances where each core contributes so many different configurations that each core is given a separate diagram. The cores are then indicated in the title and corner labels for the diagrams. Where different cores occur, they have been specifically defined on the diagrams.

Each level diagram contains a key which defines the various symbols. We include, as well, information on the ionization level, taken from Ref. 4, and the ground configurations of the ion and the next higher ion. The j,l and j,j intermediate coupling schemes are represented by different brackets, namely, [ ] and ⁊ ⁊, respectively. *All level listings are in order of increasing excitation, and all j-values are in order of increasing excitation.* When intermediate coupling occurs, the intermediate-coupling angular momenta are shown in a vertical array, again, in order of increasing excitation, from bottom to top. While j-values are shown inside parentheses, we had to account for the fact that sometimes a single energy is listed for two or more different values of j. For example, the 3d ^{4}D term of Na V is listed as:

$$\begin{matrix} 797270 \\ 797060 \end{matrix} \text{———— 3d ([5/2,3/2], 1/2)} ,$$

which means that the lower number is common to the first two j-values and the upper number belongs to $j = 1/2$.

The ionization level is shown as a horizontal dashed line. The ionization level is simply the energy difference between the ground term and the bottom of the continuum for the terms having the ground-term core.

b) Grotrian Diagrams

Diagrams showing transitions fron one spectroscopic term to another are called, "Grotrian Diagrams". We have shown most of the transitions listed by Kelly and Palumbo[4] and by Striganov and Sventitskii,[5] as well as a few others taken from more recent publications. Sometimes the line density is too great for every line to be drawn with clarity. In such cases, we have shown the line of shortest wavelength, and as many of the others as could be done conveniently. The decision as to what to omit was arbitrary. In a number of instances, the Grotrian diagram was divided into two or more. In most instances, but not all, the same energy scale has been used for the Grotrian and energy-level diagrams.

The precision with which wavelengths have been measured is often gratifyingly high, but it was impossible to incorporate the full precision into our pictures. We have usually given wavelengths with one significant figure less than that in the literature, but there are a few cases, especially for systems drawn in the early days of this project, where all decimals have been omitted. When there are two lines in a multiplet, two numbers are given, separated by a comma; when there are more than two lines, the extreme wavelengths are given, separated by a dash. Wavelengths are in vacuum for values shorter than 2000 Å, and in air for longer values.

FUTURE WORK

We are extending the present work into the elements of higher atomic number. Drawings for the second volume are now in preparation. In addition, we hope to make revisions of the present drawings as corrections, new information, and other improvements are brought to our attention. We look forward to receiving comments from the scientific community.

SOURCES

The bibliography which is given for each stage of ionization is to be taken as an addition to the general references which are listed below. For those stages of ionization for which no bibliography appears, the information was taken entirely from the following references. Occasionally, a research paper has been identified on a particular energy-level or Grotrian diagram.

The general references are:

1. W. Grotrian, *Graphische Darstellung der Spektren*, J. Springer (Berlin 1928).
2. C.E. Moore, *Atomic Energy Levels*, Circular 467, National Bureau of Standards, Vol. 1 (1949), reprinted as NSRDS-NBS 35, Vol. 1.
3. R.L. Kelly, *Tabulation of Energy Levels for Atoms and Ions*, unpublished.
4. R.L. Kelly and L.J. Palumbo, *Atomic and Ionic Emission Lines Below 2000 Angstroms, Hydrogen Through Krypton*, (U.S. Govt. Printing Office, Washington, D.C., Stock No. 0851-00061, 1973).
5. A.R. Striganov and N.S. Sventitskii, *Tables of Spectral Lines of Neutral and Ionized Atoms*, IFI/Plenum (New York, 1968).
6. C.E. Moore and P.W. Merrill, *Partial Grotrian Diagrams of Astrophysical Interest*, Appendix A of *Lines of the Chemical Elements in Astronomical Spectra*, (Carnegie Institution of Washington Publication 610, 1958 - reprinted as NSRDS-NBS-23).

ACKNOWLEDGEMENTS

Most of the drawing and much of the assembly of bibliographies were carried out by undergraduate students and drafting assistants, nearly all of them as part-time helpers. Chief among those who had significant responsibility for the final preparation of the drawings were Christopher Hogg, John Howe, and Frank Ripley. Others in the group included: D. Abels, R. Bright, F. Camacho, A. Carlin, J. Enz, M. Ferrer, M. Gizzi, S. Heising, E. Ikeda, R. Johnson, E. Jones, G. Lim, B. Littlefield, R. Lundin, E. Meinel, R. Sherry, J. Slightom, W. Tilton, D. Toy, G. Westland, P. Wittman, R. Wong. The cover was designed by J. Howe.

We owe special thanks to the many investigators who sent us line lists and partial diagrams, often in advance of publication. Our debt to Dr. Raymond Kelly in this respect is large indeed.

Useful suggestions were also received from W.S. Bickel, J.D. Garcia, J.A. Leavitt, W.C. Martin, L.J. Radziemski, Jr., and others. All of the secretarial work was done by Mrs. Lois Davis.

Financial support was received from the U.S. Air Force, NASA, ONR, and NSF.

List of Illustrations

Please note that there has been some deliberate duplication of energy level and Grotrian diagrams. This has been done especially when the transitions have been divided among a number of Grotrian diagrams, for this arrangement permits one to see the level scheme that is appropriate for each of the Grotrian diagrams. In one case in N IV it was found convenient to use separate drawings for the singlet energy levels but one Grotrian diagram. Consequently, that Grotrian diagram appears twice.

X

XIII

XIV

XV

XVII

XIX

Hydrogen (H)

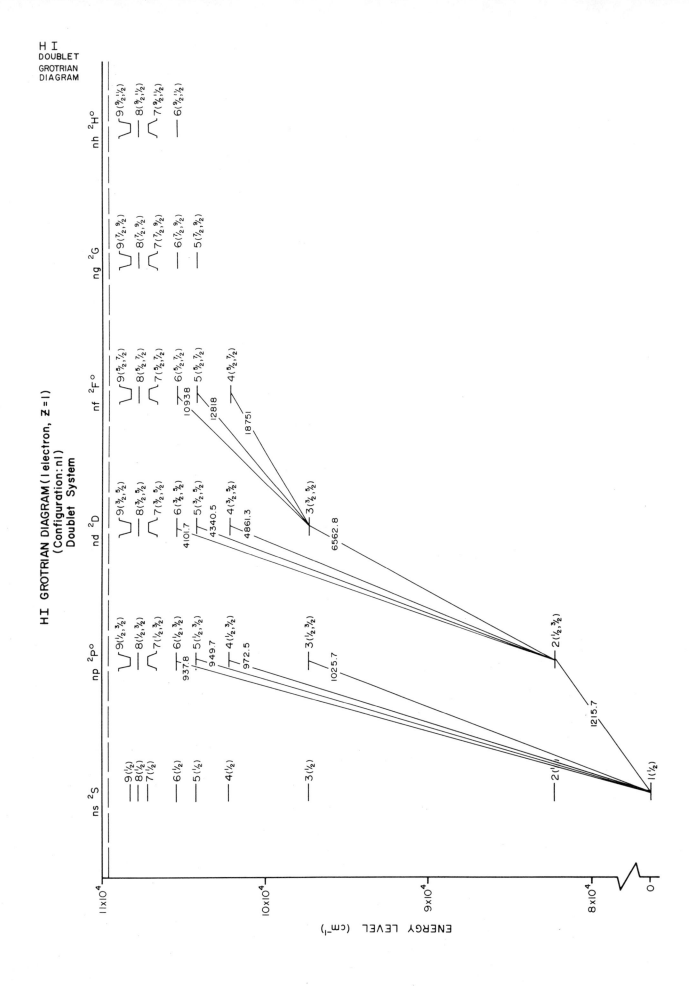

H I ENERGY LEVELS (1 electron, Z=1)
(Configuration: nl)
Doublet System

H I
DOUBLET

ns ^{2}S

108324.713 — 9(½)
107965.042 — 8(½)
107440.432 — 7(½)
106632.143 — 6(½)
105291.624 — 5(½)

102823.846 — 4(½)

97492.215 — 3(½)

82258.949 — 2(½)

0.000 — 1(½)

np ^{2}P°

108324.717 / 108324.713 — 9(½,³⁄₂)
107965.047 / 107965.042 — 8(½,³⁄₂)
107440.440 / 107440.431 — 7(½,³⁄₂)
106632.155 / 106632.141 — 6(½,³⁄₂)
105291.645 / 105291.621 — 5(½,³⁄₂)

102823.887 / 102823.842 — 4(½,³⁄₂)

97492.313 / 97492.205 — 3(½,³⁄₂)

82259.279 / 82258.913 — 2(½,³⁄₂)

nd ^{2}D

108324.718 / 108324.717 — 9(³⁄₂,⁵⁄₂)
107965.049 / 107965.047 — 8(³⁄₂,⁵⁄₂)
107440.442 / 107440.440 — 7(³⁄₂,⁵⁄₂)
106632.159 / 106632.155 — 6(³⁄₂,⁵⁄₂)
105291.653 / 105291.645 — 5(³⁄₂,⁵⁄₂)

102823.902 / 102823.887 — 4(³⁄₂,⁵⁄₂)

97492.349 / 97492.313 — 3(³⁄₂,⁵⁄₂)

nf ^{2}F°

108324.719 / 108324.718 — 9(⁵⁄₂,⁷⁄₂)
107965.050 / 107965.049 — 8(⁵⁄₂,⁷⁄₂)
107440.444 / 107440.442 — 7(⁵⁄₂,⁷⁄₂)
106632.162 / 106632.160 — 6(⁵⁄₂,⁷⁄₂)
105291.657 / 105291.653 — 5(⁵⁄₂,⁷⁄₂)

102823.910 / 102823.902 — 4(⁵⁄₂,⁷⁄₂)

ng ^{2}G

108324.719 — 9(⁷⁄₂,⁹⁄₂)
107965.051 / 107965.050 — 8(⁷⁄₂,⁹⁄₂)
107440.445 / 107440.444 — 7(⁷⁄₂,⁹⁄₂)
106632.163 / 106632.161 — 6(⁷⁄₂,⁹⁄₂)
105291.659 / 105291.656 — 5(⁷⁄₂,⁹⁄₂)

nh ^{2}H°

108324.719 — 9(⁹⁄₂,¹¹⁄₂)
107965.051 / 107965.051 — 8(⁹⁄₂,¹¹⁄₂)
107440.445 / 107440.444 — 7(⁹⁄₂,¹¹⁄₂)
106632.164 / 106632.163 — 6(⁹⁄₂,¹¹⁄₂)

KEY

105291.659 }
105291.656 } — 5(⁷⁄₂,⁹⁄₂)

J, Lowest to Highest
n
Energy Level, (cm^{-1})

Ionization Level
109678.764 cm^{-1}
(13.595 electron volts)
[H I is ^{2}S$_{1/2}$ → NUCLEUS]

ENERGY LEVEL (cm^{-1})

11×10^4
10×10^4
9×10^4
8×10^4
0

H I ENERGY LEVELS (1 electron, Z=1)
F.S., H.F.S., LAMB SHIFTS
(NOT TO SCALE)

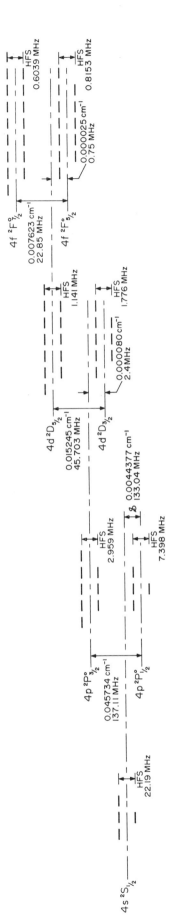
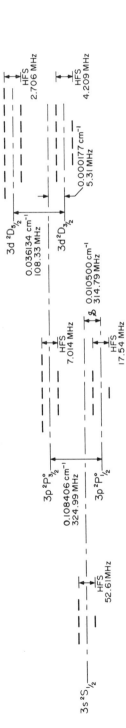
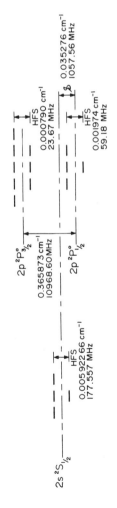

REFERENCES:
J. D. Garcia and J. E. Mack, J. Opt. Soc. Amer. 55, 654 (1965)
B. N. Taylor, W. H. Parker, D. N. Langenberg, Rev. Mod. Phys. 41 (1969)

Bibliography

H I $Z = 1$ 1 electron

Please see the general references.

Note that levels with the same value of principal quantum number (n) but different values of l have nearly the same energies. Thus all wavelengths for transitions between $n = n_1$ and $n = n_2$ are nearly independent of the values of l involved, and only one transition is shown for any pair (n_1, n_2).

Helium (He)

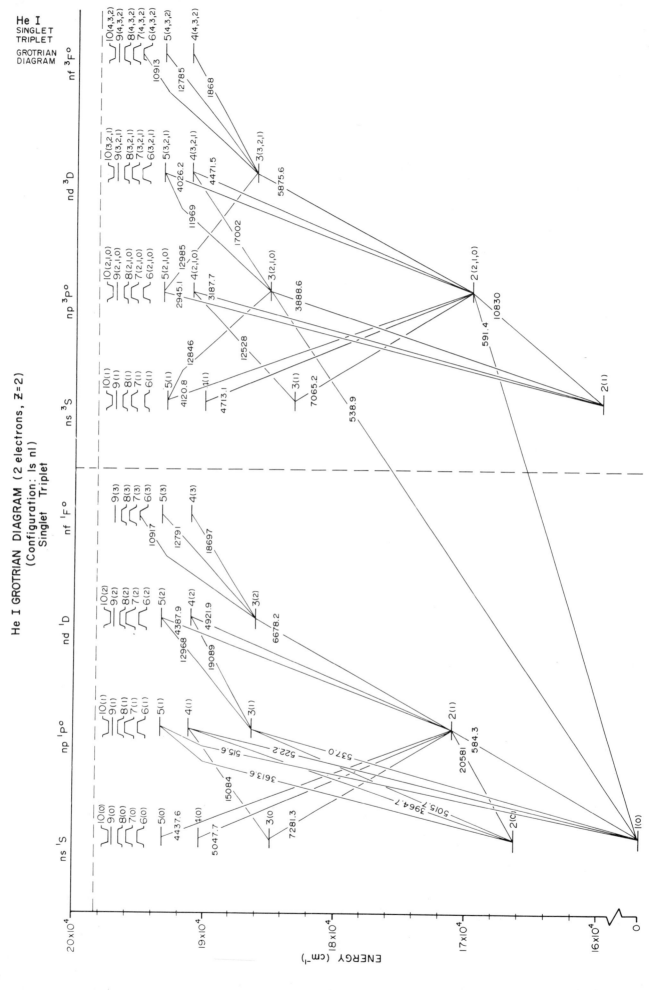

He I GROTRIAN DIAGRAM (2 electrons, Z=2)
(Configuration: 1s nl)
Singlet   Triplet

He I
SINGLET
TRIPLET

GROTRIAN
DIAGRAM

ENERGY (cm⁻¹)

8

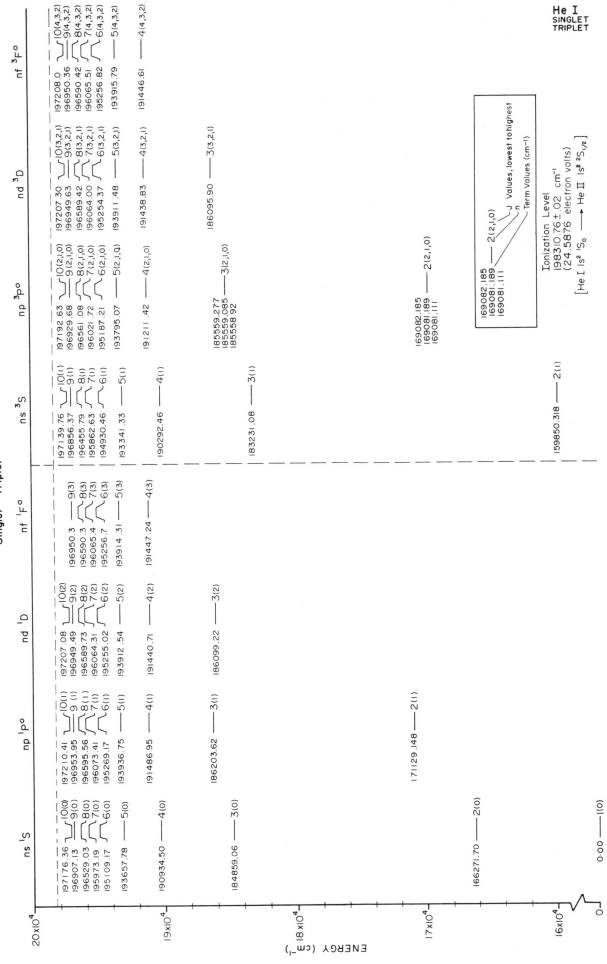

He I ENERGY LEVELS (2 electrons, Z=2)
(Configuration: 1s nl)
Singlet Triplet

He I
SINGLET
TRIPLET

6

9

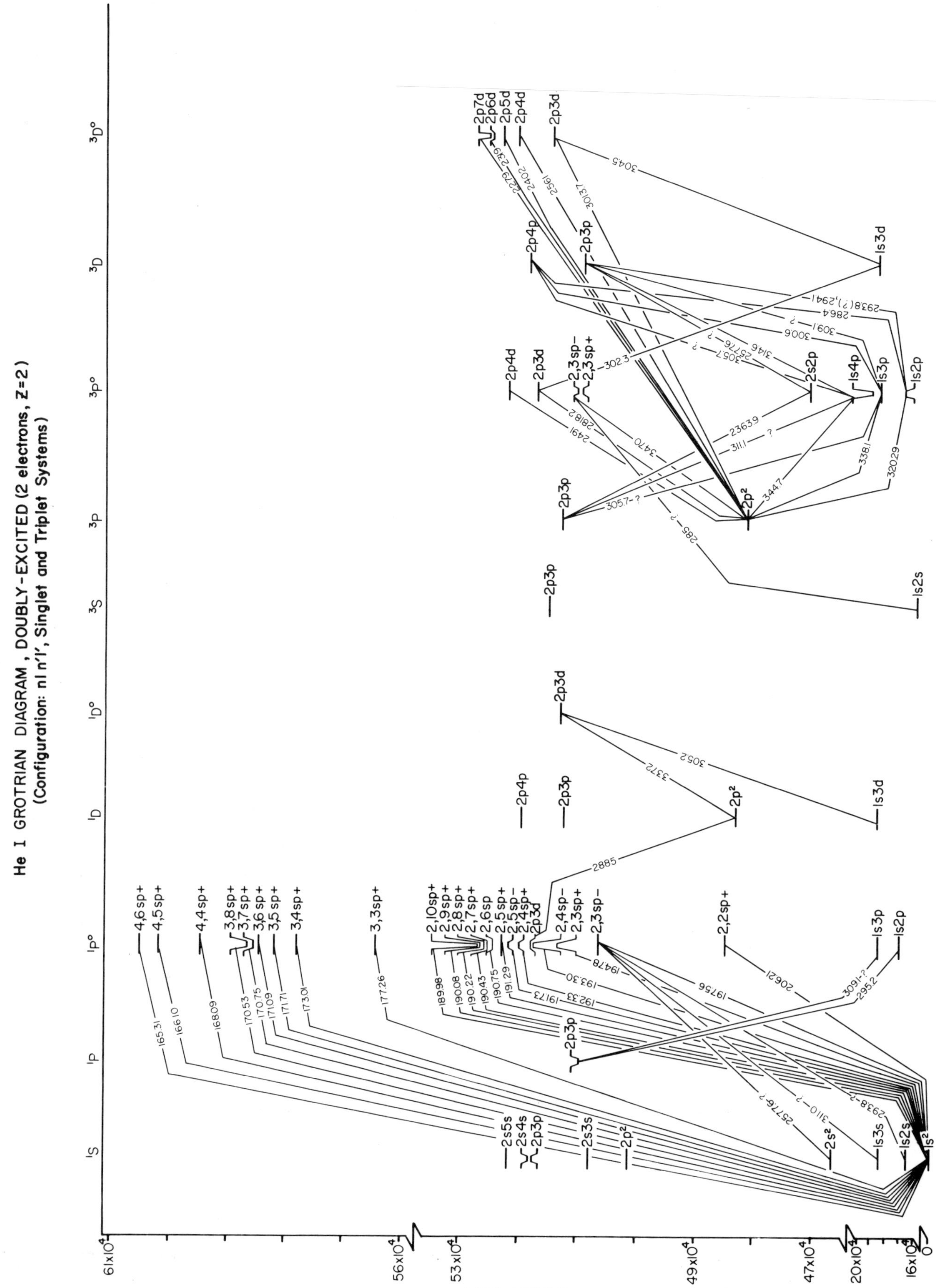

He I
DOUBLY-EXCITED
SINGLET, TRIPLET
GROTRIAN DIAGRAM

He I GROTRIAN DIAGRAM, DOUBLY-EXCITED (2 electrons, Z=2)
(Configuration: n l n' l', Singlet and Triplet Systems)

ENERGY (cm⁻¹)

10

He I ENERGY LEVELS, DOUBLY-EXCITED (2 electrons, Z=2)
(Configuration: nl n'l', Singlet and Triplet Systems)

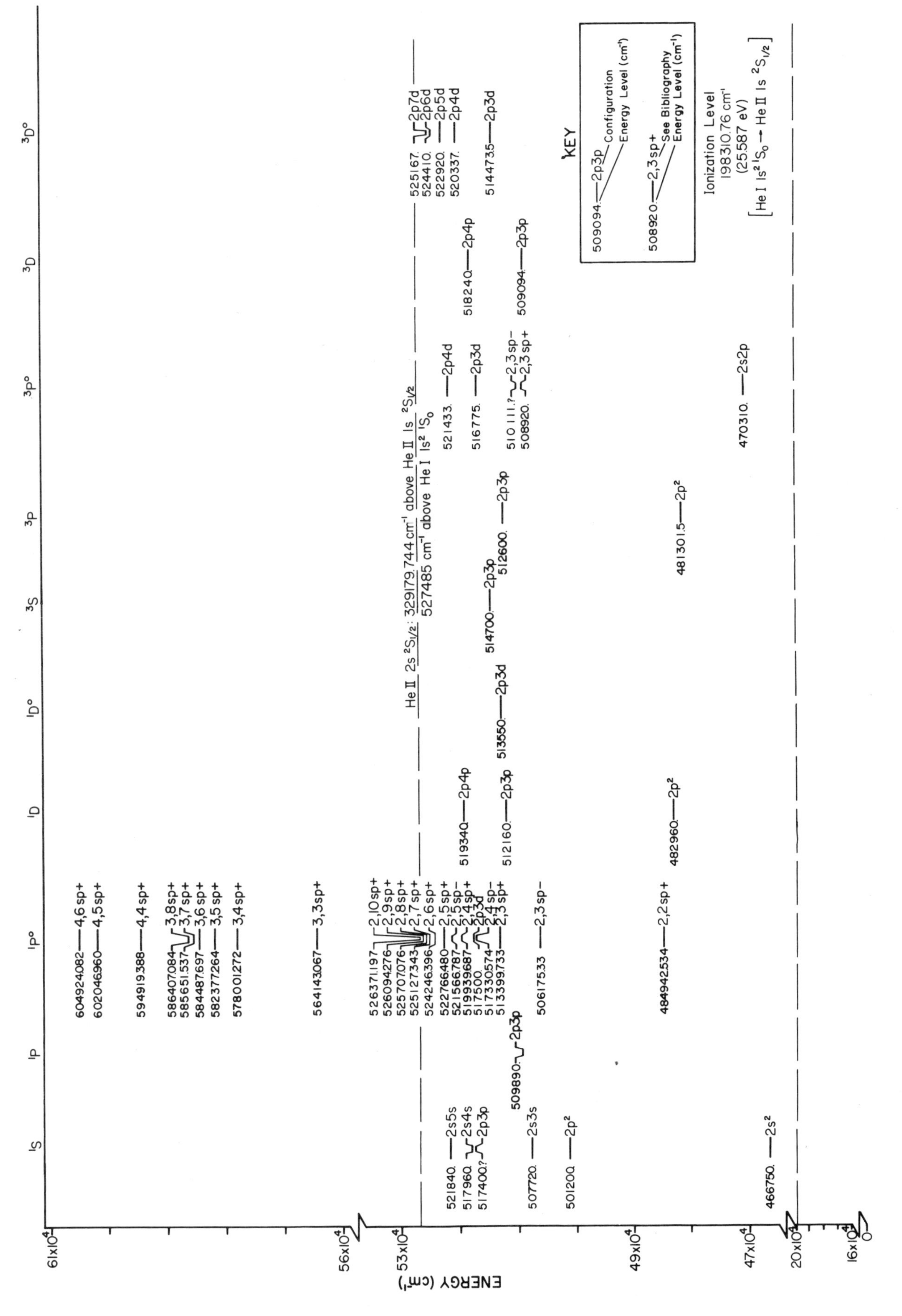

He I
DOUBLY-EXCITED
SINGLET, TRIPLET

11

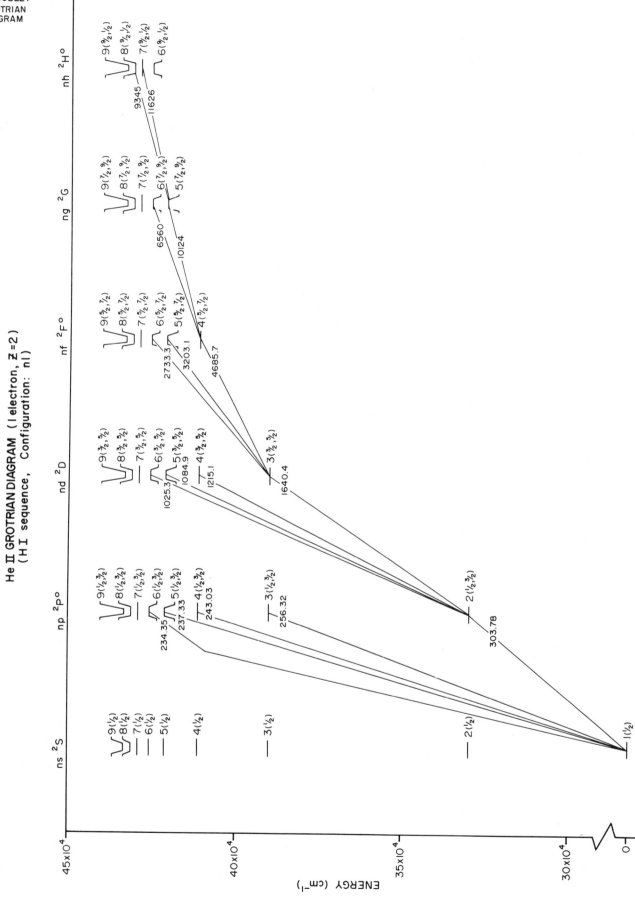

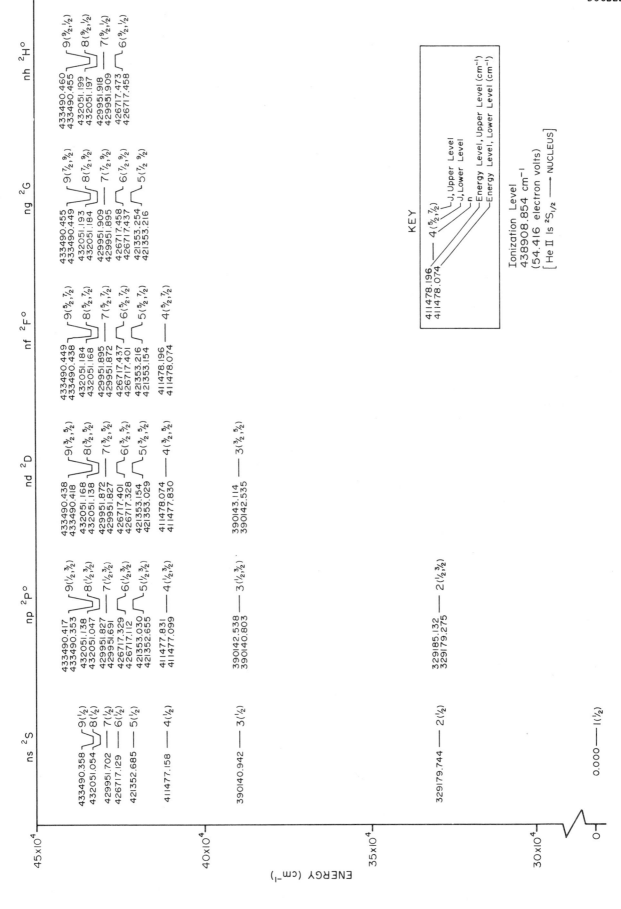

He II ENERGY LEVELS (1 electron, Z=2)
(HI sequence, Configuration: nl)

He II
DOUBLET

He I $Z = 2$ 2 electrons

R. Arrathoon, J. Opt. Soc. Amer. **61**, 332 (1971).

Author gives transition wavelengths and probabilities.

H.G. Berry, J. Desesquelles, and M. Dufay, Phys. Rev. **A6**, 600 (1972).

H.G. Berry, I. Martinson, L.J. Curtis, and L. Lundin, Phys. Rev. **A3**, 1934 (1971).

Authors give energies and transitions for doubly-excited states.

J. Humphreys and H.J. Kostkowski, J. Research NBS **49**, 73 (1952).

Authors give observed infrared spectra.

A.N. Ivanova, U.I. Safronova, and V.N. Kharitonova, Opt. and Spectros. **24**, 55 (1968).

Authors give an energy level table of calculated and experimental values.

E.J. Knystautus and R. Drovin, Nucl.Instrum.Methods **110**, 95 (1973)

U. Litzén, Physica Scripta **2**, 103 (1970).

Author gives improved infrared wavelengths.

R. Madden and K. Codling, Ap. J. **141**, 364 (1965).

Authors give line and energy level tables for resonant absorption lines observed in the region 165–200 Å.

W.C. Martin, J. Opt. Soc. Amer. **50**, 174 (1960).

Author gives observed wavelengths.

W.C. Martin, J. Phys. Chem. Ref. Data **2**, 257 (1973).

This is a compilation of all levels observed to date with one and two electrons excited.

On doubly-excited terms, alternative transitions are given for some spectral lines. Thus Berry *et al.* (1972) suggest 1s 2s ^{1}S–sp23–^{1}P° and/or 1s 2p ^{3}P°–pp23–^{3}D for λ 293.8 Å, and 2s^2 ^{1}S–sp23–^{1}P and/or 2s 2p ^{3}P°–pp23–^{3}D for λ 2577.6 Å. We have shown all of these lines with question marks. Berry *et al.* (1971) assign λ 311.0 Å to 1s 3s ^{1}S–sp23 ^{1}P° whereas Knystautus and Drovin assign λ 311.1 Å to 2p 3p ^{3}P–1s 4p ^{3}P°. Knystautus and Drovin also list λ 305.7 Å as belonging to 2p 4p ^{3}D–1s 4p ^{3}P° and/or 2p 3p ^{1}D–1s 3p ^{1}P°, and λ 309.1 Å as coming from 2p 3p ^{1}P–1s 3p ^{1}P° and/or 2p 3p ^{3}D–1s 3p ^{3}P°. These possibilities are all shown with question marks.

The (±) notation comes from Madden and Codling. Briefly, the wavefunction ψ is written in terms of single-particle wavefunctions U as

$$\psi(\text{sp}2n\pm) = (1/\sqrt{2})\,[U(2s n\text{p}) \pm U(2p n\text{s})\,]$$

He II $Z = 2$ 1 electron

E.G. Kessler, Jr., and F.L. Roesler, J. Opt. Soc. Amer. **62**, 440 (1972).

Authors study fine structure of $n = 4$–5 and $n = 4$–6 transitions in He II at low temperature.

Lithium (Li)

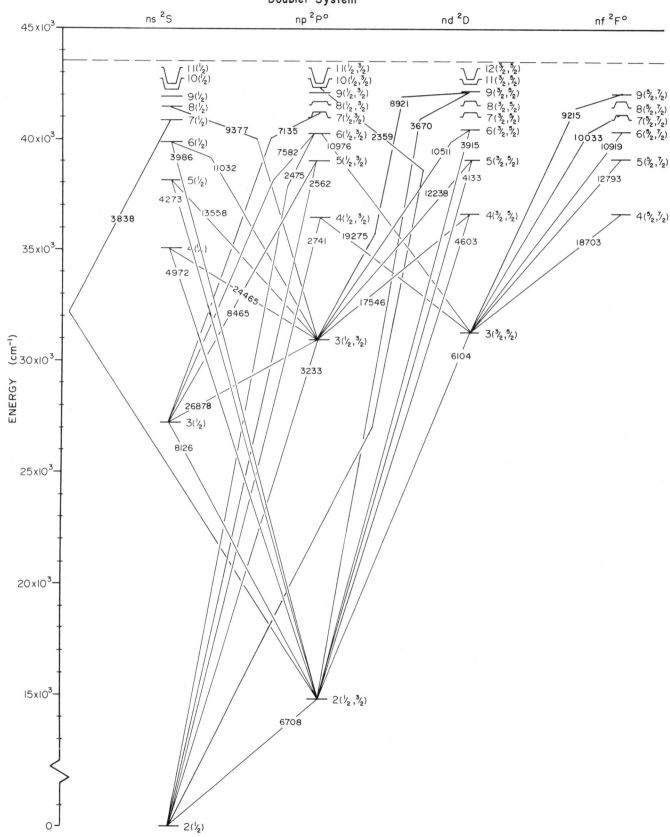

Li I
DOUBLET
GROTRIAN
DIAGRAM

Li I GROTRIAN DIAGRAM (3 electrons, Z = 3)
(Configuration: 1s² nl)
Doublet System

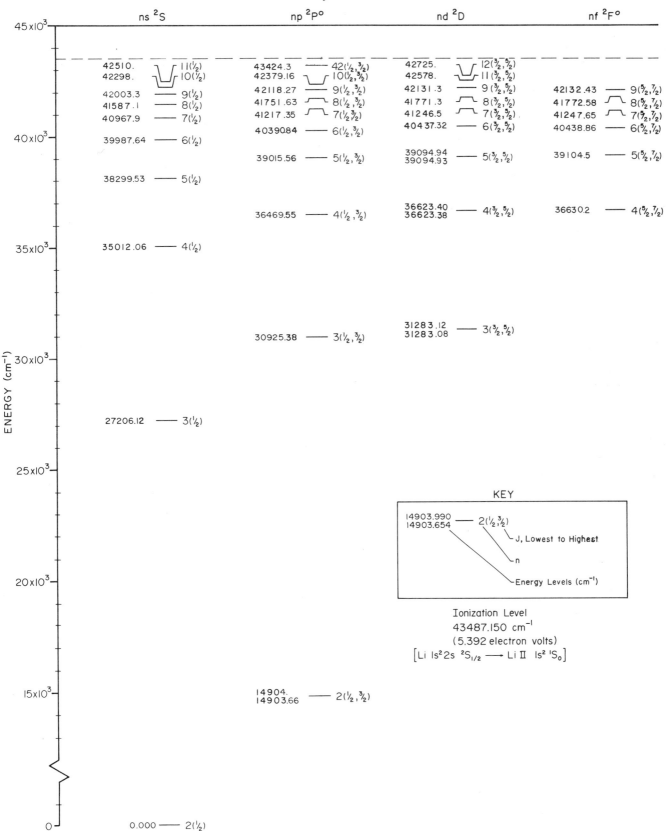

Li I ENERGY LEVELS (3 electrons, Z̄ = 3)
(Configuration 1s² nl)
Doublet System

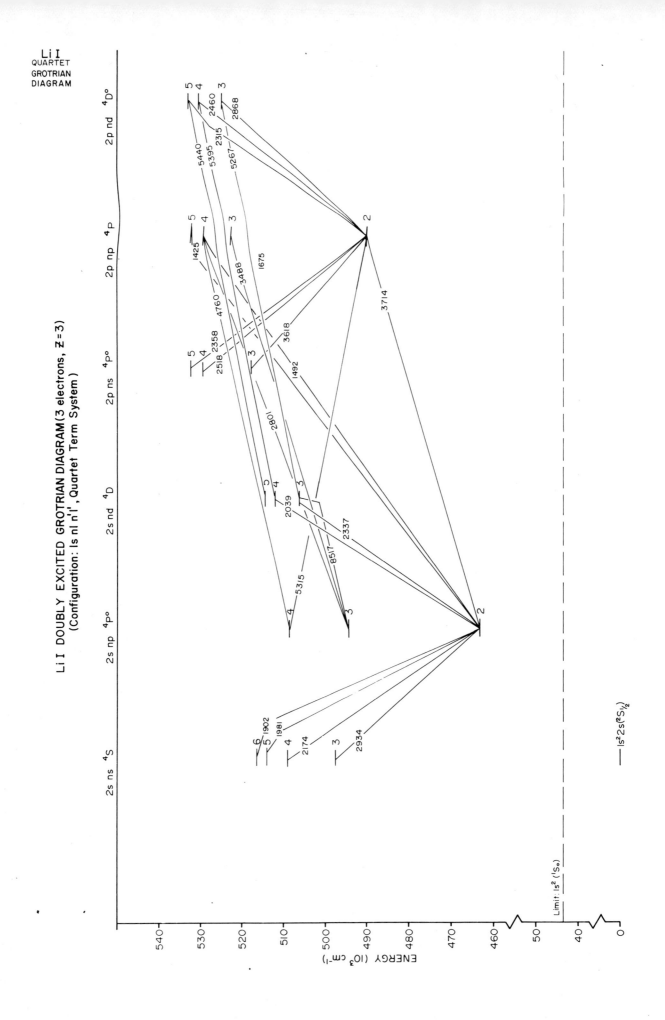

Li I DOUBLY EXCITED ENERGY LEVELS (3 electrons, Z=3)
(Configuration: 1s nl n'l', Quartet Term System)

2s ns ^{4}S	2s np ^{4}P°	2s nd ^{4}D	2p ns ^{4}P°	2p np ^{4}P	2p nd ^{4}D°
				535380 — 6	
				533690 — 5	533360 — 5
			532570 — 5	530550 — 4	530810 — 4
			529880 — 4		525280 — 3
				523230 — 3	
518850 — 7(2)	518070 — 4	514980 — 5	518080 — 3		
516110 — 6(2)	515620	512450 — 4			
514010 — 5(2)		506300 — 3			
509580 — 4(2)	506270 — 3				
	494570				
497600 — 3(2)				490440 — 2	
	463520 — 2				

Li I
QUARTET

KEY

490450 — 2 — n
Energy Level (cm^{-1})

Ionization Level
43487.150 cm^{-1}
(5.392 eV)

[Li I 2s ^{2}S$_{1/2}$ → Li II 1s^2 ^{1}S$_0$]

Limit: 1s^2 (^{1}S$_0$)   43487.150

0.00 — 1s^{2}2s(^{2}S$_{1/2}$)

ENERGY (10^3 cm^{-1})

540 530 520 510 500 490 480 470 460 50 40 0

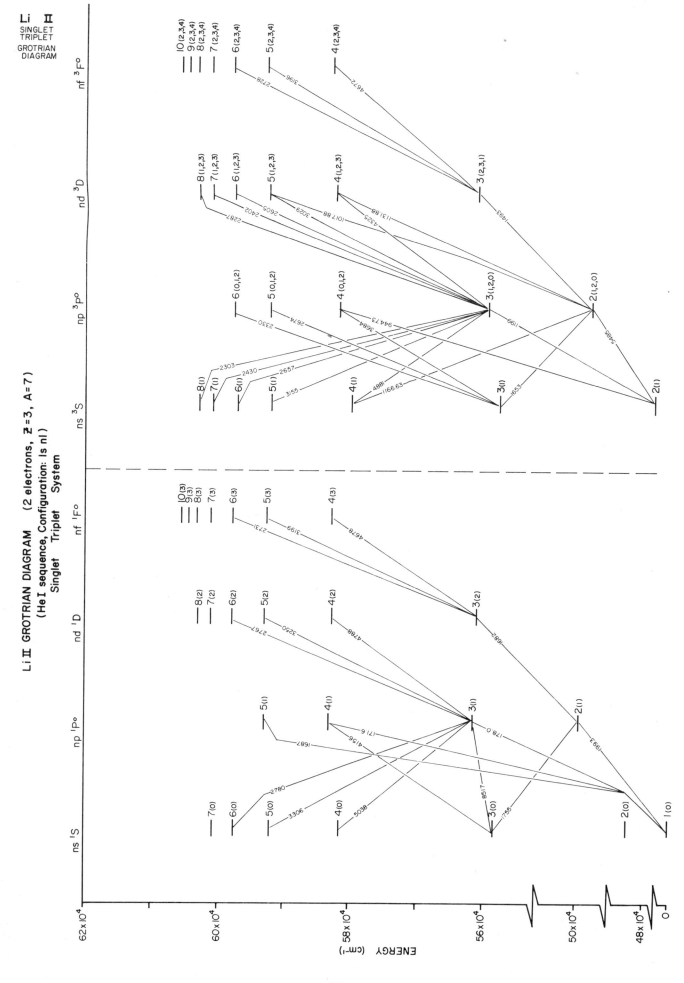

20

Li II ENERGY LEVELS (2 electrons, Z=3, A=7)
(He I sequence, Configuration: 1s nl)
Singlet   Triplet   System

Li II
SINGLET
TRIPLET

ns ¹S

600930. —— 7(0)
597580.53 —— 6(0)
591989.55 —— 5(0)
581596.77 —— 4(0)
558777.88 —— 3(0)
491374.6 —— 2(0)
0.(±3) —— 1(0)

np ¹P°

592634.91 —— 5(1)
582830.11 —— 4(1)
561752.82 —— 3(1)
501808.59 —— 2(1)

nd ¹D

603219.50 —— 8(2)
601119.02 —— 7(2)
597882.52 —— 6(2)
592514.43 —— 5(2)
582630.95 —— 4(2)
561273.62 —— 3(2)

nf ¹F°

605690.5 —— 10(3)
604660.70 —— 9(3)
603221.21 —— 8(3)
603221.21 —— 7(3)
601121.55 —— 6(3)
592521.11 —— 5(3)
582644.04 —— 4(3)

ns ³S

602902.98 —— 8(1)
600643.90 —— 7(1)
597121.95 —— 6(1)
591184.26 —— 5(1)
579981.33 —— 4(1)
554754.45 —— 3(1)
476034.98 —— 2(1)

np ³P°

597663.40
597662.73 —— 6 (0,1,2)
597662.35
592134.70
592134.03 —— 5 (0,1,2)
592133.65
581886.70
581885.98 —— 4 (0,1,2)
581885.58
559502.32
559501.42 —— 3 (1,2,0)
559500.35
494266.57
494263.44 —— 2 (1,2,0)
494261.17

nd ³D

603216.87
603216.21 —— 8 (1,2,3)
603215.82
601115.11
601114.45 —— 7 (1,2,3)
601114.06
597876.60
597875.94 —— 6 (1,2,3)
597875.55
592504.75
592504.09 —— 5 (1,2,3)
592503.70
582614.07
582613.41 —— 4 (1,2,3)
582613.02
561244.30
561243.77 —— 3 (2,3,1)
561243.15

nf ³F°

605689.4 —— 10 (2,3,4)
604659.55 —— 9 (2,3,4)
603220.0 —— 8 (2,3,4)
601120.4 —— 7 (2,3,4)
597885.43 —— 6 (2,3,4)
592520.11 —— 5 (2,3,4)
582642.97 —— 4 (2,3,4)

KEY

494266.57
494263.44 —— 2 (1,2,0)
494261.17

J Lowest to highest
n
Energy Level (cm⁻¹)

Ionization Level
610079.0 cm⁻¹
(75.638 eV)
[Li II 1s² ¹S₀ → Li III 1s ²S₁/₂]

ENERGY (cm⁻¹)

62×10⁴
60×10⁴
58×10⁴
56×10⁴
50×10⁴
48×10⁴
0

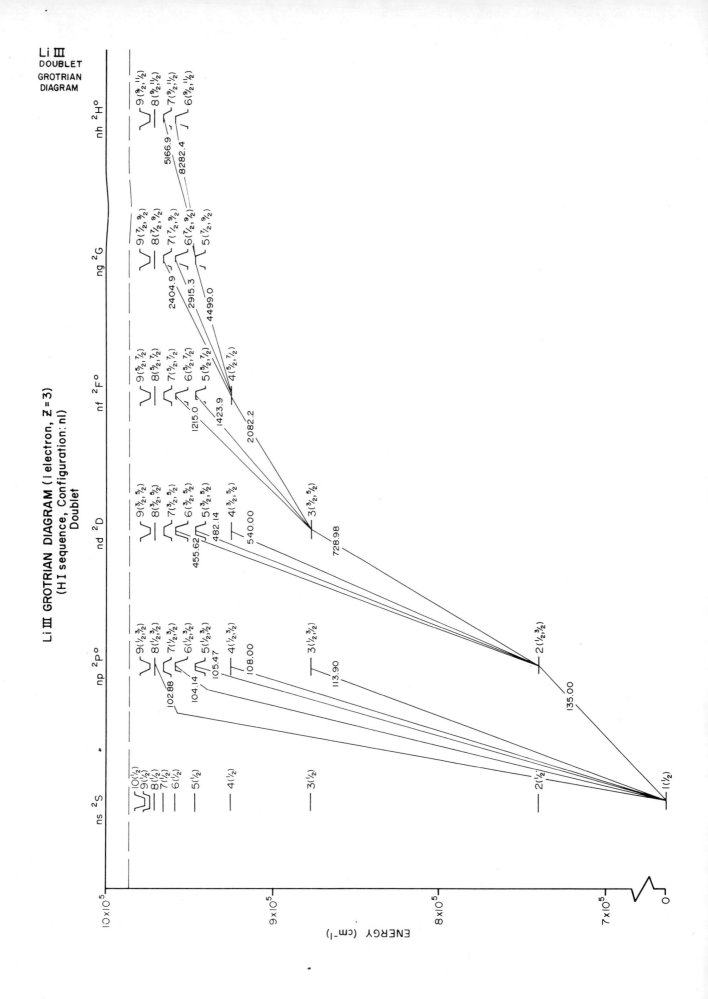

Li III ENERGY LEVELS (1 electron, Z = 3) Mass Number 7
(H I sequence, Configuration: nl) Doublet
(NOTE: MAGNETIC HYPERFINE STRUCTURE HAS BEEN IGNORED)

Li III
DOUBLET

ns ²S

977784.938 ⌐ 10(½)
975468.292 ⌐ 9(½)
972229.537 — 8(½)
967505.505 — 7(½)
960226.923 — 6(½)

948155.517 — 5(½)

925932.786 — 4(½)

877919.682 — 3(½)

740736.381 — 2(½)

0.000 — 1(½)

np ²P°

975468.594 ⌐ 9(½,³⁄₂)
975468.269 ⌐
972229.968 ⌐ 8(½,³⁄₂)
972229.504 ⌐
967506.148 ⌐ 7(½,³⁄₂)
967505.456 ⌐
960227.943 ⌐ 6(½,³⁄₂)
960226.844 ⌐
948157.280 ⌐ 5(½,³⁄₂)
948155.382 ⌐

925936.229 ⌐ 4(½,³⁄₂)
925932.522 ⌐

877927.845 ⌐ 3(½,³⁄₂)
877919.058 ⌐

740763.945 ⌐ 2(½,³⁄₂)
740734.288 ⌐

nd ²D

975468.702 ⌐ 9(³⁄₂,⁵⁄₂)
975468.593 ⌐
972230.121 ⌐ 8(³⁄₂,⁵⁄₂)
972229.967 ⌐
967506.377 ⌐ 7(³⁄₂,⁵⁄₂)
967506.146 ⌐
960228.307 ⌐ 6(³⁄₂,⁵⁄₂)
960227.941 ⌐
948157.909 ⌐ 5(³⁄₂,⁵⁄₂)
948157.277 ⌐

925937.458 ⌐ 4(³⁄₂,⁵⁄₂)
925936.223 ⌐

877930.759 ⌐ 3(³⁄₂,⁵⁄₂)
877927.831 ⌐

nf ²F°

975468.756 ⌐ 9(⁵⁄₂,⁷⁄₂)
975468.702 ⌐
972230.198 ⌐ 8(⁵⁄₂,⁷⁄₂)
972230.121 ⌐
967506.492 ⌐ 7(⁵⁄₂,⁷⁄₂)
967506.376 ⌐
960228.489 ⌐ 6(⁵⁄₂,⁷⁄₂)
960228.306 ⌐
948158.224 ⌐ 5(⁵⁄₂,⁷⁄₂)
948157.908 ⌐

925938.074 ⌐ 4(⁵⁄₂,⁷⁄₂)
925937.456 ⌐

ng ²G

975468.788 ⌐ 9(⁷⁄₂,⁹⁄₂)
975468.756 ⌐
972230.244 ⌐ 8(⁷⁄₂,⁹⁄₂)
972230.198 ⌐
967506.561 ⌐ 7(⁷⁄₂,⁹⁄₂)
967506.491 ⌐
960228.599 ⌐ 6(⁷⁄₂,⁹⁄₂)
960228.489 ⌐
948158.414 ⌐ 5(⁷⁄₂,⁹⁄₂)
948158.224 ⌐

nh ²H°

975468.810 ⌐ 9(⁹⁄₂,¹¹⁄₂)
975468.788 ⌐
972230.275 ⌐ 8(⁹⁄₂,¹¹⁄₂)
972230.244 ⌐
967506.607 ⌐ 7(⁹⁄₂,¹¹⁄₂)
967506.561 ⌐
960228.672 ⌐ 6(⁹⁄₂,¹¹⁄₂)
960228.598 ⌐

KEY

877930.759 ⌐ 3(³⁄₂,⁵⁄₂)
877927.831 ⌐

J, Upper Level
J, Lower Level
n
Energy, Upper Level (cm⁻¹)
Energy, Lower Level (cm⁻¹)

Ionization Level
987660.945 cm⁻¹
(122.451 electron volts)
[Li III 1s ²S₁/₂ ⟶ NUCLEUS]

ENERGY (cm⁻¹)

10×10⁵
9×10⁵
8×10⁵
7×10⁵
0

23

Bibliography

Li I $Z = 3$ 3 electrons

N. Andersen, W.S. Bickel, G.W. Carriveau, K. Jensen, and E. Veje, Physica Scripta **4**, 113 (1971).

Authors extend the spectra of Li I and Li II.

H.G. Berry, J. Bromander, I. Martinson, and R. Buchta, Physica Scripta **3**, 63 (1971).

Authors give observed energy levels and wavelengths of doubly-excited Li.

H.G. Berry, E.H. Pinnington, and J.L. Subtil, J. Opt. Soc. Amer. **62**, 767 (1972).

Authors give a line table and a Grotrian diagram from lines observed in doubly-excited lithium.

J.P. Buchet, M.C. Buchet-Poulizac, and H.G. Berry, Phys. Rev. **A7**, 922 (1973).

Authors present classifications of transitions in doubly-excited Li and Li II.

J.P. Buchet, A. Denis, J. Désesquelles, and M. Dufay, Phys. Letters **28A**, 529 (1969).

Authors treat new lines plausibly attributed to doubly-excited states of lithium.

J.W. Cooper, M.J. Conneely, K. Smith, and S. Ormonde, Phys. Rev. Letters **25**, 1540 (1970).

Authors calculate resonant energy levels above the $2\ ^3S$ threshold and compare values to observed values.

D.L. Ederer, T. Lucatorto, and R.P. Madden, Phys. Rev. Letters **25**, 1537 (1970).

Authors give line and energy level tables for observed resonance lines due to an excitation of a K-shell electron or due to a K-shell electron and an outer electron.

D.L. Ederer, T. Lucatorto, and R.P. Madden, Journal de Physique **32**, C4 (1971).

Authors compare energy level values from observations and calculations for resonances due to an excitation of a K-shell electron or due to a K-shell electron and an outer electron.

P. Feldman, M. Levitt, and R. Novick, Phys. Rev. Letters **21**, 331 (1968).

Authors use energy values for definitive identification of lithium spectral lines.

J.D. Garcia and J.E. Mack, Phys. Rev. **138A**, 987 (1965).

Authors give energy levels and spectral lines of doubly-excited lithium.

Li II $Z = 3$ 2 electrons

H.G. Berry, E.H. Pinnington, and J.L. Subtil, J. Opt. Soc. Amer. **62**, 767 (1972).

Authors give a table of lines observed in doubly-excited lithium.

J.P. Buchet, A. Denis, J. Desesquelles, and M. Dufay, Phys. Letters **28A**, 529 (1969).

Authors treat new lines plausibly attributed to doubly-excited states of lithium.

E. Holøien and J. Midtdal, J. Phys. B: Atom. Molec. Phys. **4**, 1243 (1971).

Authors calculate energy levels nonrelativistically for He I isoelectronic sequence.

C.L. Pekeris, Phys. Rev. **126**, 143 (1962).

Author presents calculations of energies of some states in Li II.

Y.G. Turesson and B. Edlén, Ark. Fys. **23**, 117 (1962).

Authors identify the transition at 9581 Å.

Note that a few doubly-excited states are known in Li II; these are described by J.P. Buchet, M.C. Buchet-Poulizac, H.G. Berry, and G.W.F. Drake, Phys. Rev. A **7**, 922 (1973).

Li III $Z = 3$ 1 electron

Please see the general references.

Beryllium (Be)

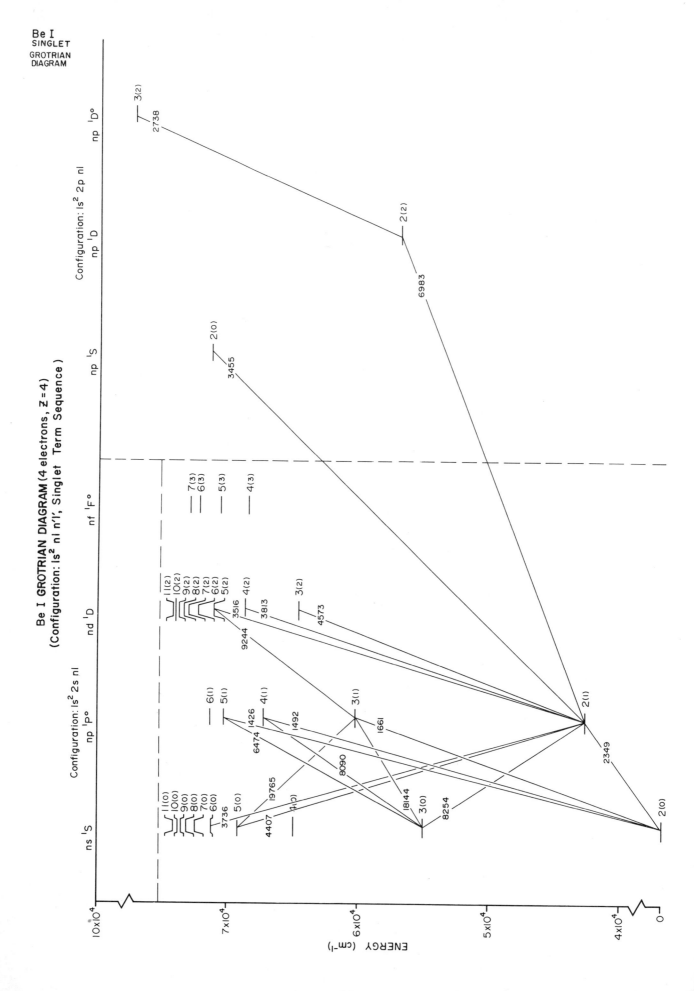

Be I
SINGLET
GROTRIAN
DIAGRAM

Be I GROTRIAN DIAGRAM (4 electrons, Z = 4)
(Configuration: 1s² nl n'l', Singlet Term Sequence)

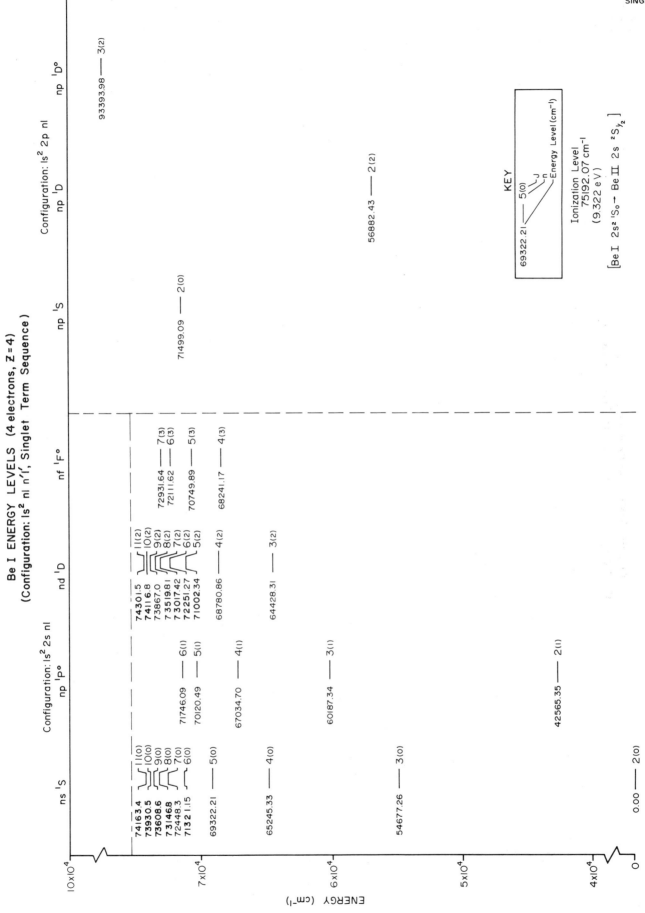

Be I ENERGY LEVELS (4 electrons, Z=4)
(Configuration: 1s² 2l nl', Singlet Term Sequence)

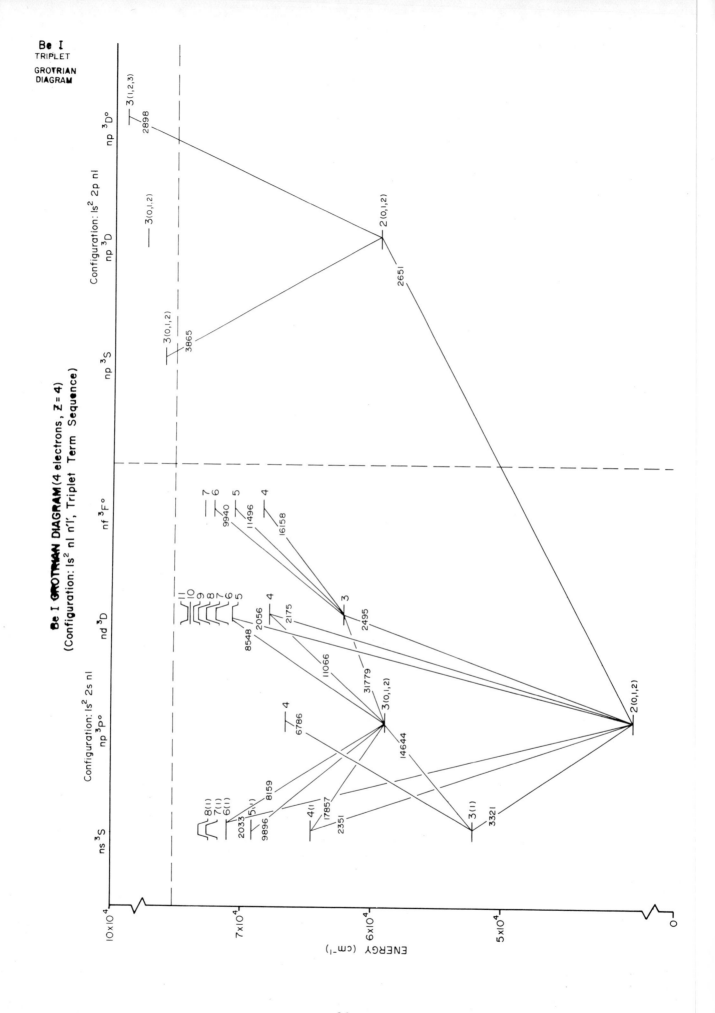

Be I
TRIPLET

GROTRIAN
DIAGRAM

Be I GROTRIAN DIAGRAM (4 electrons, Z = 4)
(Configuration: ls² nl n'l', Triplet Term Sequence)

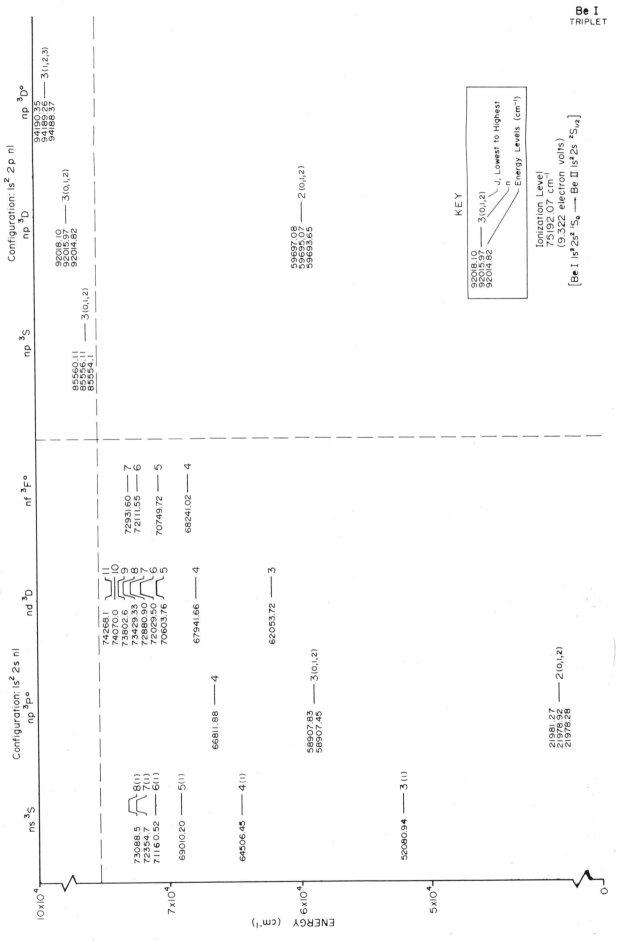

Be I ENERGY LEVELS (4 electrons, Z = 4)
(Configuration: 1s² nl n'l', Triplet Term Sequence)

Be I
TRIPLET

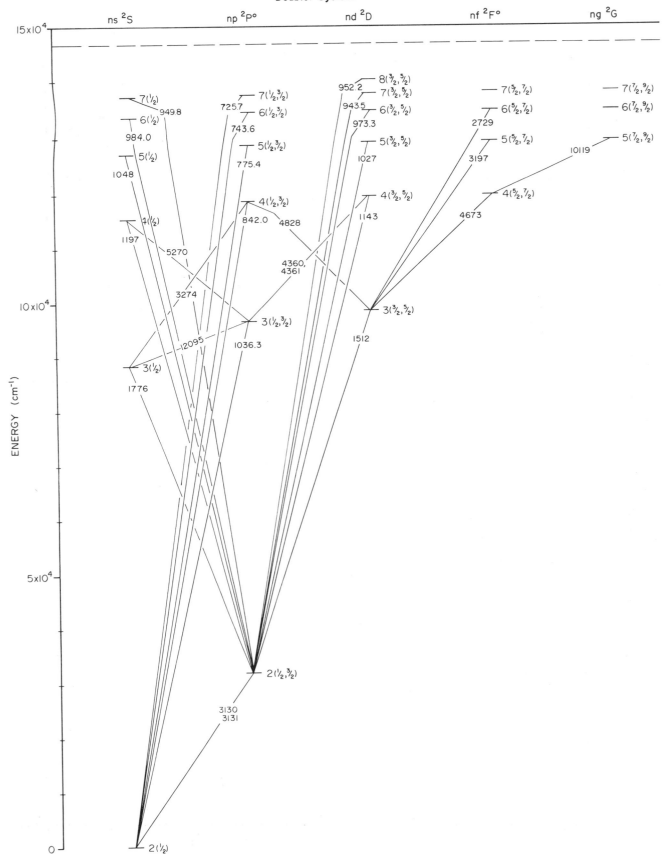

Be II
DOUBLET
GROTRIAN
DIAGRAM

Be II GROTRIAN DIAGRAM (3 electrons, Z=3)
(Li I sequence, Configuration: ls² nl)
Doublet System

ns ²S   np ²P°   nd ²D   nf ²F°   ng ²G

ENERGY (cm⁻¹)

Be II ENERGY LEVELS (3 electrons, Z=3)
(Li I sequence, Configuration: ls² nl)
Doublet System

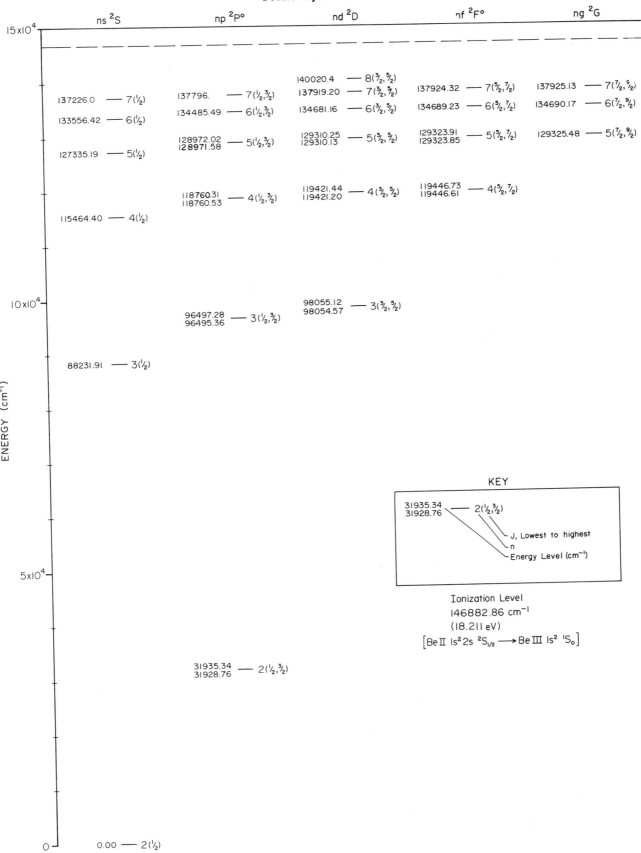

ns ²S   np ²P°   nd ²D   nf ²F°   ng ²G

140020.4 — 8(³/₂,⁵/₂)

137226.0 — 7(¹/₂)   137796. — 7(¹/₂,³/₂)   137919.20 — 7(³/₂,⁵/₂)   137924.32 — 7(⁵/₂,⁷/₂)   137925.13 — 7(⁷/₂,⁹/₂)

133556.42 — 6(¹/₂)   134485.49 — 6(¹/₂,³/₂)   134681.16 — 6(³/₂,⁵/₂)   134689.23 — 6(⁵/₂,⁷/₂)   134690.17 — 6(⁷/₂,⁹/₂)

128972.02
128971.58 — 5(¹/₂,³/₂)   129310.25
129310.13 — 5(³/₂,⁵/₂)   129323.91
129323.85 — 5(⁵/₂,⁷/₂)   129325.48 — 5(⁷/₂,⁹/₂)

127335.19 — 5(¹/₂)

118760.31
118760.53 — 4(¹/₂,³/₂)   119421.44
119421.20 — 4(³/₂,⁵/₂)   119446.73
119446.61 — 4(⁵/₂,⁷/₂)

115464.40 — 4(¹/₂)

98055.12
98054.57 — 3(³/₂,⁵/₂)

96497.28
96495.36 — 3(¹/₂,³/₂)

88231.91 — 3(¹/₂)

ENERGY (cm⁻¹)

KEY

31935.34
31928.76 — 2(¹/₂,³/₂)

J, Lowest to highest
n
Energy Level (cm⁻¹)

Ionization Level
146882.86 cm⁻¹
(18.211 eV)
[Be II ls²2s ²S₁/₂ ⟶ Be III ls² ¹S₀]

31935.34
31928.76 — 2(¹/₂,³/₂)

0.00 — 2(¹/₂)

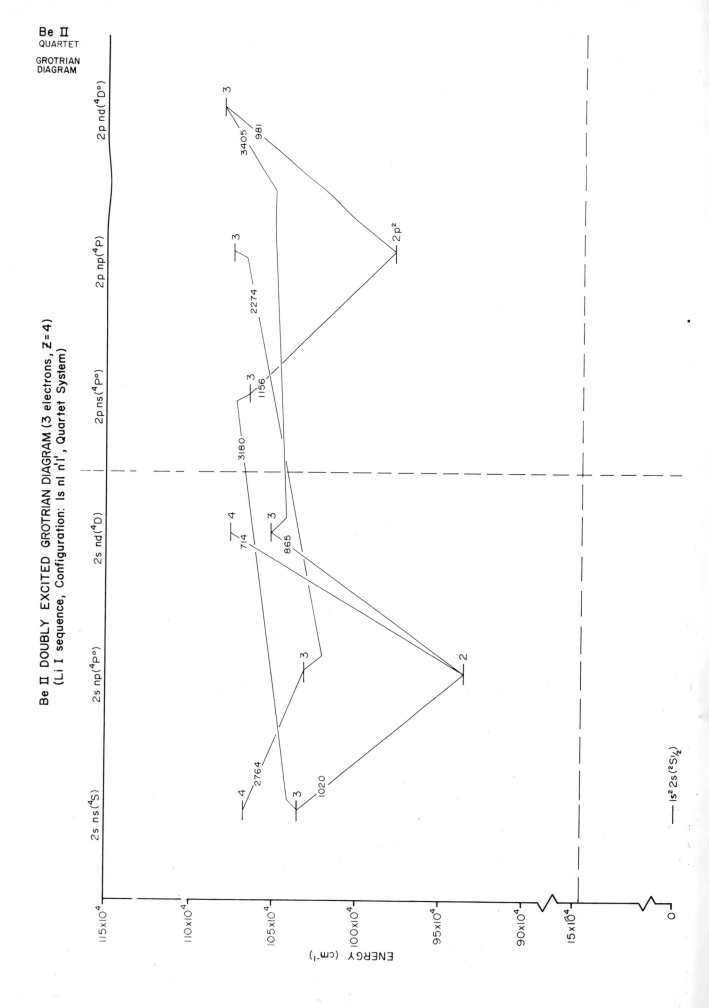

Be II DOUBLY EXCITED GROTRIAN DIAGRAM (3 electrons, Z = 4)
(Li I sequence, Configuration: ls nl n'l', Quartet System)

Be II
QUARTET

GROTRIAN
DIAGRAM

Be II Doubly Excited Energy Levels (3 electrons, Z=4) (Li I sequence, Configuration: 1s nl n'l', Quartet System)

Be II
QUARTET

35

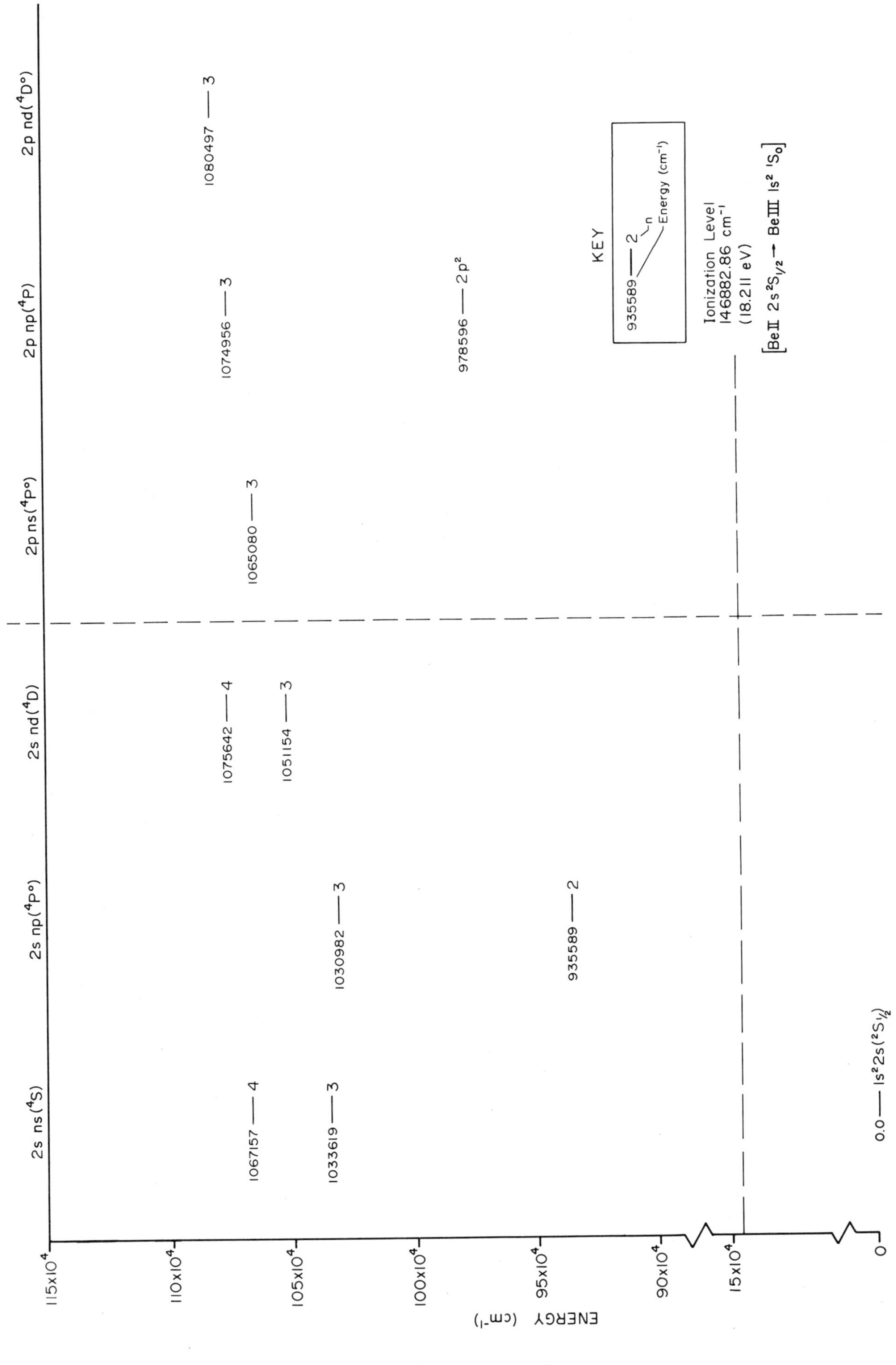

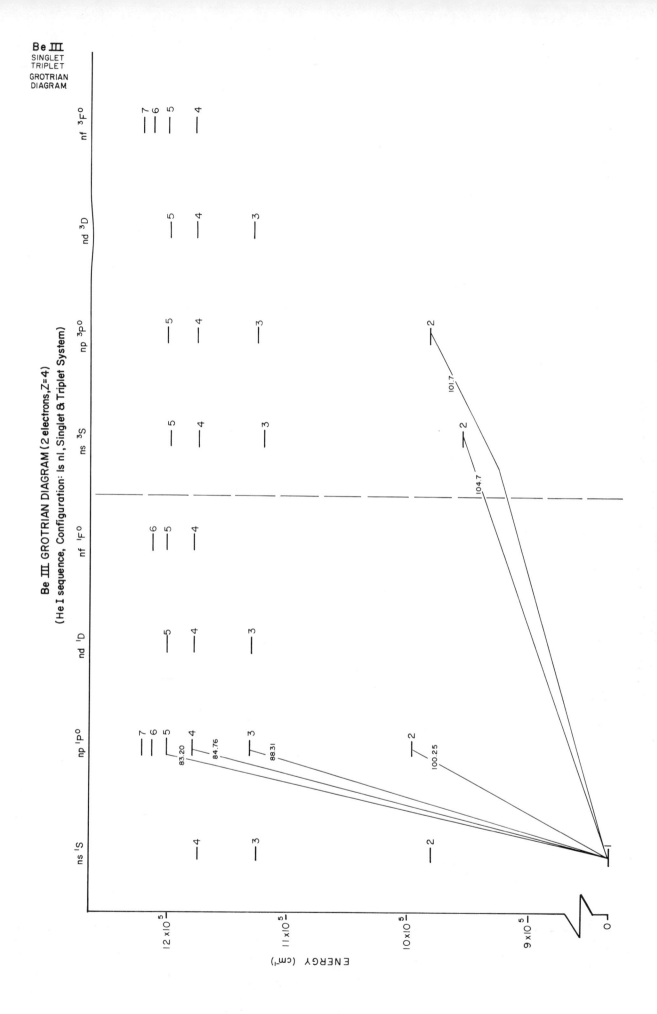

Be III
SINGLET
TRIPLET
GROTRIAN
DIAGRAM

Be III GROTRIAN DIAGRAM (2 electrons, Z=4)
(He I sequence, Configuration: 1s nl, Singlet & Triplet System)

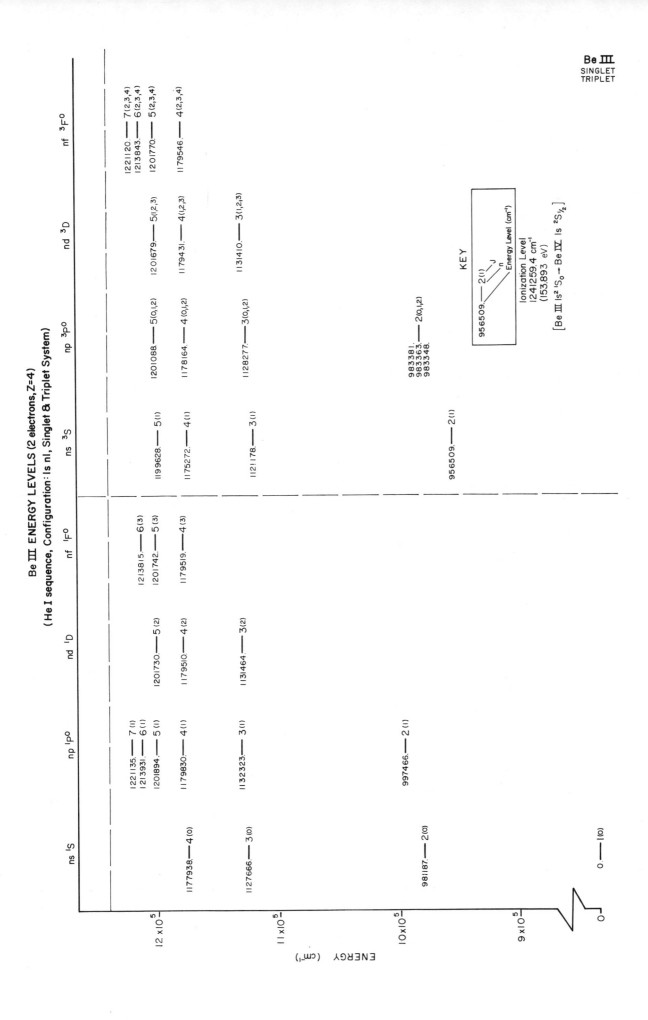

Be III ENERGY LEVELS (2 electrons, Z=4)
(He I sequence, Configuration: 1s nl, Singlet & Triplet System)

Be III
SINGLET
TRIPLET

KEY

956509.——— 2(1)

Ionization Level
1241259.4 cm⁻¹
(153.893 eV)

[Be III 1s² ¹S₀ → Be IV 1s ²S½]

Energy Level (cm⁻¹)
n
J

ns ¹S

1177938.——4(0)

1127666.——3(0)

981187.——2(0)

0 ——1(0)

np ¹P⁰

1221135.——7(1)
1213931.——6(1)
1201894.——5(1)
1179830.——4(1)

1132323.——3(1)

997466.——2(1)

nd ¹D

1201730.——5(2)
1179510.——4(2)

1131464.——3(2)

nf ¹F⁰

1213815.——6(3)
1201742.——5(3)
1179519.——4(3)

ns ³S

1199628.——5(1)
1175272.——4(1)

1121178.——3(1)

956509.——2(1)

np ³P⁰

1201088.——5(0,1,2)
1178164.——4(0,1,2)

1128277.——3(0,1,2)

983381.
983363.——2(0,1,2)
983348.

nd ³D

1201679.——5(1,2,3)
1179431.——4(1,2,3)

1131410.——3(1,2,3)

nf ³F⁰

1221120.——7(2,3,4)
1213843.——6(2,3,4)
1201770.——5(2,3,4)
1179546.——4(2,3,4)

ENERGY (cm⁻¹)

12 x10⁵

11 x10⁵

10x10⁵

9 x10⁵

0

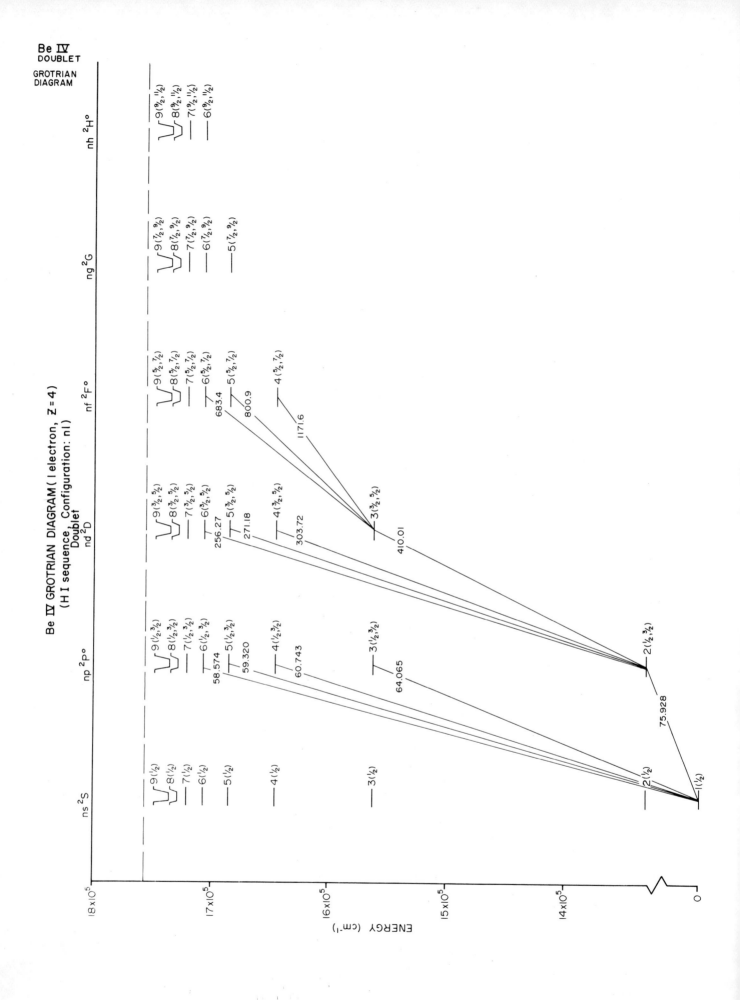

38

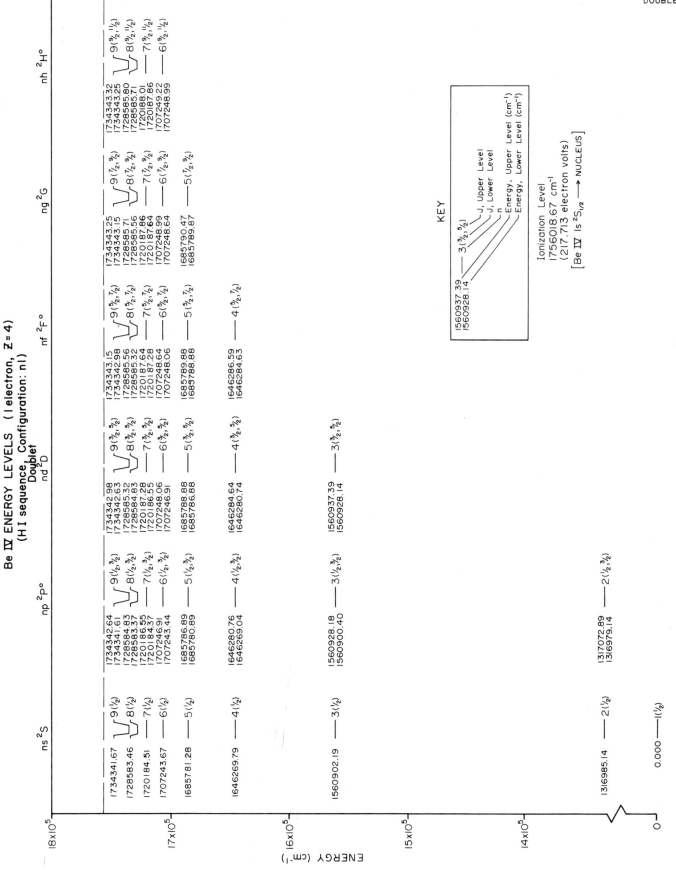

Be IV ENERGY LEVELS (1 electron, Z=4)
(HI sequence, Configuration: nl)
Doublet

Be IV
DOUBLET

39

Be I $Z = 4$ 4 electrons

H.G. Berry, J. Bromander, I. Martinson, and R. Buchta, Physica Scripta **3**, 63 (1971).

Authors suggest classifications for spectral lines.

J.-E. Holmström and L. Johansson, Ark. Fys. **40**, 133 (1969).

Authors give line table observed in the range 35 000–3800 Å with Ritz formulae.

E.W.H. Selwyn, Proc. Phys. Soc. **41**, 392 (1929).

Author gives a line table for lines observed in the region 1600–2100 Å.

Note that a few doubly-excited levels not shown on the accompanying diagrams are described in the above reference by Berry *et al.*

Be II $Z = 4$ 3 electrons

H.G. Berry, J. Bromander, I. Martinson, and R. Buchta, Physica Scripta **3**, 63 (1971).

Authors summarize available experimental information on the doubly-excited levels in the Li I isoelectronic sequence.

J.-E. Holmström and L. Johansson, Ark. Fys. **40**, 133 (1969).

Authors give line table and term table for wavelengths observed in range 35 000–3800 Å.

Be III $Z = 4$ 2 electrons

H.G. Berry, J. Désesquelles, and M. Dufay, Phys. Rev. **A6**, 600 (1972).

This paper summarizes experiment and theory for doubly-excited states in the He I isoelectronic sequence.

E. Holøien and J. Midtdal, J. Phys. B: Atom. Molec. Phys. **4**, 1243 (1971).

Authors calculated energy levels nonrelativistically for the He I isoelectronic sequence.

Be IV $Z = 4$ 1 electron

Please see the general references.

Borium (B)

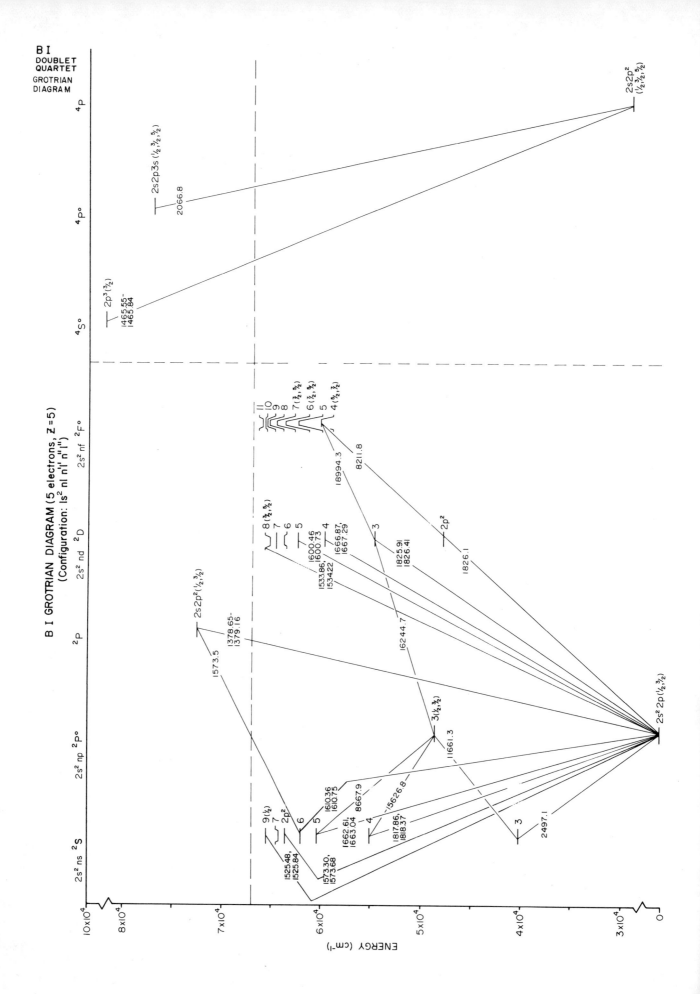

B I ENERGY LEVELS (5 electrons, Z=5)
(Configuration: 1s² nl n'l' n''l'' Doublet, Quartet)

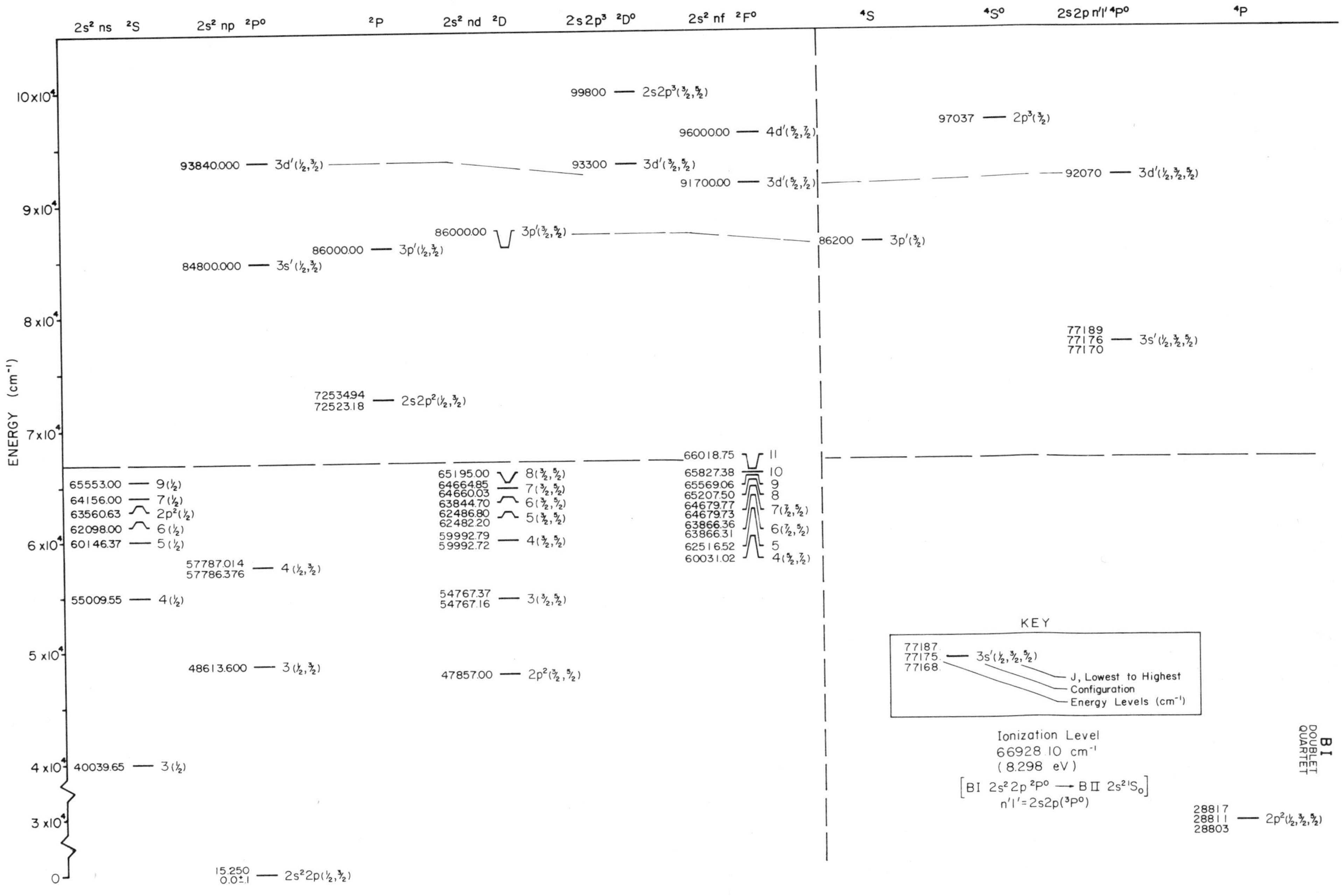

ENERGY (cm⁻¹)

B I
DOUBLET
QUARTET

Column headings:
2s² ns ²S | 2s² np ²P° | ²P | 2s² nd ²D | 2s2p³ ²D° | 2s² nf ²F° | ⁴S | ⁴S° | 2s2p nl' ⁴P° | ⁴P

KEY
77187
77175 —— 3s'(½, 3/2, 5/2)
77168

J, Lowest to Highest
Configuration
Energy Levels (cm⁻¹)

Ionization Level
66928.10 cm⁻¹
(8.298 eV)
[B I 2s²2p ²P° → B II 2s² ¹S₀]
n'l' = 2s2p(³P°)

²S column:
6555300 —— 9 (½)
6415600 —— 7 (½)
6356063 —— 2p²(½)
6209800 —— 6 (½)
6014637 —— 5 (½)
5500955 —— 4 (½)
4003965 —— 3 (½)

²P° column:
93840000 —— 3d'(½, 3/2)
84800000 —— 3s'(½, 3/2)
57787.014
57786.376 —— 4 (½, 3/2)
48613600 —— 3 (½, 3/2)
15.250
0.0±.1 —— 2s²2p(½, 3/2)

²P column:
8600000 —— 3p'(½, 3/2)
7253494
7252318 —— 2s2p²(½, 3/2)

²D column:
9800000 —— 3d'(½, 3/2)
8600000 —— 3p'(½, 3/2)
65195.00 —— 8 (½, 5/2)
6466485
6466003 —— 7 (3/2, 5/2)
6384470 —— 6 (3/2, 5/2)
6248680 —— 5 (3/2, 5/2)
6248220
5999279
5999272 —— 4 (3/2, 5/2)
5476737
5476716 —— 3 (3/2, 5/2)
4785700 —— 2p²(3/2, 5/2)

²D° column:
99800 —— 2s2p³(3/2, 5/2)
93300 —— 3d'(3/2, 5/2)

²F° column:
9600000 —— 4d'(5/2, 7/2)
9170000 —— 3d'(3/2, 5/2)
86200 —— 3p'(½)
66018.75 —— 11
6582738 —— 10
6556906 —— 9
6520750 —— 8
6479977
6479973 —— 7 (7/2, 5/2)
6386636
6386631 —— 6 (7/2, 5/2)
6251652
6003102 —— 5
—— 4 (5/2, 7/2)

⁴S column:
97037 —— 2p³(3/2)

⁴P° column:
92070 —— 3d'(½, 3/2, 5/2)
77189
77176 —— 3s'(½, 3/2, 5/2)
77170

⁴P column:
28817
28811 —— 2p²(½, 3/2, 5/2)
28803

Energy axis (left):
10x10⁴
9x10⁴
8x10⁴
7x10⁴
6x10⁴
5x10⁴
4x10⁴
3x10⁴
0

43

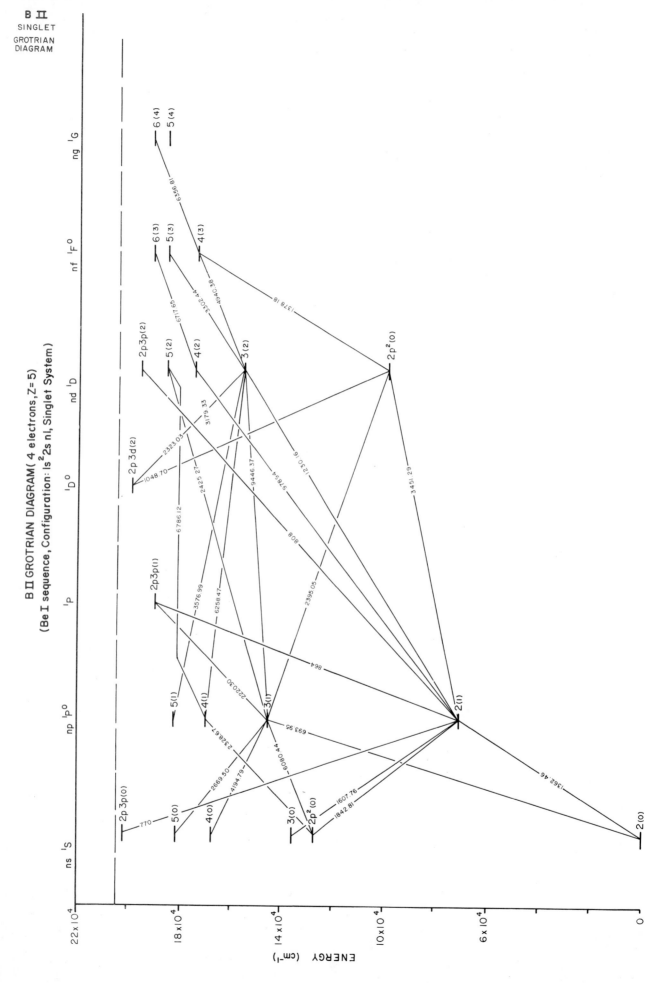

B II

SINGLET

GROTRIAN

DIAGRAM

B II GROTRIAN DIAGRAM (4 electrons, Z = 5)

(Be I sequence, Configuration: 1s²2s nl, Singlet System)

44

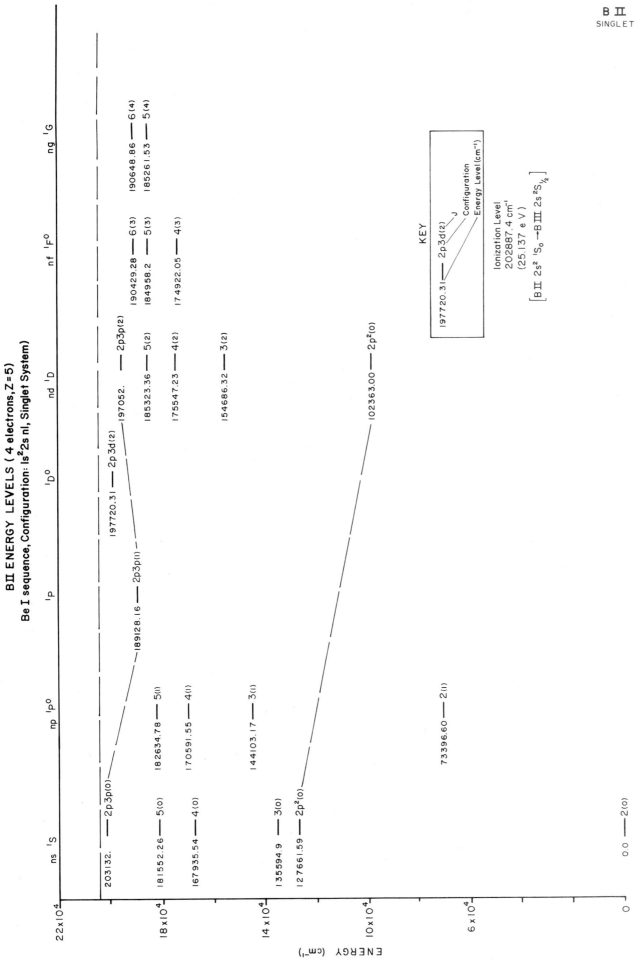

B II
SINGLET

B I ENERGY LEVELS (4 electrons, Z=5)
Be I sequence, Configuration: ls²2s nl, Singlet System)

45

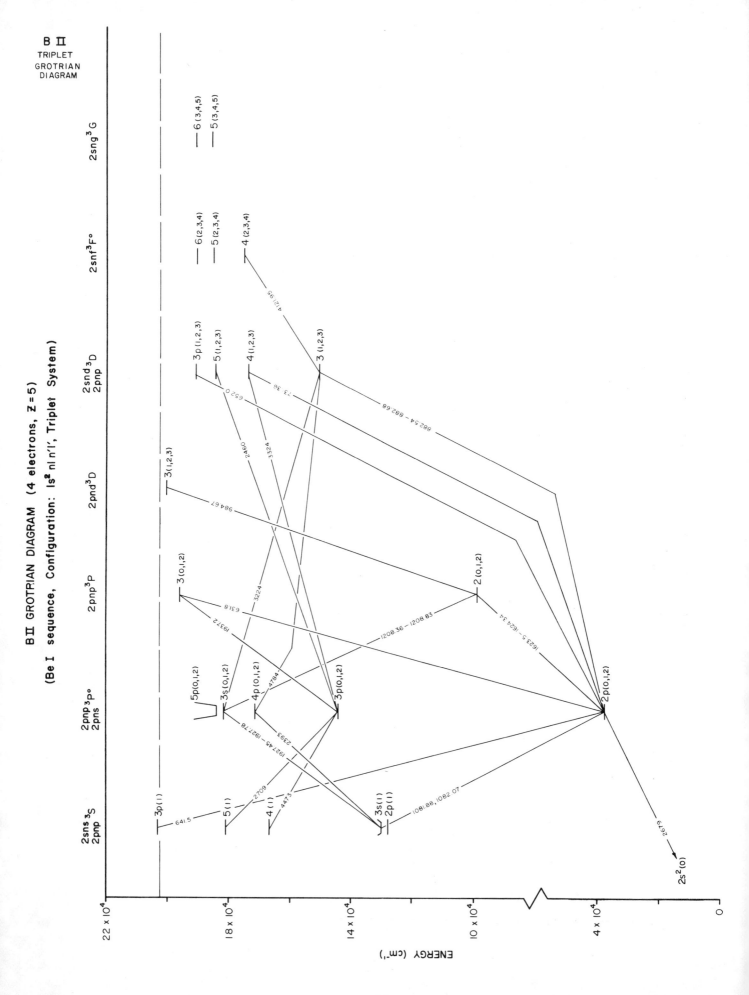

BI GROTRIAN DIAGRAM (4 electrons, Z = 5)

(Be I sequence, Configuration: 1s² nl n'l', Triplet System)

B II
TRIPLET
GROTRIAN
DIAGRAM

ENERGY (cm⁻¹)

46

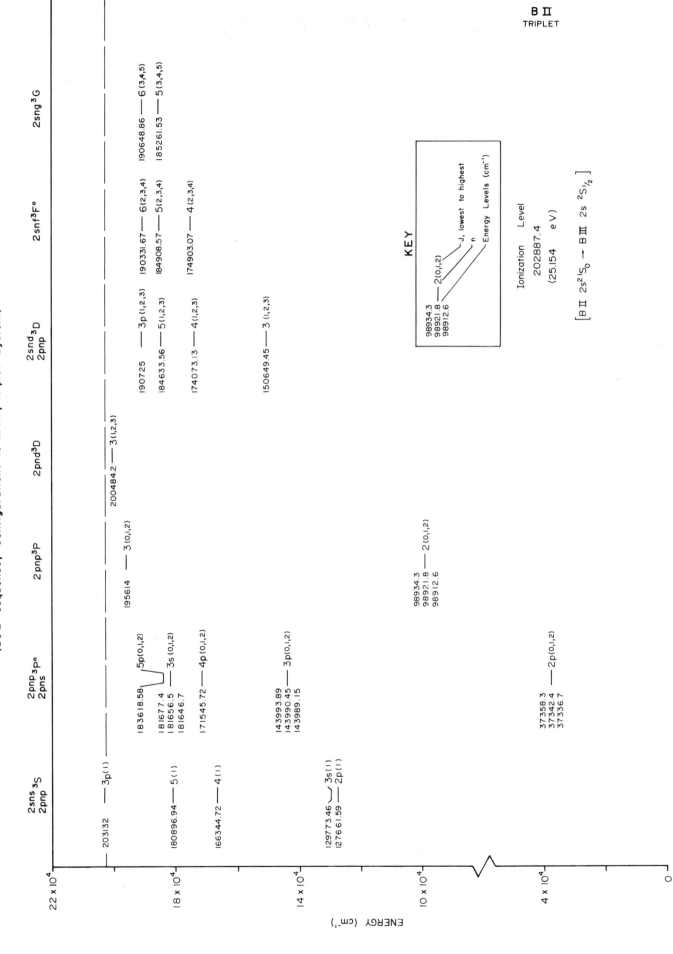

B II ENERGY LEVELS (4 electrons, Z = 5)

(Be I sequence, Configuration: 1s²nl n'l', Triplet System)

B II
TRIPLET

2sns³S / 2pnp

203132 —— 3p(1)

180896.94 —— 5(1)

166344.72 —— 4(1)

2pnp³P°/2pns

183618.58 —— 5p(0,1,2)
181677.4 —— 3s(0,1,2)
181656.5
181646.7
171545.72 —— 4p(0,1,2)

143993.89 —— 3p(0,1,2)
143990.45
143989.15

37358.3 —— 2p(0,1,2)
37342.4
37336.7

2pnp³P

195614 —— 3(0,1,2)

98934.3 —— 2(0,1,2)
98921.8
98912.6

2pnd³D

200484.2 —— 3(1,2,3)

2snd³D / 2pnp

190725 —— 3p(1,2,3)
184633.56 —— 5(1,2,3)
174073.13 —— 4(1,2,3)

150649.45 —— 3(1,2,3)

2snf³F°

190031.67 —— 6(2,3,4)
184908.57 —— 5(2,3,4)
174903.07 —— 4(2,3,4)

2sng³G

190648.86 —— 6(3,4,5)
18526l.53 —— 5(3,4,5)

KEY

98934.3
98921.8 —— 2(0,1,2)
98912.6

J, lowest to highest
n
Energy Levels (cm⁻¹)

Ionization Level
202887.4
(25.154 eV)

[B II 2s² ¹S₀ → B III 2s ²S½]

ENERGY (cm⁻¹)

22 x 10⁴

18 x 10⁴

14 x 10⁴

10 x 10⁴

4 x 10⁴

0

47

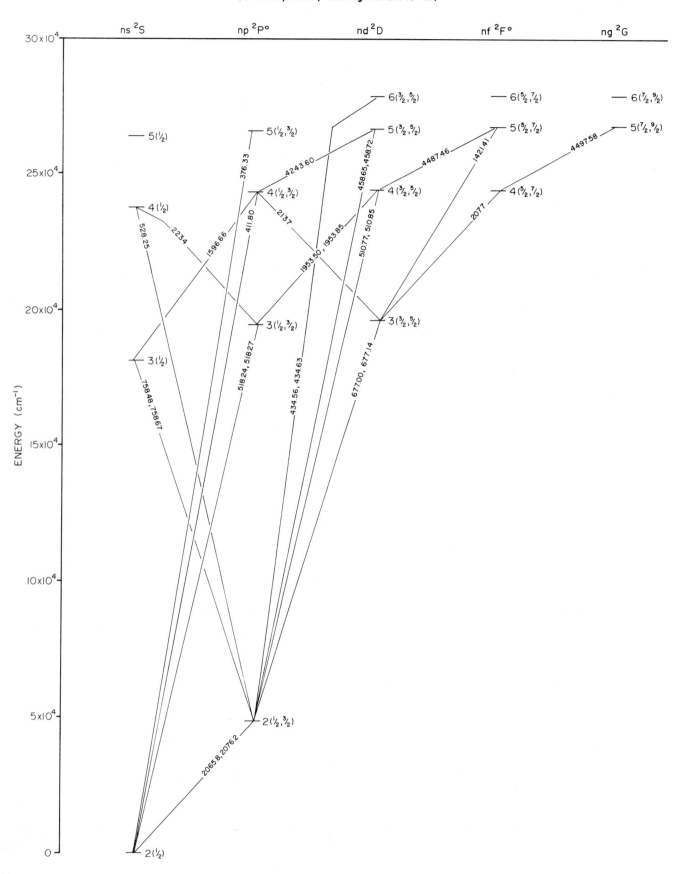

B III GROTRIAN DIAGRAM (3 electrons, Z=5)
(Li I sequence, Configuration: $1s^2\,nl$)

ns ^{2}S   np ^{2}P°   nd ^{2}D   nf ^{2}F°   ng ^{2}G

ENERGY (cm^{-1})

B III ENERGY LEVELS (3 electrons, Z=5)
(Li I sequence, Configuration: 1s² nl)

ns ²S	np ²P°	nd ²D	nf ²F°	ng ²G

278473.7 — 6(³/₂,⁵/₂)   278491.7 — 6(⁵/₂,⁷/₂)   278497.5 — 6(⁷/₂,⁹/₂)

263156.2 — 5(½)   265719.7 — 5(½,³/₂)   266389.5 — 5(³/₂,⁵/₂)   266416.5 — 5(⁵/₂,⁷/₂)   266427.2 — 5(⁷/₂,⁹/₂)

242832.4 / 242828.1 — 4(½,³/₂)   244138.9 / 244138.6 — 4(³/₂,⁵/₂)   244199.2 — 4(⁵/₂,⁷/₂)

237695.5 — 4(½)

192959.4 / 192949.2 — 3(½,³/₂)   196071.2 / 196068.1 — 3(³/₂,⁵/₂)

180201.8 — 3(½)

KEY

48392.50 / 48358.40 — 2(½,³/₂)
J, Lowest to Highest
n
Energy Levels (cm⁻¹)

Ionization Level
305931.10 cm⁻¹
(37.930 electron volts)
[B III 1s²2s ²S₁/₂ ⟶ B IV 1s² ¹S₀]

48392.6 / 48358.5 — 2(½,³/₂)

0.0 — 2(½)

ENERGY (cm⁻¹)

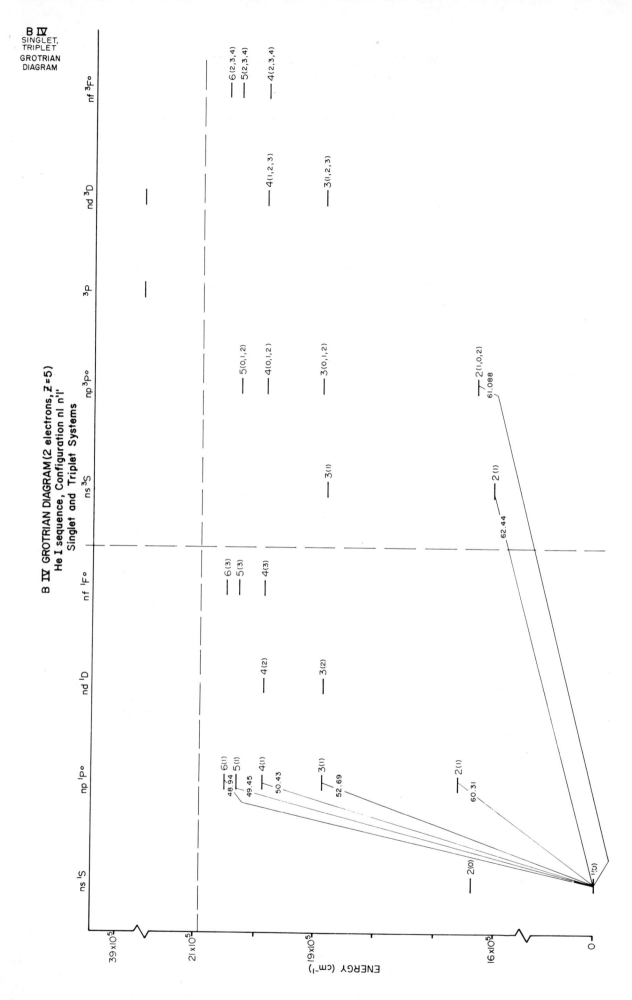

B IV
SINGLET,
TRIPLET
GROTRIAN
DIAGRAM

B IV GROTRIAN DIAGRAM (2 electrons, Z = 5)
He I sequence, Configuration nl n'l'
Singlet and Triplet Systems

ENERGY (cm⁻¹)

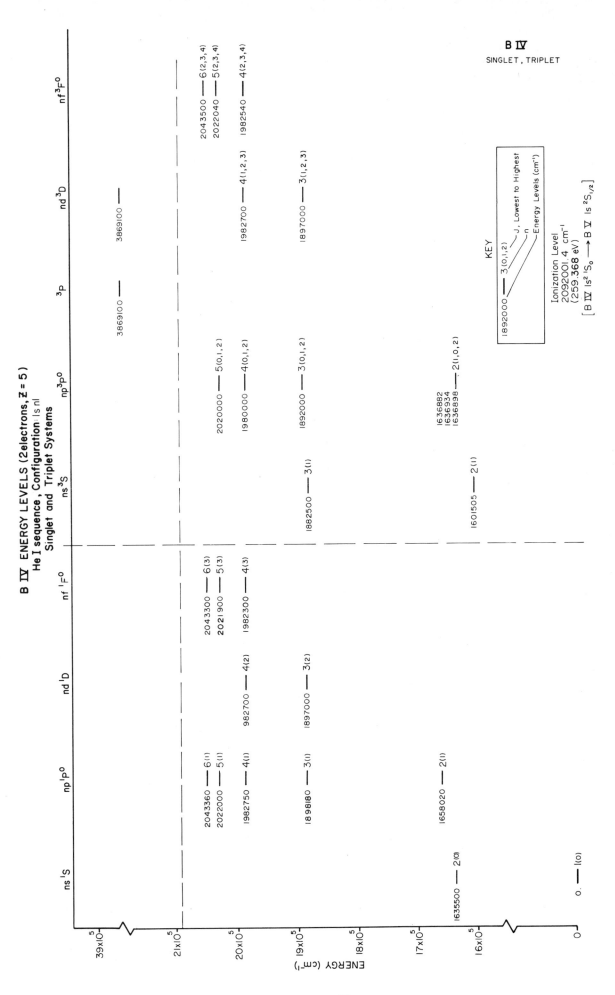

B IV ENERGY LEVELS (2electrons, Z = 5)
He I sequence, Configuration: 1s nl
Singlet and Triplet Systems

51

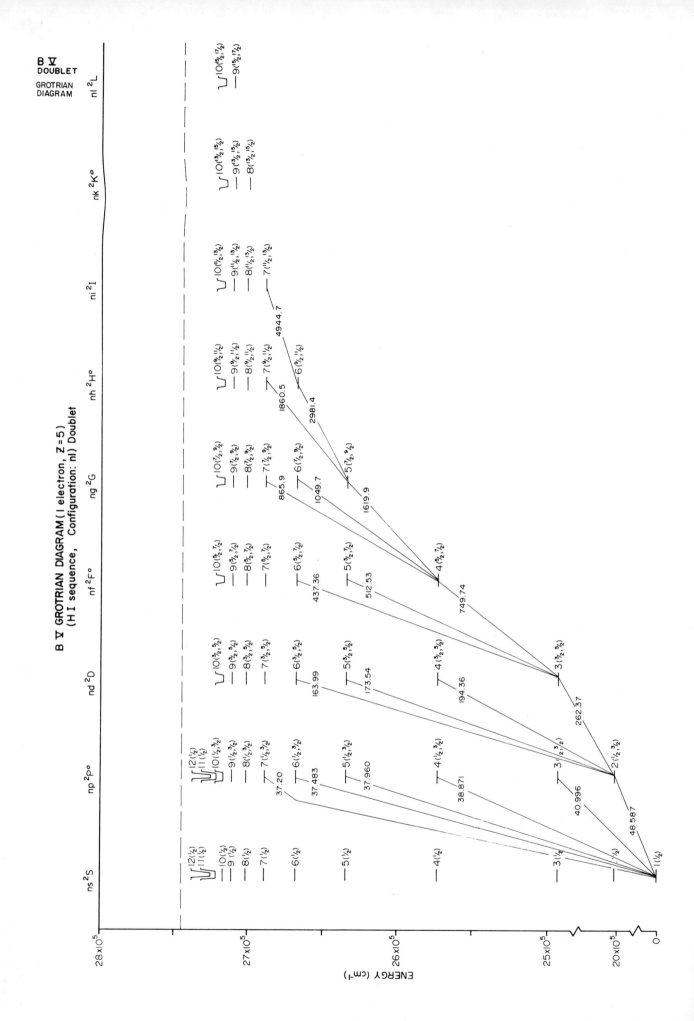

B Ⅴ
DOUBLET

GROTRIAN
DIAGRAM

B Ⅴ GROTRIAN DIAGRAM (1 electron, Z = 5)
(H I sequence, Configuration: nl) Doublet

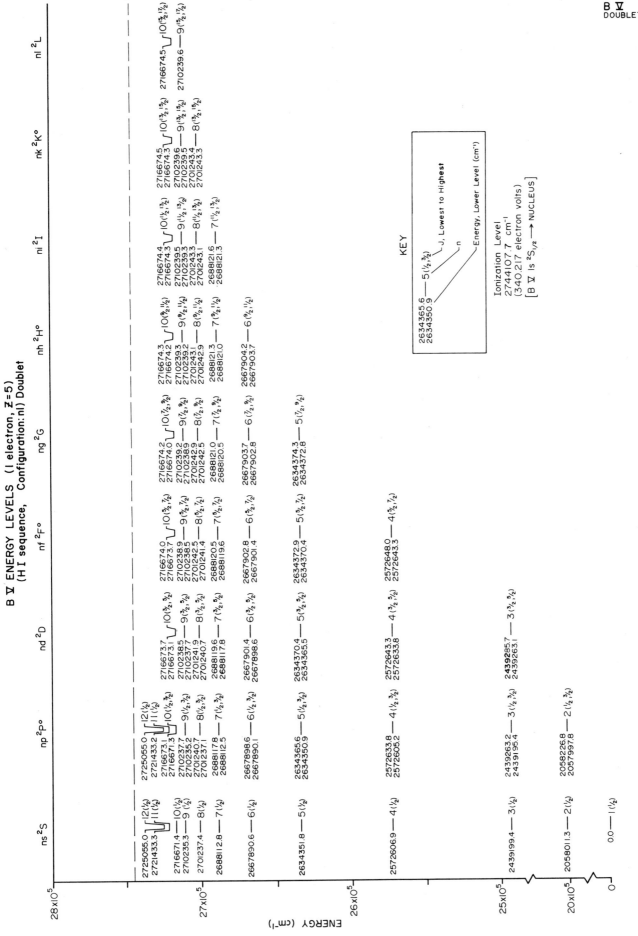

B V ENERGY LEVELS (1 electron, Z=5)
(H I sequence, Configuration: nl) Doublet

B V
DOUBLET

ENERGY (cm⁻¹)

Bibliography

B I $Z = 5$ 5 electrons

B. Edlén, A. Ölme, G. Herzberg, and J.W.C. Johns, J. Opt. Soc. Amer. **60**, 889 (1970).

  Authors give line tables from observed spectra.

P. Gunnvald and L. Minnhagen, Ark. Fys. **22**, 327 (1962).

  Authors give line and energy level tables and a Grotrian diagram for the transitions observed.

B II $Z = 5$ 4 electrons

E.W.H. Selwyn, Proc. Phys. Soc. **41**, 392 (1929).

  Author gives observed lines of wavelengths in the region 1600–2000 Å.

B III $Z = 5$ 3 electrons

  Please see the general references.

B IV $Z = 5$ 2 electrons

E. Holøien and J. Midtdal, J. Phys. B: Atom. Molec. Phys. **4**, 1243 (1971).

  Authors calculate energy levels nonrelativistically for the He I isoelectronic sequence.

B V $Z = 5$ 1 electron

  Please see the general references.

Carbon (C)

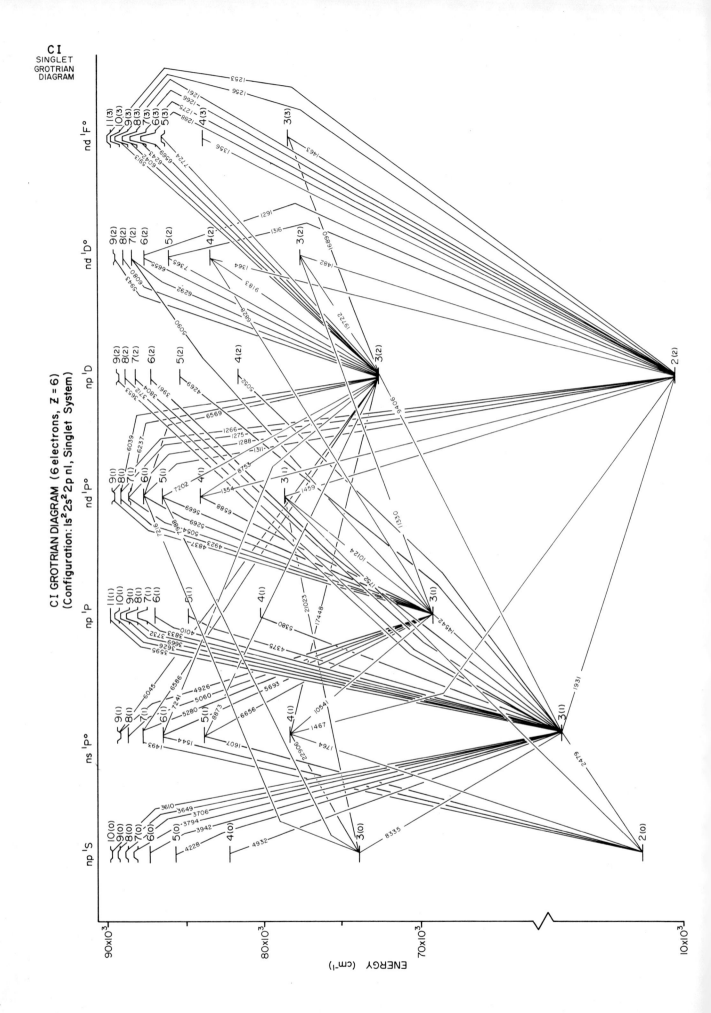

C I
SINGLET
GROTRIAN
DIAGRAM

C I GROTRIAN DIAGRAM (6 electrons, Z = 6)
(Configuration: ls²2s²2p nl, Singlet System)

ENERGY (cm⁻¹)

56

CI ENERGY LEVELS (6 electrons, Z = 6)
(Configuration: 1s² 2s² 2p nl, Singlet System)

np ¹S
- 89678.11 — 10(0)
- 89381.61 — 9(0)
- 88960.64 — 8(0)
- 88333.98 — 7(0)
- 87341.04 — 6(0)
- 85625.18 — 5(0)
- 82251.71 — 4(0)
- 73975.91 — 3(0)
- 21648.01 — 2(0)

ns ¹P°
- 89149.35 — 9(1)
- 88615.01 — 8(1)
- 87789.63 — 7(1)
- 86416.55 — 6(1)
- 83877.31 — 5(1)
- 78340.28 — 4(1)
- 61981.82 — 3(1)

np ¹P
- 89789.21 — 11(1)
- 89555.53 — 10(1)
- 89232.41 — 9(1)
- 88766.98 — 8(1)
- 88061.28 — 7(1)
- 88912.86 — 6(1)
- 84851.53 — 5(1)
- 80562.85 — 4(1)
- 68856.33 — 3(1)

nd ¹P°
- 89525.54 — 9(1)
- 89164.74 — 8(1)
- 88639.02 — 7(1)
- 87830.17 — 6(1)
- 86491.41 — 5(1)
- 84032.15 — 4(1)
- 78731.27 — 3(1)

np ¹D
- 89350.10 — 9(2)
- 88913.56 — 8(2)
- 88260.37 — 7(2)
- 87218.26 — 6(2)
- 85399.81 — 5(2)
- 81769.79 — 4(2)
- 72610.72 — 3(2)
- 10192.63 — 2(2)

nd ¹D°
- 89431.48 — 9(2)
- 89054.16 — 8(2)
- 88498.62 — 7(2)
- 87633.75 — 6(2)
- 86185.20 — 5(2)
- 83497.62 — 4(2)
- 77679.82 — 3(2)

nd ¹F°
- 89971.3 — 11(3)
- 89778.9 — 10(3)
- 89519.13 — 9(3)
- 89155.70 — 8(3)
- 88625.00 — 7(3)
- 87806.93 — 6(3)
- 86449.19 — 5(3)
- 83947.43 — 4(3)
- 78529.62 — 3(3)

KEY

72610.72 — 3(2)
J
n
Energy Level (cm⁻¹)

Ionization Level
90820.42 cm⁻¹
(11.260 eV)
[C I 2s² 2p² ³P₀ → C II 2s² 2p ²P°₁/₂]

ENERGY (cm⁻¹)
90×10³
80×10³
70×10³
10×10³

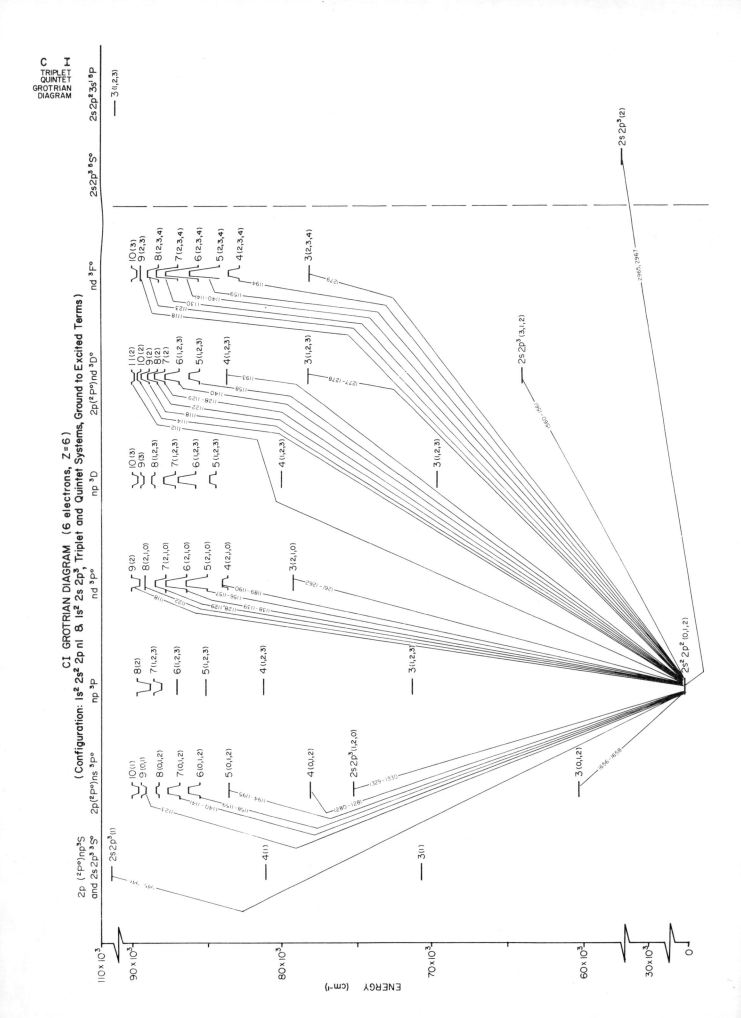

58

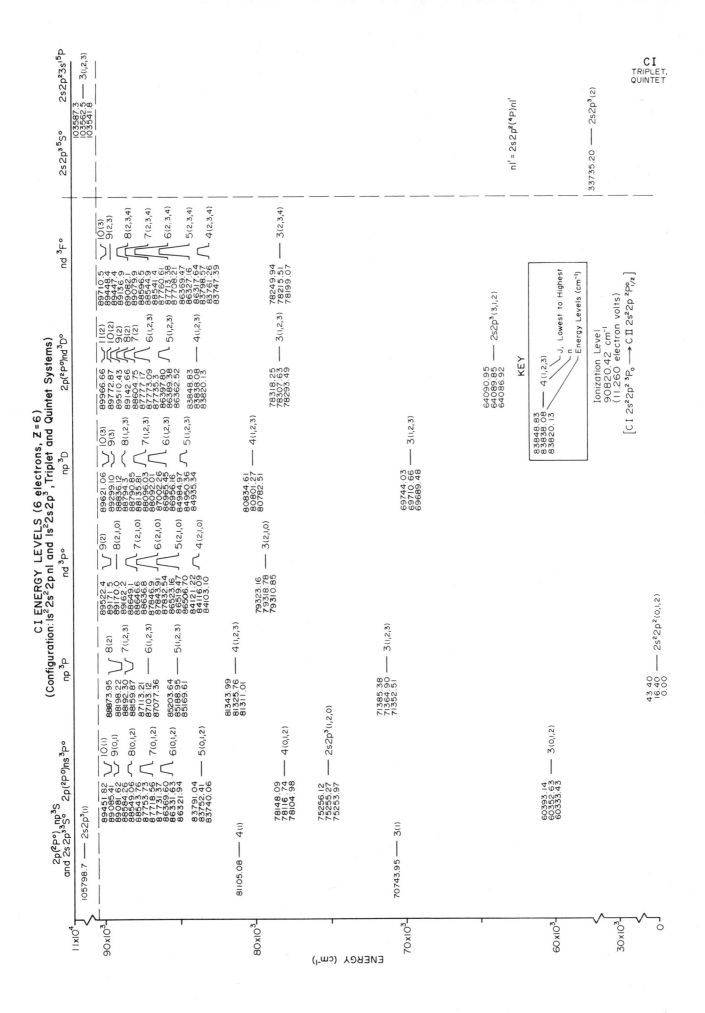

CI ENERGY LEVELS (6 electrons, Z=6)
(Configuration: $1s^2 2s^2 2pnl$ and $1s^2 2s2p^3$, Triplet and Quintet Systems)

59

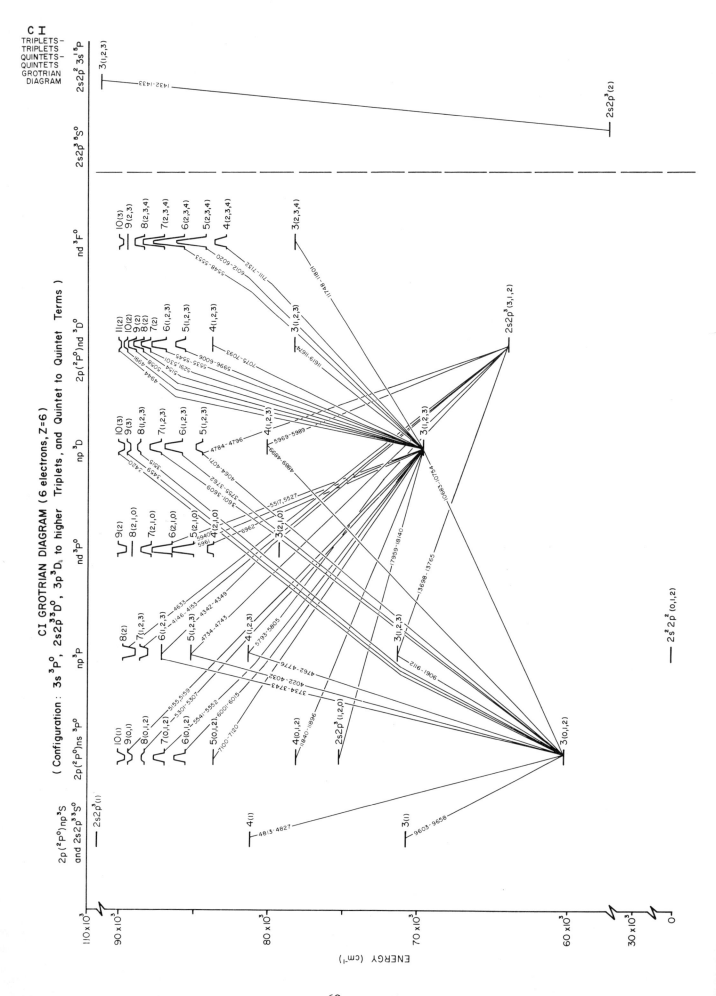

CI
TRIPLETS—
TRIPLETS
QUINTETS—
QUINTETS
GROTRIAN
DIAGRAM

CI GROTRIAN DIAGRAM (6 electrons, Z=6)

(Configuration : 3s³Pᵒ, 2s2p³Dᵒ, 3p³Dₐ, to higher Triplets, and Quintet to Quintet Terms)

60

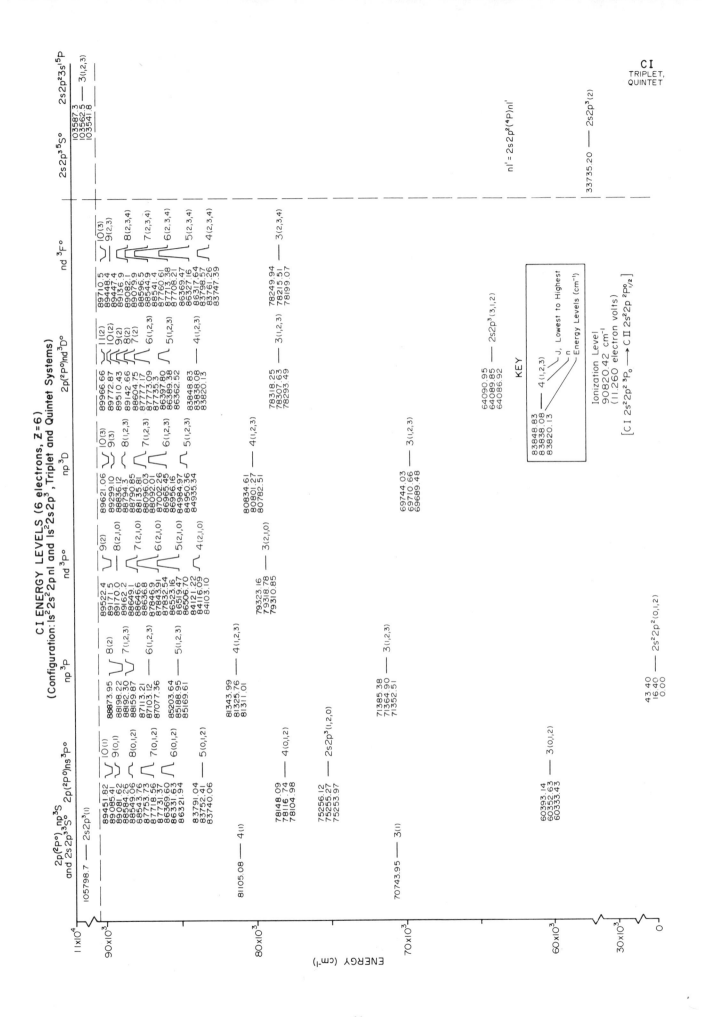

C I ENERGY LEVELS (6 electrons, Z=6)
(Configuration: 1s²2s²2pnl and 1s²2s2p³, Triplet and Quintet Systems)

C I
TRIPLET,
QUINTET

ENERGY (cm⁻¹)

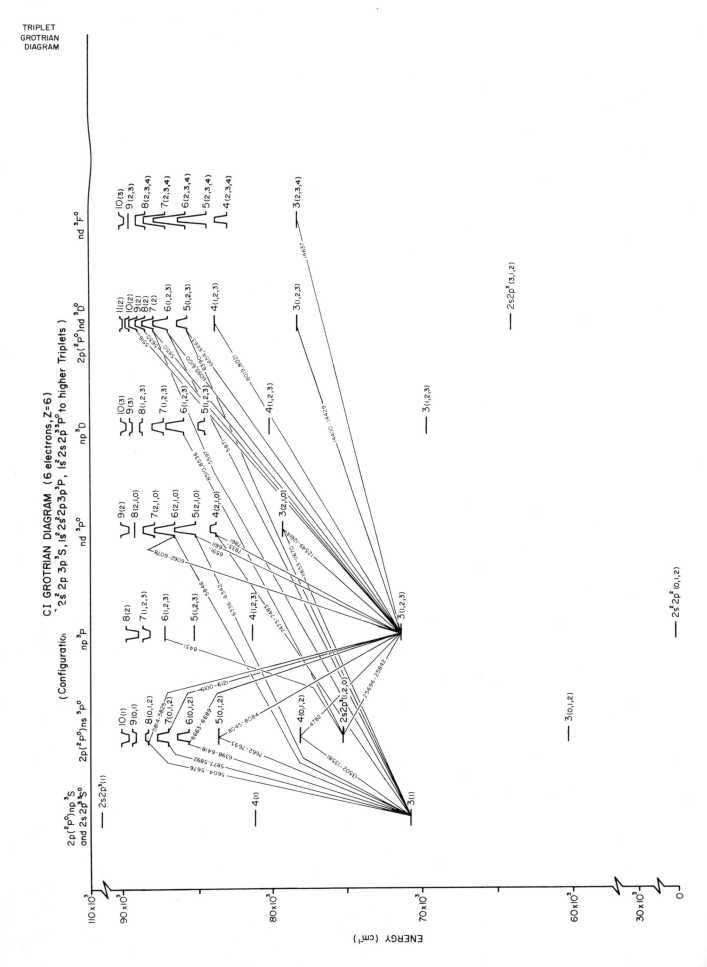

TRIPLET
GROTRIAN
DIAGRAM

CI GROTRIAN DIAGRAM (6 electrons, Z=6)
2s² 2p 3p ³S, 1s²2s²2p3p³P, 1s²2s 2p³³P⁰ to higher Triplets)

(Configuration

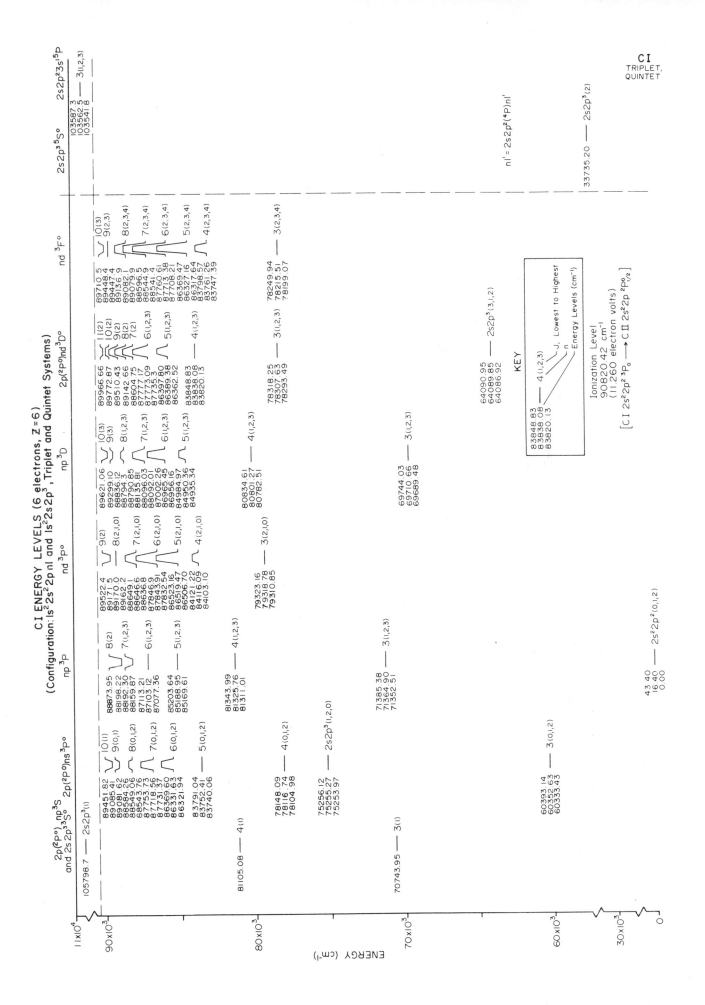

C I ENERGY LEVELS (6 electrons, Z=6)
(Configuration: 1s²2s²2pnl and 1s²2s2p³, Triplet and Quintet Systems)

C I ENERGY LEVELS (6 electrons, Z=6)

(Configurations: $1s^2 2s^2 2p\,(^2P^o)nf\,L_j(J)$ and $1s^2 2s2p^2\,(^4P)\,3s\;^5P$)

Intermediate Coupling and Autoionizing Terms

	D	F	G	5p

5p

102587.3 —— (1,2,3)
102562.5
102541.8

F column (G terms):

89162.10 ⌇ $8\left[\tfrac{7}{2}\right]$ (4)

88634.07 —— $7\left[\tfrac{7}{2}\right]$ (3,4)
88633.98

87820.00 —— $6\left[\tfrac{7}{2}\right]$ (3,4)
87819.90

86469.66 —— $5\left[\tfrac{7}{2}\right]$ (3,4)
86469.51

84016.25 —— $9\left[\tfrac{9}{2}\right]$ (5,4)
84015.86 —— $4\left[\tfrac{7}{2}\right]$ (3,4)
83986.45
83986.22

F column (F terms):

89101.82 ⌇ $8\left[\tfrac{7}{2}\right]$ (3,4)
89101.79 ⌇ $8\left[\tfrac{5}{2}\right]$ (2)
89101.76

88575.31 —— $7\left[\tfrac{7}{2}\right]$ (4)
88574.87 —— $7\left[\tfrac{5}{2}\right]$ (2,3)
88574.85

87763.24 —— $7\left[\tfrac{7}{2}\right]$ (4,3)
87763.10 —— $6\left[\tfrac{5}{2}\right]$ (3,2)
87762.22
87762.12

86414.69 —— $5\left[\tfrac{7}{2}\right]$ (3,4)
86414.49 —— $5\left[\tfrac{5}{2}\right]$ (3,2)
86412.05
86411.98

83926.37 —— $7\left[\tfrac{7}{2}\right]$ (3,4)
83926.20 —— $4\left[\tfrac{5}{2}\right]$ (3,2)
83919.76
83919.65

D column:

88645.33 ⌇ $7\left[\tfrac{3}{2}\right]$ (2)
88638.30 ⌇ $\left[\tfrac{5}{2}\right]$ (2)

87837.74 —— $6\left[\tfrac{3}{2}\right]$ (3,2)
87822.02 —— $\left[\tfrac{5}{2}\right]$ (3,2)
87826.94

86498.64 —— $5\left[\tfrac{3}{2}\right]$ (1,2)
86498.55 —— $\left[\tfrac{5}{2}\right]$ (3,2)
86482.78
86482.66

84036.40 —— $4\left[\tfrac{3}{2}\right]$ (1,2)
84036.29 —— $\left[\tfrac{5}{2}\right]$ (3,4)
84013.40
84013.25

KEY

$$
\begin{array}{l}
87837.74 \\
87827.02 \\
87826.94
\end{array}
\;\; 6\left[\begin{array}{l}\tfrac{3}{2}\\ \tfrac{5}{2}\end{array}\right] (3,2)
$$

Final J, lowest to highest
Intermediate J, lowest to highest
n
Energy level (cm^{-1})

Ionization Level
9820.42 cm^{-1}
(11.260 eV)

$\left[\mathrm{C\,I}\,(s^2)s^2 2p^2\,{}^3P_o \rightarrow \mathrm{C\,II}\,(s^2)2s^2 2p\;{}^2P^o_{1/2}\right]$

Y-axis labels: 105 × 10^3, 102 × 10^3, 90 × 10^3, 85 × 10^3, 80 × 10^3

CI INTERCOMBINATION GROTRIAN DIAGRAM (6 electrons, Z = 6)
(Configuration: 1s²2s²2pnl, Singlets to Triplets)

CI
SINGLETS to
TRIPLETS
INTERCOMBINATION
GROTRIAN DIAGRAM

ENERGY (cm⁻¹)

66

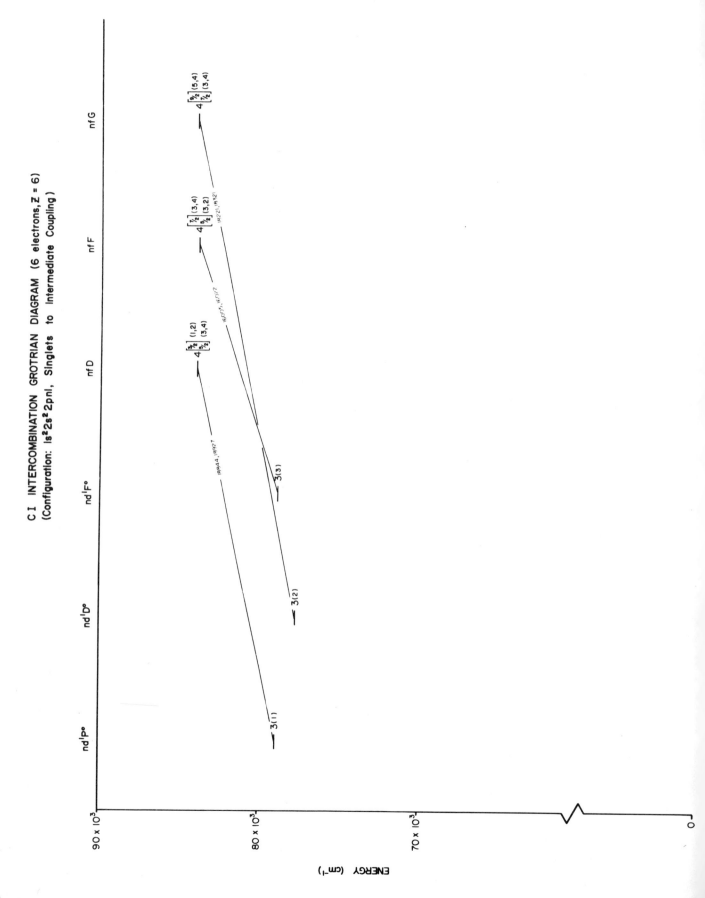

C I INTERCOMBINATION GROTRIAN DIAGRAM (6 electrons, Z = 6)
(Configuration: $1s^2 2s^2 2pnl$, Singlets to Intermediate Coupling)

68

C I INTERCOMBINATION GROTRIAN DIAGRAM (6 electrons, Z = 6)

(Configuration: 1s²2s²2pnl, Ground term (2p³P) to Singlet, Quintet Terms)

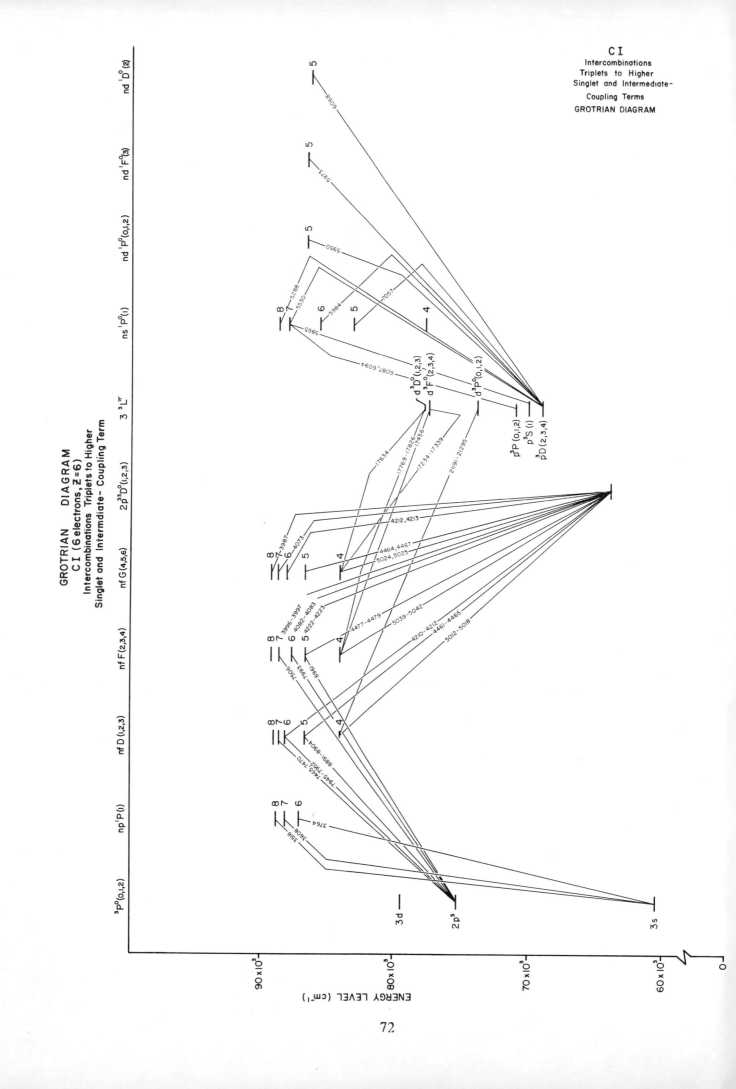

GROTRIAN DIAGRAM
C I (6 electrons, Z = 6)
Intercombinations Triplets to Higher
Singlet and Intermdiate- Coupling Term

ENERGY LEVEL (cm⁻¹)

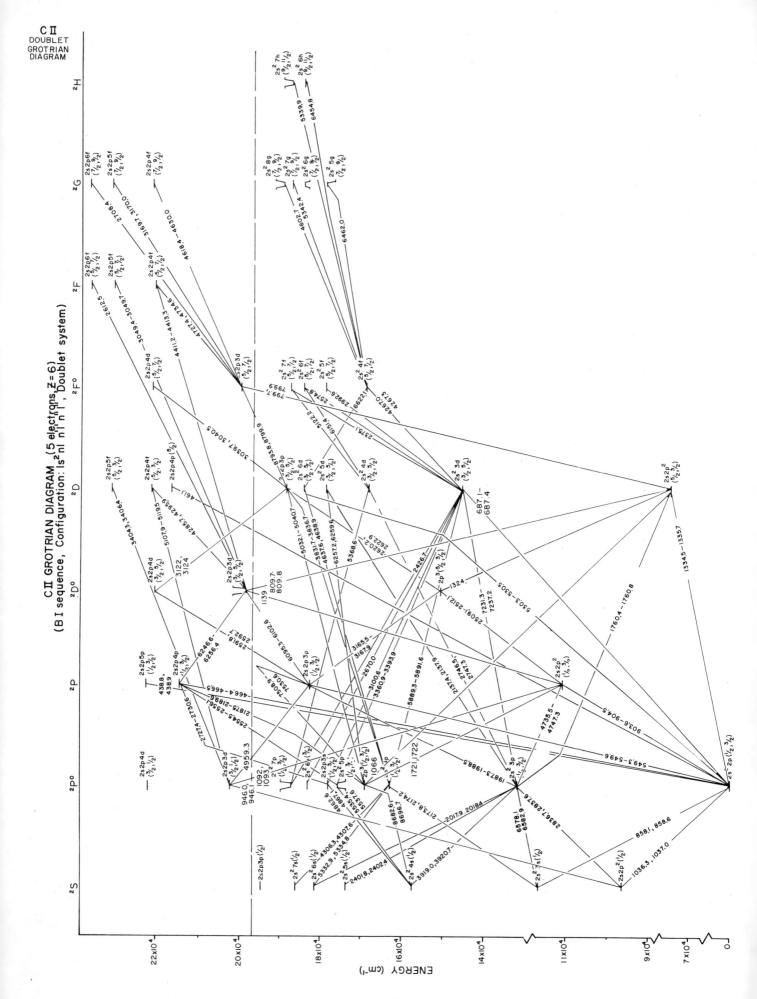

CII
DOUBLET
GROTRIAN
DIAGRAM

CII GROTRIAN DIAGRAM (5 electrons, Z=6)
(BI sequence, Configuration: 1s²nl n'l'n''l'', Doublet system)

ENERGY (cm⁻¹)

74

C II ENERGY LEVELS (5 electrons, Z = 6)

(BI Sequence, Configuration $1s^2 nl\ n'l'\ n''l''$, Doublet Term System)

C II
DOUBLET

Column headers: ²S ²Pº ²P ²Dº ²D ²Fº ²F ²G ²H

ENERGY LEVELS (cm⁻¹)

22 × 10⁴
20 × 10⁴
18 × 10⁴
16 × 10⁴
14 × 10⁴
11 × 10⁴
9 × 10⁴
7 × 10⁴

²S column:
194571.9 — 2s2p3p (½)
185733.36 — 2s²7s (½)
181264.67 — 2s²6s (½)
173348.27 — 2s²5s (½)
157234.50 — 2s²4s (½)
116538.08 — 2s²3s (½)
96494.17 — 2s2p² (½)

²Pº column:
222286.1 — 2s2p4d (³⁄₂, ½)
222259.2
202204.95 — 2s2p3d (³⁄₂, ½)
202180.28
186746.3 — 2s²7p (½, ³⁄₂)
182993.66 — 2s²6p (³⁄₂)
177793.97 — 2s²5p (½, ³⁄₂)
177775.02
175295.18 — 2s²5p (½, ³⁄₂)
175287.82
168748.73 — 2p³ (½, ³⁄₂)
168729.96
162525.00 — 2s²4p (½, ³⁄₂)
162518.32
131735.95 — 2s²3p (½, ³⁄₂)
131724.80

²P column:
227908.14 — 2s2p5p (½, ³⁄₂)
227881.64
214430.38 — 2s2p4p (½, ³⁄₂)
214407.76
182043.84 — 2s2p3p (½, ³⁄₂)
182024.29
150467.12 — 2p³(½, ³⁄₂)
150462.01
110665.99 — 2s²p² (½, ³⁄₂)
110624.60
63.42 — 2s²2p (½, ⁷⁄₂)
0.0

²Dº column:
220614.94 — 2s2p4d (³⁄₂, ½)
220601.96
198436.74 — 2s2p3d (³⁄₂, ½)
198425.86
150467.12 — 2p³(⁵⁄₂, ³⁄₂)
150462.01

²D column:
231570.8 — 2s2p5p (⁵⁄₂, ³⁄₂)
231528.82
221708.14 — 2s2p4f (⁵⁄₂, ³⁄₂)
221752.69
216927 — 2s2p4p (⁵⁄₂)
188615.50 — 2s2p3p (³⁄₂, ⁵⁄₂)
188581.68
184075.71 — 2s²6d (³⁄₂, ⁵⁄₂)
184075.02
178496.14 — 2s²5d (³⁄₂, ⁵⁄₂)
178495.54
168124.88 — 2s²4d (³⁄₂, ⁵⁄₂)
168124.17
144551.13 — 2s²3d (³⁄₂, ⁵⁄₂)
144549.70
74933.10 — 2s²p² (⁵⁄₂, ³⁄₂)
74930.60

²Fº column:
199983.67 — 2s2p3d (⁵⁄₂, ⁷⁄₂)
199941.84
187642.0 — 2s²7f (⁵⁄₂, ⁷⁄₂)
184376.49 — 2s²6f (⁵⁄₂, ⁷⁄₂)
178956.37 — 2s²5f (⁵⁄₂, ⁷⁄₂)
168978.77 — 2s²4f (⁵⁄₂, ⁷⁄₂)

²F column:
236703 — 2s2p6f (⁵⁄₂, ⁷⁄₂)
236693
231217.63 — 2s2p5f (⁵⁄₂, ⁷⁄₂)
231209.63
221098.35 — 2s2p4f (⁵⁄₂, ⁷⁄₂)
221089.31

²G column:
236890.1 — 2s2p6f (⁷⁄₂, ⁹⁄₂)
236857.0
231519.6 — 2s2p5f (⁷⁄₂, ⁹⁄₂)
231481.1
221626.15 — 2s2p4f (⁷⁄₂, ⁹⁄₂)
221587.55
189794.6 — 2s²8g (⁷⁄₂, ⁹⁄₂)
187691.8 — 2s²7g (⁷⁄₂, ⁹⁄₂)
184449.70 — 2s²6g (⁷⁄₂, ⁹⁄₂)
179073.48 — 2s²5g (⁷⁄₂, ⁹⁄₂)

²H column:
187701 — 2s²7h (⁹⁄₂, ¹¹⁄₂)
184529.9 — 2s²6h (⁹⁄₂, ¹¹⁄₂)

KEY

150467.12 ⎫ 2p³(⁵⁄₂, ³⁄₂)
150462.01 ⎭

J VALUES – LOWER TO
UPPER
CONFIGURATION
ENERGY LEVELS (cm⁻¹)

IONIZATION LEVEL
196664.7 cm⁻¹
(24.383 eV)
[C II 2s²2p ²Pº₁/₂ → C III 2s² ¹S₀]

75

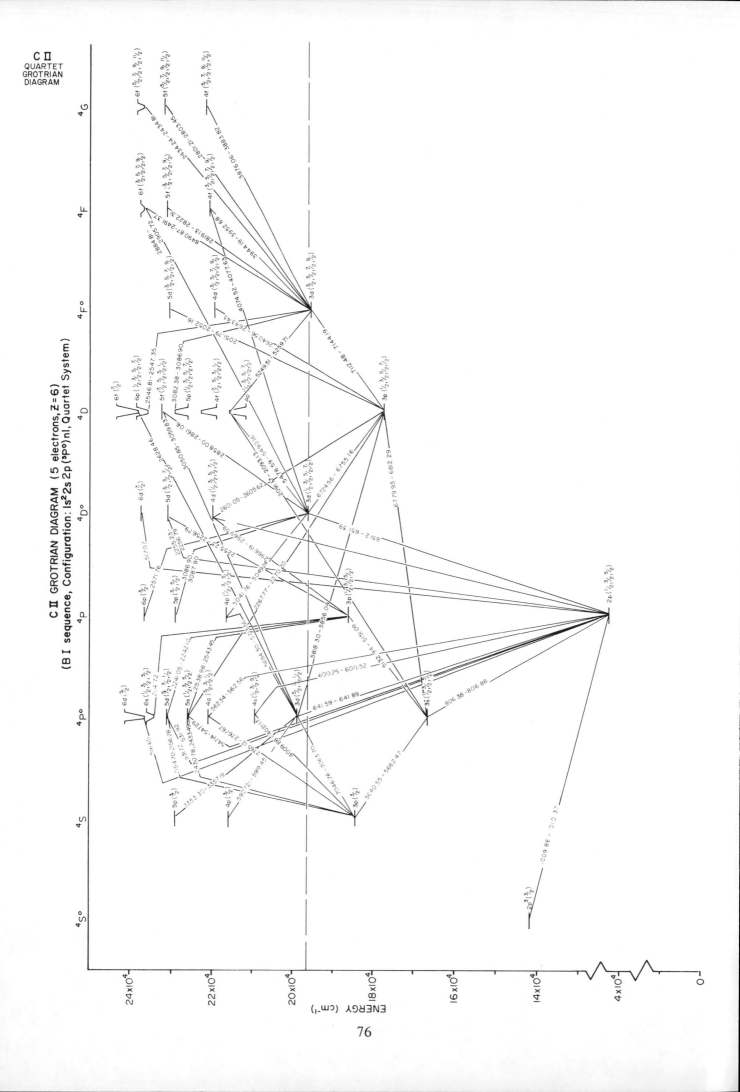

C II ENERGY LEVELS (5 electrons, Z=6)
(BI Sequence, Configuration, 1s² 2s 2p (³P°) nl, Quartet Term System)

77

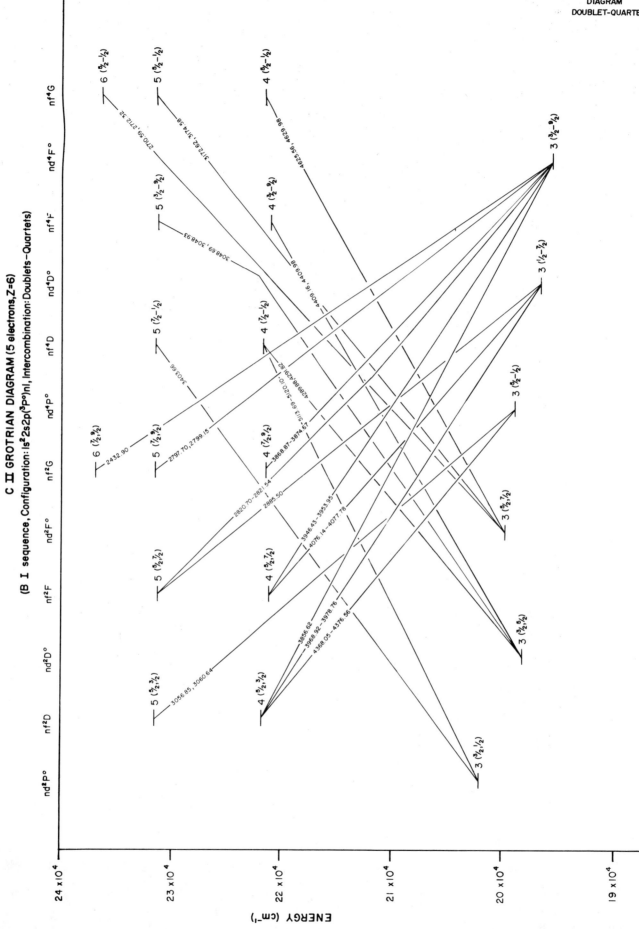

C II GROTRIAN DIAGRAM (5 electrons, Z=6)

(B I sequence, Configuration: $1s^2 2s 2p(^3P^o)nl$, Intercombination: Doublets–Quartets)

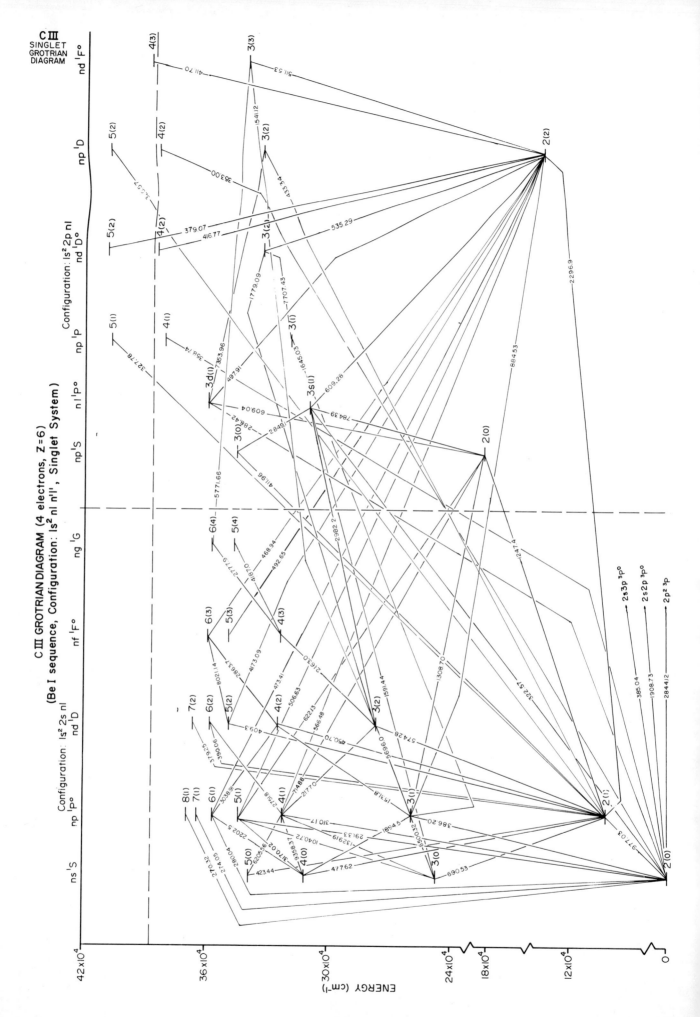

C III
SINGLET
GROTRIAN
DIAGRAM

C III GROTRIAN DIAGRAM (4 electrons, Z=6)
(Be I sequence, Configuration: 1s² nl n'l', Singlet System)

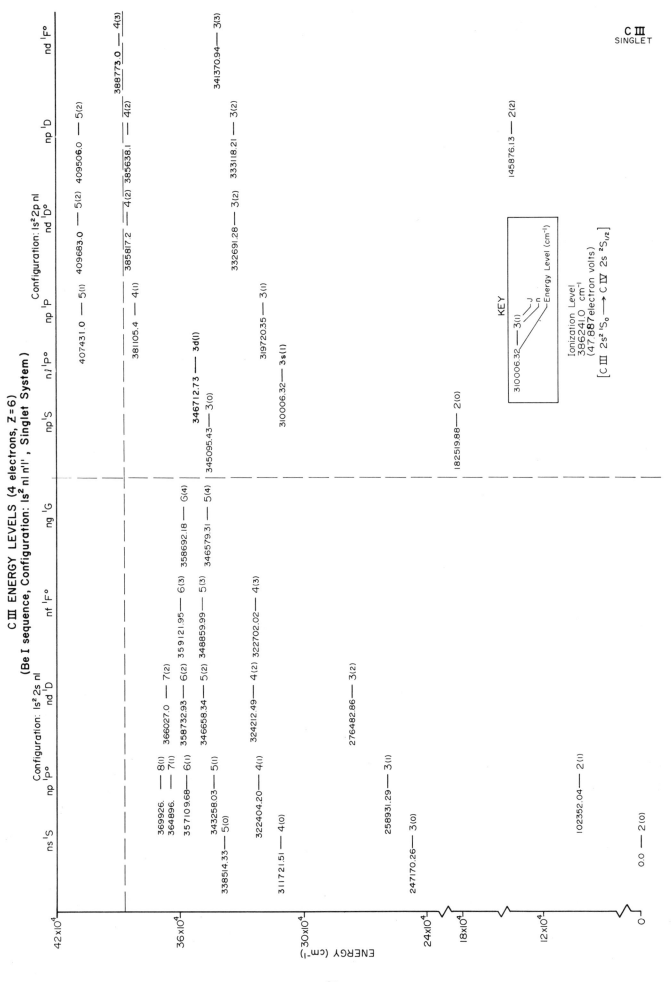

C III
SINGLET

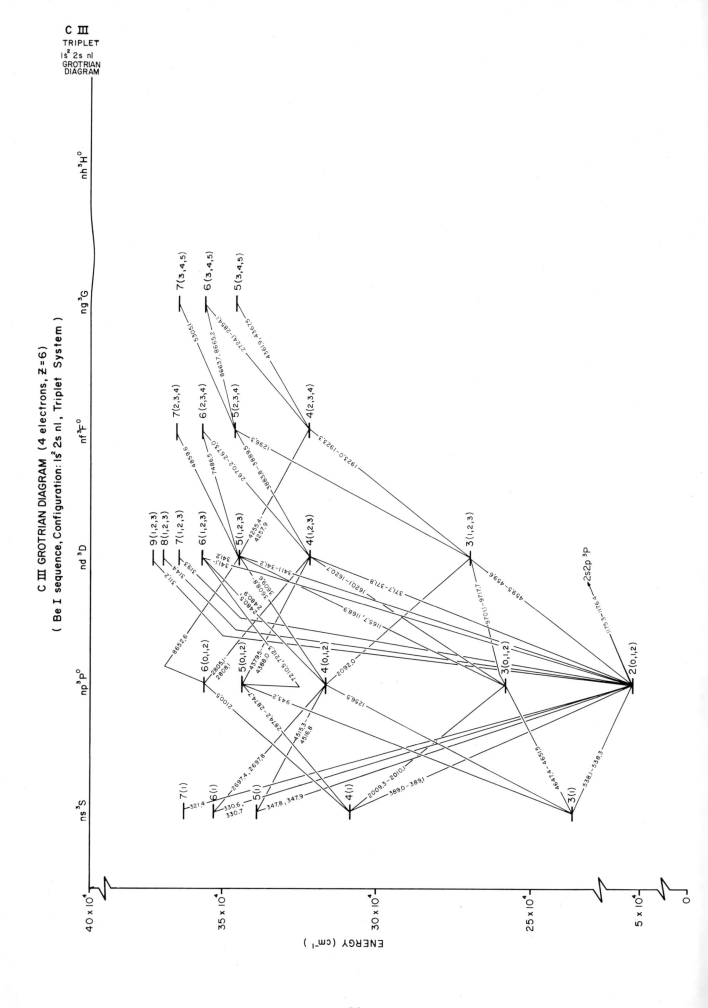

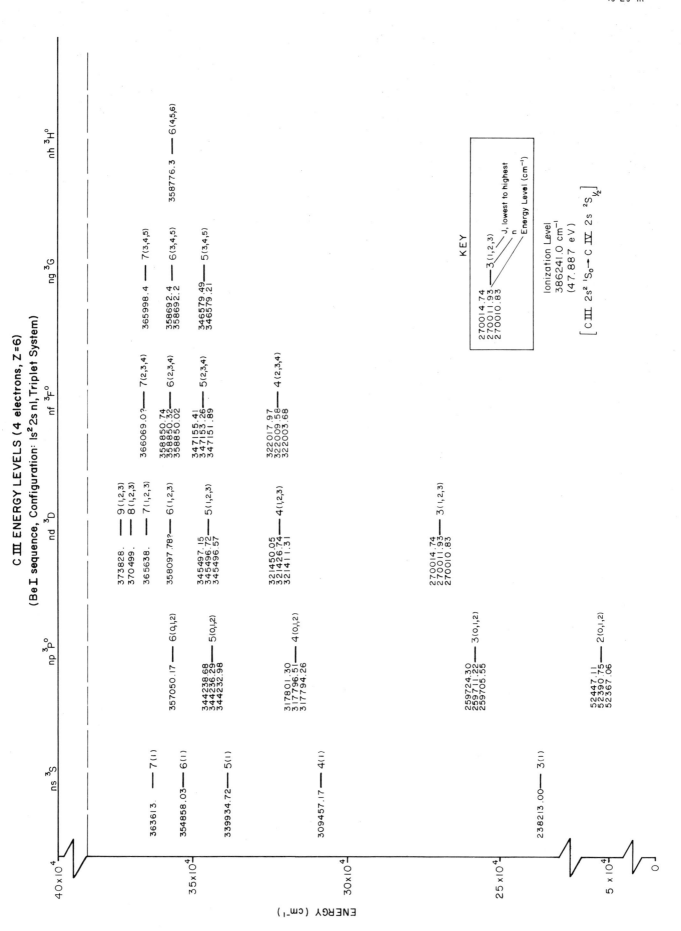

CIII ENERGY LEVELS (4 electrons, Z=6)
(Be I sequence, Configuration: 1s²2s nl, Triplet System)

C III
TRIPLET
1s²2s nl

ns ³S

363613. —— 7(1)

354858.03 —— 6(1)

339934.72 —— 5(1)

309457.17 —— 4(1)

238213.00 —— 3(1)

np ³P°

357050.17 —— 6(0,1,2)

344238.68
344236.29 —— 5(0,1,2)
344232.98

317801.30
317796.51 —— 4(0,1,2)
317794.26

259724.30
259711.22 —— 3(0,1,2)
259705.55

52447.11
52390.75 —— 2(0,1,2)
52367.06

nd ³D

373828.
370499. —— 9(1,2,3)
        —— 8(1,2,3)

365638. —— 7(1,2,3)

358097.78? —— 6(1,2,3)

345497.15
345496.72 —— 5(1,2,3)
345496.57

321450.05
321426.74 —— 4(1,2,3)
321411.31

270014.74
270011.93 —— 3(1,2,3)
270010.83

nf ³F°

366069.0? —— 7(2,3,4)

358850.74
358850.32 —— 6(2,3,4)
358850.02

347155.41
347153.26 —— 5(2,3,4)
347151.89

322017.97
322009.58 —— 4(2,3,4)
322003.68

ng ³G

365998.4 —— 7(3,4,5)

358692.4
358692.2 —— 6(3,4,5)

346579.49
346579.21 —— 5(3,4,5)

nh ³H°

358776.3 —— 6(4,5,6)

KEY

270014.74
270011.93 —— 3(1,2,3)
270010.83

3(1,2,3) ←— J, lowest to highest
         ←— n
         ←— Energy Level (cm⁻¹)

Ionization Level
386241.0 cm⁻¹
(47.887 eV)

[C III 2s² ¹S₀ → C IV 2s ²S₁/₂]

ENERGY (cm⁻¹)

40×10⁴

35×10⁴

30×10⁴

25×10⁴

5×10⁴

0

83

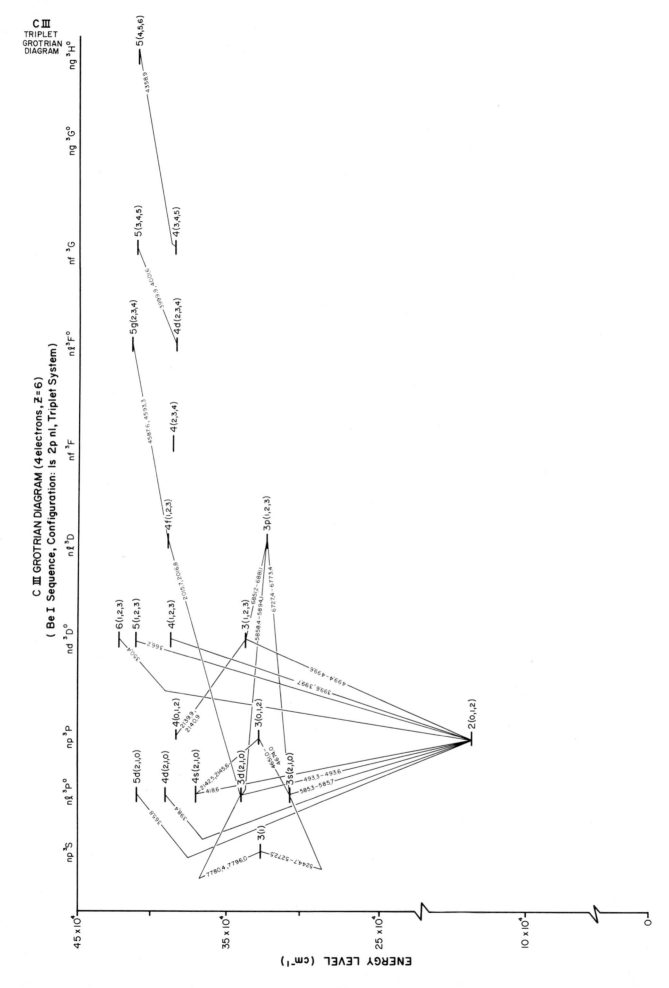

C III GROTRIAN DIAGRAM (4 electrons, Z=6)

(Be I Sequence, Configuration: 1s 2p nl, Triplet System)

C III
TRIPLET
GROTRIAN
DIAGRAM

ENERGY LEVEL (cm⁻¹)

84

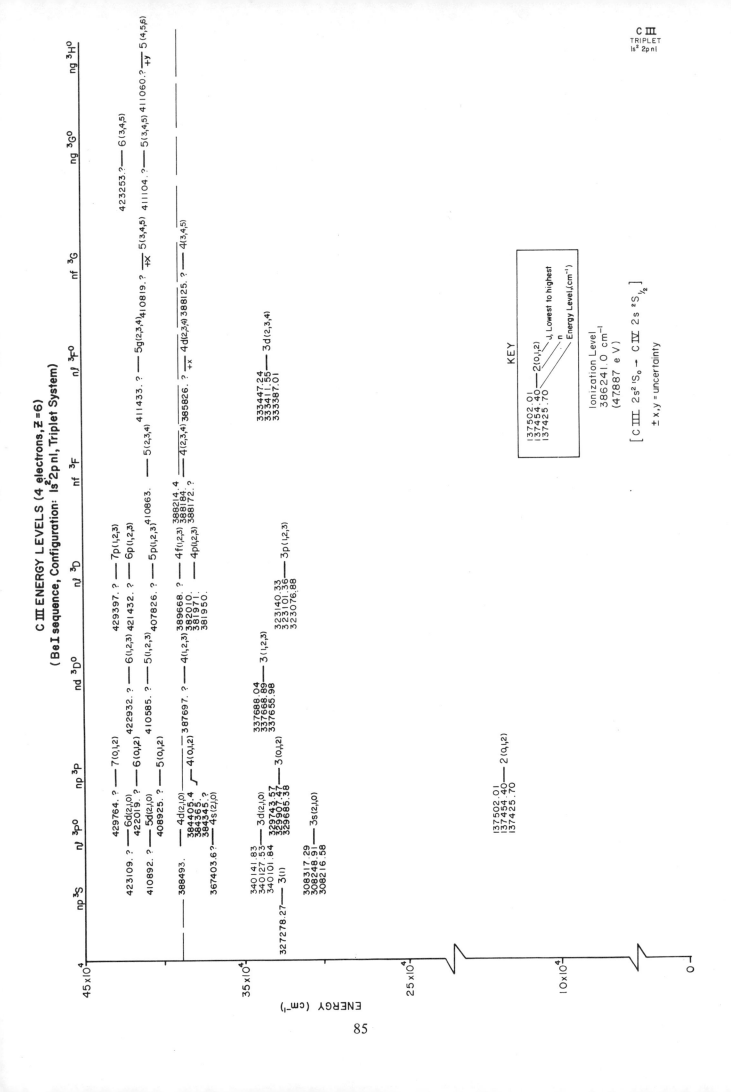

C III ENERGY LEVELS (4 electrons, Z = 6)
(Be I sequence, Configuration: $1s^2 2pnl$, Triplet System)

C III
TRIPLET
$1s^2 2pnl$

KEY

137502.01
137454.40 ——— 2(0,1,2) — J, Lowest to highest
137425.70 — n
——— Energy Level, (cm⁻¹)

Ionization Level
386241.0 cm⁻¹
(47.887 eV)

[C III $2s^2 \, ^1S_0 \rightarrow$ C IV $2s \, ^2S_{1/2}$]

± x, y = uncertainty

ENERGY (cm⁻¹)

45×10^4

35×10^4

25×10^4

10×10^4

0

np ³S

423109. ? ———
410892. ? ———
388493. ———
367403.6? ———
327278.27 ——— 3(1)

nl ³P⁰

429764. ? ——— 7(0,1,2)
6d(2,1,0)
422019. ? ——— 6(0,1,2)
5d(2,1,0)
410892. ? ——— 5(0,1,2)
4d(2,1,0)
384405.4
384365. ——— 4(0,1,2)
384345. ?
4s(2,1,0)
340141.83
340127.53 3d(2,1,0)
340101.84
329743.57
329907.47 ——— 3(0,1,2)
329685.38
308317.29
308248.91 ——— 3s(2,1,0)
308216.58

np ³P

429397. ? ——— 7(1,2,3)
6p(1,2,3)
422932. ? ——— 6(1,2,3)
5p(1,2,3)
410585. ? ——— 5(1,2,3)
4f(1,2,3)
389668. ? ——— 4(1,2,3)
382010.
381971.
381950.
337688.04
337668.89 ——— 3(1,2,3)
337655.98
323140.33
323101.36 ——— 3p(1,2,3)
323076.88
137502.01
137454.40 ——— 2(0,1,2)
137425.70

nd ³D⁰

nl ³D

429397. ? ——— 7p(1,2,3)
6p(1,2,3)
421432. ? —— 6p(1,2,3)
407826. ? ——— 5p(1,2,3)
388214.4
388184. ——— 4f(1,2,3)
388010. ———
381971. 4p(1,2,3)
388172. ?

nf ³F

388125. ? ——— 4(3,4,5)
385826. ? ——— 4(2,3,4)
+x ——— 5(3,4,5)
411433. ? ——— 5g(2,3,4)
410819. ? ———
333447.24
333411.55 ——— 3d(2,3,4)
333387.01

nJ ³F⁰

ng ³G⁰

423253. ? ——— 6(3,4,5)
411104. ? ——— 5(3,4,5)
411060. ? +y ——— 5(4,5,6)

ng ³H⁰

85

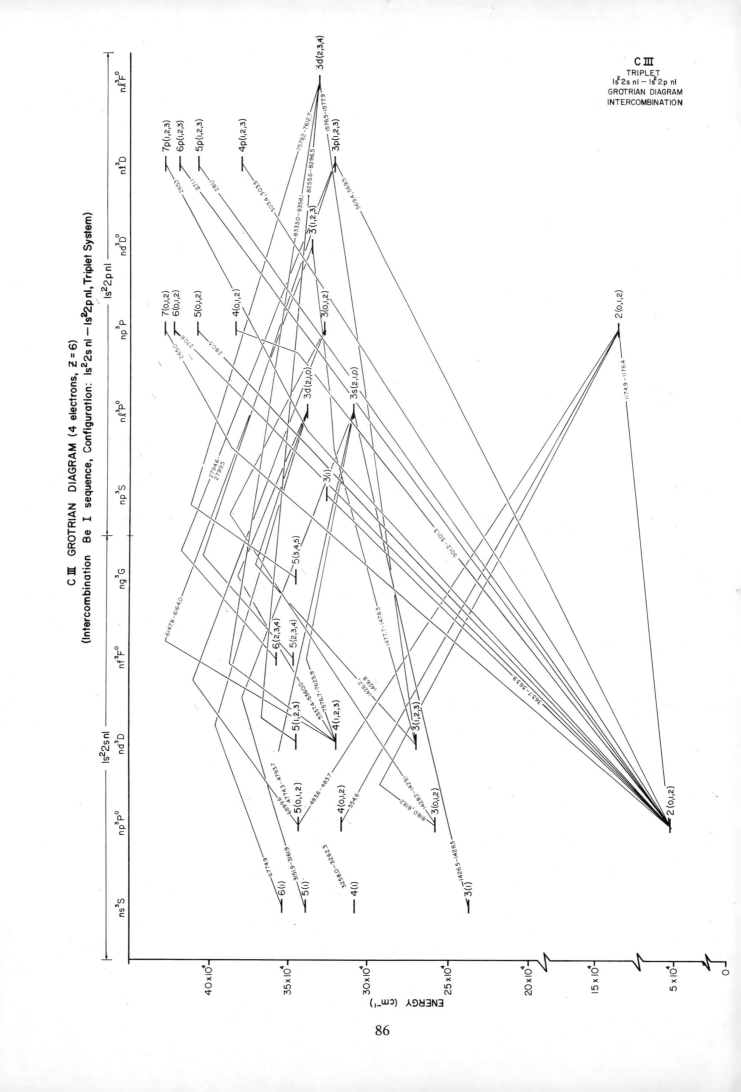

C III GROTRIAN DIAGRAM (4 electrons, Z = 6)

(Intercombination Be I sequence, Configuration: $1s^2 2s\,nl - 1s^2 2p\,nl$, Triplet System)

C III
TRIPLET
$1s^2 2s\,nl - 1s^2 2p\,nl$
GROTRIAN DIAGRAM
INTERCOMBINATION

98

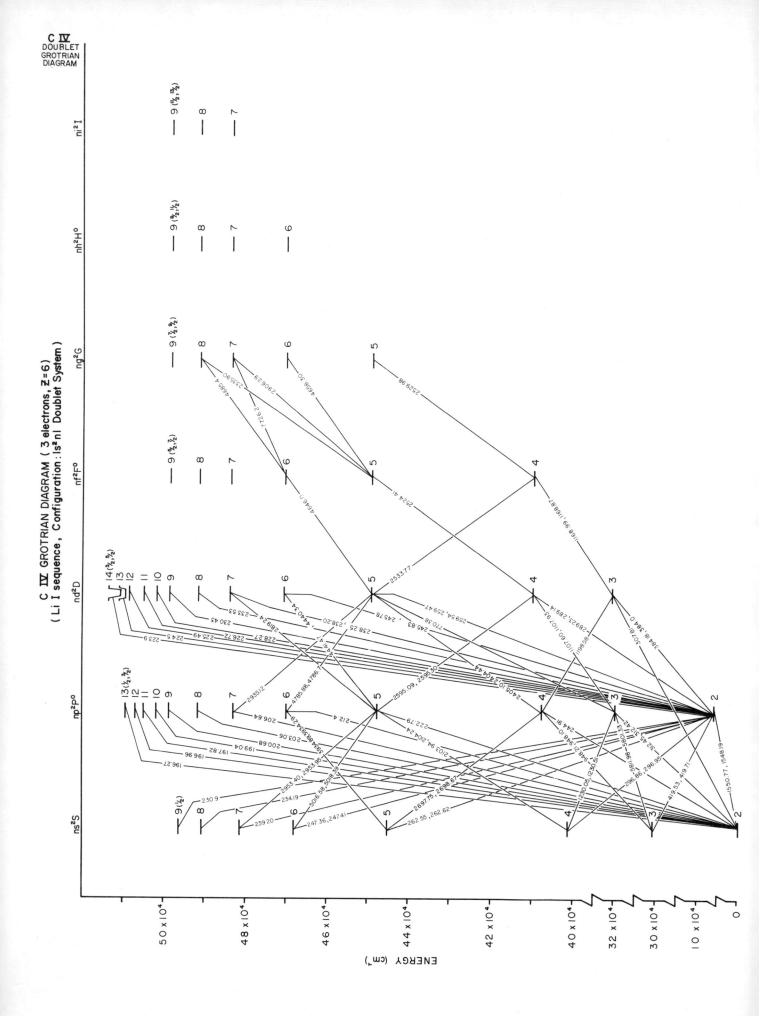

C IV GROTRIAN DIAGRAM (3 electrons, Z̄=6)
(Li I sequence, Configuration : ls²nl Doublet System)

C IV
DOUBLET
GROTRIAN
DIAGRAM

88

C IV ENERGY LEVELS (3 electrons, Z=6)
(Li I sequence, Configuration: 1s²nl Doublet System)

KEY

44994840 —— 5 (7/2, 9/2)
| J, Lowest to Highest
| Configuration
| Energy Level (cm⁻¹)

Ionization Level
520178.4 cm⁻¹
(64.492 eV)

[C IV 1s²2s ²S₁/₂ — C IV 1s² ¹S₀]

ni ²I

49850l.44 —— 9 (¹¹/₂, ¹³/₂)
492743.86 —— 8 (¹¹/₂, ¹³/₂)
48434596 —— 7 (¹¹/₂, ¹³/₂)

nh ²H°

49850l.37 —— 9 (⁹/₂, ¹¹/₂)
492743.78 —— 8 (⁹/₂, ¹¹/₂)
48434584 —— 7 (⁹/₂, ¹¹/₂)
471406.80 —— 6 (⁹/₂, ¹¹/₂)

ng ²G

49850l.17 —— 9 (⁷/₂, ⁹/₂)
49274349 —— 8 (⁷/₂, ⁹/₂)
48434542 —— 7 (⁷/₂, ⁹/₂)
471406l6 —— 6 (⁷/₂, ⁹/₂)
44994840 —— 5 (⁷/₂, ⁹/₂)

nf ²F°

4985003 —— 9 (⁵/₂, ⁷/₂)
4927422 —— 8 (⁵/₂, ⁷/₂)
4843435 —— 7 (⁵/₂, ⁷/₂)
4714032 —— 6 (⁵/₂, ⁷/₂)
449939.8 —— 5 (⁵/₂, ⁷/₂)
4104342 —— 4 (⁵/₂, ⁷/₂)

nd ²D

51l2540 —— 14 (³/₂, ⁵/₂)
509821.0 —— 13 (³/₂, ⁵/₂)
508018.0 —— 12 (³/₂, ⁵/₂)
5056960 —— 11 (³/₂, ⁵/₂)
502398.0 —— 10 (³/₂, ⁵/₂)
4984906 —— 9 (³/₂, ⁵/₂)
4927285 —— 8 (³/₂, ⁵/₂)
4843206
4843000 —— 7 (³/₂, ⁵/₂)
47l37l5
47l3703 —— 6 (³/₂, ⁵/₂)
4498899
4498882 —— 5 (³/₂, ⁵/₂)
4103401
410336.1 —— 4 (³/₂, ⁵/₂)
3248903
324879.8 —— 3 (³/₂, ⁵/₂)

np ²P°

5097280 —— 13 (½, ³/₂)
507906.0 —— 12 (½, ³/₂)
505510.0 —— 11 (½, ³/₂)
5024l2.0 —— 10 (½, ³/₂)
49831l5.7
49831l4.6 —— 9 (½, ³/₂)
4924793
49247l7.7 —— 8 (½, ³/₂)
4839508
483948.4 —— 7 (½, ³/₂)
4707789
470775.0 —— 6 (½, ³/₂)
4488629
448855.8 —— 5 (½, ³/₂)
4083242
408311.1 —— 4 (½, ³/₂)
3200817
320050.1 —— 3 (½, ³/₂)
64591.7
64484.0 —— 2 (½, ³/₂)

ns ²S

4977367 —— 9 (½)
49l650.8 —— 8 (½)
482706.0 —— 7 (½)
468784.0 —— 6 (½)
445368.5 —— 5 (½)
401348.1 —— 4 (½)
302849.0 —— 3 (½)
0.0 —— 2 (½)

ENERGY (cm⁻¹)

50 x 10⁴
48 x 10⁴
46 x 10⁴
44 x 10⁴
42 x 10⁴
40 x 10⁴
32 x 10⁴
30 x 10⁴
10 x 10⁴
0

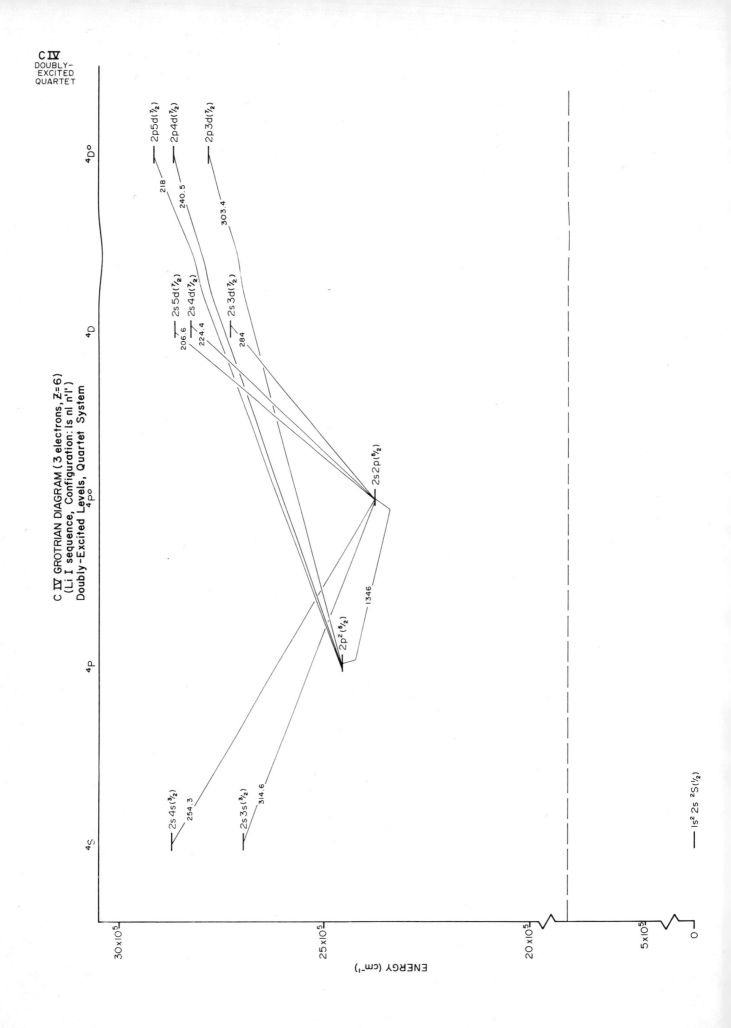

C Ⅳ GROTRIAN DIAGRAM (3 electrons, Z=6)
(Li I sequence, Configuration: ls nl n'l')
Doubly-Excited Levels, Quartet System

C Ⅳ
DOUBLY-
EXCITED
QUARTET

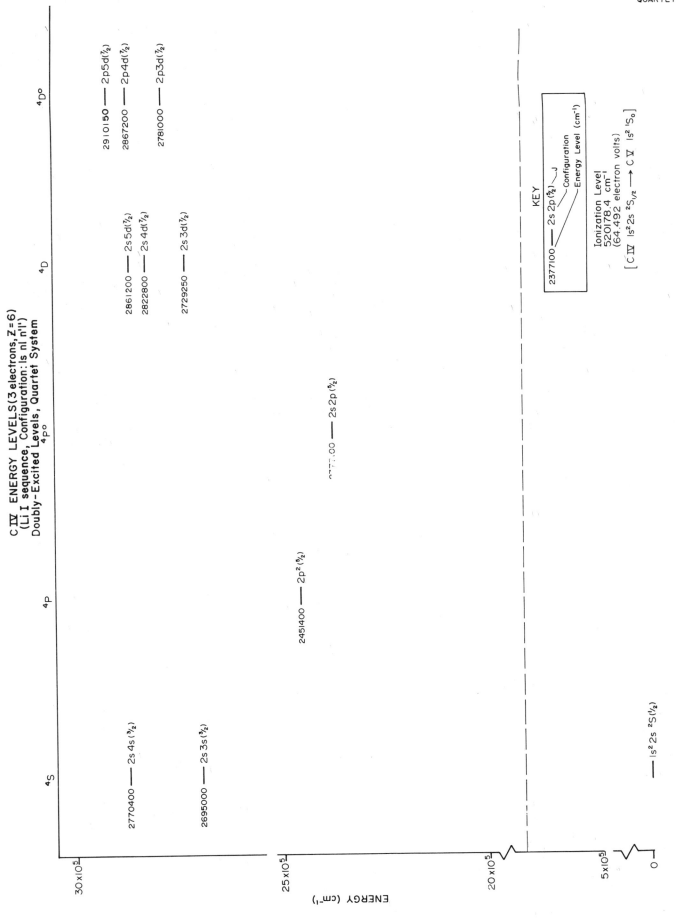

C IV ENERGY LEVELS (3 electrons, Z = 6)
(Li I sequence, Configuration: 1s nl n'l')
Doubly-Excited Levels, Quartet System

ENERGY (cm⁻¹)

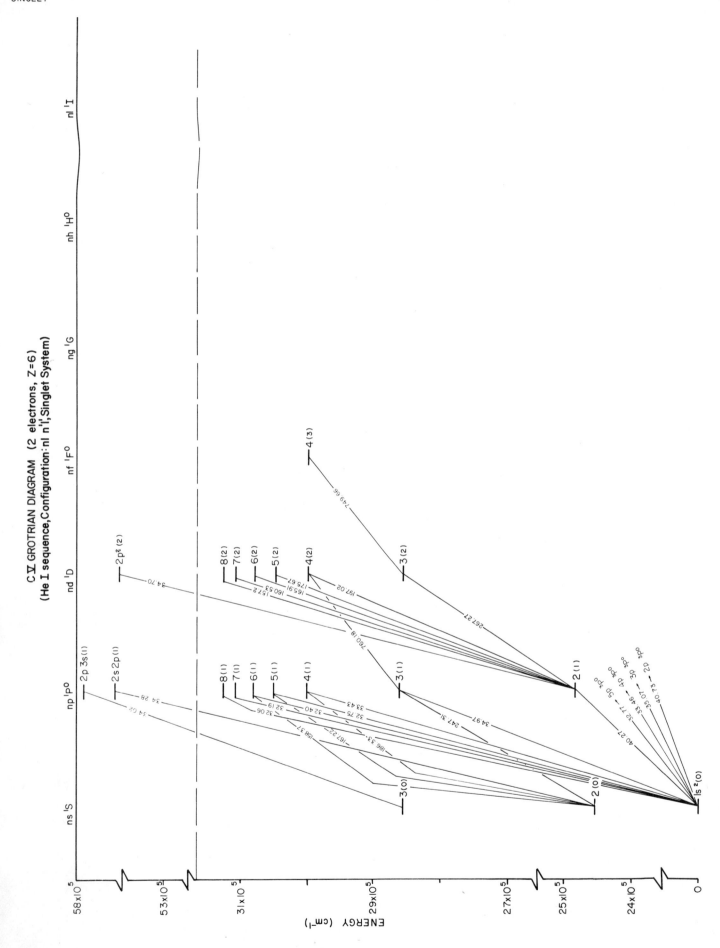

C V GROTRIAN DIAGRAM (2 electrons, Z=6)
(He I sequence, Configuration: nl n'l', Singlet System)

C V
SINGLET

92

C Ⅴ ENERGY LEVELS (2 electrons, Z=6)
(He I Sequence, Configuration; nl n'l', Singlet System)

C Ⅴ
SINGLET

Columns: ns¹S | np¹P° | nd¹D | nf¹F° | ng¹G | nh¹H° | ni¹I

ns¹S:
- 2851180 — 3(0)
- 2455024 — 2(0)
- 0 — 1s²(0)

np¹P°:
- 5789300 — 2p3s(1)
- 5371400 — 2s2p(1)
- 3119619 — 8(1)
- 3106541 — 7(1)
- 3086439 — 6(1)
- 3053044 — 5(1)
- 2991710 — 4(1)
- 2859375 — 3(1)
- 2483771 — 2(1)

nd¹D:
- 5365100 — 2p²(2)
- 3119530 — 8(2)
- 3106407 — 7(2)
- 3086189 — 6(2)
- 3052656 — 5(2)
- 2990923 — 4(2)
- 2857529 — 3(2)

nf¹F°:
- 3086186 — 6(3)
- 3052653 — 5(3)
- 2990923 — 4(3)

ng¹G:
- 31064074 — 7(4)
- 3086l908 — 6(4)
- 3052654.4 — 5(4)

nh¹H°:
- 31064082 — 7(5)
- 3086l908 — 6(5)

ni¹I:
- 31064086 — 7(6)

KEY

2859360 — 3(1)
J
n
Energy Level (cm⁻¹)

Ionization Level
3162395 cm⁻¹
(392.077 eV)
[C Ⅴ 1s² ¹S₀ → C Ⅵ 1s ²S₁/₂]

ENERGY (cm⁻¹)

58×10⁵
53×10⁵
31×10⁵
29×10⁵
27×10⁵
25×10⁵
24×10⁵
0

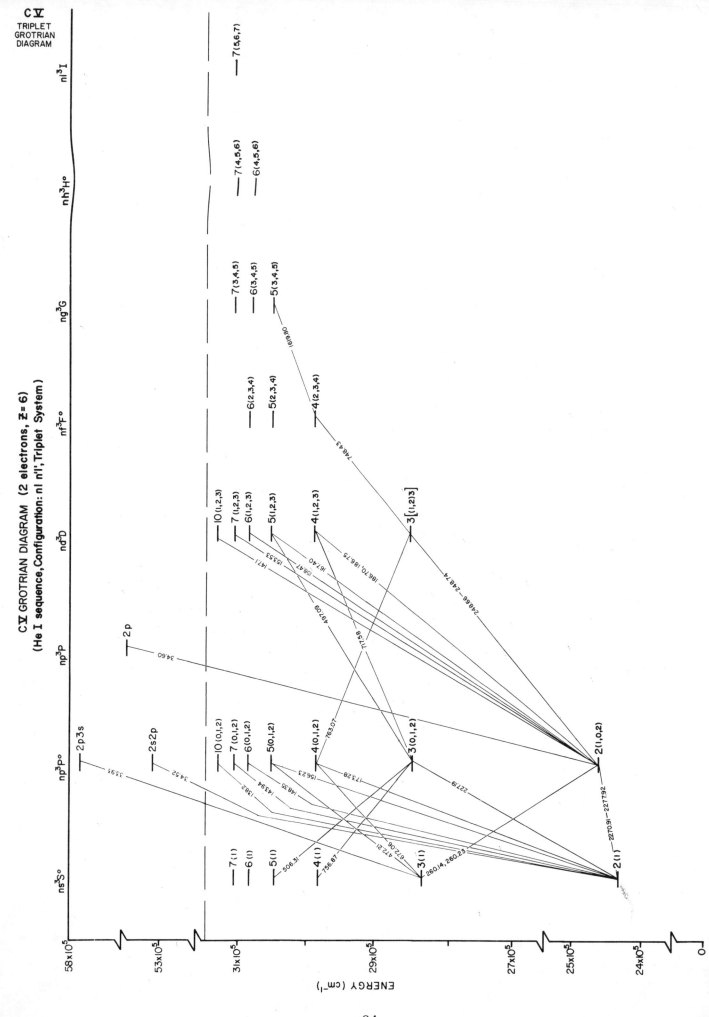

C V ENERGY LEVELS (2 electrons, $Z=6$)
(He I sequence, Configuration; nl n'l'; Triplet System)

C V
TRIPLET

Column headings: $ns\,^3S^\circ$ $np\,^3P$ $np\,^3P^\circ$ $nd\,^3D$ $nf\,^3F^\circ$ $ng\,^3G$ $nh\,^3H^\circ$ $ni\,^3I$

$ns\,^3S^\circ$:
5786700 —— 2p3s

$np\,^3P$:
5345500 —— 2p

$np\,^3P^\circ$:
5308200 —— 2s2p

$ni\,^3I$:
3106408.6 —— 7(5,6,7)

$nh\,^3H^\circ$:
3106408.2 —— 7(4,5,6)
3086190.8 —— 6(4,5,6)

$ng\,^3G$:
3106407.4 —— 7(3,4,5)
3086189.6 —— 6(3,4,5)
3052659.4 —— 5(3,4,5)

$nf\,^3F^\circ$:
3086186 —— 6(2,3,4)
3052653 —— 5(2,3,4)
2990923 —— 4(2,3,4)

$nd\,^3D$:
3135060 —— 10(1,2,3)
3106374 —— 7(1,2,3)
3086138 —— 6(1,2,3)
3052589 —— 5(1,2,3)
2990776 —— 4(1,2,3)
2857315
2857305 —— 3[(1,2)3]

$np\,^3P^\circ$:
3134860 —— 10(0,1,2)
3105933 —— 7(0,1,2)
3085435 —— 6(0,1,2)
3051332 —— 5(0,1,2)
2988359 —— 4(0,1,2)
2851418 —— 3(0,1,2)
2455284
2455161 —— 2(1,0,2)
2455148

$ns\,^3S^\circ$:
3105066 —— 7(1)
3084048 —— 6(1)
3048927 —— 5(1)
2998541 —— 4(1)
2839562 —— 3(1)
2411262 —— 2(1)

KEY

2859360 —— 3(1)
J
n
Energy Level (cm⁻¹)

Ionization Level
3162395 cm^{-1}
(392.077 eV)
[C V 1s$^2\,^1S_0$ → C VI 1s $^2S_{1/2}$]

ENERGY (cm⁻¹)

58×10⁵
53×10⁵
31×10⁵
29×10⁵
27×10⁵
25×10⁵
24×10⁵
0

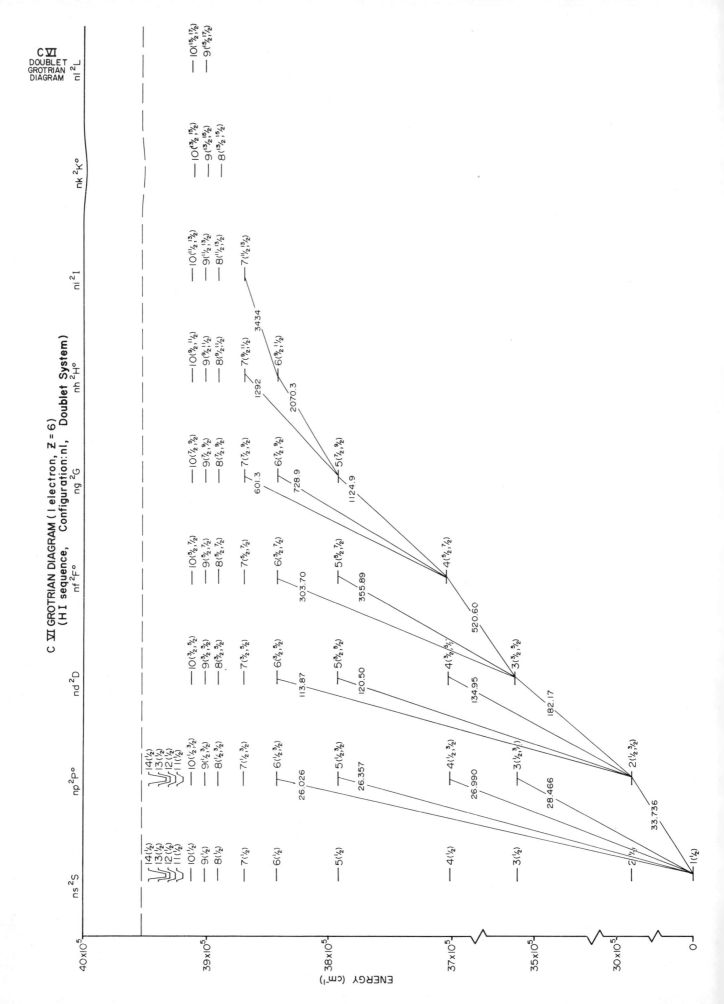

C XI GROTRIAN DIAGRAM (1 electron, Z = 6)
(H I sequence, Configuration:nl, Doublet System)

C XI
DOUBLET
GROTRIAN
DIAGRAM

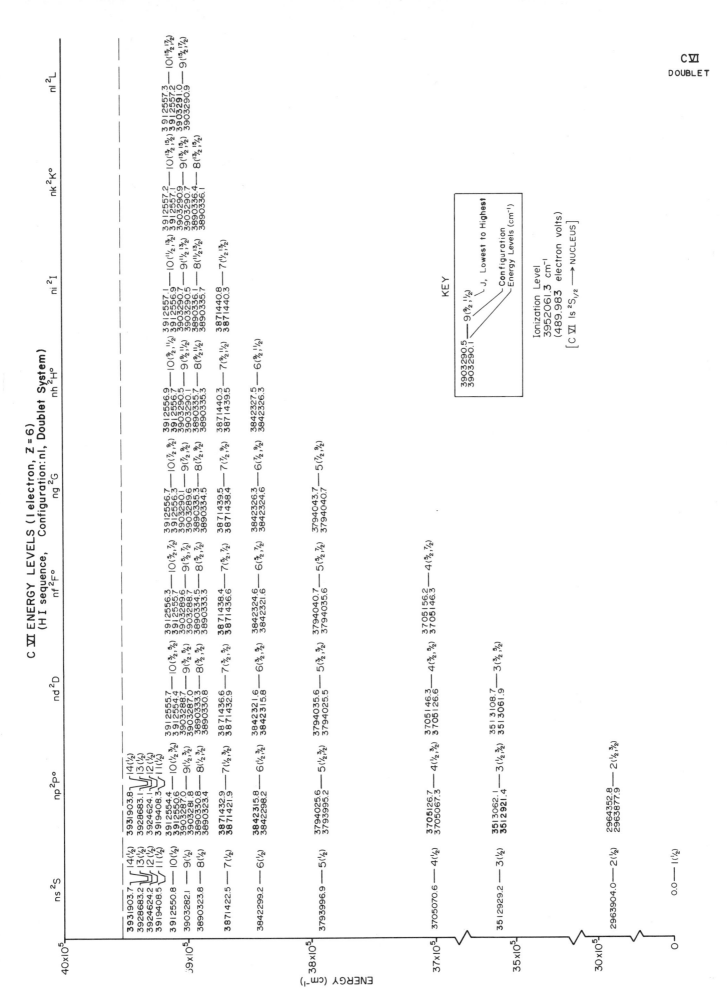

C VI ENERGY LEVELS (1 electron, Z = 6)
(H I sequence, Configuration: nl, Doublet System)

C VI
DOUBLET

97

C I $\qquad$ $Z = 6$ $\qquad$ 6 electrons

I.S. Bowen, Ap. J. **132**, 1 (1960).

Author gives a line table from empirical data of forbidden transitions.

L. Johansson, Ark. Fys. **25**, 425 (1963).

Author gives line tables for spectral lines observed in the range 3420–9659 Å.

Comment on intermediate coupling: According to L. Johansson, Ark. Fys. **31**, 201 (1966), the pair coupling with f electrons lies between

$$[\{(l_1,s_1)j_1,l_2\}K_1s_2]J \quad \text{and} \quad [\{(l_1,l_2)L,s_1\}K_1s_2]J .$$

In our key, we have called Johansson's "K", "intermediate J", and we have not used the symbol, "K", at all.

C II $\qquad$ $Z = 6$ $\qquad$ 5 electrons

Please see the general references.

C III $\qquad$ $Z = 6$ $\qquad$ 4 electrons

B. Edlén, Handbuch der Phys. **27**, 172 (1964).

Author gives corrections to energy-level values by K. Bockasten, Ark. Fys. **9**, 457 (1955).

C IV $\qquad$ $Z = 6$ $\qquad$ 3 electrons

H.G. Berry, M.C. Buchet-Poulizac, and J.P. Buchet, J. Opt. Soc. Amer. **63**, 240 (1973).

Lines arising from transitions within the quartet system are given.

C V $\qquad$ $Z = 6$ $\qquad$ 2 electrons

H.G. Berry, J. Désesquelles, and M. Dufay, Phys. Rev. **A6**, 600 (1972).

These authors summarize information on doubly-excited states in the He I isoelectronic sequence.

B.C. Fawcett, A.H. Gabriel, W.G. Griffin, B.B. Jones, and R. Wilson, Nature **200**, 1303 (1963).

Authors identify wavelengths observed in zeta spectrum in range 16–400 Å.

E. Holøien and J. Midtdal, J. Phys. B: Atom. Molec. Phys. **4**, 1243 (1971).

Authors give nonrelativistic calculations of energies in treatment of the helium isoelectronic sequence.

C VI $\qquad$ $Z = 6$ $\qquad$ 1 electron

B.C. Fawcett, A.H. Gabriel, W.G. Griffin, B.B. Jones, and R. Wilson, Nature **200**, 1303 (1963).

Authors give observed spectrum in range 16–400 Å.

Nitrogen (N)

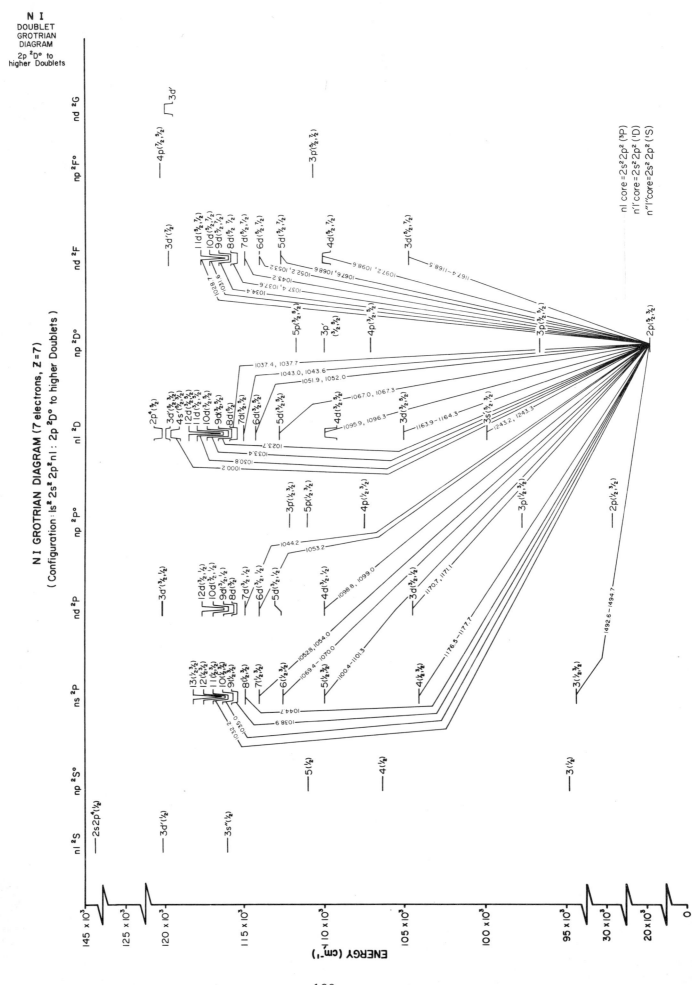

N I
DOUBLET
GROTRIAN
DIAGRAM
2p ²D° to
higher Doublets

N I GROTRIAN DIAGRAM (7 electrons, Z = 7)
(Configuration : 1s² 2s² 2p²nl : 2p ²D° to higher Doublets)

nl core = 2s² 2p² (³P)
n'l' core = 2s² 2p² (¹D)
n''l''core = 2s² 2p² (¹S)

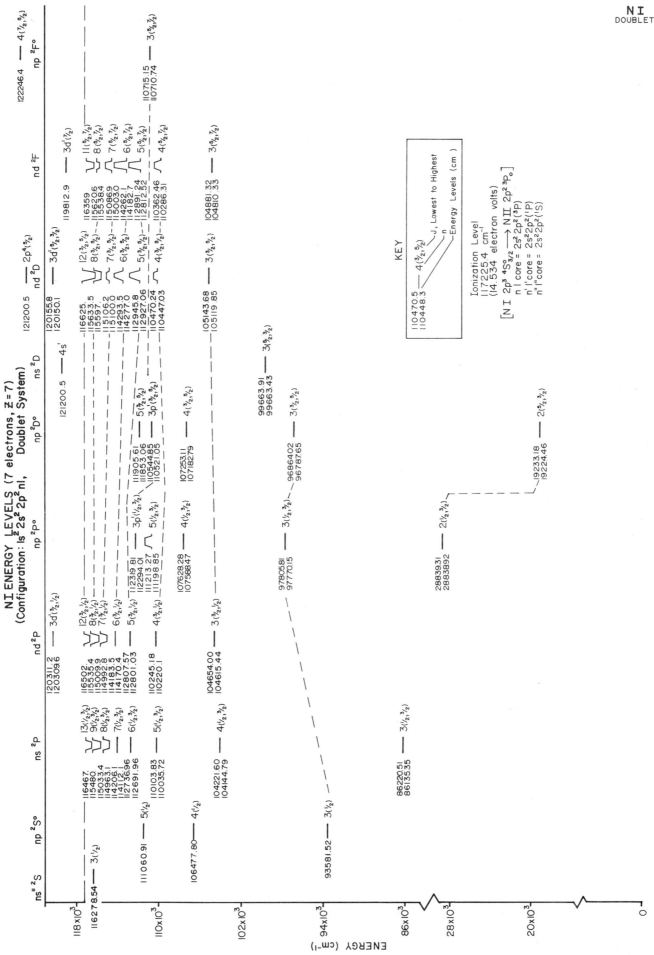

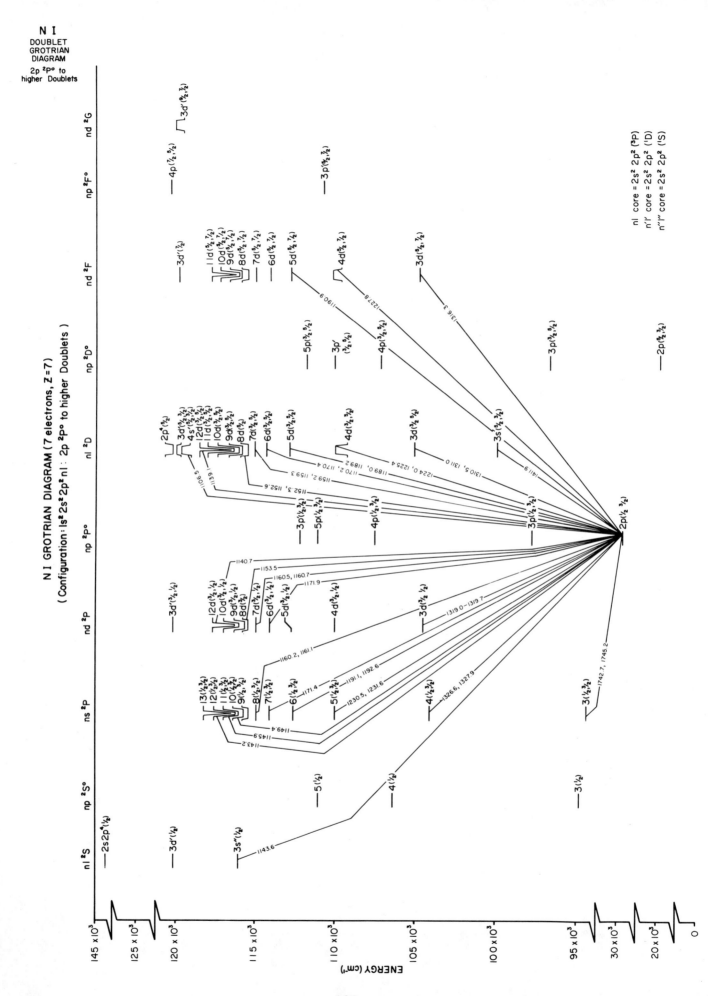

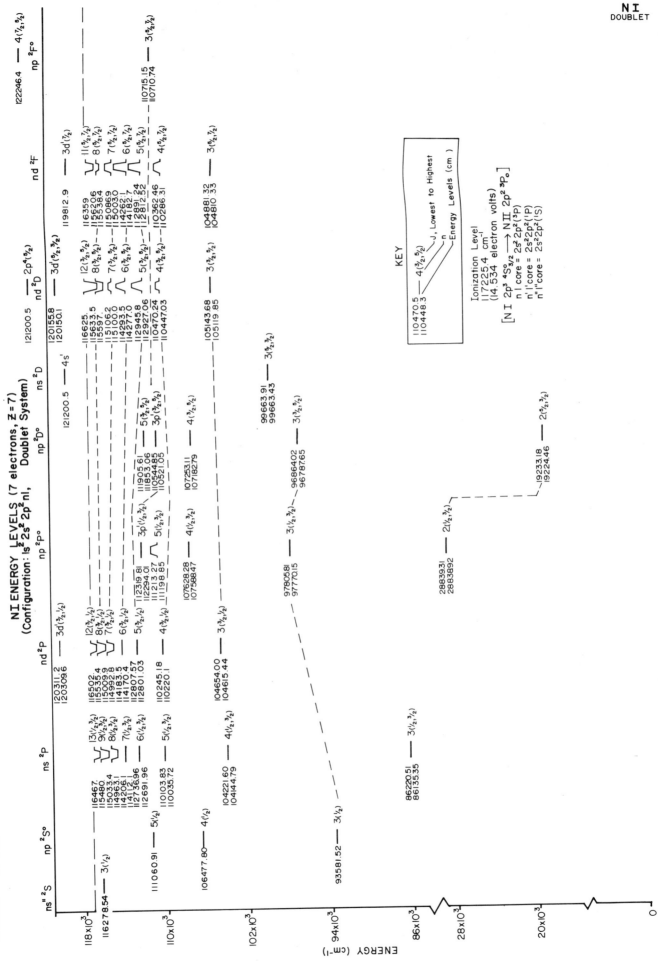

NI ENERGY LEVELS (7 electrons, Z=7)
(Configuration: 1s² 2s² 2p² nl, Doublet System)

N I
DOUBLET

103

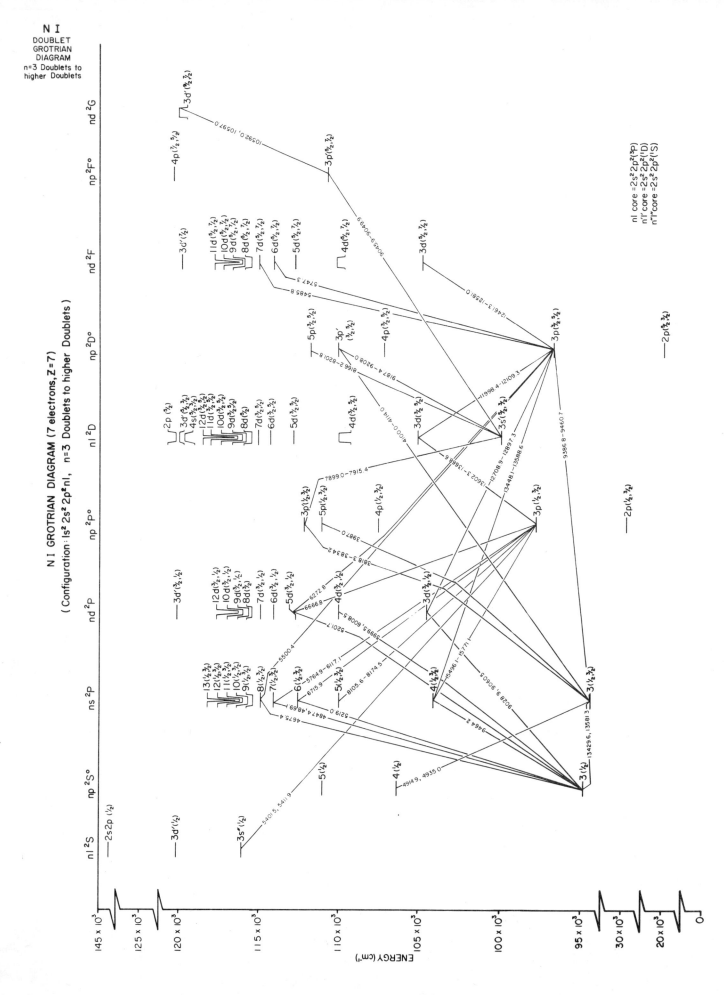

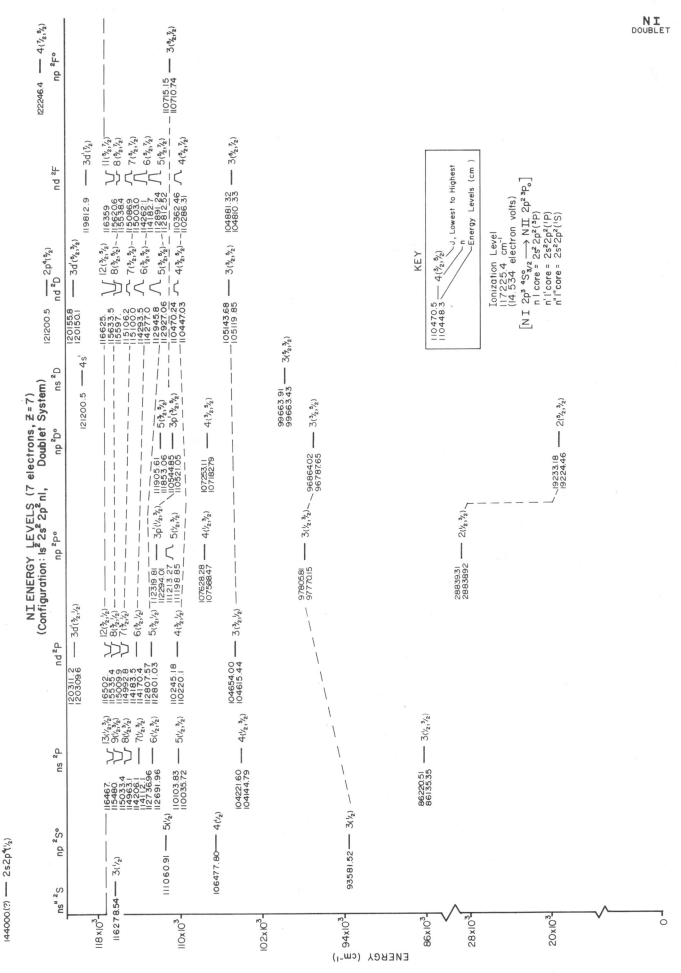

N I
DOUBLET

NI ENERGY LEVELS (7 electrons, Z = 7)
(Configuration: 1s² 2s² 2p² nl,   Doublet System)

ENERGY (cm⁻¹)

105

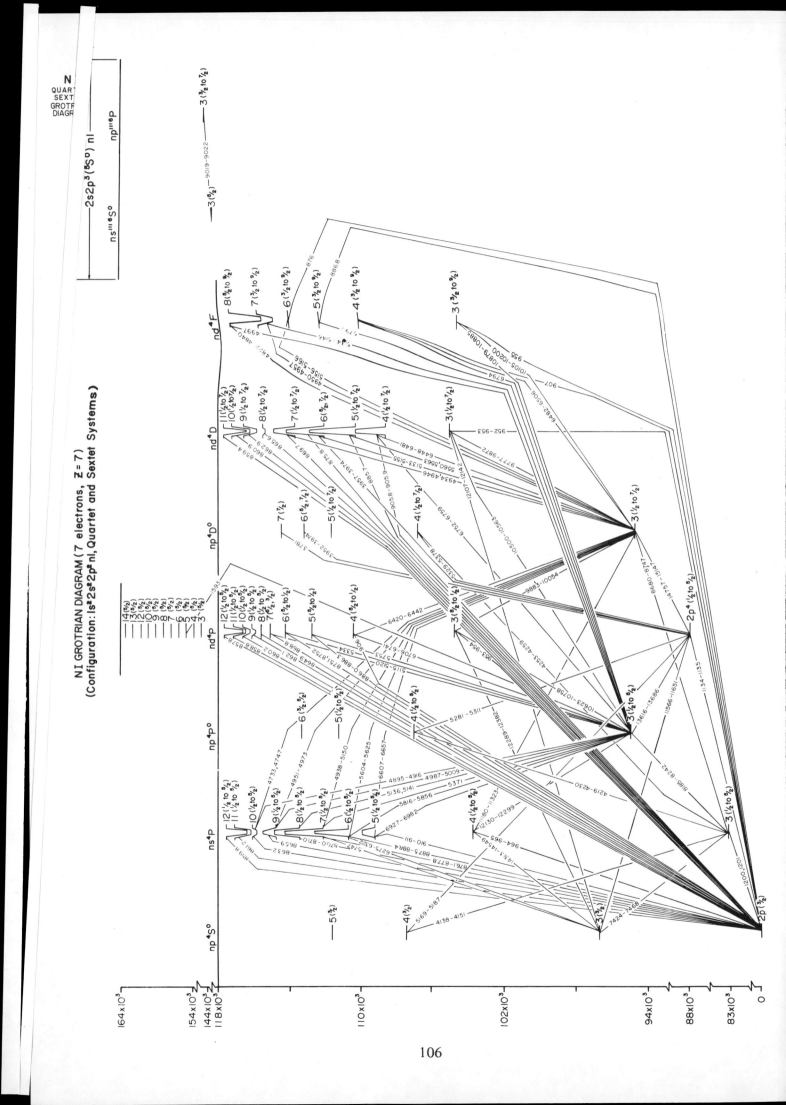

NI GROTRIAN DIAGRAM (7 electrons, Z = 7)
(Configuration: 1s²2s²2p⁴ nl, Quartet and Sextet Systems)

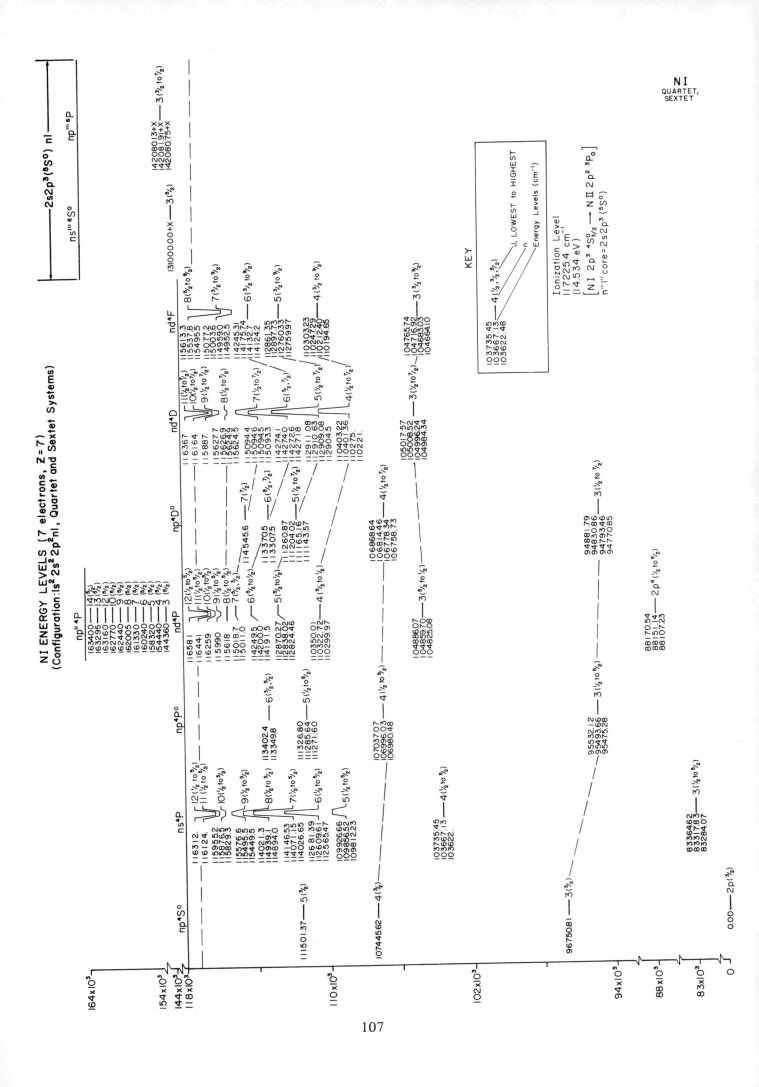

NI ENERGY LEVELS (7 electrons, Z = 7)
(Configuration: 1s² 2s² 2p²nl, Quartet and Sextet Systems)

NI
QUARTET,
SEXTET

107

NI ENERGY LEVELS (7 electrons, Z=7)
(Configuration: 1s² 2s² 2p²(ᵐL) nf, Intermediate Coupling)

(³P)D°nf   (³P)F°nf   (³P)G°nf   (¹D)H°nf

125688O7 —— 4f' [5] (¹¹/₂, ⁹/₂)

(³P)G°nf column:

1143OO11O   [5] (¹¹/₂)
1142220B 6 [4] (⁷/₂, ⁹/₂)
114216S3   [3] (⁵/₂)

112953SO   5 (¹¹/₂, ⁹/₂)
11295340
112877B9   5 4 (⁹/₂, ⁷/₂)
1128687O   3 (⁷/₂)

11O473Z1   5 (¹¹/₂, ⁹/₂)
11O47306
11O4OZ1S   4 (⁷/₂, ⁹/₂)
11O4O2O6
11O3BS33   3 (⁹/₂, ⁷/₂)
11O3BSZ6

KEY

(³P)G°nf ——— Core
           l₁+l₂ (L)
           Final J
5 [4] (⁹/₂, ⁷/₂)   j-l Coupling
  [3] (⁷/₂)   n
                    Energy Level (cm⁻¹)

112967I7
11296706 5
11296S33

Ionization Level
117225.4 cm⁻¹
(14534 eV)
[N I 2p³ ⁴S°₃/₂ —— N II 2p² ³P₀]

(³P)F°nf column:

1143O782   [4] (⁹/₂)
114306?? 6 [3] (⁷/₂)
1143O2O6   [2] (⁵/₂)

11296717   5 4 (⁹/₂, ⁷/₂)
112967O6   3 (⁷/₂)
1129653B

11O5O18O   [4] (⁹/₂, ⁷/₂)
11O5O16S
11O4984O   4 3 (⁷/₂, ⁵/₂)
11O49B39
11O4B6O1   2 (³/₂, ⁵/₂)
11O4B593

(³P)D°nf column:

1142961S   [1] (³/₂)
1142S378 6 2 (⁵/₂, ⁷/₂)
11417199   3 (⁵/₂, ⁷/₂)

112BBOSO 5 [2] (⁵/₂)

11O4S976   [1]
11O4O4S2   4 2 (³/₂, ⁵/₂)
11O4O447
11O349I4   3 (⁵/₂, ⁷/₂)
11O349O6

ENERGY (cm⁻¹)

13×10⁴
12×10⁴
11×10⁴
10×10⁴

109

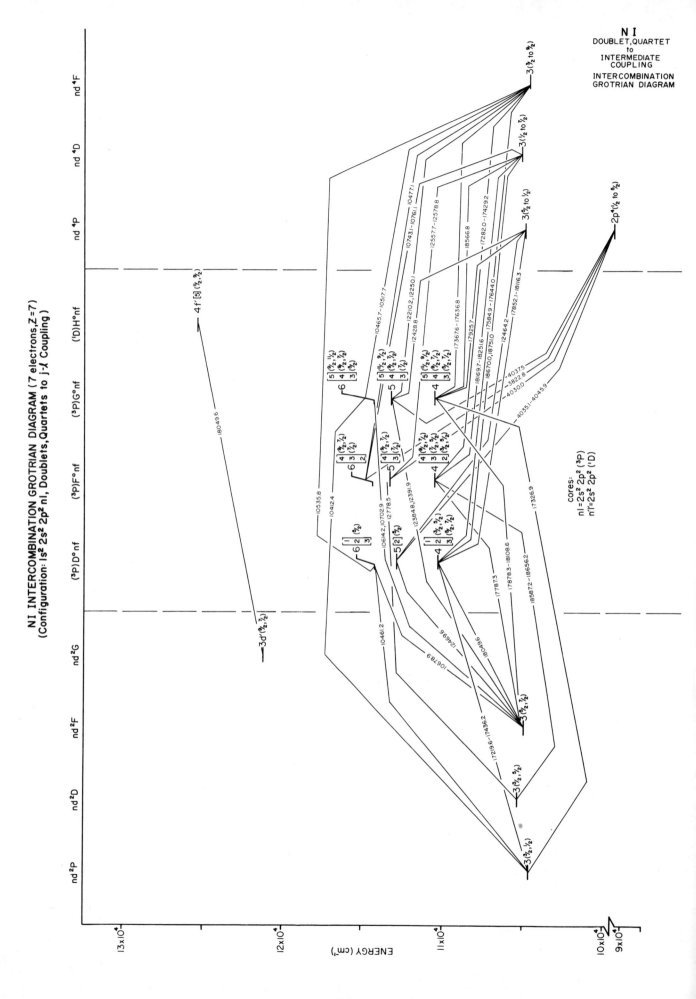

N I INTERCOMBINATION GROTRIAN DIAGRAM (7 electrons, Z=7)
(Configuration: 1s² 2s² 2p² nl, Doublets, Quartets to j-λ Coupling)

N I
DOUBLET, QUARTET
to
INTERMEDIATE
COUPLING

INTERCOMBINATION
GROTRIAN DIAGRAM

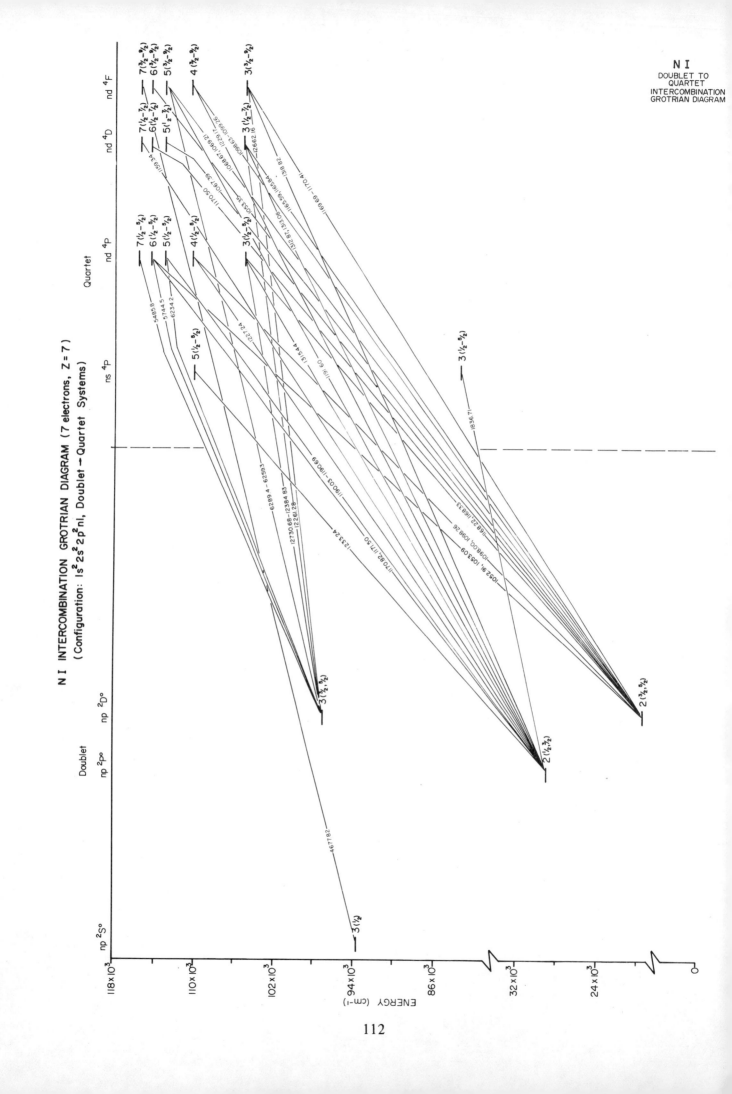

N I INTERCOMBINATION GROTRIAN DIAGRAM (7 electrons, Z = 7)
(Configuration: 1s² 2s² 2p² nl, Doublet → Quartet Systems)

N I
DOUBLET TO
QUARTET
INTERCOMBINATION
GROTRIAN DIAGRAM

ENERGY (cm-1)

112

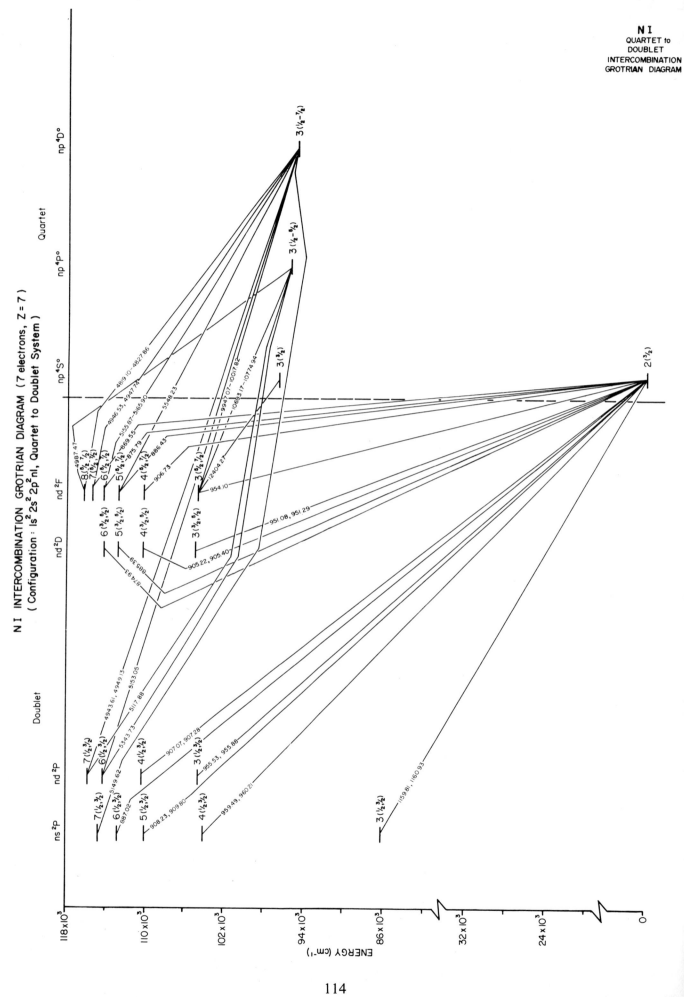

N I INTERCOMBINATION GROTRIAN DIAGRAM (7 electrons, Z = 7)
(Configuration : $1s^2 2s^2 2p^2 nl$, Quartet to Doublet System)

ENERGY (cm⁻¹)

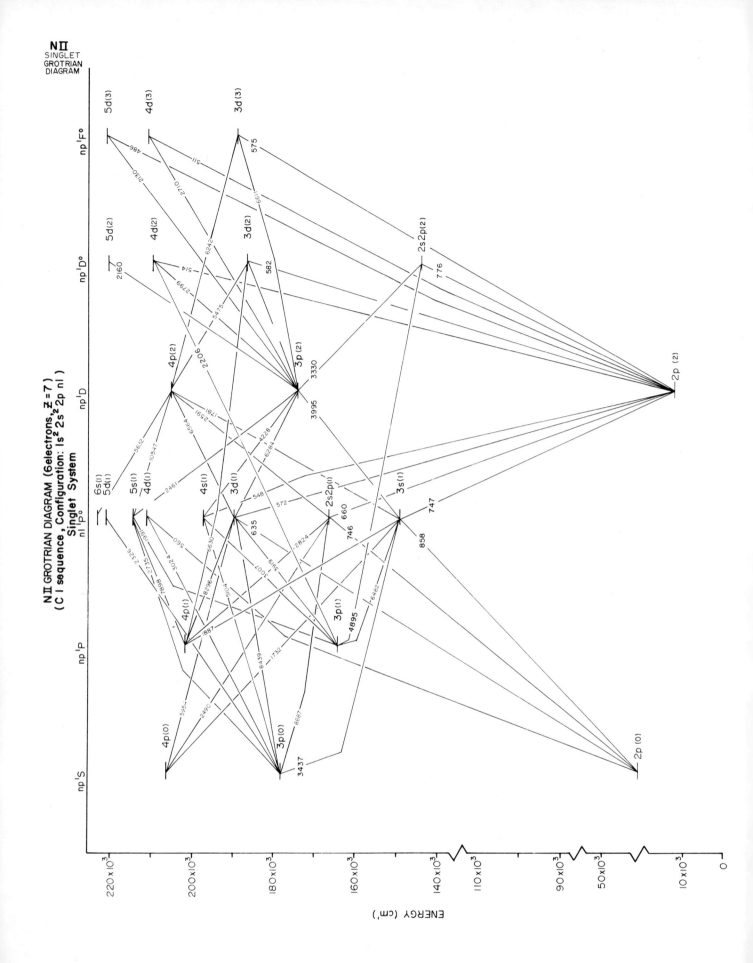

N II GROTRIAN DIAGRAM (6electrons, $\bar{z}$ =7)
(C I sequence, Configuration: $1s^2\ 2s^2 2p\ nl$)
Singlet System

N II
SINGLET
GROTRIAN
DIAGRAM

ENERGY (cm⁻¹)

N II ENERGY LEVELS (6 electrons, Z=7)
(CI sequence, Configuration: 1s² 2s² 2p nl , Singlet System)

N II
SINGLET

np¹S   np¹P   nl¹P°   np¹D   np¹D°   np¹F°

206910.24 —— 4p(0)

202170.63 —— 4p(1)

223101.82 —— 6s(1)
221246.17 —— 5d(1)

214829.79 —— 5s(1)
211336.16 —— 4d(1)

197858.69 —— 4s(1)

190120.24 —— 3d(1)

178273.38 —— 3p(0)

174212.03 —— 3p(2)

220495.36 —— 5d(2)

209925.76 —— 4d(2)

187091.37 —— 3d(2)

221141.61 —— 5d(3)

211103.63 —— 4d(3)

189335.16 —— 3d(3)

205350.18 —— 4p(2)

164610.76 —— 3p(1)

166765.66 —— 2s2p³(1)

144187.94 —— 2s2p³(2)

149187.80 —— 3s(1)

15316.2 —— 2p(2)

32688.8 —— 2p(0)

KEY

149187 80 —— 2s3s(1)

Configuration
J
Energy Level (cm⁻¹)

Ionization Level
238750.5±1.3(cm⁻¹)
(29.601 electron volts)
[N II 2p² ³P₀ → N III 2p ²P°₁/₂]

ENERGY (cm⁻¹)

220×10³
200×10³
180×10³
160×10³
140×10³
110×10³
90×10³
50×10³
10×10³
0

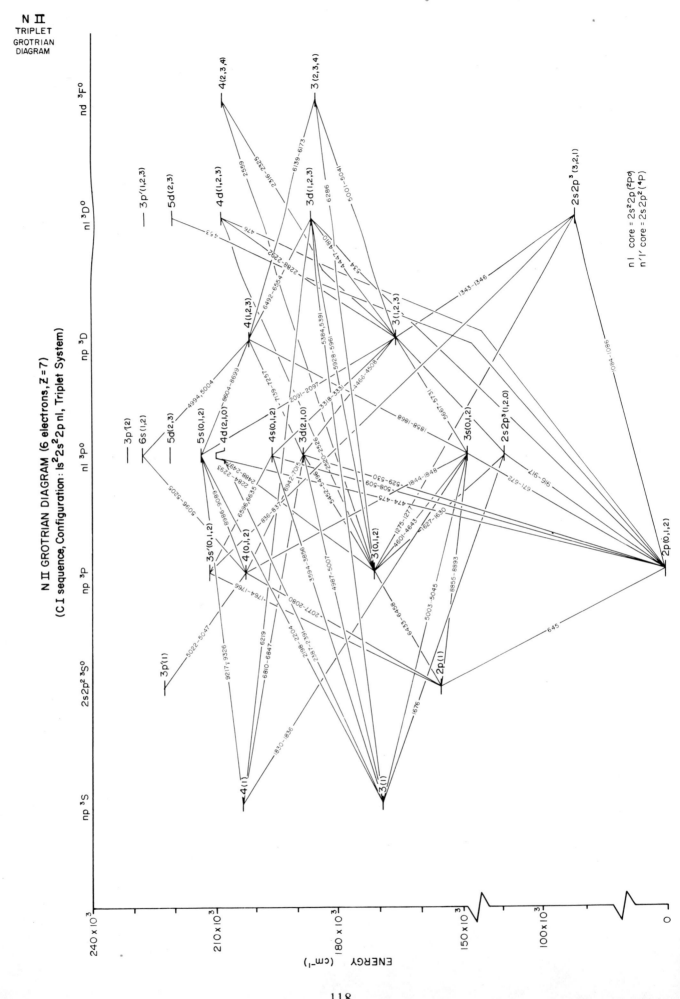

N II
TRIPLET
GROTRIAN
DIAGRAM

N II GROTRIAN DIAGRAM (6 electrons, Z = 7)
(C I sequence, Configuration : 1s²2s²2p nl, Triplet System)

ENERGY (cm⁻¹)

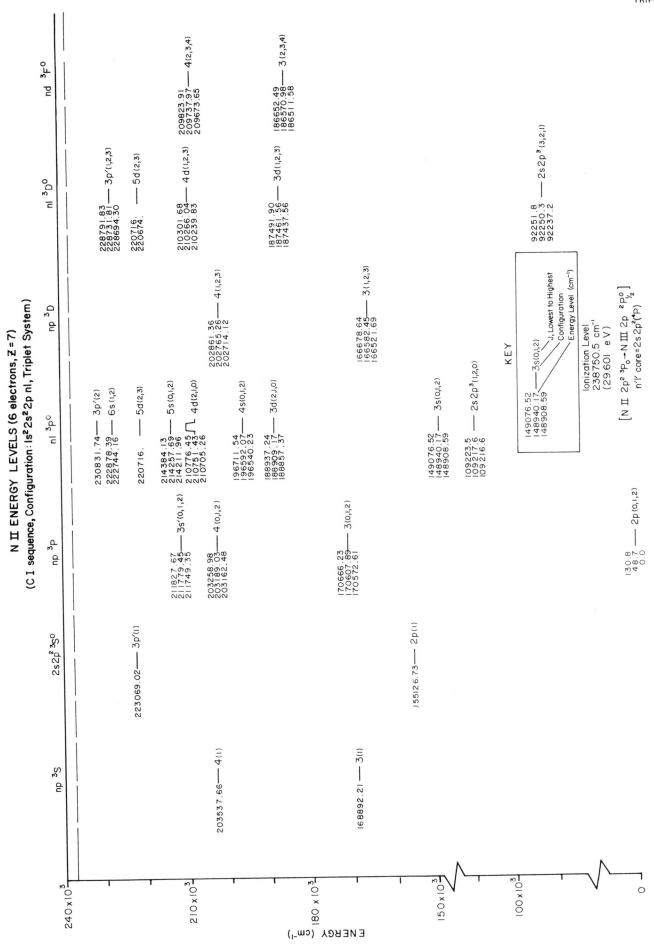

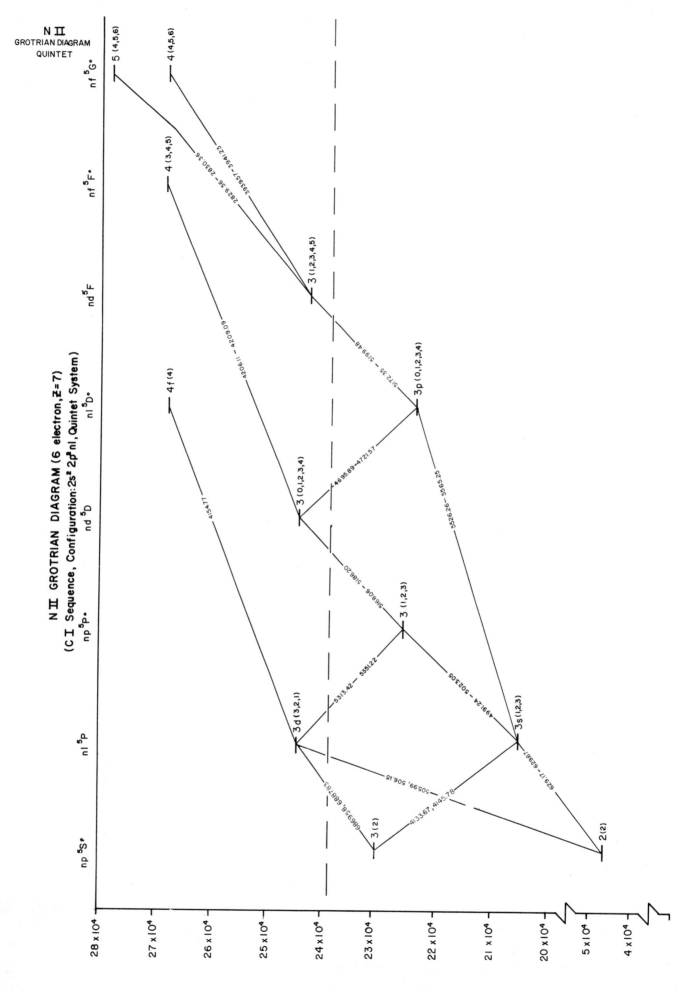

N II
GROTRIAN DIAGRAM
QUINTET

N II GROTRIAN DIAGRAM (6 electron, Z = 7)
(C I Sequence, Configuration: 2s² 2p² nl, Quintet System)

120

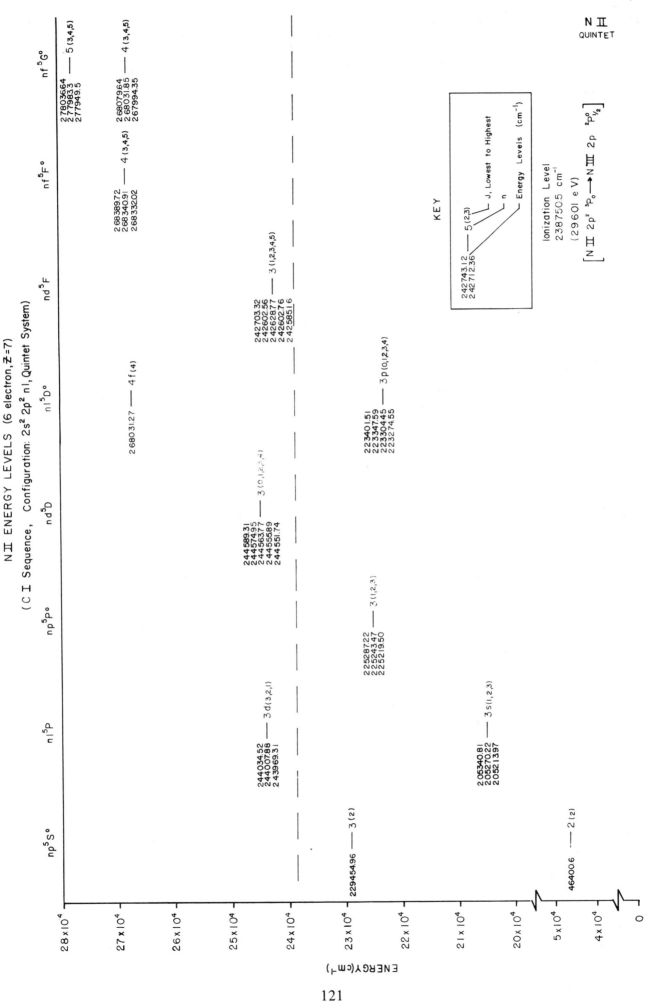

N II ENERGY LEVELS (6 electron, Z̄ = 7)

(C I Sequence, Configuration: 2s² 2p² nl, Quintet System)

N II
QUINTET

KEY

Ionization Level
2387505 cm⁻¹
(296.01 eV)

$\left[N\ II\ 2p^2\ ^3P_o \longrightarrow N\ III\ 2p\ \ ^2P^o_{1/2} \right]$

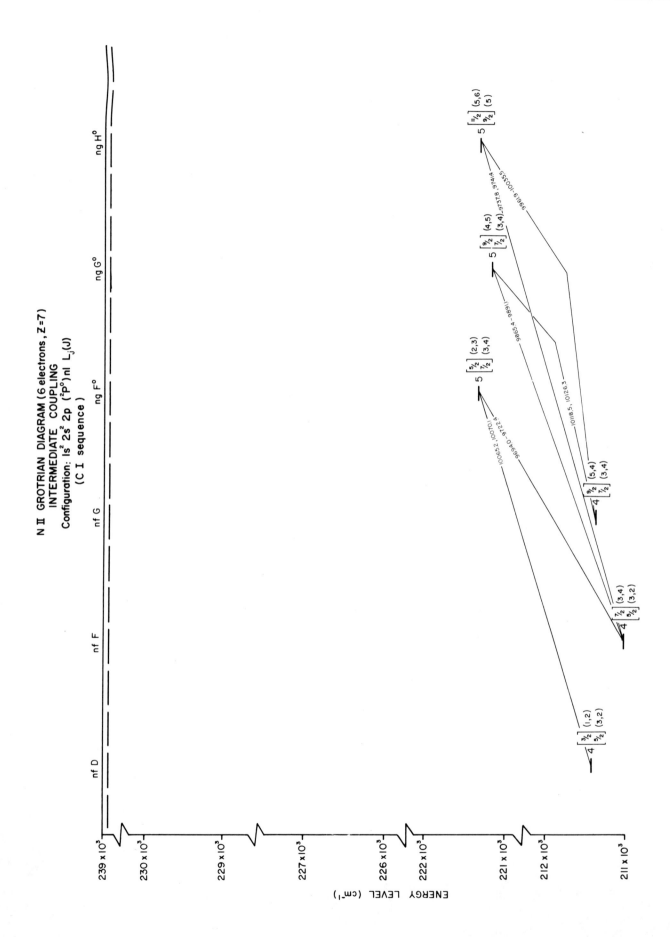

N II GROTRIAN DIAGRAM (6 electrons, Z=7)
INTERMEDIATE COUPLING
Configuration: $1s^2\,2s^2\,2p\,(^2P^o)nl\ L_J(J)$
(C I sequence)

ENERGY LEVEL (cm⁻¹)

N II ENERGY LEVELS (6 electrons, Z=7)
INTERMEDIATE COUPLING
(C I sequence, Configuration: 1s² 2s² 2p (²P°)nl L_j(J))

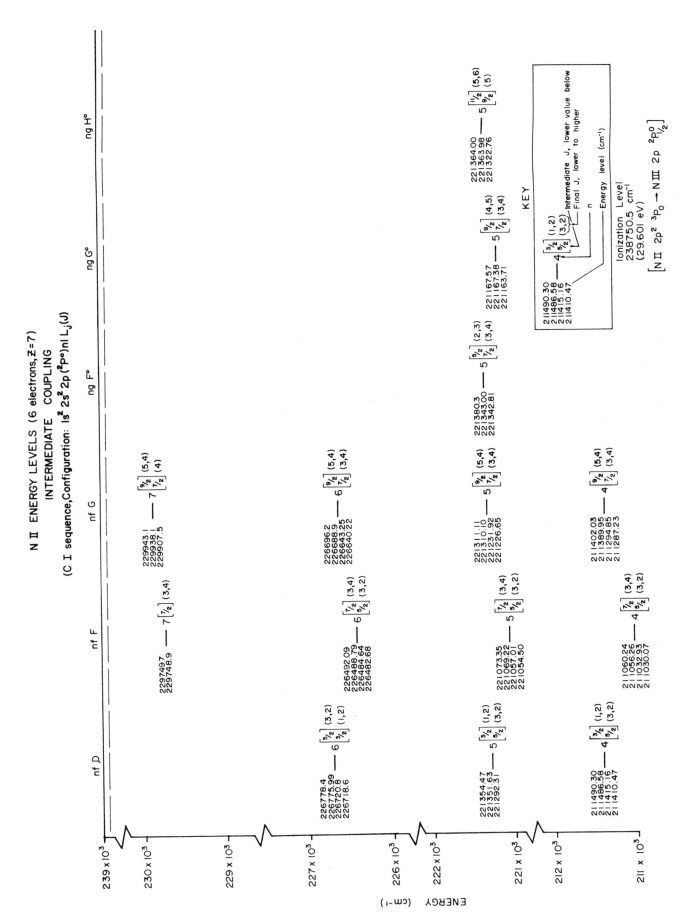

ENERGY (cm⁻¹)

123

N II INTERCOMBINATIONS (6 electrons, Z = 7)

C I sequence, Singlet to Triplet Systems

GROTRIAN DIAGRAM

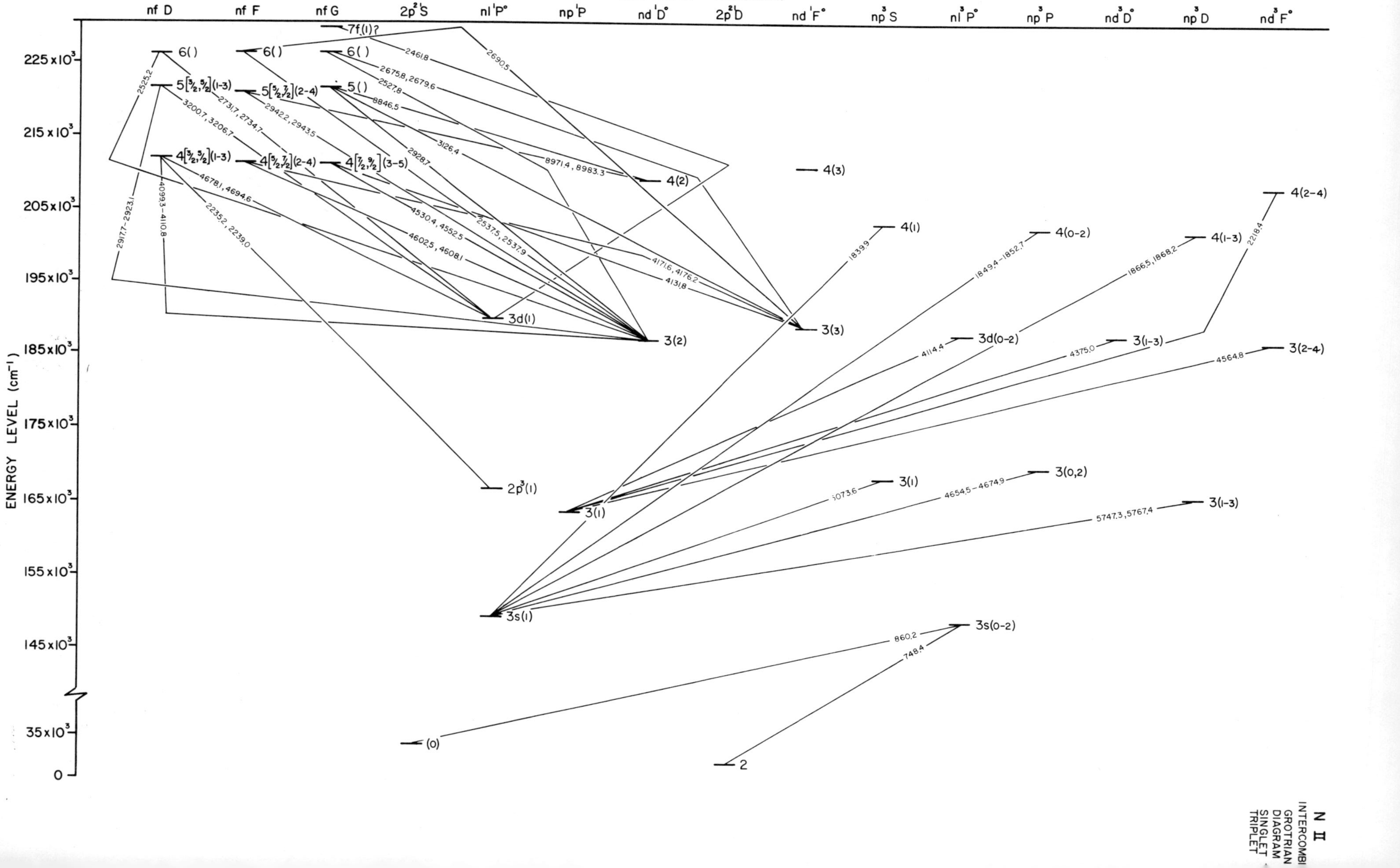

N II
INTERCOMBI
GROTRIAN
DIAGRAM
SINGLET
TRIPLET

N II

INTERCOMBINATIONS
GROTRIAN DIAGRAM
SINGLETS–
INTERMEDIATE
COUPLING

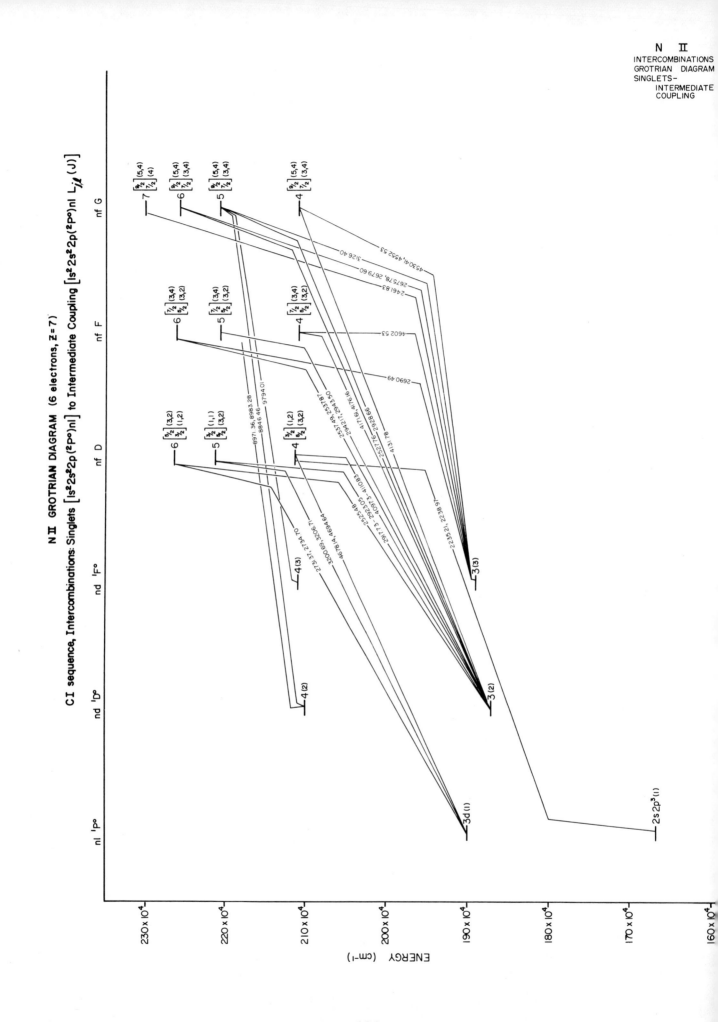

N II GROTRIAN DIAGRAM (6 electrons, Z = 7)

CI sequence, Intercombinations: Singlets $[1s^2 2s^2 2p(^2P^o)nl]$ to Intermediate Coupling $[1s^2 2s^2 2p(^2P^o)nl] L_{j\ell} (J)$

N II GROTRIAN DIAGRAM (6 electrons, Z = 7)
(C I sequence, Intercombinations: Triplets to Singlets)

N II
GROTRIAN DIAGRAM
INTERCOMBINATION
TRIPLET to SINGLET

128

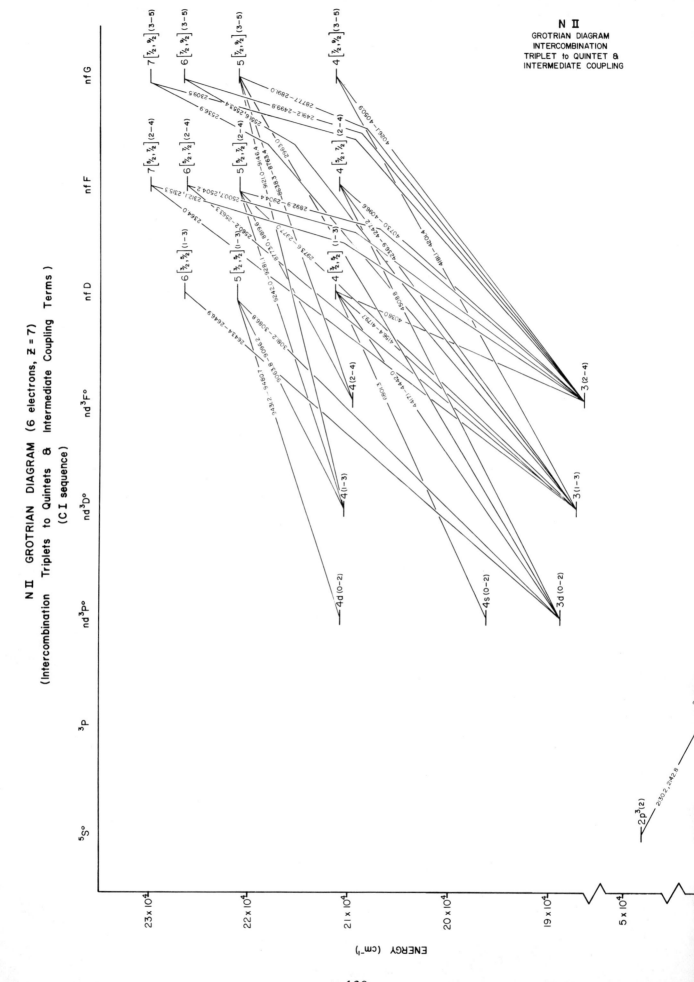

N II GROTRIAN DIAGRAM (6 electrons, Z = 7)

(Intercombination Triplets to Quintets & Intermediate Coupling Terms)

(C I sequence)

N II
GROTRIAN DIAGRAM
INTERCOMBINATION
TRIPLET to QUINTET &
INTERMEDIATE COUPLING

ENERGY (cm⁻¹)

130

N III GROTRIAN DIAGRAM (5 electrons, Z=7)

(BI Sequence, Configuration Is² 2s nl nl' Doublet Term System)

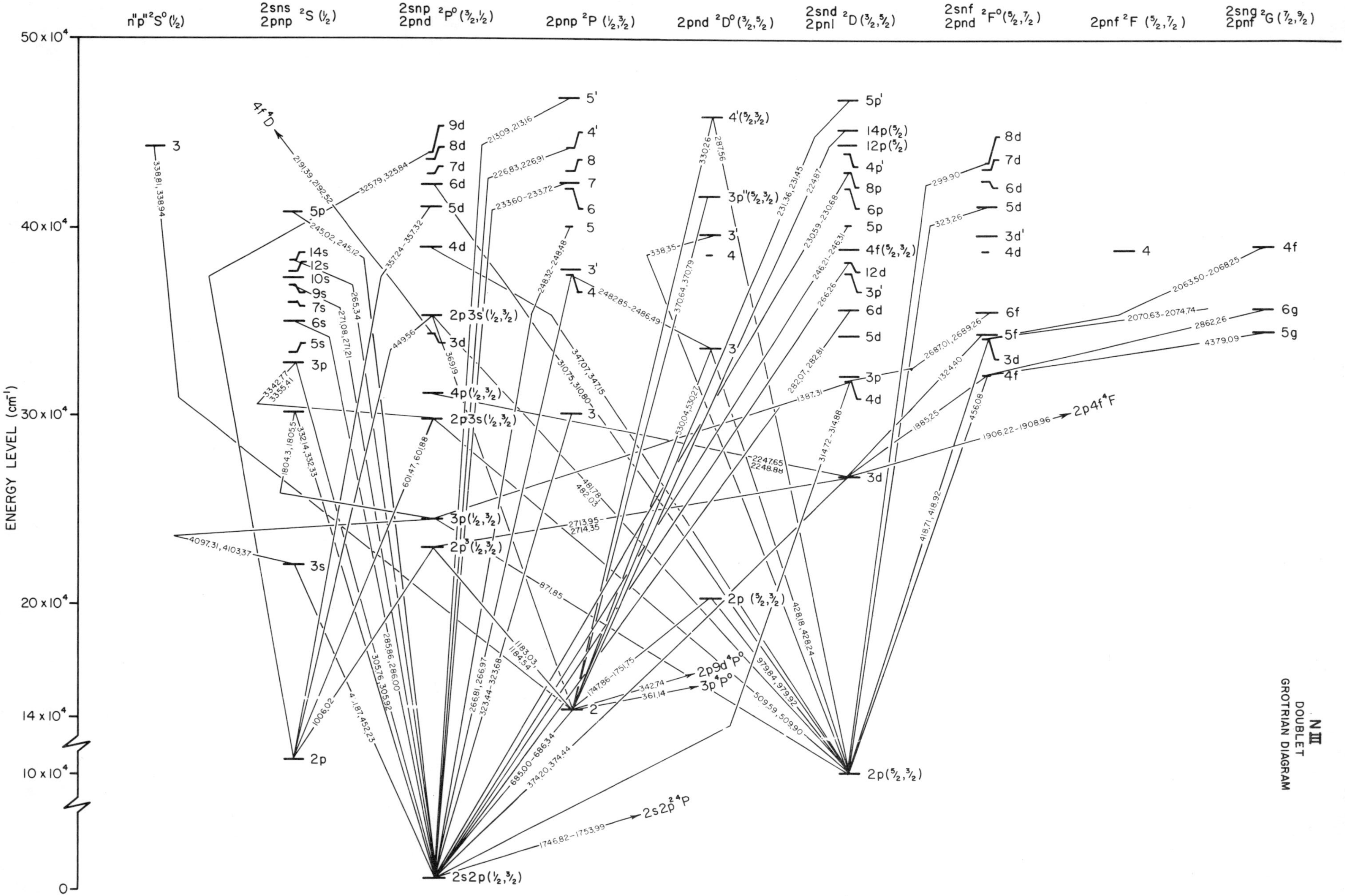

N III
DOUBLET
GROTRIAN DIAGRAM

N III ENERGY LEVELS (5 electrons, Z=7)

(BI Sequence, Configuration 1s² 2s nl n'l' Doublet Term System)

N III
DOUBLET

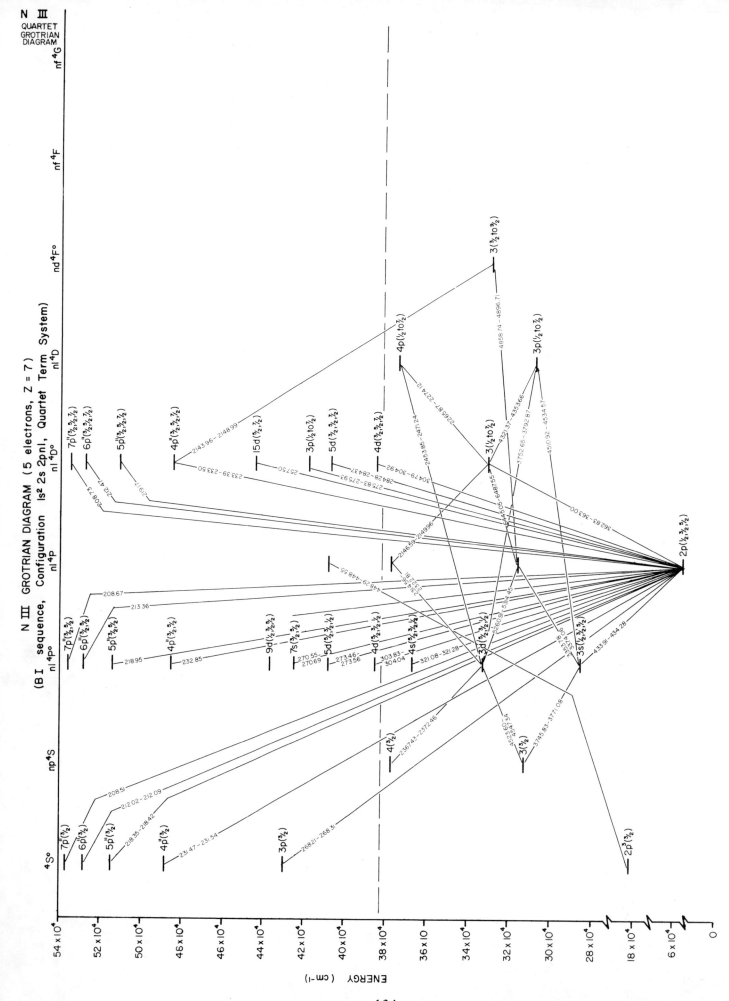

N III

QUARTET GROTRIAN DIAGRAM

N III GROTRIAN DIAGRAM (5 electrons, Z = 7)

(B I sequence, Configuration 1s² 2s 2pnl, Quartet Term System)

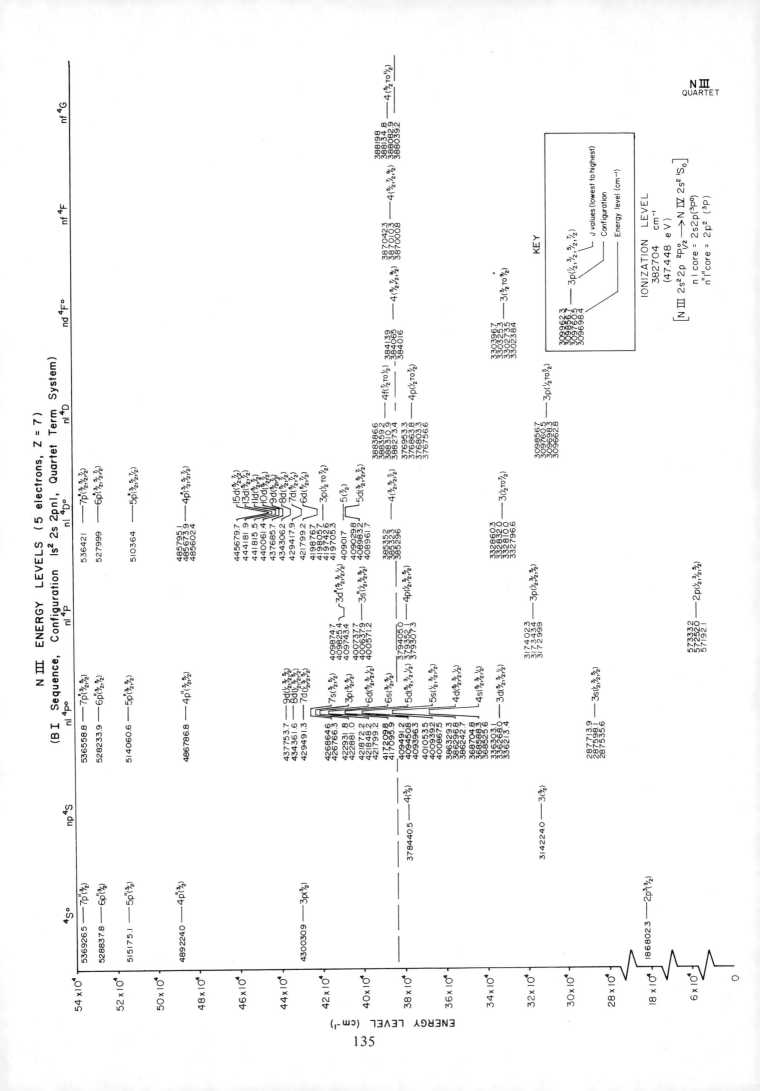

N III ENERGY LEVELS (5 electrons, Z = 7)

(B I Sequence, Configuration 1s² 2s 2pnl, Quartet Term System)

N III
QUARTET

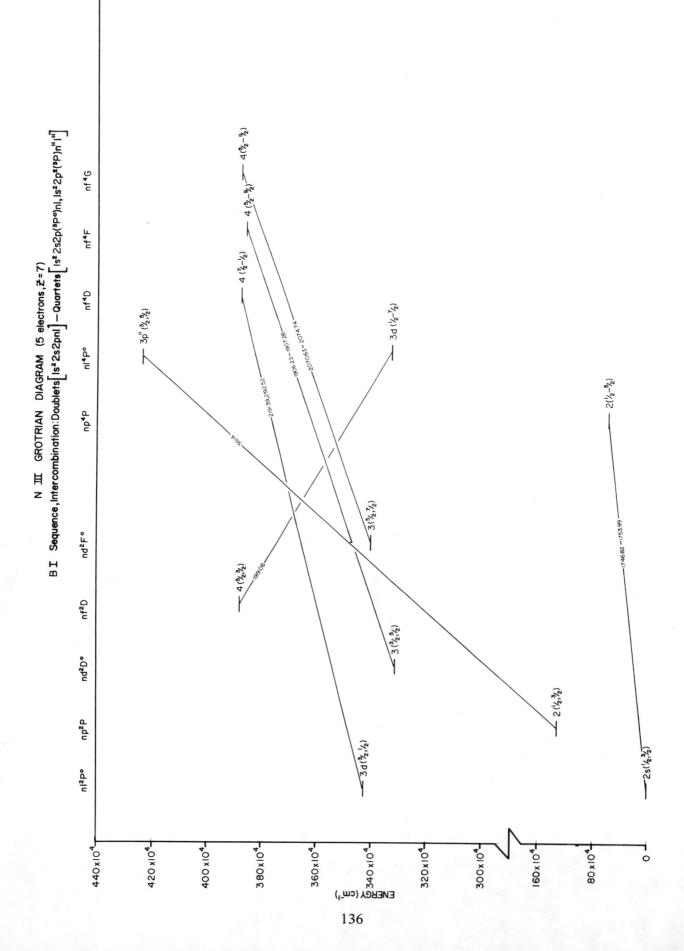

N III GROTRIAN DIAGRAM (5 electrons, Z=7)

B I Sequence, Intercombination: Doublets $\left[1s^2 2s2pnl\right]$ — Quartets $\left[1s^2 2s2p(^3P^\circ)nl, 1s^2 2p^2(^3P)n'l'\right]$

N III
INTERCOMBINATION
GROTRIAN DIAGRAM
DOUBLETS–
QUARTETS

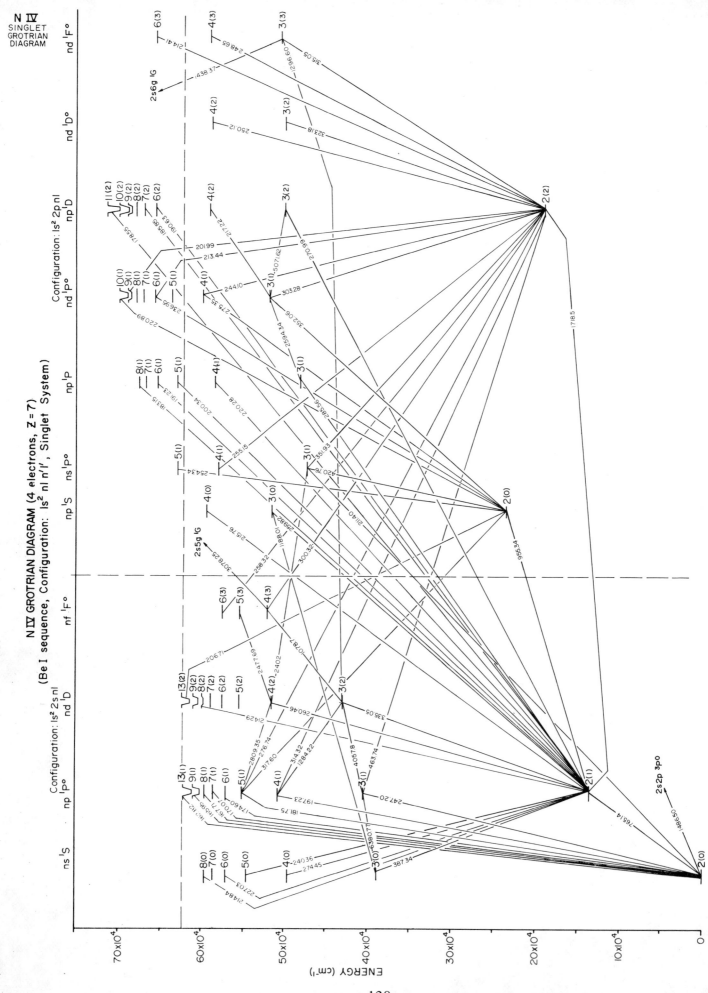

N IV
SINGLET
GROTRIAN
DIAGRAM

N IV GROTRIAN DIAGRAM (4 electrons, Z = 7)
(Be I sequence, Configuration: Is² nl n'l', Singlet System)

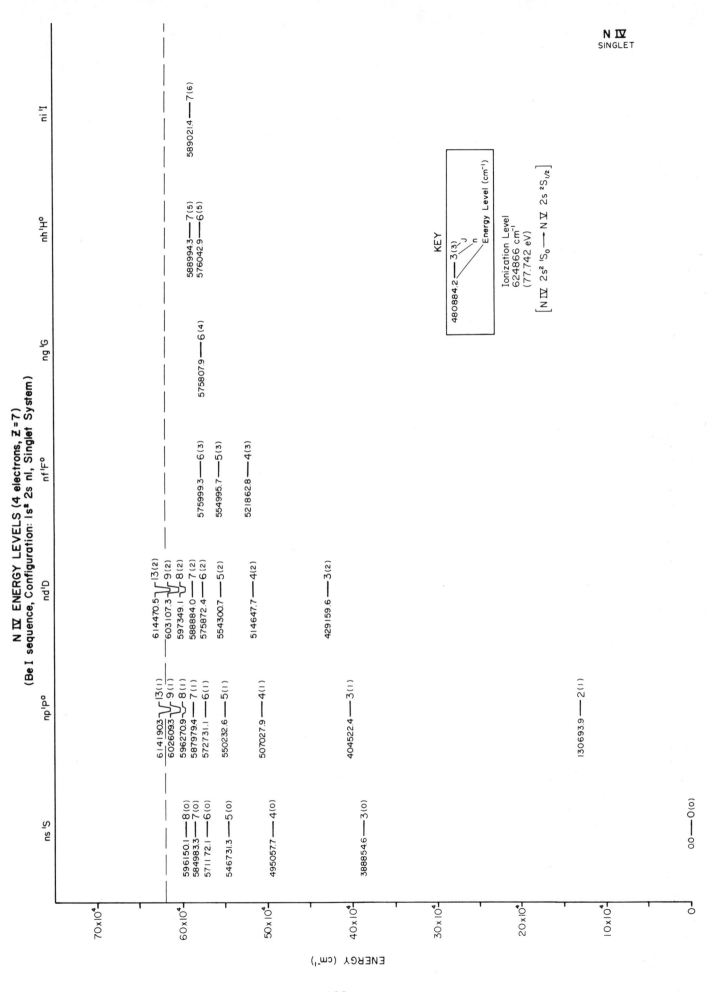

N IV ENERGY LEVELS (4 electrons, Z = 7)
(Be I sequence, Configuration: 1s² 2s nl, Singlet System)

N IV
SINGLET

139

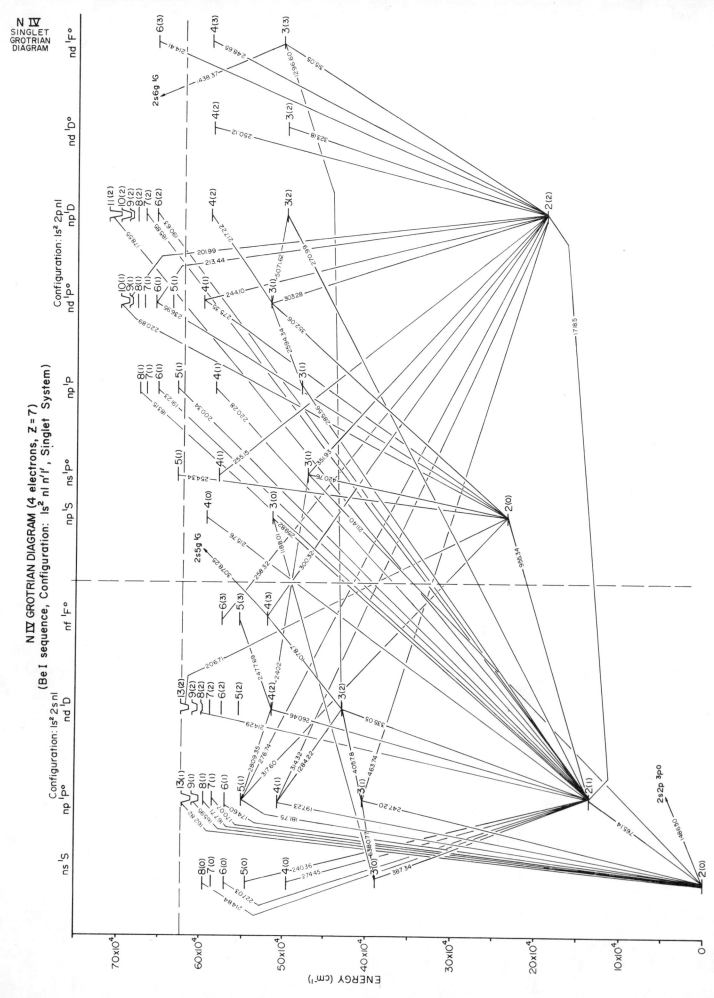

N IV
SINGLET
GROTRIAN
DIAGRAM

N IV GROTRIAN DIAGRAM (4 electrons, Z = 7)
(Be I sequence, Configuration: 1s² nl n'l', Singlet System)

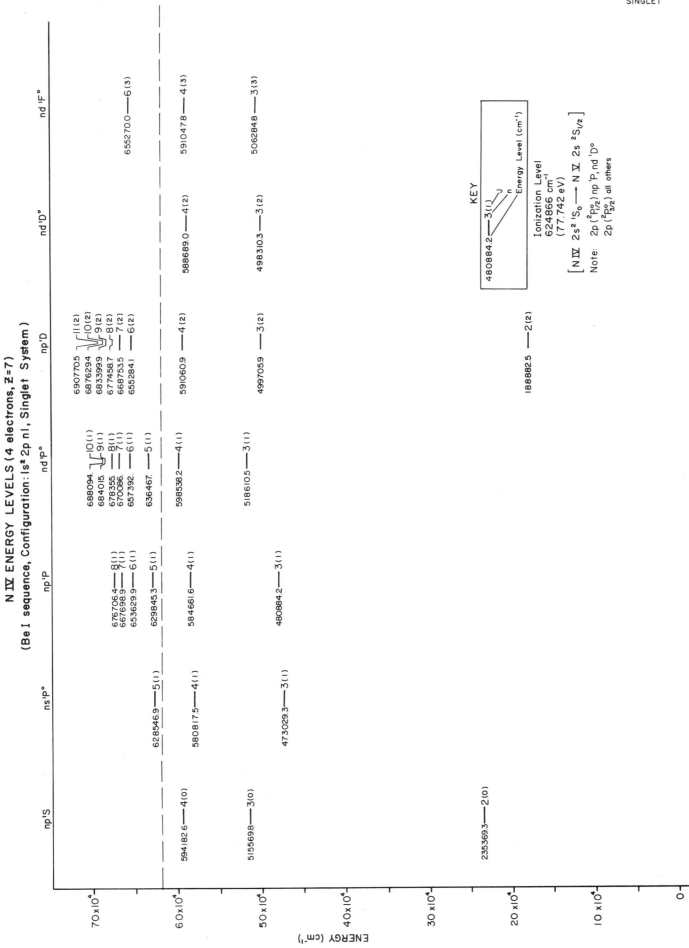

N IV ENERGY LEVELS (4 electrons, Z=7)

(Be I sequence, Configuration: 1s² 2p nl, Singlet System)

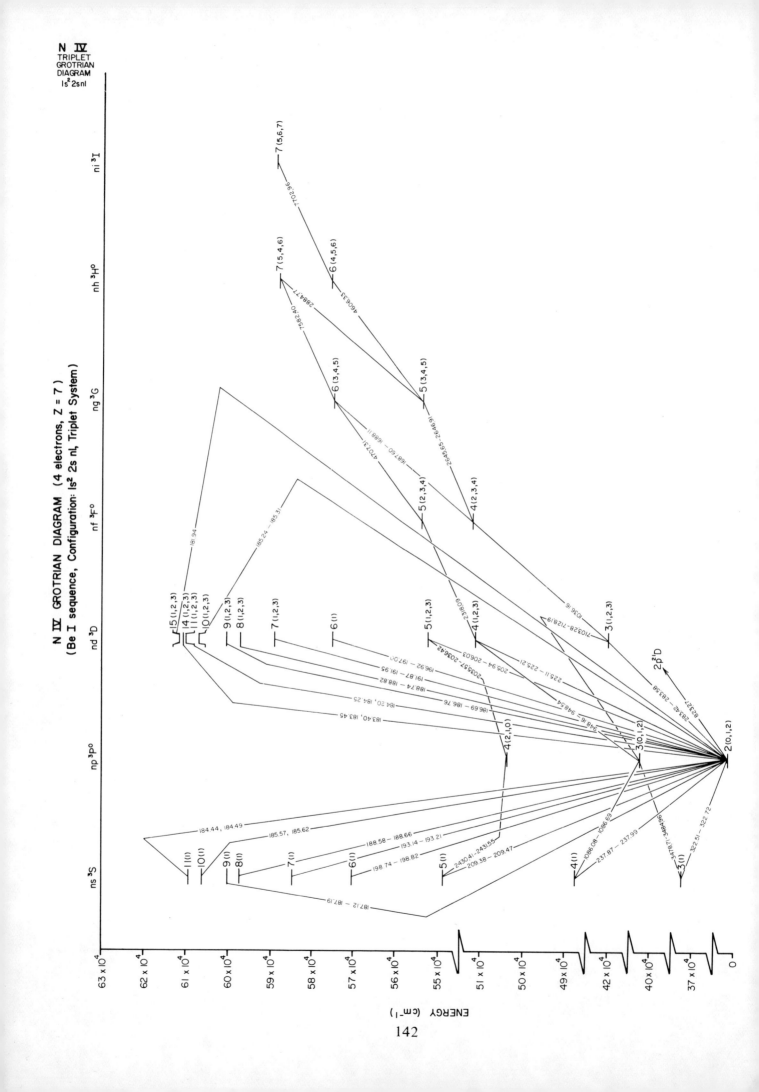

N IV GROTRIAN DIAGRAM (4 electrons, Z = 7)
(Be I sequence, Configuration: Is² 2s nl, Triplet System)

ENERGY (cm⁻¹)

142

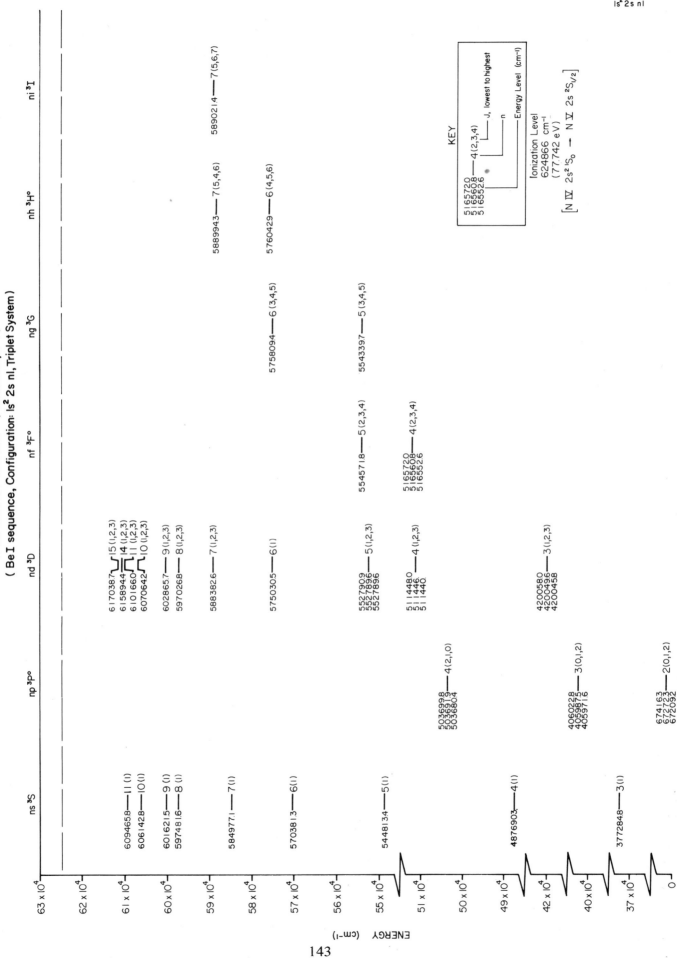

N IV ENERGY LEVELS (4 electrons, Z=7)
(Be I sequence, Configuration: 1s² 2s nl, Triplet System)

N IV
TRIPLET
1s² 2s nl

ENERGY (cm⁻¹)

143

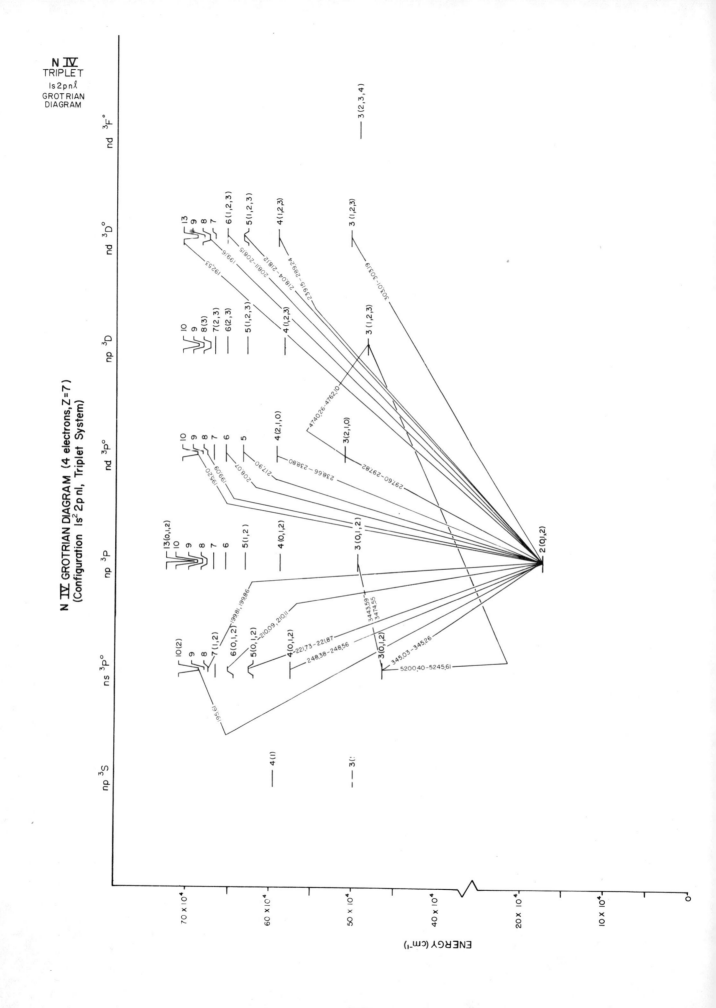

N IV GROTRIAN DIAGRAM (4 electrons, Z=7)
(Configuration 1s² 2p nl, Triplet System)

N IV
TRIPLET
1s2pnℓ
GROTRIAN
DIAGRAM

ENERGY (cm⁻¹)

144

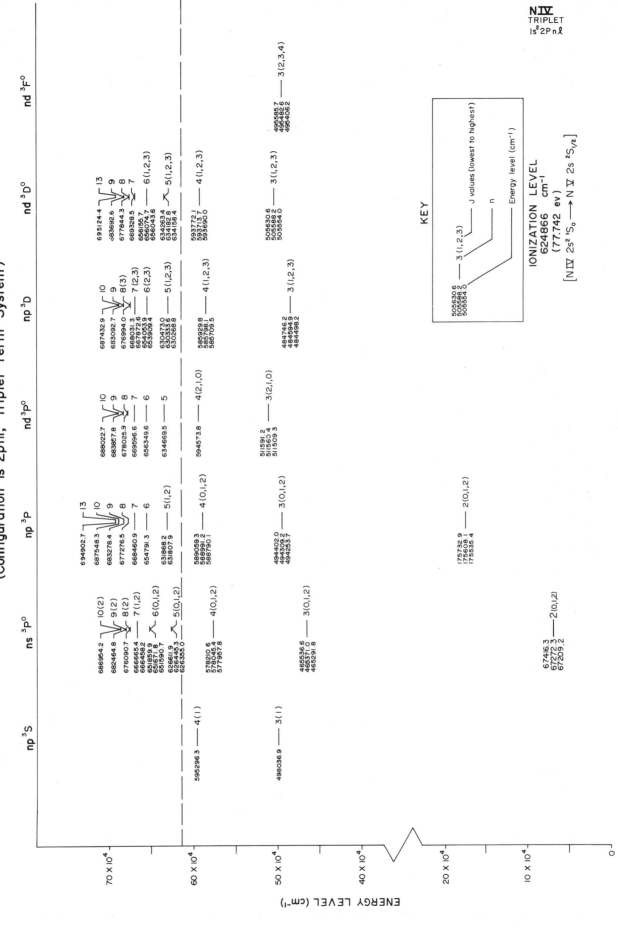

N IV ENERGY LEVELS (4 electrons, Z = 7)

(Configuration 1s²2pnl, Triplet Term System)

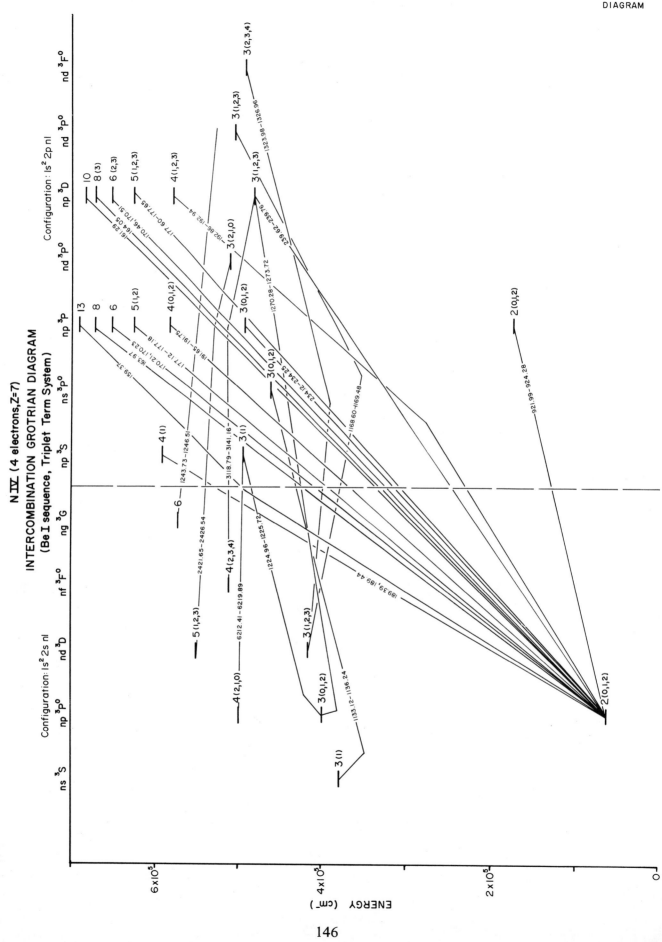

N IV (4 electrons, Z=7)
INTERCOMBINATION GROTRIAN DIAGRAM
(Be I sequence, Triplet Term System)

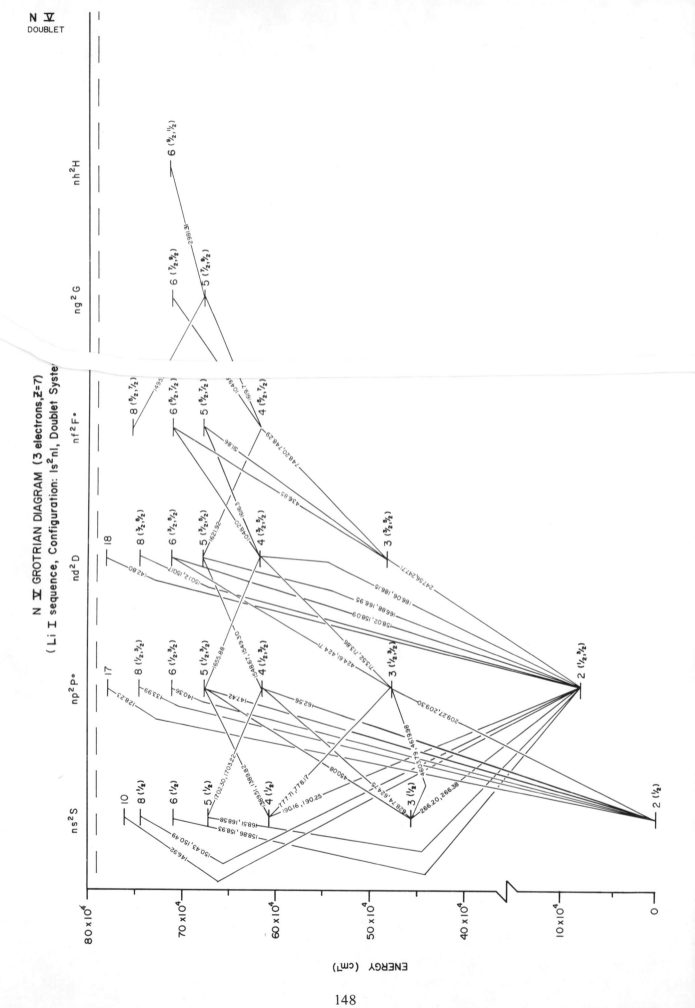

N V GROTRIAN DIAGRAM (3 electrons, Z=7)

(Li I sequence, Configuration: 1s²nl, Doublet System

N V
DOUBLET

ENERGY (cm⁻¹)

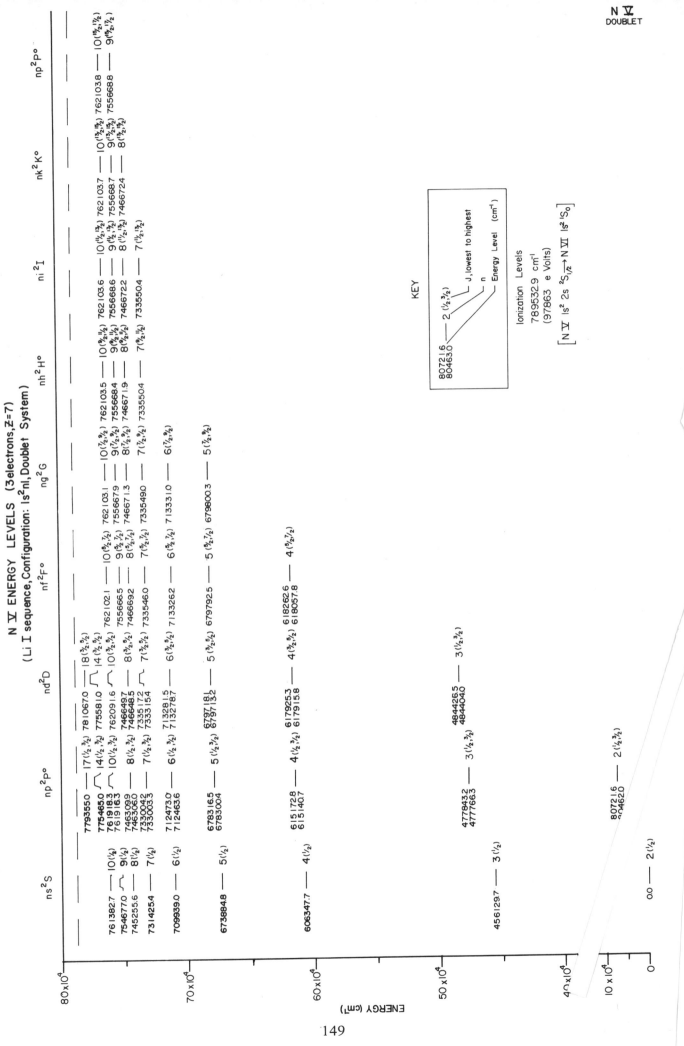

N V ENERGY LEVELS (3 electrons, Z=7)
(Li I sequence, Configuration: 1s²nl, Doublet System)

149

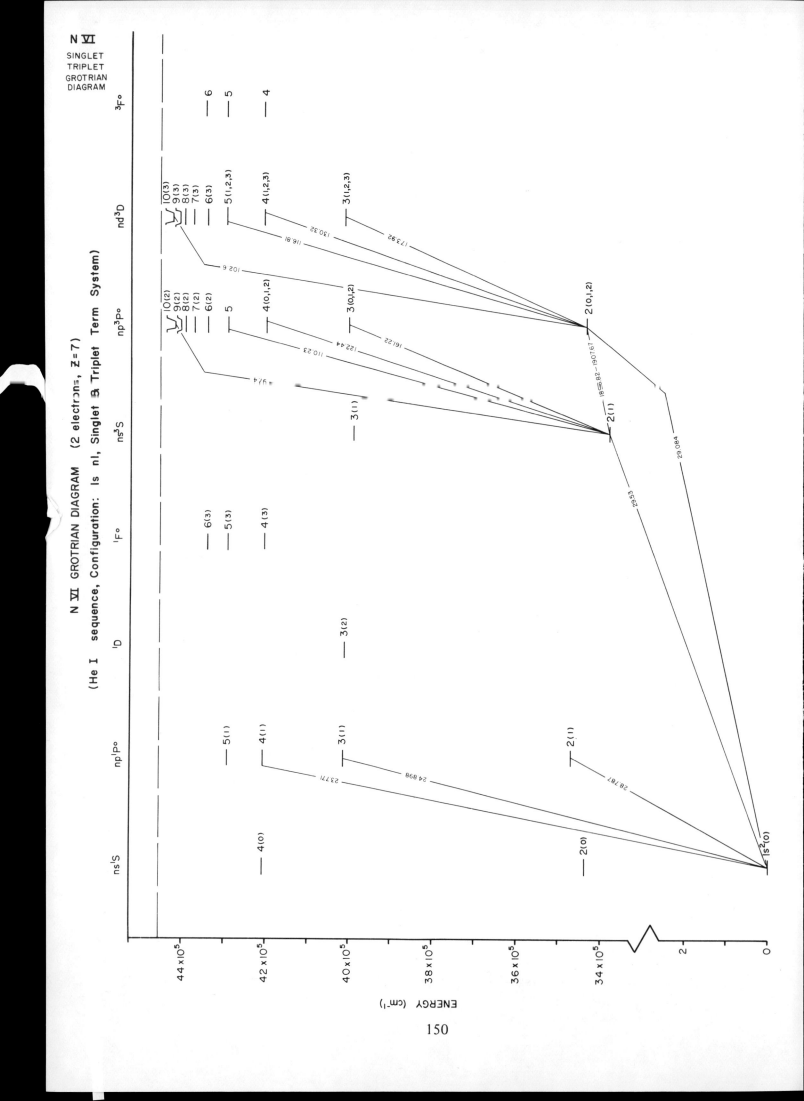

N $\overline{\text{VI}}$ GROTRIAN DIAGRAM (2 electrons, Z=7)

(He I sequence, Configuration: Is nl, Singlet & Triplet Term System)

N $\overline{\text{VI}}$

SINGLET
TRIPLET
GROTRIAN
DIAGRAM

ENERGY (cm⁻¹)

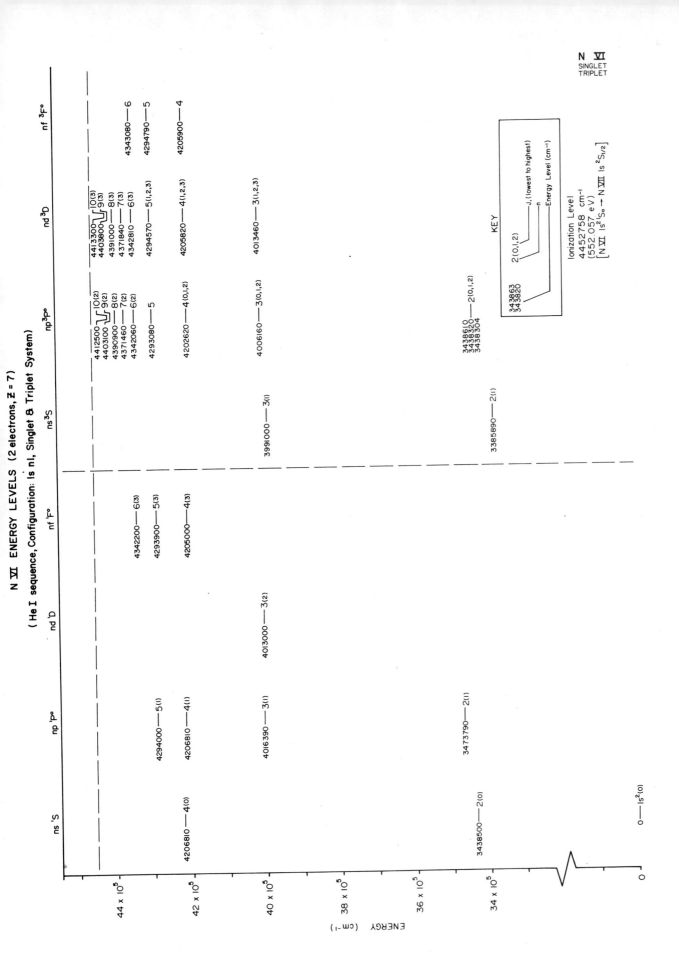

N VI ENERGY LEVELS (2 electrons, Z = 7)

(He I sequence, Configuration: Is nl, Singlet & Triplet System)

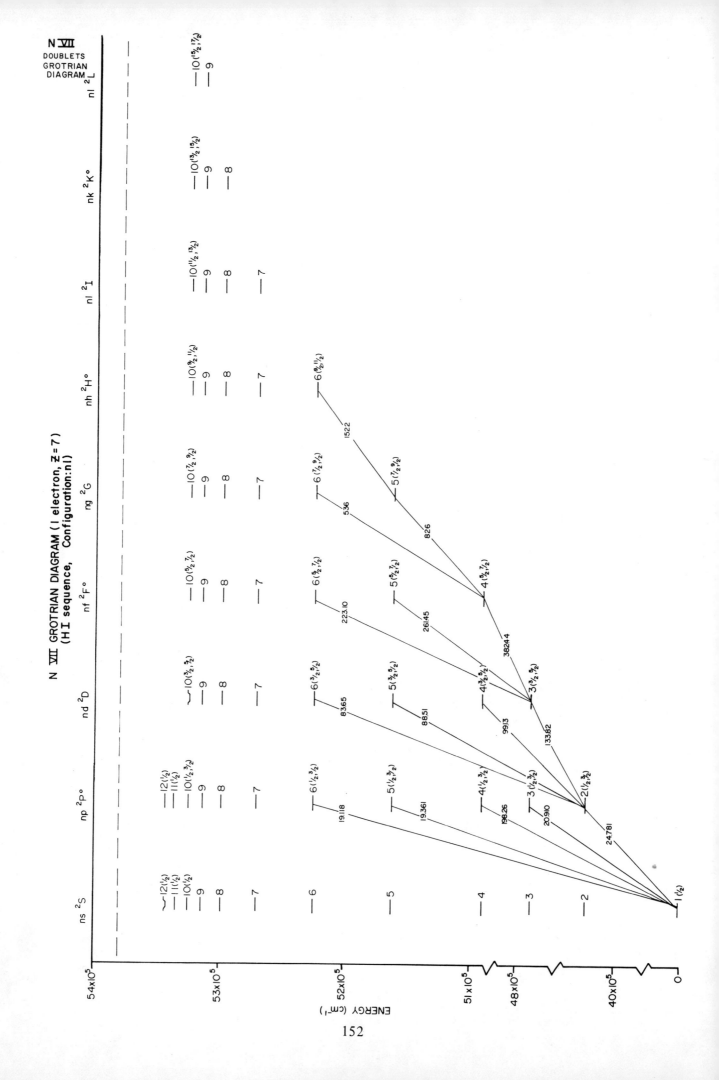

N VII ENERGY LEVELS (1 electron, Z=7)
(H I sequence, Configuration: nl)

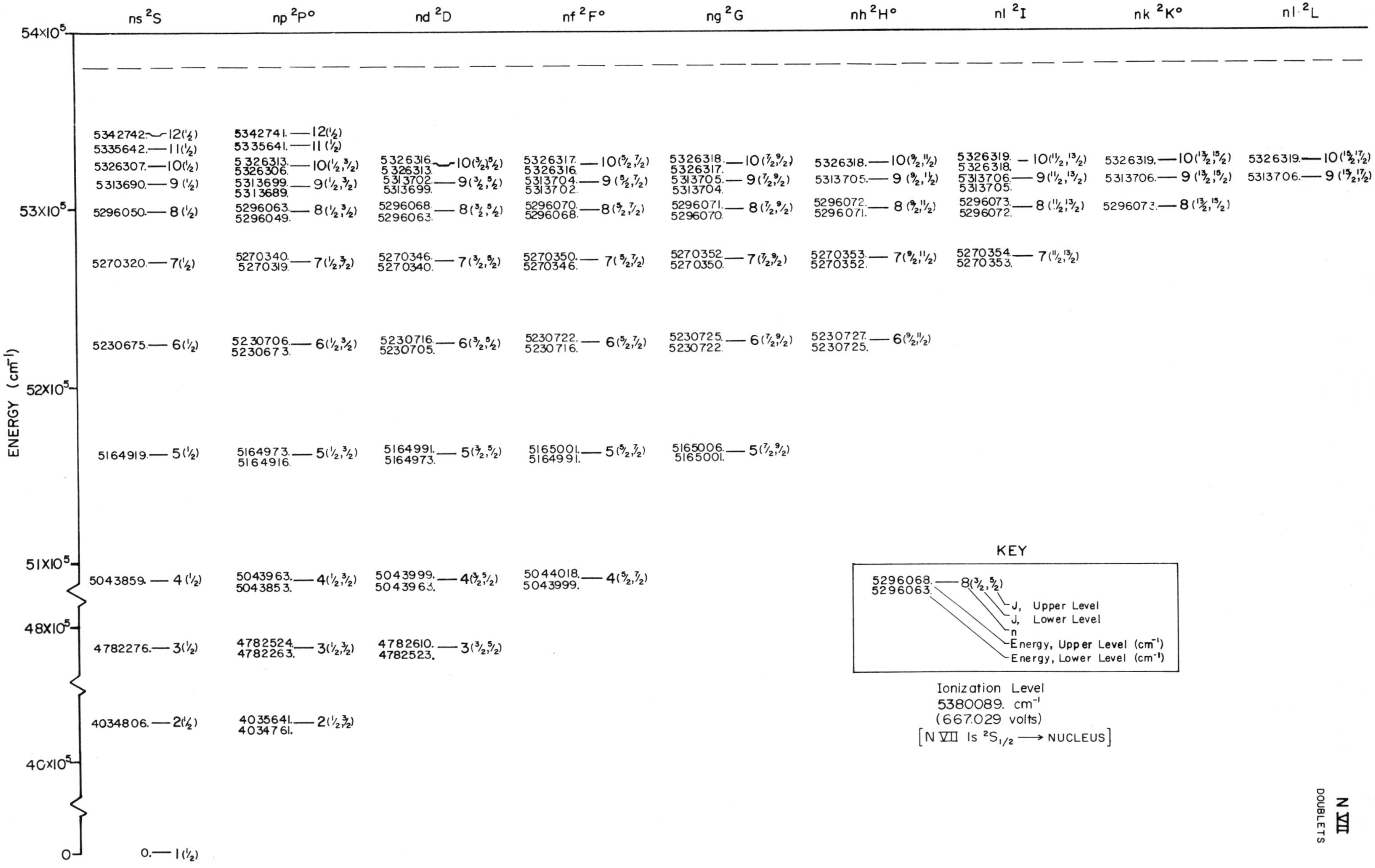

N VII
DOUBLETS

Column headers (left to right):

ns ²S | np ²Pᵒ | nd ²D | nf ²Fᵒ | ng ²G | nh ²Hᵒ | nl ²I | nk ²Kᵒ | nl ²L

ns ²S

5342742.—12(½)
5335642.—11(½)
5326307.—10(½)
5313690.—9 (½)
5296050.—8 (½)
5270320.—7(½)
5230675.—6(½)
5164919.—5 (½)
5043859.—4 (½)
4782276.—3(½)
4034806.—2(½)
0.—1 (½)

np ²Pᵒ

5342741.—12(½)
5335641.—11 (½)
5326313.—10(½,³/₂)
5326306.
5313699.—9(½,³/₂)
5313689.
5296063.—8(½,³/₂)
5296049.
5270340.—7(½,³/₂)
527039.
5230706.—6(½,³/₂)
5230673.
5164973.—5(½,³/₂)
5164916.
5043963.—4(½,³/₂)
5043853.
4782524.—3(½,³/₂)
4782263.
4035641.—2(½,³/₂)
4034761.

nd ²D

5326316.—10(³/₂,⁵/₂)
5326313.
5313702.—9(³/₂,⁵/₂)
5313699.
5296068.—8(³/₂,⁵/₂)
5296063.
5270346.—7(³/₂,⁵/₂)
5270340.
5230716.—6(³/₂,⁵/₂)
5230705.
5164991.—5(³/₂,⁵/₂)
5164973.
5043999.—4(³/₂,⁵/₂)
5043963.
4782610.—3(³/₂,⁵/₂)
4782523.

nf ²Fᵒ

5326317.—10(³/₂,⁵/₂)
5326316.
5313704.—9(⁵/₂,⁷/₂)
5313702.
5296070.—8(⁵/₂,⁷/₂)
5296068.
5270350.—7(⁵/₂,⁷/₂)
5270346.
5230722.—6(⁵/₂,⁷/₂)
5230716.
5165001.—5(⁵/₂,⁷/₂)
51649 91.
5044018.—4(⁵/₂,⁷/₂)
5043999.

ng ²G

5326318.—10(⁷/₂,⁹/₂)
5326317.
5313705.—9(⁷/₂,⁹/₂)
5313704.
5296071.—8(⁷/₂,⁹/₂)
5296070.
5270352.—7(⁷/₂,⁹/₂)
5270350.
5230725.—6(⁷/₂,⁹/₂)
5230722.
5165006.—5(⁷/₂,⁹/₂)
5165001.

nh ²Hᵒ

5326318.—10(⁹/₂,¹¹/₂)
5313705.—9 (⁹/₂,¹¹/₂)
5296072.—8 (⁹/₂,¹¹/₂)
5296071.
5270353.—7(⁹/₂,¹¹/₂)
5270352.
5230727.—6(⁹/₂,¹¹/₂)
5230725.

nl ²I

5326319.—10(¹¹/₂,¹³/₂)
5326318.
5313706.—9(¹¹/₂,¹³/₂)
5313705.
5296073.—8(¹¹/₂,¹³/₂)
5296072.
5270354.—7(¹¹/₂,¹³/₂)
5270353.

nk ²Kᵒ

5326319.—10(¹³/₂,¹⁵/₂)
5313706.—9(¹³/₂,¹⁵/₂)
5296072.—8(¹³/₂,¹⁵/₂)

nl ²L

5326319.—10(¹⁵/₂,¹⁷/₂)
5313706.—9(¹⁵/₂,¹⁷/₂)

KEY

5296068.——8(³/₂,⁵/₂)
5296063.

J, Upper Level
J, Lower Level
n
Energy, Upper Level (cm⁻¹)
Energy, Lower Level (cm⁻¹)

Ionization Level
5380089. cm⁻¹
(667.029 volts)
[N VII 1s ²S₁/₂ ⟶ NUCLEUS]

ENERGY (cm⁻¹)

54×10⁵
53×10⁵
52×10⁵
51×10⁵
48×10⁵
40×10⁵
0

153

N I $Z = 7$ 7 electrons

I.S. Bowen, Ap. J. **132**, 1 (1960).

Author gives transition wavelengths from nebular observations.

K.B.S. Eriksson, Physica Scripta **9**, 151 (1974).

Line and energy-level tables based on wavelengths from 860—9100 Å.

J.W. McConkey, D.J. Burns, and J.A. Kernahan, J. Quant. Spectrosc. Radiat. Transfer **8**, 823 (1968).

Authors give a spectrogram and a line table for lines observed in the region 10 105—10 775 Å.

Comment on intermediate coupling: Here an f-electron ($l = 3$) is coupled to a ^{3}P core ($l = 1$), whence a total orbital L-value of 2, 3, or 4 can be made. Terms with these values are represented by the symbols D, F, and G, respectively.

Following the symbol for L is a bracketed number formed from a core j-value and the l-value of the f-electron. The spin of the f-electron is then coupled to the bracketed number to yield the final J-value of the level.

See Eriksson and Johansson, Ark. Fys. **19**, 235 (1961).

Note: New Tables of Energy Levels and Multiplets for N I, N II, and N III have recently appeared.
The reference is: C.E. Moore, Selected Tables of Atomic Spectra, N I, N II, N III, NSRDS-NBS 3, Sec. 5 (U.S. Govt. Printing Office, 1975).
The difference between the new compilation and our own is small

N II $Z = 7$ 6 electrons

I.S. Bowen, Ap. J. **132**, 1 (1960).

Author gives transition wavelengths from nebular observations.

W.B. Bridges and A.N. Chester, IEEE J. Qu. Electronics **1**, 66 (1965).

Authors give a line table for transitions calculated and observed in ion lasers.

B. Edlén, Handbuch der Phys. **27**, 172 (1964).

Author gives corrections to Eriksson's energy level values.

K.B.S. Eriksson, Phys. Rev. **102**, 102 (1956).

Theory of coupling of ^{2}P core with f and g electrons, applied to N II.

T. Sasaki, N. Kaifu, N. Itoh, K. Sakai, and I. Shimada, Sci. of Light **14**, 142 (1965).

Authors give Grotrian diagrams.

J.B. Tatus, Mon. Not. R. Astr. Soc. **140**, 87 (1968).

Author gives a line table of calculated and observed wavelengths.

Comment on intermediate coupling: We have to deal with the coupling of f- and g-electrons to a core. Following Eriksson, Phys. Rev. **102**, 102 (1956), the pair-coupling approximation is used to link a p-electron with f or g, as required. A total L-value appears, represented by D, F, G, or H, along with an intermediate-coupling angular momentum formed from L and j_l. We call this value, $[j,l]$ in our key; Eriksson calls it K. There is then a total angular momentum formed from $[j,l]$ and the spin of the f- or g-electron.

The table below, adapted from Eriksson's paper, summarizes the coupling assigned to a ^{2}P core with an f- or g-electron.

	J_P	L	$J_{P}, l_2 = 3$:	J_{final}
^{2}P + f	3/2	G	9/2	5,4
			7/2	4,3
	1/2	F	7/2	4,3
			5/2	3,2
	3/2	D	5/2	3,2
			3/2	2,1
			$J_{P}, l_2 = 4$:	
^{2}P + g	3/2	H	11/2	6,5
			9/2	5,4
	1/2	G	9/2	5,4
			7/2	4,3
	3/2	F	7/2	4,3
			5/2	3,2

See Note in bibliography of N I

N III $\qquad\qquad$ $Z = 7$ $\qquad\qquad$ 5 electrons

H.G. Berry, W.S. Bickel, S. Bashkin, J. Désesquelles, and R.M. Schectman, J. Opt. Soc. Amer. **61**, 947 (1971).

   Authors give transition wavelengths and Grotrian diagrams.

W.B. Bridges and A.N. Chester, IEEE J. Qu. Electronics **1**, 66 (1965).

   Authors give wavelengths calculated and observed in ion gas lasers.

See Note in bibliography of N I

N IV $\qquad\qquad$ $Z = 7$ $\qquad\qquad$ 4 electrons

W.B. Bridges and A.N. Chester, IEEE J. Qu. Electronics **1**, 66 (1965).

   Authors give wavelengths calculated and observed in ion gas lasers.

T. Sasaki, N. Kaifu, N. Itoh, K. Sakai, and I. Shimada, Sci. of Light **14**, 142 (1965).

   Authors give Grotrian diagrams.

N V $\qquad\qquad$ $Z = 7$ $\qquad\qquad$ 3 electrons

   Please see the general references.

B.C. Fawcett, A.H. Gabriel, W.G. Griffin, B.B. Jones, and R. Wilson, Nature **200**, 1303 (1963).

Authors identify lines attributed to N VI in the range 16—400 Å.

E. Holøien and J. Midtdal, J. Phys. B: Atom. Molec. Phys. **4**, 1243 (1971).

Authors calculate energies nonrelativistically in a treatment of the He I isoelectronic sequence.

R.L. Blake, T.A. Chubb, H. Friedman, and A.E. Unzicker, Science **146**, 1037 (1964).

Authors give a line table of lines observed in the solar spectrum below 25 Å.

B.C. Fawcett, A.H. Gabriel, W.G. Griffin, B.B. Jones, and R. Wilson, Nature **200**, 1303 (1963).

Authors give observed spectrum in range 16—400 Å.

Oxygen (O)

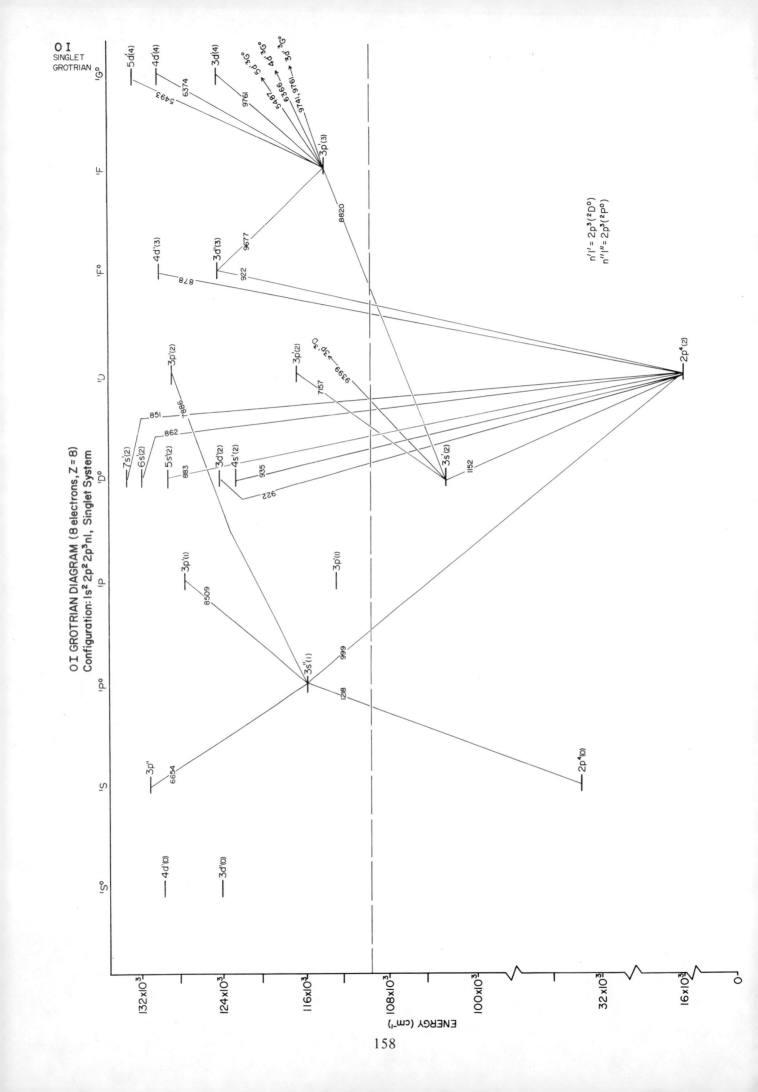

O I
SINGLET
GROTRIAN

O I GROTRIAN DIAGRAM (8 electrons, Z = 8)
Configuration: 1s² 2p² 2p³ nl, Singlet System

n′l′ = 2p³(²D°)
n″l″ = 2p³(²P°)

158

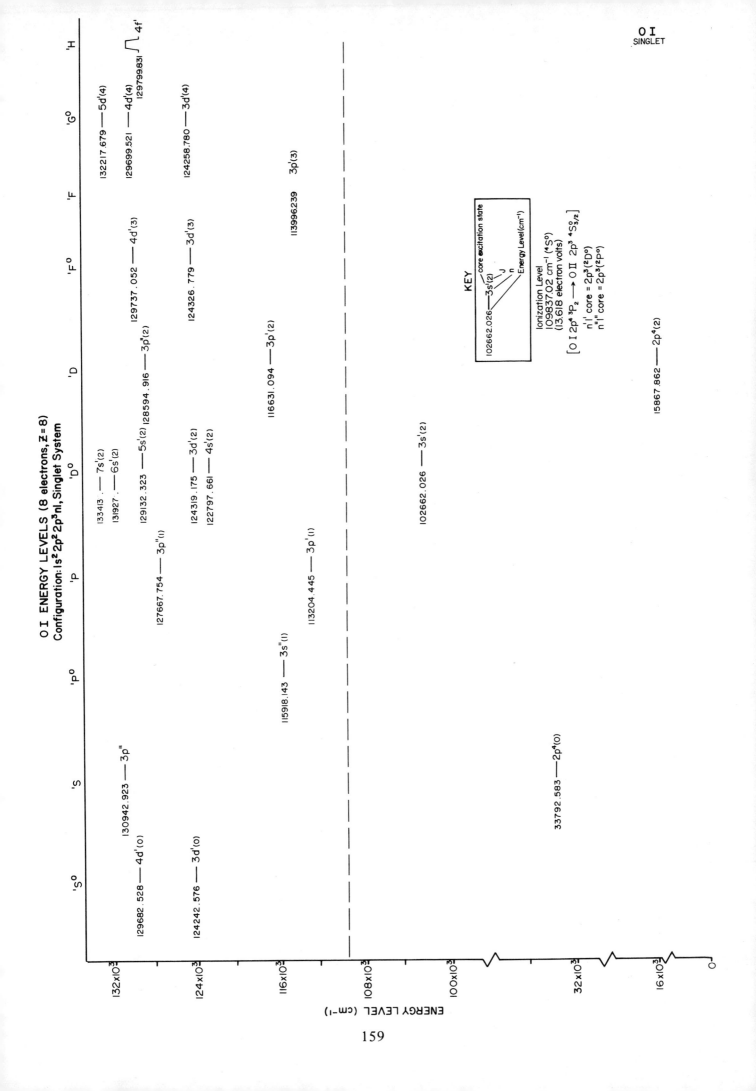

O I ENERGY LEVELS (8 electrons, Z = 8)
Configuration: 1s² 2p² 2p³ nl, Singlet System

O I
SINGLET

KEY

core excitation state
102662.026 — 3s'(2)
J
n
Energy Level (cm⁻¹)

Ionization Level
109837.02 cm⁻¹ (⁴S°)
(13.618 electron volts)
[O I 2p⁴ ³P₂ → O II 2p³ ⁴S°₃/₂]
n'l' core = 2p³(²D°)
n"l" core = 2p³(²P°)

'S°
129682.528 — 4d'(0)
124242.576 — 3d'(0)

'S
130942.923 — 3p"
33792.583 — 2p⁴(0)

'P°
115918.143 — 3s"(1)

'P
127667.754 — 3p"(1)
113204.445 — 3p'(1)

'D°
133413. — 7s'(2)
131927. — 6s'(2)
129132.323 — 5s'(2)
124319.175 — 3d'(2)
122797.661 — 4s'(2)
102662.026 — 3s'(2)

'D
128594.916 — 3p"(2)
116631.094 — 3p'(2)
15867.862 — 2p⁴(2)

'F°
129737.052 — 4d'(3)
124326.779 — 3d'(3)

'F
113996239   3p'(3)

'G°
132217.679 — 5d'(4)
129699.521 — 4d'(4)
129799831 ⌐ 4f'
124258.780 — 3d'(4)

'H°

ENERGY LEVEL (cm⁻¹)

132x10³
124x10³
116x10³
108x10³
100x10³
32x10³
16x10³
0

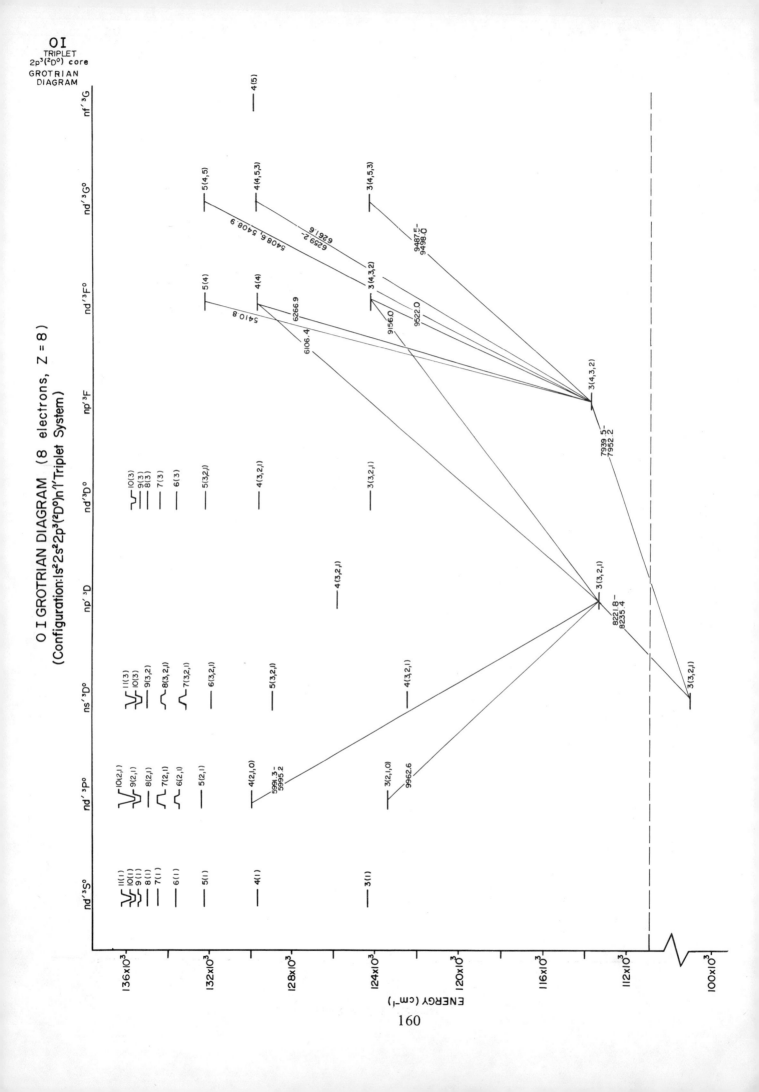

O I GROTRIAN DIAGRAM (8 electrons, Z = 8)
(Configuration: ls²2s²2p³(²D°)nℓ'Triplet System)

OI
TRIPLET
2p³(²D°) core
GROTRIAN
DIAGRAM

ENERGY (cm⁻¹)

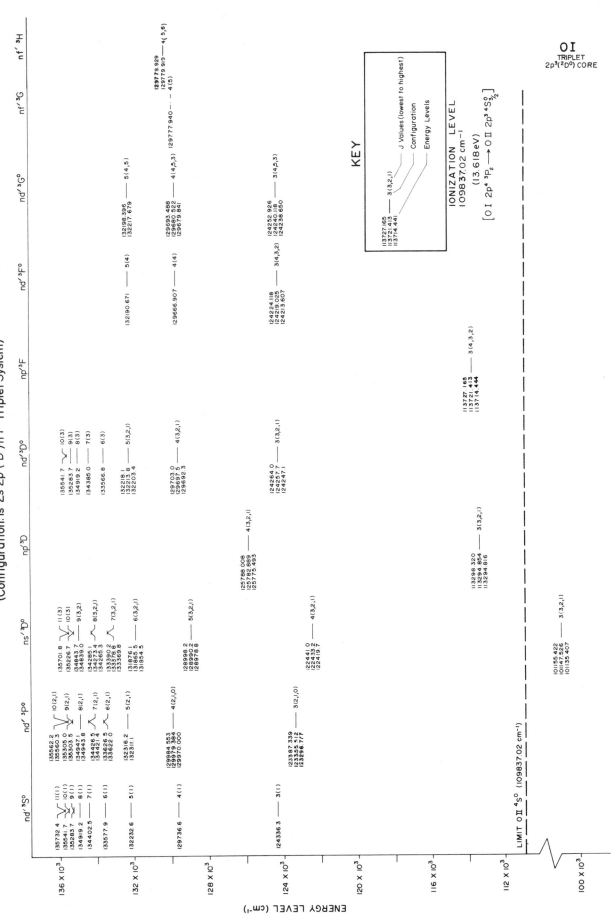

O I ENERGY LEVELS (8 electrons, Z = 8)
(Configuration: ls²2s²2p³(²D⁰)n′l′ Triplet System)

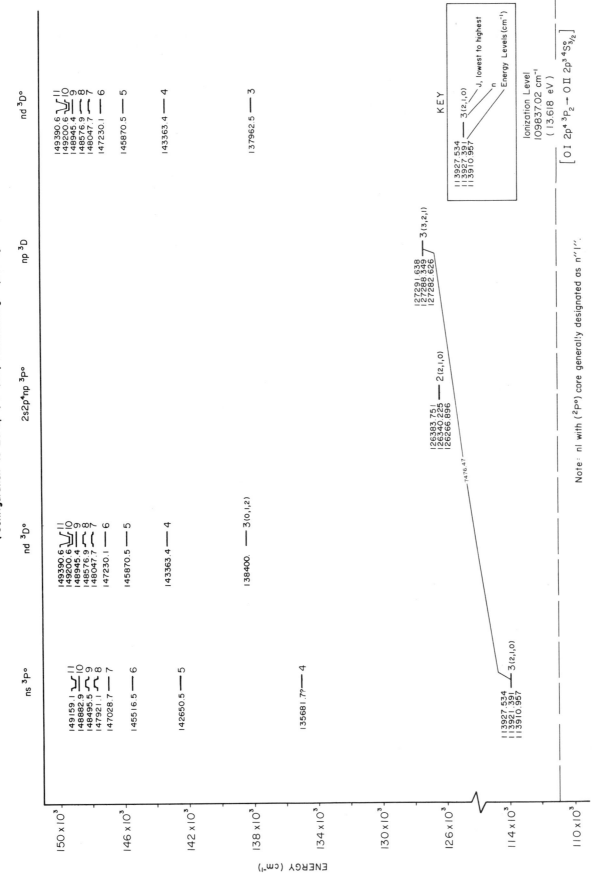

OI ENERGY LEVELS (8 electrons, Z=8)

(Configuration: 1s²2s²2p³(²P°)nl, Autoionizing Triplet System)

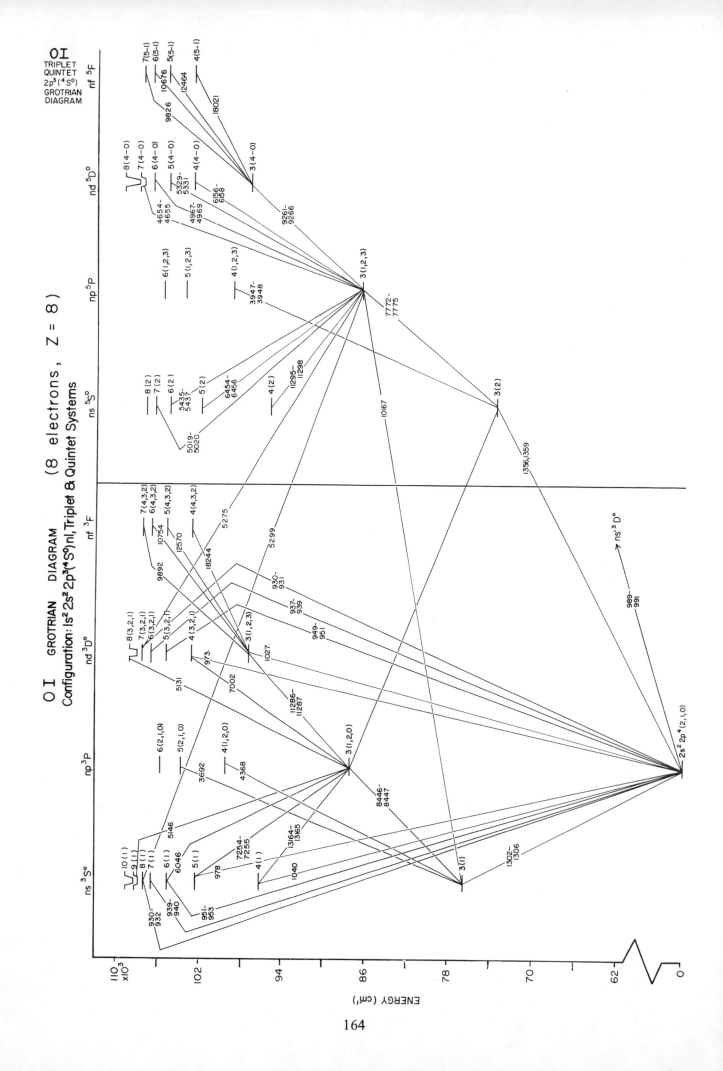

O I GROTRIAN DIAGRAM (8 electrons, Z = 8)
Configuration: 1s² 2s² 2p³(⁴S°)nl, Triplet & Quintet Systems

OI
TRIPLET
QUINTET
2p³(⁴S°)
GROTRIAN
DIAGRAM

ENERGY (cm⁻¹)

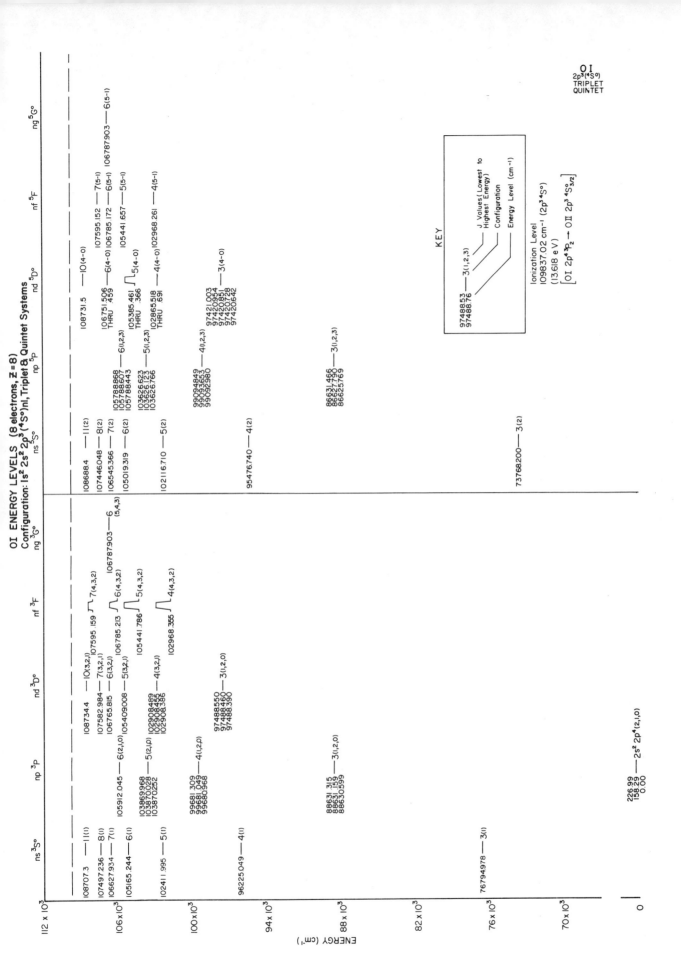

OI ENERGY LEVELS (8 electrons, Z=8)
Configuration: 1s² 2s² 2p³(⁴S°)nl, Triplet & Quintet Systems

ENERGY (cm⁻¹)

165

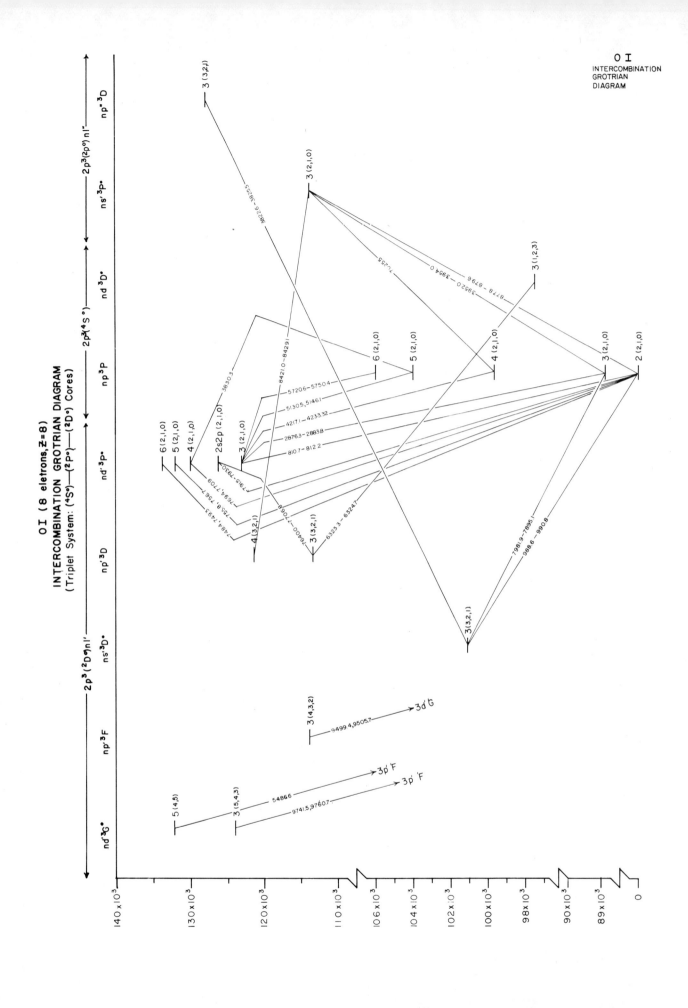

OI
INTERCOMBINATION
GROTRIAN
DIAGRAM

OI (8 eletrons, Z=8)
INTERCOMBINATION GROTRIAN DIAGRAM
(Triplet System: (⁴S°)—(²P°)—(²D°) Cores)

166

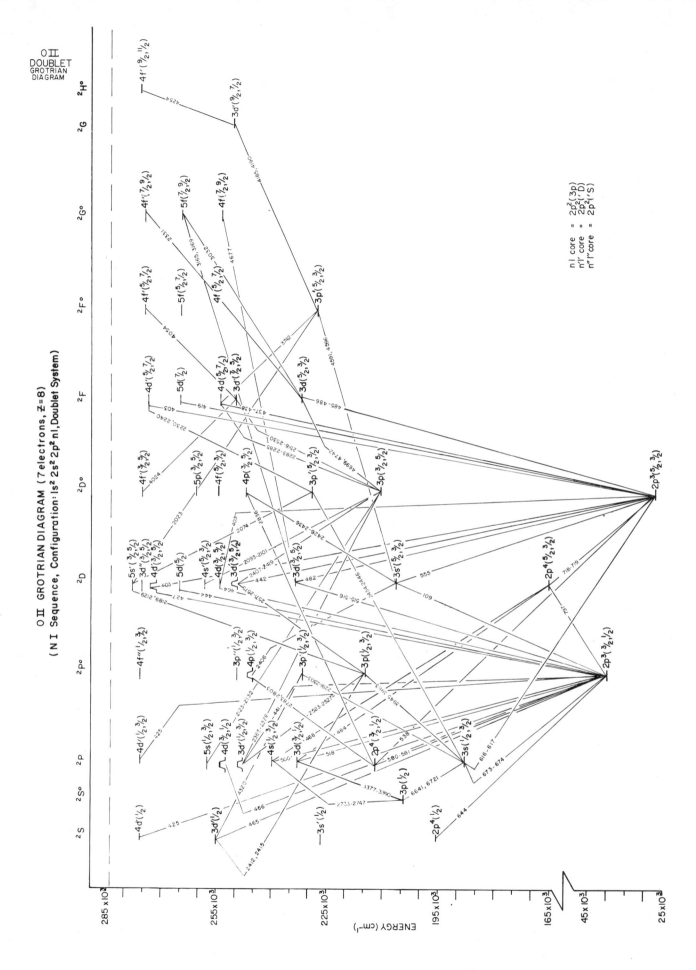

O II GROTRIAN DIAGRAM (7 electrons, Z=8)
(N I Sequence, Configuration: 1s² 2s² 2p² nl, Doublet System)

O II
DOUBLET
GROTRIAN
DIAGRAM

nl core = 2p²(³P)
n'l' core = 2p²('D)
n"l"core = 2p²('S)

168

O II ENERGY LEVELS (7 electrons, Z = 8)
(N I Sequence, Configuration: 1s² 2s² 2p² nl, Doublet System)

O II
DOUBLET

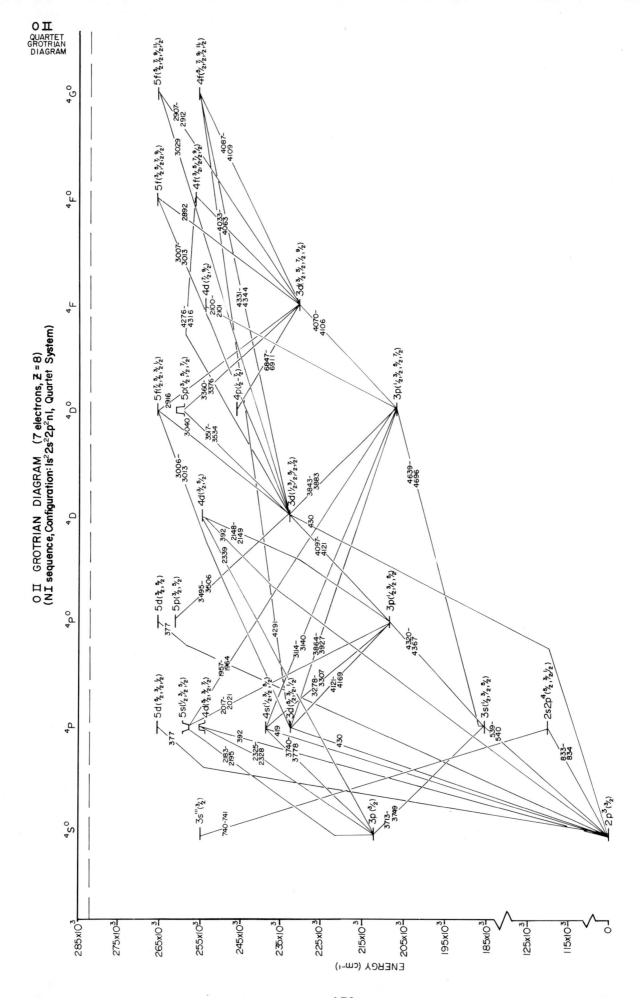

O II GROTRIAN DIAGRAM (7 electrons, Z = 8)
(N I sequence, Configuration: 1s²2s²2p²nl, Quartet System)

O II
QUARTET
GROTRIAN
DIAGRAM

170

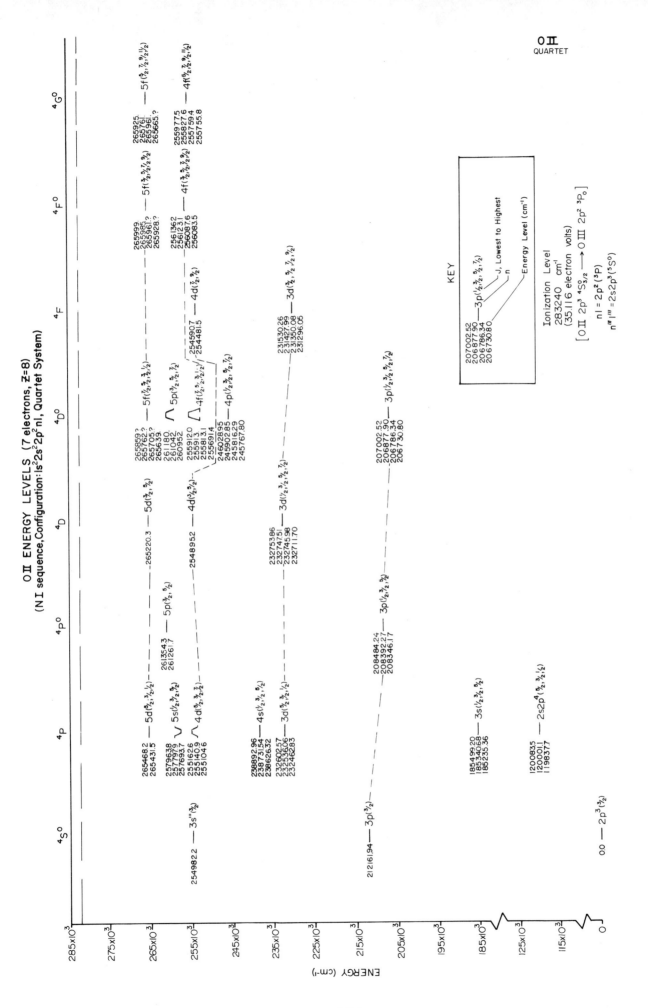

OII ENERGY LEVELS (7 electrons, Z=8)
(N I sequence, Configuration: 1s²2s²2p²nl, Quartet System)

O II
QUARTET

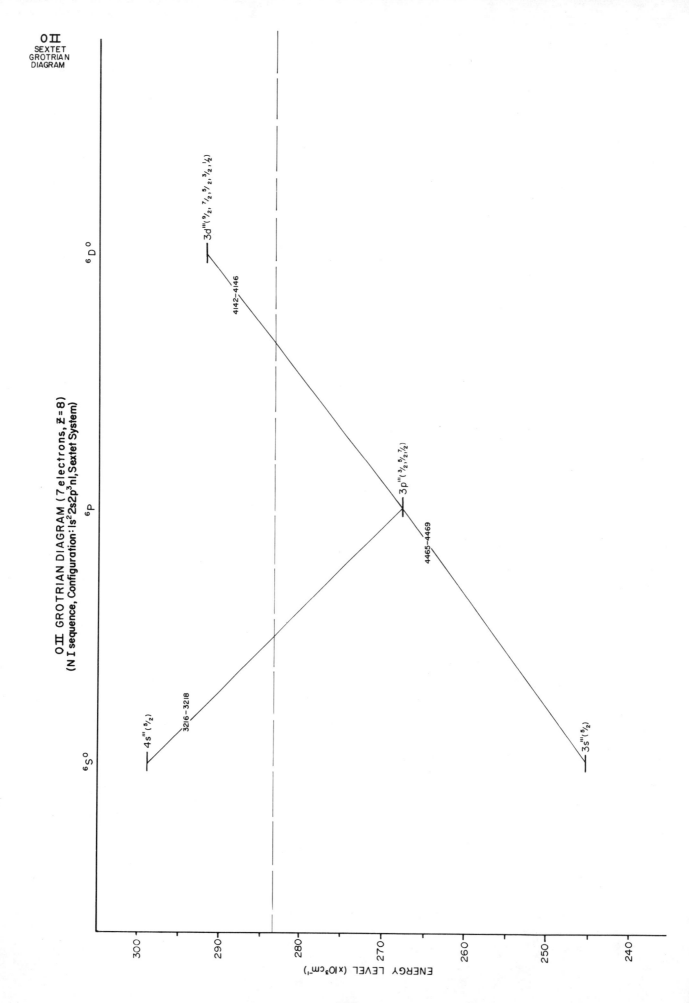

O II
SEXTET
GROTRIAN
DIAGRAM

O II GROTRIAN DIAGRAM (7 electrons, Z = 8)
(N I sequence, Configuration: 1s²2s2p³nl, Sextet System)

$^6S^o$ ^6P $^6D^o$

$4s'''(^5/_2)$

$3216-3218$

$3p'''(^3/_2, ^5/_2, ^7/_2)$

$4465-4469$

$4142-4146$

$3d'''(^9/_2, ^7/_2, ^5/_2, ^3/_2, ^1/_2)$

$3s'''(^5/_2)$

ENERGY LEVEL (x10³cm⁻¹)

300   290   280   270   260   250   240

172

O II ENERGY LEVELS (7 electrons, Z=8)
(N I Sequence, Configuration: 1s² 2s2p³ nl, Sextet Term System)

$^6S^o$

298849.2 + x ———— $4s'''(^5/_2)$

245395.5 + x ———— $3s'''(^5/_2)$

^6P

267783.40 + x
267770.85 + x
267763.39 + x

———— $3p'''(^3/_2, ^5/_2, ^7/_2)$

$^6D^o$

291899.81 + x
291899.01 + x
291898.01 + x
291896.78 + x
291895.90 + x

———— $3d'''(^9/_2, ^7/_2, ^5/_2, ^3/_2, ^1/_2)$

KEY

Core Excitation State

$3p'''(^3/_2, ^5/_2, ^7/_2)$

J, Lowest to Highest

n

Energy Level (cm⁻¹)

Uncertainty
(x=± few hundred cm⁻¹)

267783.40 + x
267770.85 + x
267763.39 + x

O II
SEXTET

IONIZATION LIMIT
285240   cm⁻¹

(35.116 ev)

[O II 2p³ ⁴S°_{3/2} → O III 2p² ³P₀]

n'''|''' = 2s2p³(⁵S°)

ENERGY LEVEL (x 10³ cm⁻¹)

173

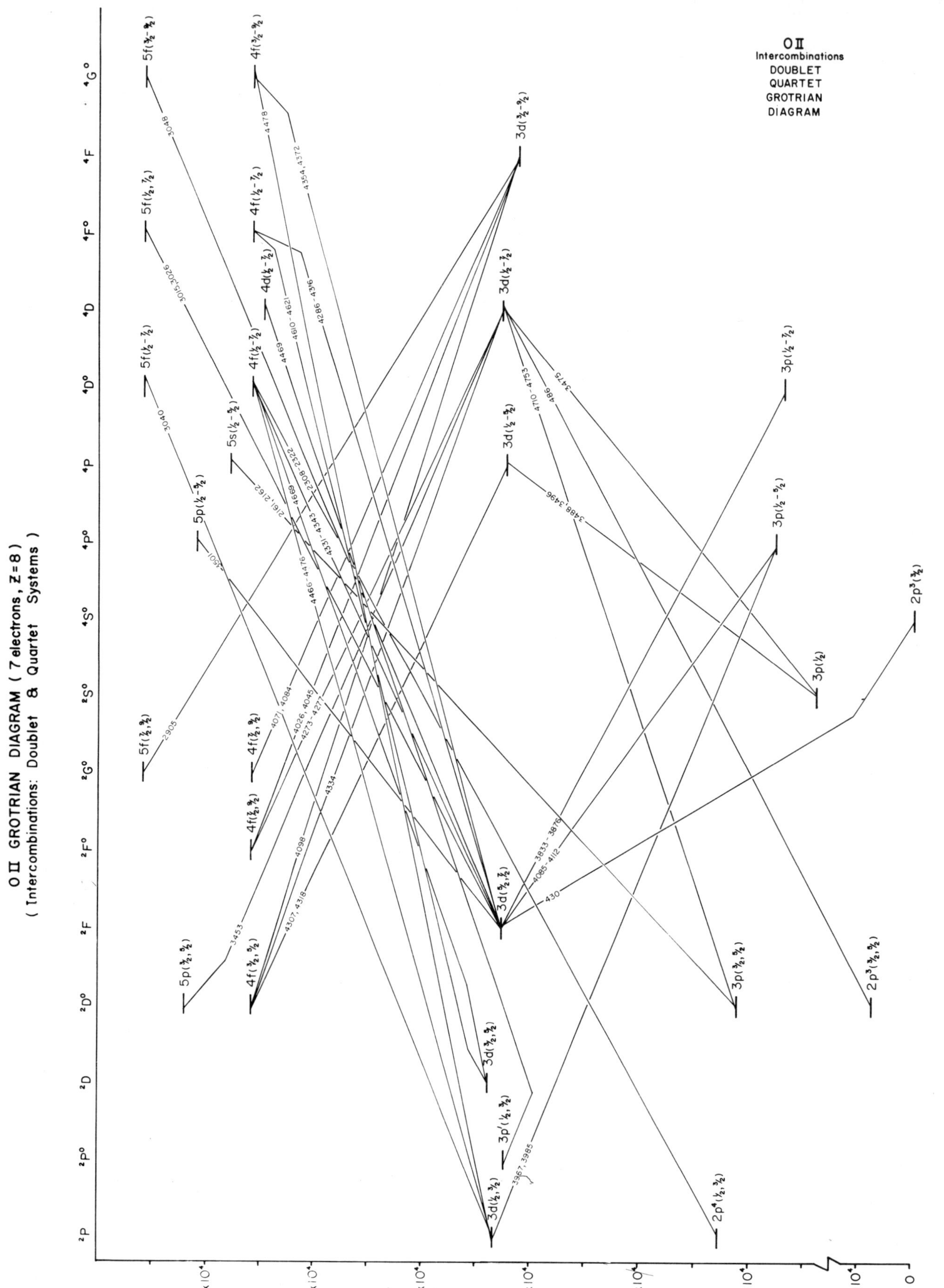

O II GROTRIAN DIAGRAM (7 electrons, Z=8)
(Intercombinations: Doublet & Quartet Systems)
O II
Intercombinations
DOUBLET
QUARTET
GROTRIAN
DIAGRAM
ENERGY (cm⁻¹)
26×10⁴
25×10⁴
24×10⁴
23×10⁴
22×10⁴
21×10⁴
3×10⁴
0

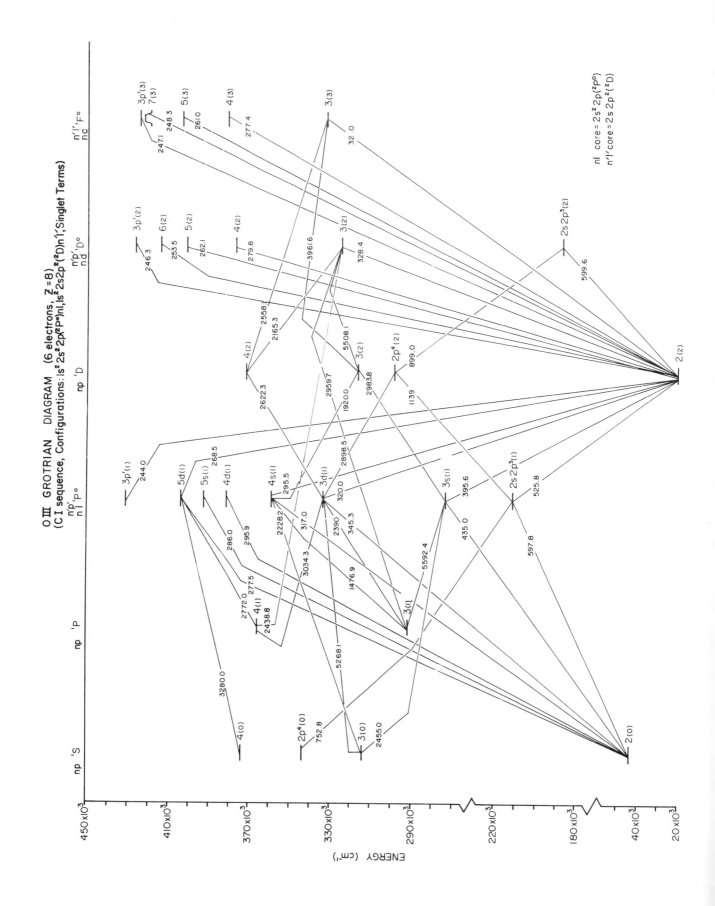

O III
SINGLET
GROTRIAN
DIAGRAM

O III GROTRIAN DIAGRAM (6 electrons, Z=8)
(C I sequence, Configurations: 1s² 2s² 2p(²P°)nl, 1s²2s2p²(²D)n'l';Singlet Terms)

nl core = 2s² 2p(²P°)
n'l' core = 2s 2p²(²D)

176

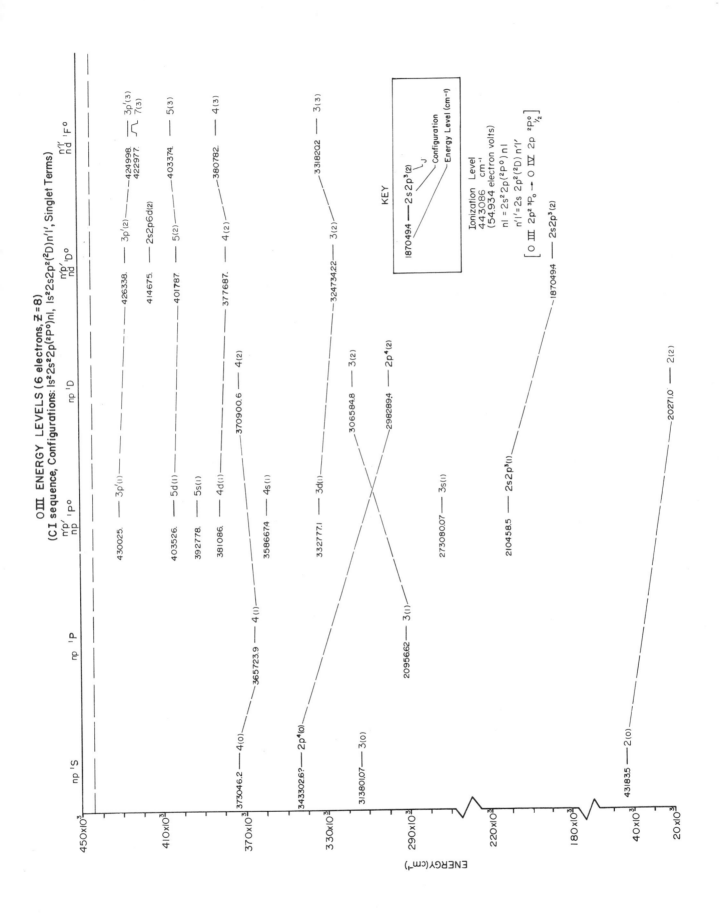

O III
SINGLET

O III ENERGY LEVELS (6 electrons, Z = 8)
(C I sequence, Configurations: 1s²2s²2p(²P°)nl, 1s²2s2p²(²D)n'l', Singlet Terms)

177

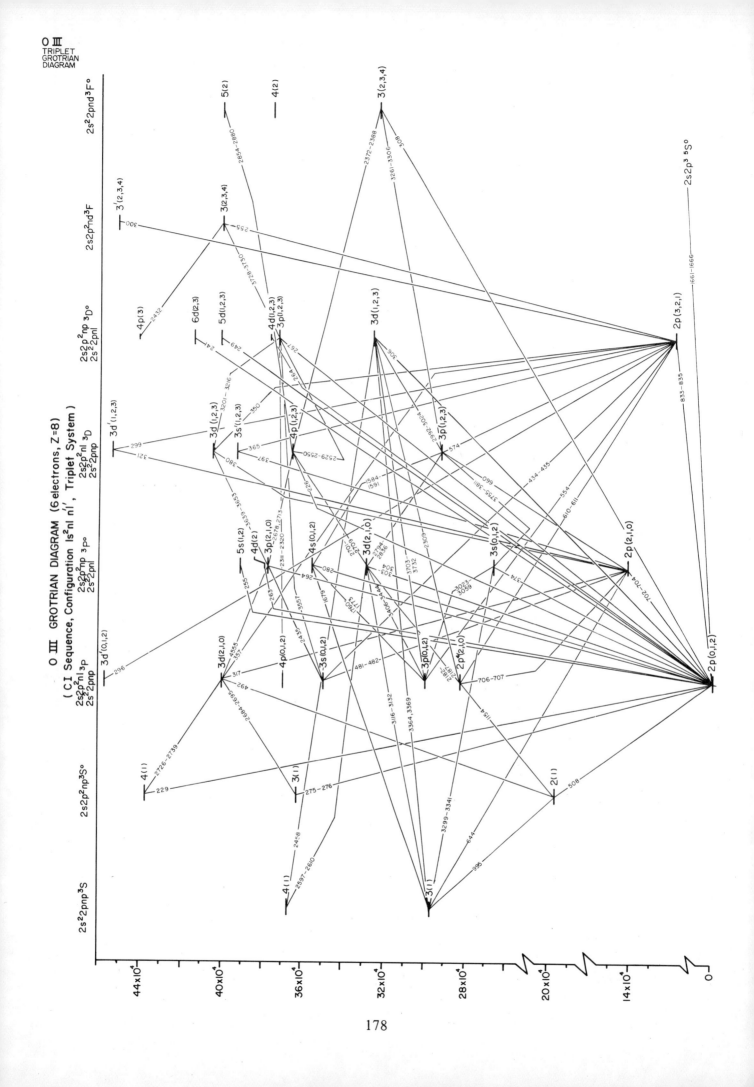

178

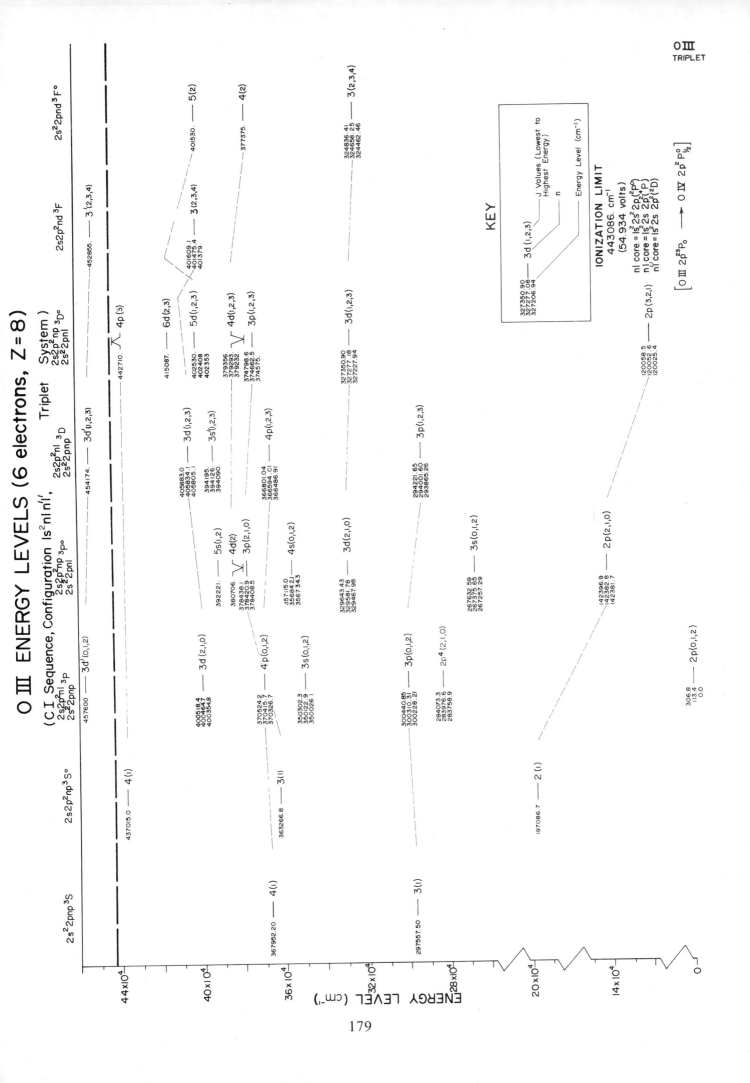

O III ENERGY LEVELS (6 electrons, Z = 8)

Triplet System

(C I Sequence, Configuration 1s²nln'l')

179

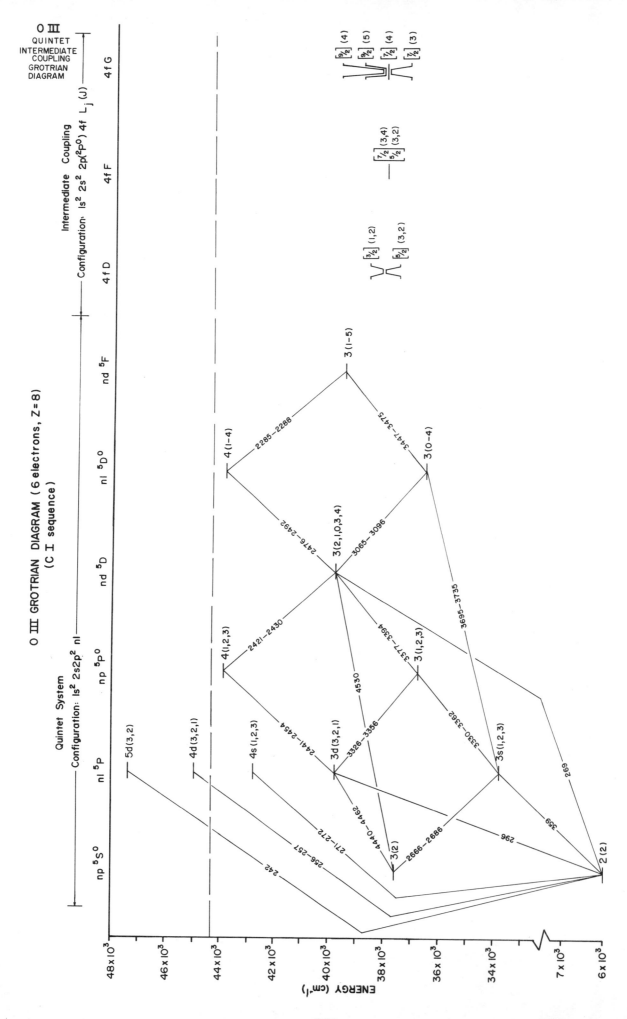

O III ENERGY LEVELS (6 electrons, Z=8)
(C I sequence)

Quintet System
—————— Configuration: $1s^2 2s2p^2$ nl ——————

Intermediate Coupling
—————— Configuration: $1s^2 2s^2 2p(^2P^o)$ 4f L_j (J) ——————

KEY

J values (lowest to highest)
n
Energy Level (cm⁻¹)

398231.9
398150.4
398148.1
398144.5
398140.4

3(2,1,0,3,4)

$[\frac{7}{2}]$ (3)

381176.9

Final J
Intermediate J

Ionization Level
443086. cm⁻¹
(54.934 eV)
$[O\ III\ 2p^2\ ^3P_0 \rightarrow O\ IV\ 2p\ ^2P^o_{1/2}]$

181

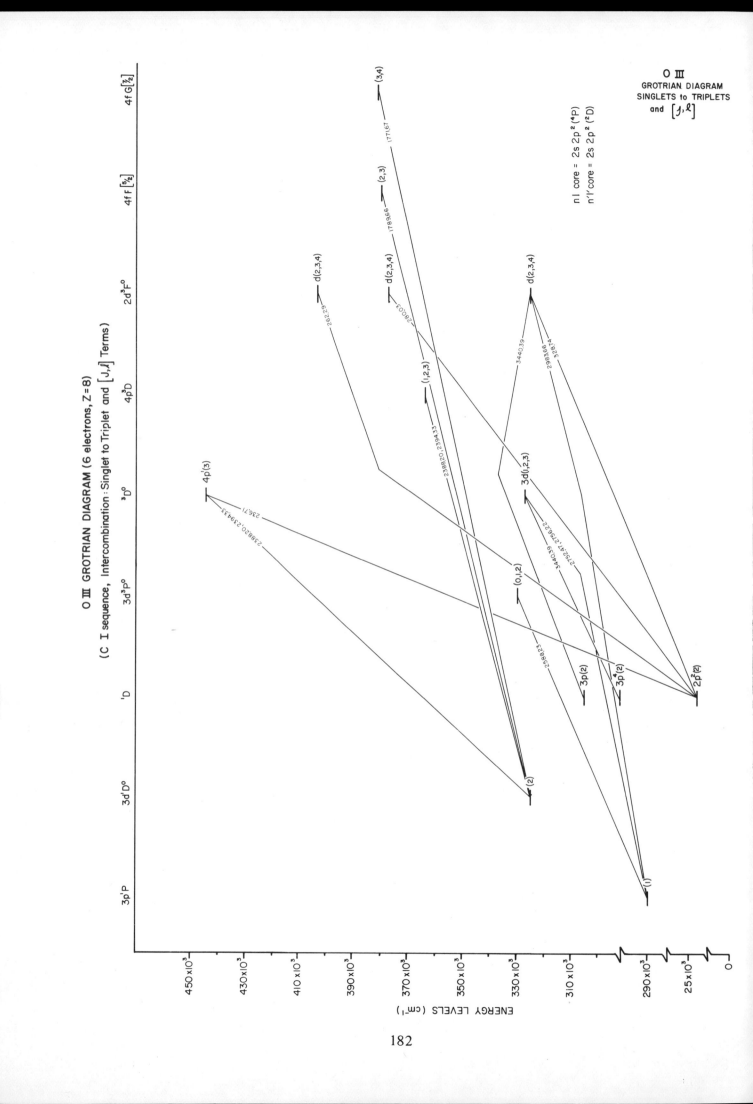

O III GROTRIAN DIAGRAM (6 electrons, Z=8)

(C I sequence, Intercombination: Singlet to Triplet and [J,ℓ] Terms)

O III
GROTRIAN DIAGRAM
SINGLETS to TRIPLETS
and [J,ℓ]

nl core = 2s 2p² (⁴P)
n'l' core = 2s 2p² (²D)

ENERGY LEVELS (cm⁻¹)

182

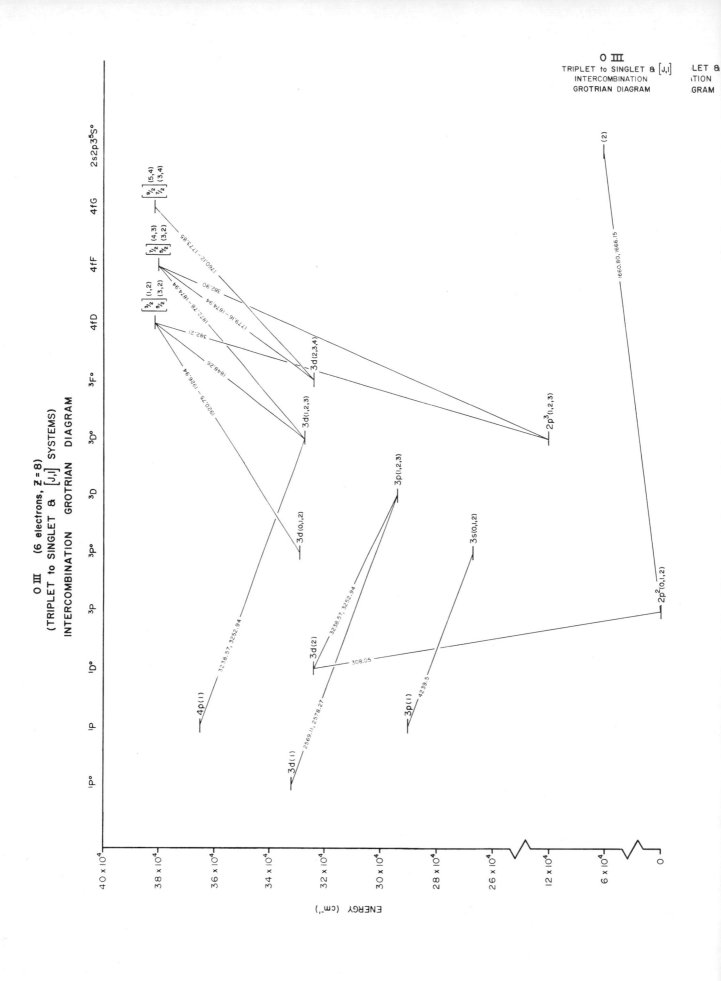

O III
(6 electrons, Z = 8)
(TRIPLET to SINGLET & [J,l] SYSTEMS)
INTERCOMBINATION GROTRIAN DIAGRAM

O III
TRIPLET to SINGLET & [J,l]
INTERCOMBINATION
GROTRIAN DIAGRAM

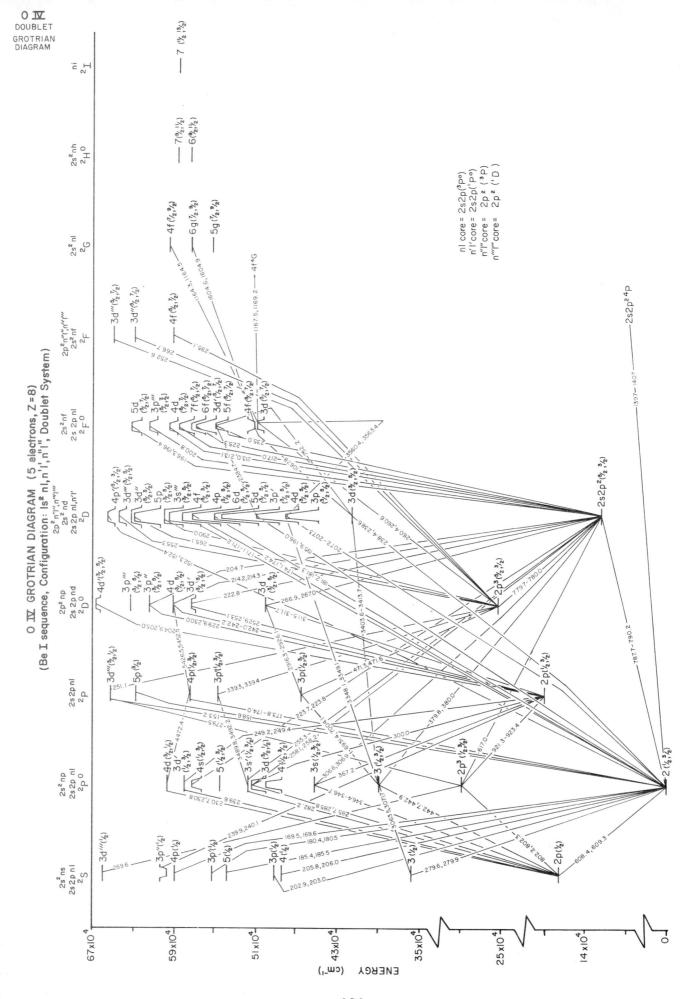

186

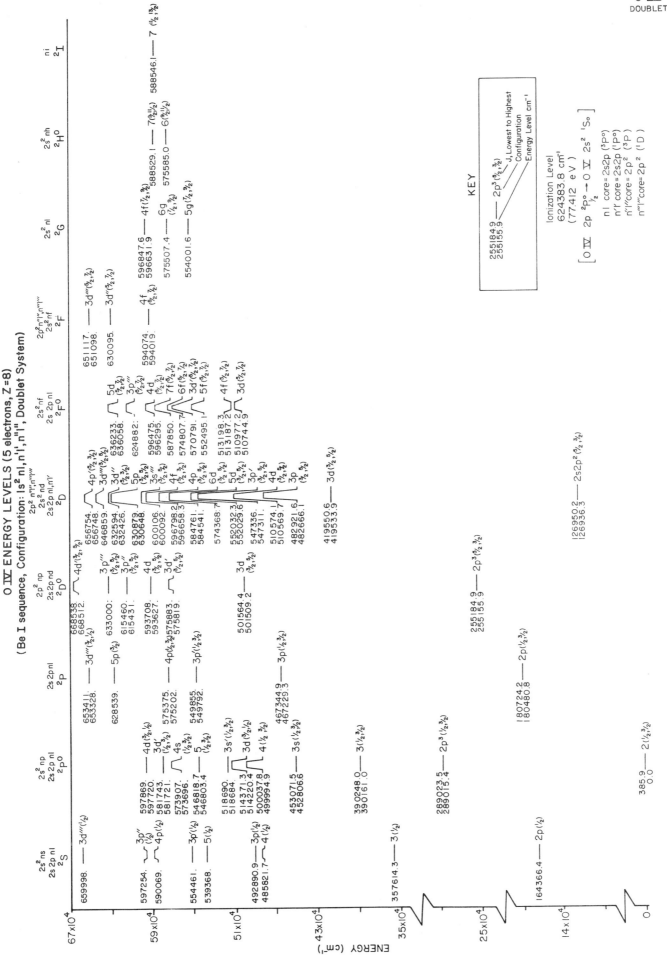

O IV ENERGY LEVELS (5 electrons, Z=8)
(Be I sequence, Configuration: 1s² nl,n'l',n"l",n'"l", Doublet System)

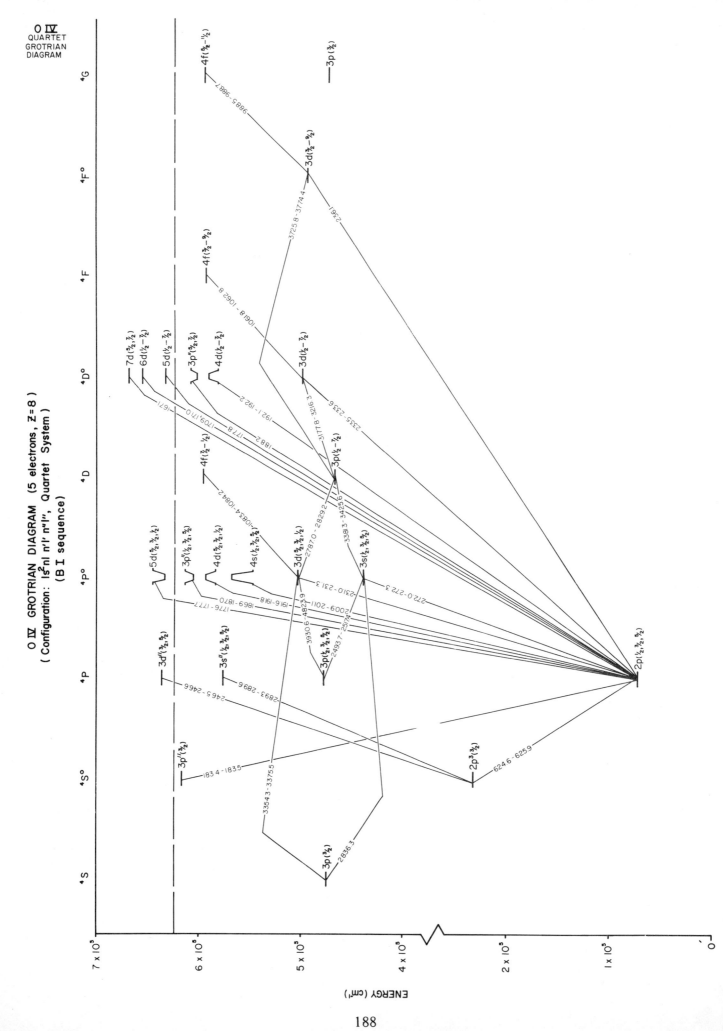

O IV QUARTET GROTRIAN DIAGRAM

O IV GROTRIAN DIAGRAM (5 electrons, Z=8)
(Configuration: 1s²nl n'l' n''l'', Quartet System)
(B I sequence)

ENERGY (cm⁻¹)

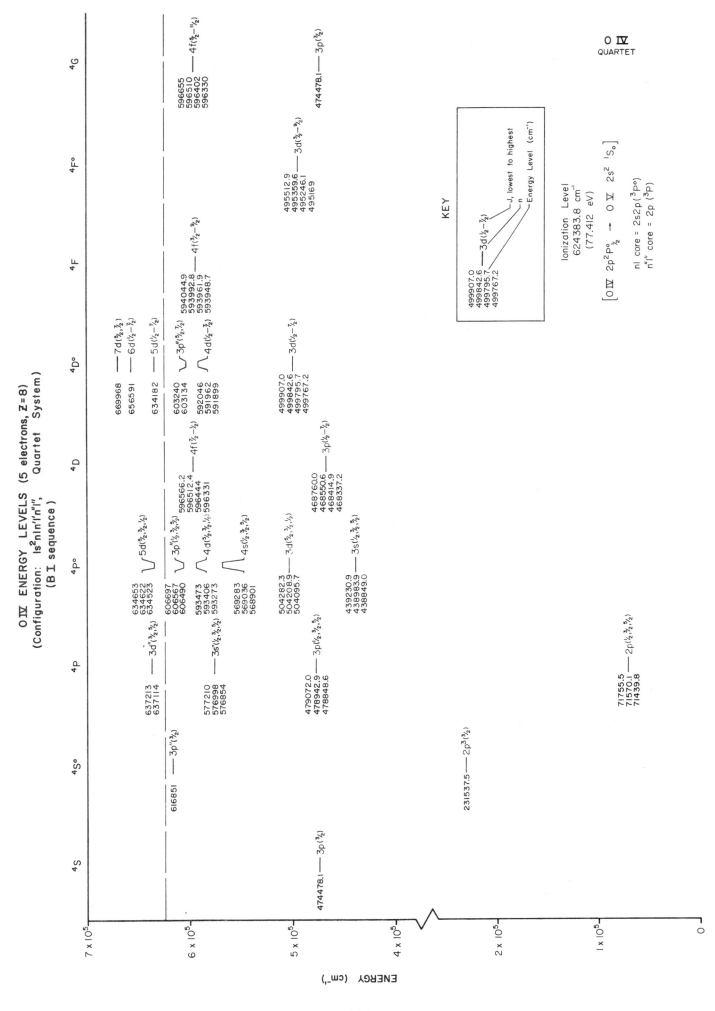

O IV ENERGY LEVELS (5 electrons, Z=8)
(Configuration: 1s²nln'l'n''l'', Quartet System)
(B I sequence)

189

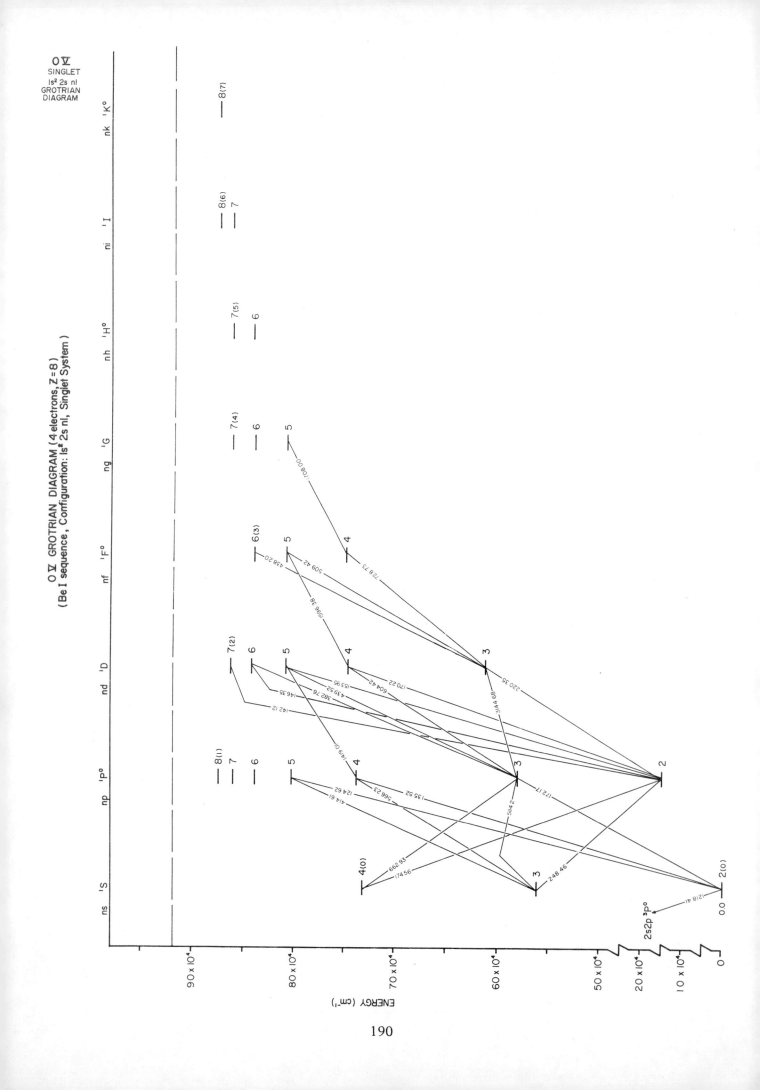

O V ENERGY LEVEL (4 electrons, Z=8)
(Be I sequence, Configuration: 1s² 2s nl, Singlet System)

ENERGY (cm⁻¹)

ns ¹S | np ¹P° | nd ¹D | nf ¹F° | ng ¹G | nh ¹H° | ni ¹I | nk ¹K°

8744470 — 8(1)
8608740 — 7(1)
8396160 — 6(1)

8757744 — 8(6)   8757858 — 8(7)
8626504 — 7(6)

8628007 — 7(4)   8626062 — 7(5)
8403852 — 6(4)   8423732 — 6(5)
8083886 — 5(4)

8624190 — 7(2)
8420870 — 6(2)
8083523 — 5(2)

8408215 — 6(3)
8089168 — 5(3)
7498405 — 4(3)

8024660 — 5(1)

7378808 — 4(1)

7462749 — 4(2)

7316705 — 4(0)

6126156 — 3(2)

5808249 — 3(1)

5612764 — 3(0)

1587977 — 2(1)

0.0 — 2(0)

90 x 10⁴
80 x 10⁴
70 x 10⁴
60 x 10⁴
50 x 10⁴
20 x 10⁴
10 x 10⁴
0

KEY

7462749 —— 4(2)
       J
   n
Energy Level (cm⁻¹)

Ionization Level
98657 cm⁻¹
(113.896 eV)
[O V 1s²2s²¹S₀→O VI 1s²2s²S₁/₂]

O V
SINGLET
1s² 2s nl

191

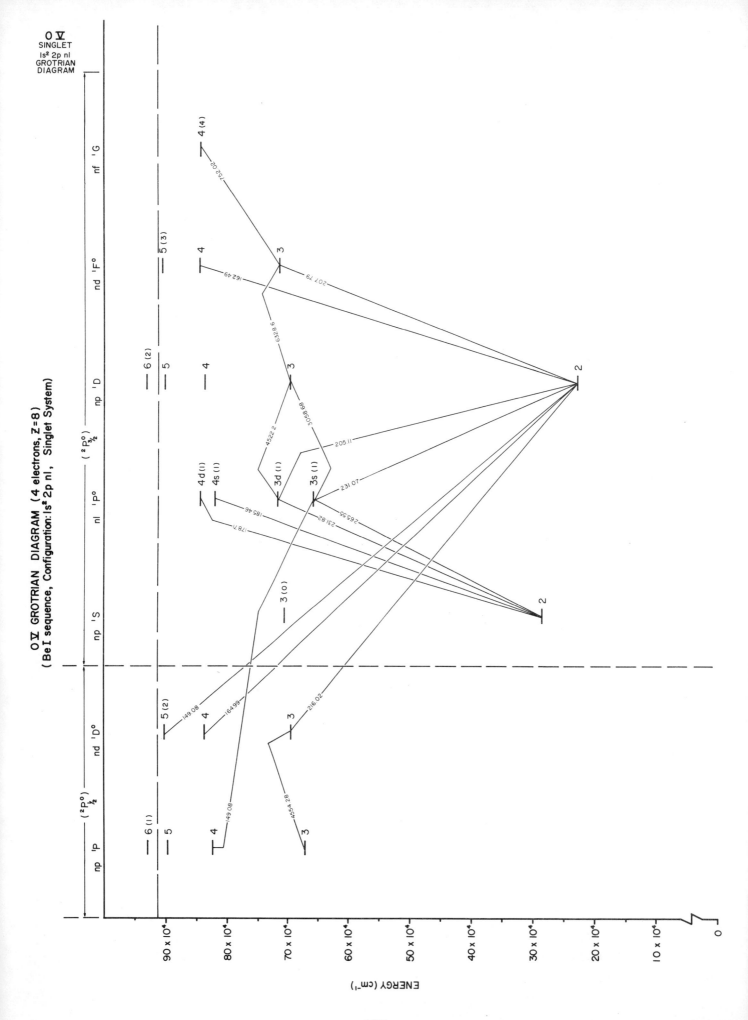

O Ⅴ GROTRIAN DIAGRAM (4 electrons, Z=8)
(Be I sequence, Configuration: 1s² 2p nl, Singlet System)

O Ⅴ
SINGLET
1s² 2p nl
GROTRIAN
DIAGRAM

O V ENERGY LEVELS (4 electrons, Z = 8)
(Be I sequence, Configuration: ls² 2p nl, Singlet System)

O V
SINGLET
ls² 2p nl

KEY

847460.0 ——— 4 d (1) ⌐J
                      └n
                Energy Level (cm⁻¹)

Ionization   Level
918657   cm⁻¹
(113.896 eV)

[O V ls² 2s² ¹S₀ → O VI ls² 2s ²S_½]

np ¹P   (²P°_½)   nd ¹D°   np ¹S   nl ¹P°   (²P°_{3/2})   np ¹D   nd ¹F°   nf ¹G

935093.0 ——— 6 (1)

898580.0 ——— 5 (1)   902691.0 ——— 5 (2)   937341.0 ——— 6 (2)

                                                                                                          902442.0 ——— 5 (2)   906403.0 ——— 5 (3)

829597.0 ——— 4 (1)   837833.0 ——— 4 (2)   847460.0 ——— 4d (1)   837855.0 ——— 4 (2)   847136.0 ——— 4 (3)   845942.6 ——— 4 (4)

                                                                    824282.0 ——— 4s (1)

707635.5 ——— 3 (0)   719274.9 ——— 3d (1)   697170.2 ——— 3 (2)   712963.5 ——— 3 (3)

672693.8 ——— 3 (1)   694643.8 ——— 3 (2)   664485.9 ——— 3s (1)

287910.3 ——— 2 (0)   231721.4 ——— 2 (2)

ENERGY (cm⁻¹)

90 × 10⁴

80 × 10⁴

70 × 10⁴

60 × 10⁴

50 × 10⁴

40 × 10⁴

30 × 10⁴

20 × 10⁴

10 × 10⁴

0

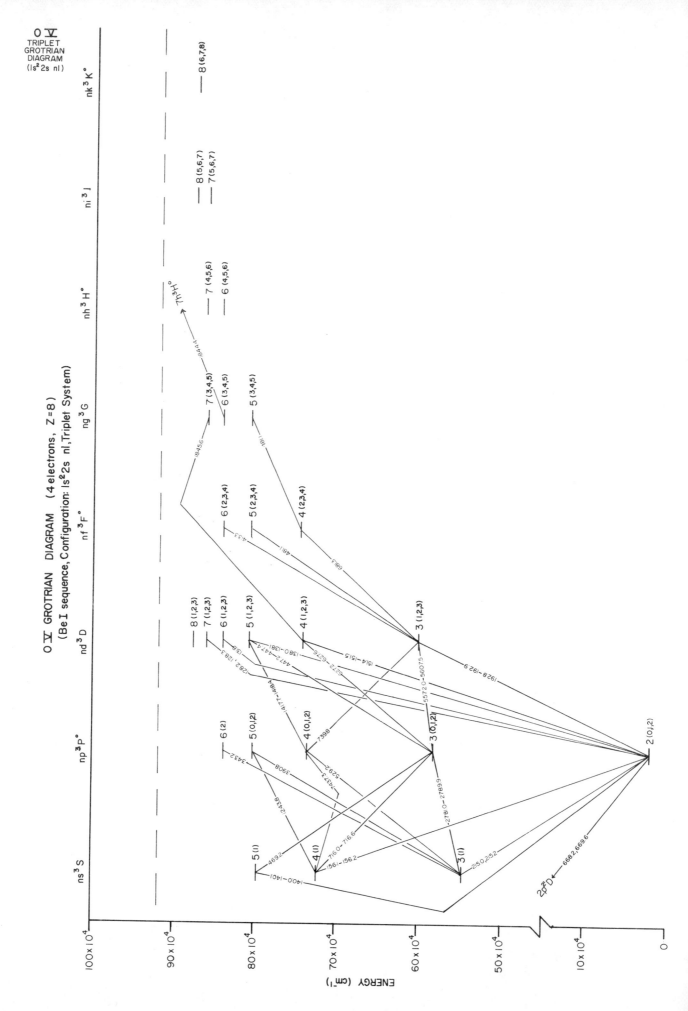

O V GROTRIAN DIAGRAM (4 electrons, Z = 8)
(Be I sequence, Configuration: 1s²2s nl, Triplet System)

O V
TRIPLET
GROTRIAN
DIAGRAM
(1s² 2s nl)

ENERGY (cm⁻¹)

194

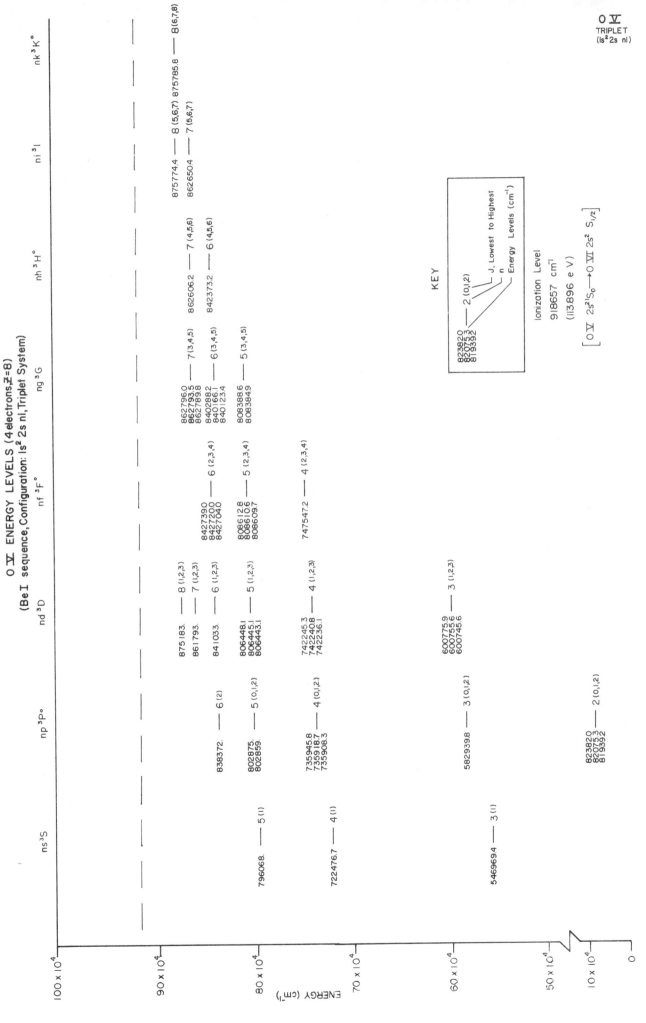

O V ENERGY LEVELS (4 electrons, Z=8)
(Be I sequence, Configuration: 1s² 2s nl, Triplet System)

ENERGY (cm⁻¹)

195

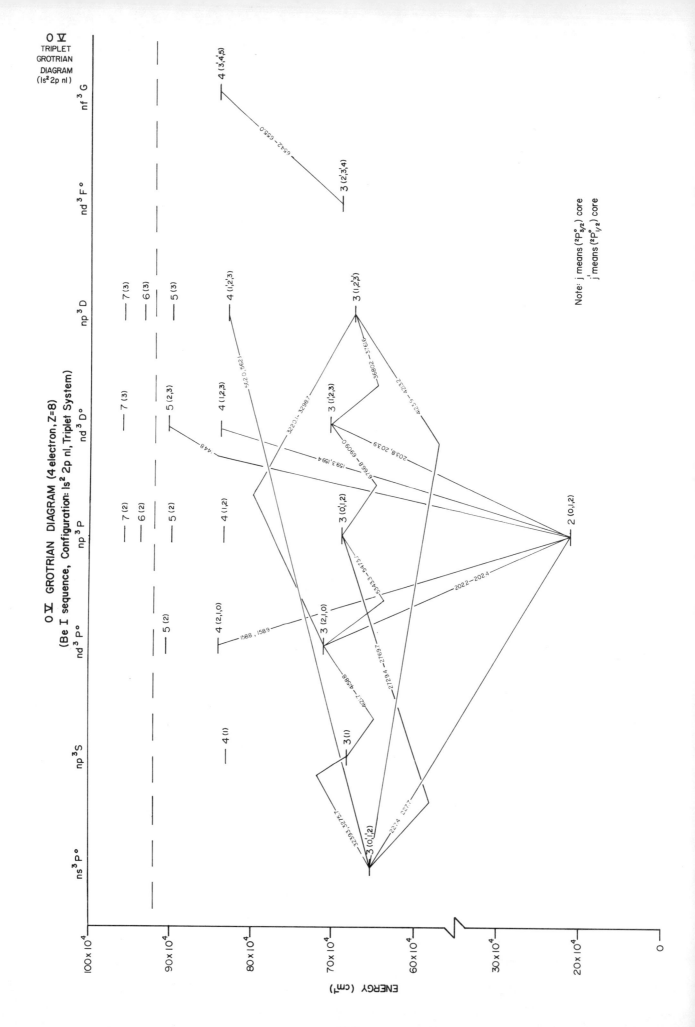

O V GROTRIAN DIAGRAM (4 electron, Z=8)
(Be I sequence, Configuration: 1s² 2p nl, Triplet System)

O V
TRIPLET
GROTRIAN
DIAGRAM
(1s² 2p nl)

Note: j means (²P°₃/₂) core
j' means (²P°₁/₂) core

196

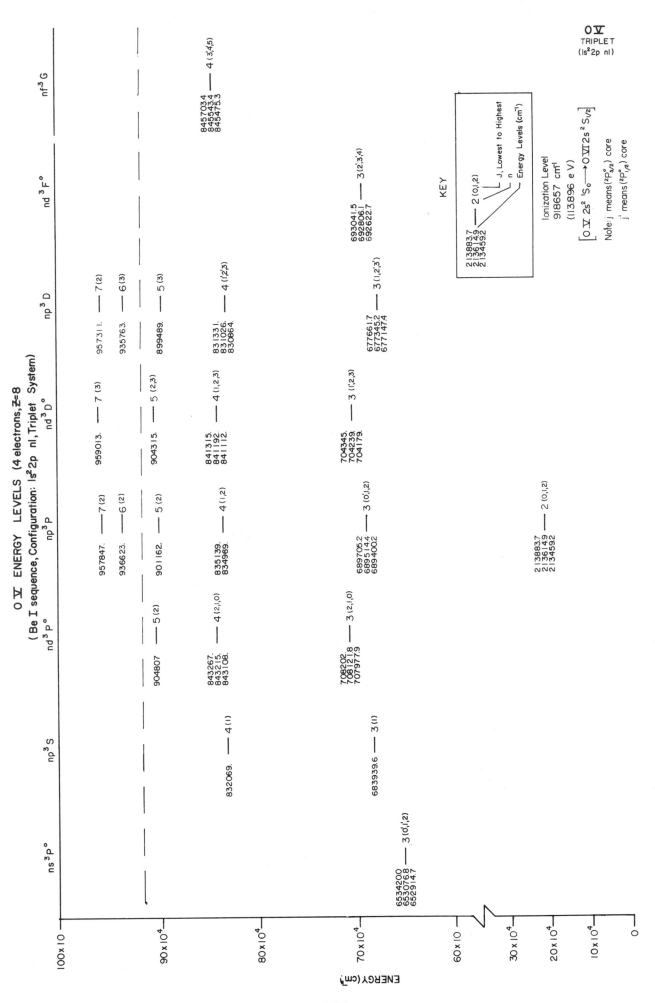

O V ENERGY LEVELS (4 electrons, Z=8)
(Be I sequence, Configuration: 1s²2p nl, Triplet System)

197

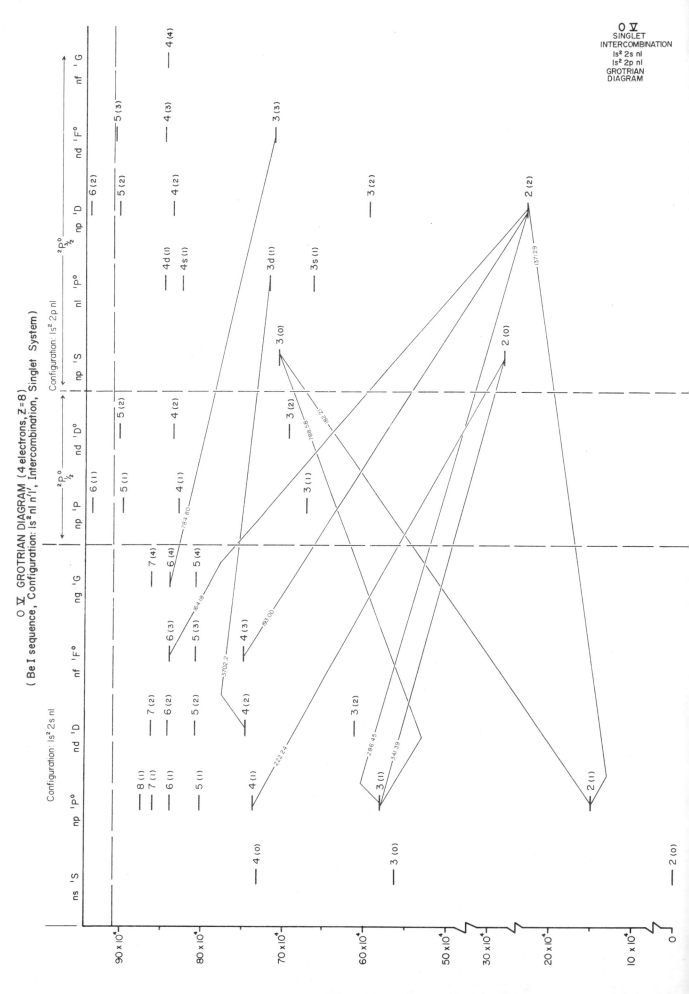

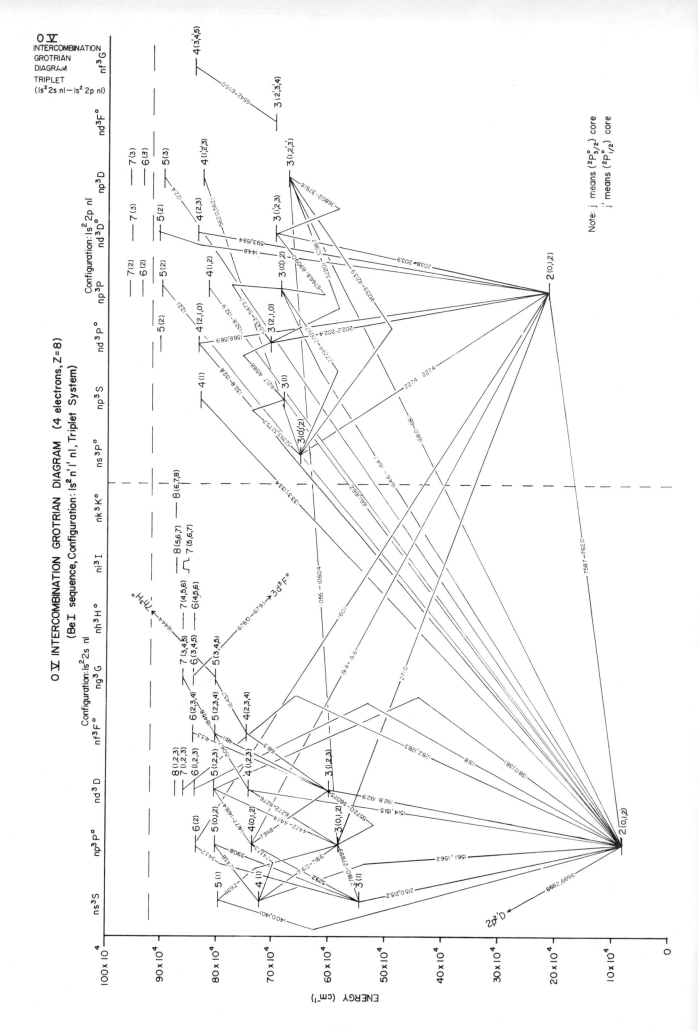

O V INTERCOMBINATION GROTRIAN DIAGRAM (4 electrons, Z=8)

(Be I sequence, Configuration: ls² n l' nl, Triplet System)

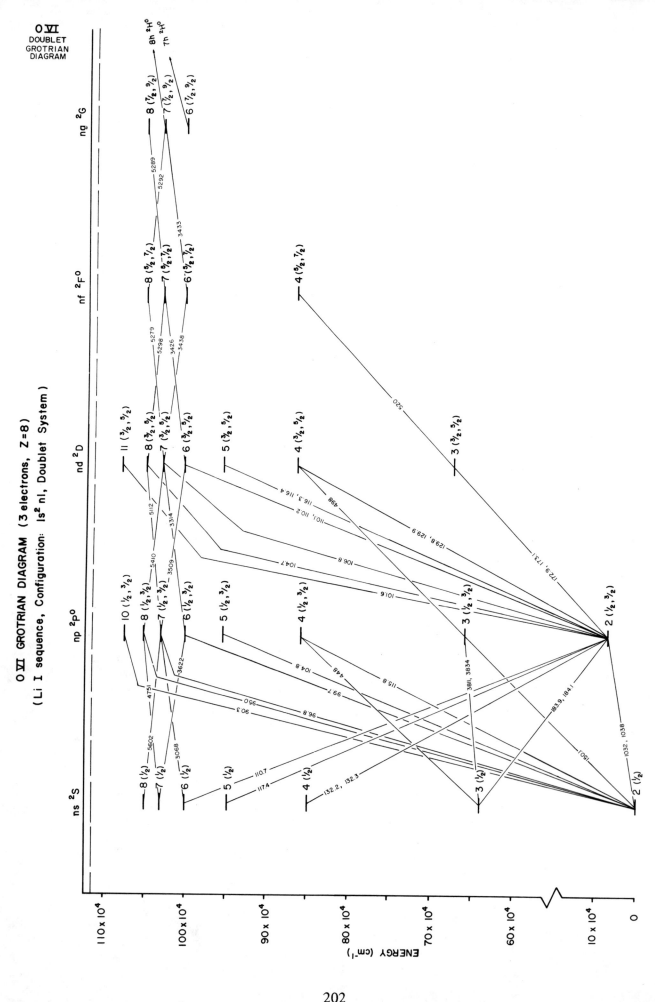

O VI GROTRIAN DIAGRAM (3 electrons, Z=8)

(Li I sequence, Configuration: Is² nl, Doublet System)

O VI
DOUBLET
GROTRIAN
DIAGRAM

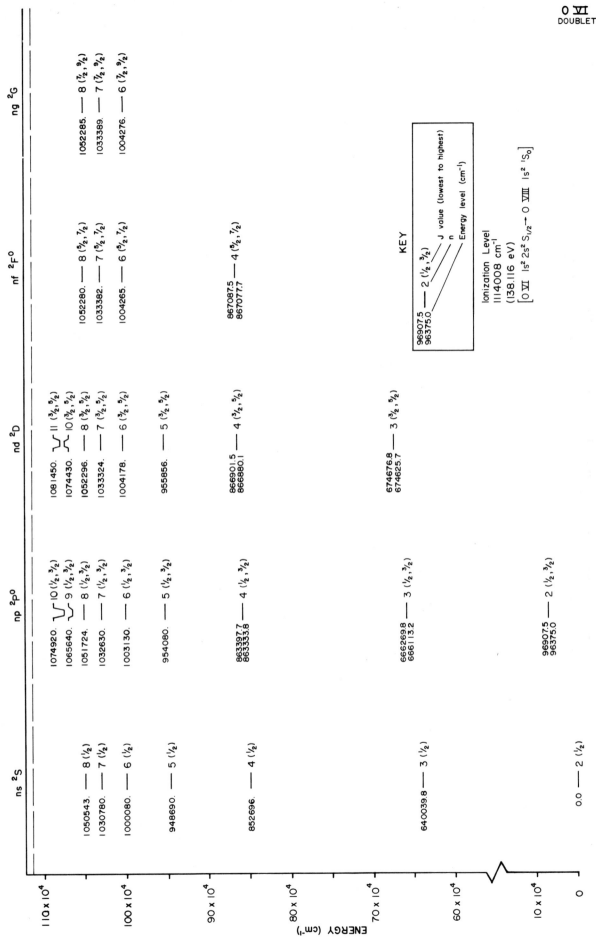

O VI ENERGY LEVELS (3 electrons, Z=8)

(Li I sequence, Configuration: 1s² nl, Doublet System)

O VI
DOUBLET

ENERGY (cm⁻¹)

203

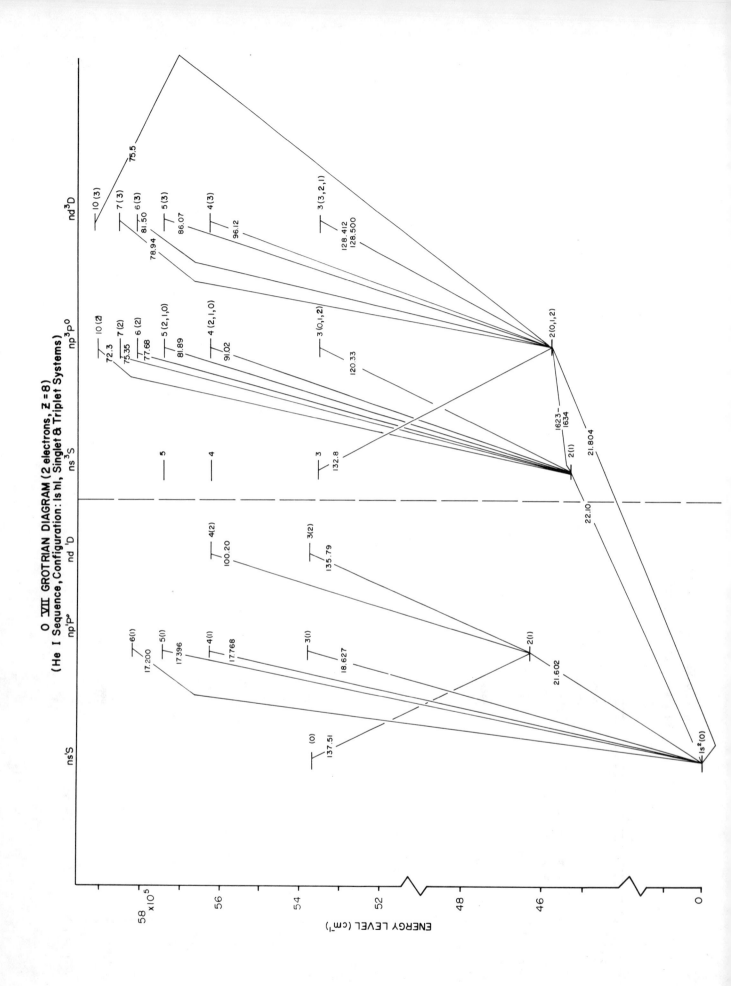

O VII GROTRIAN DIAGRAM (2 electrons, Z = 8)
(He I Sequence, Configuration: 1s nl, Singlet & Triplet Systems)

204

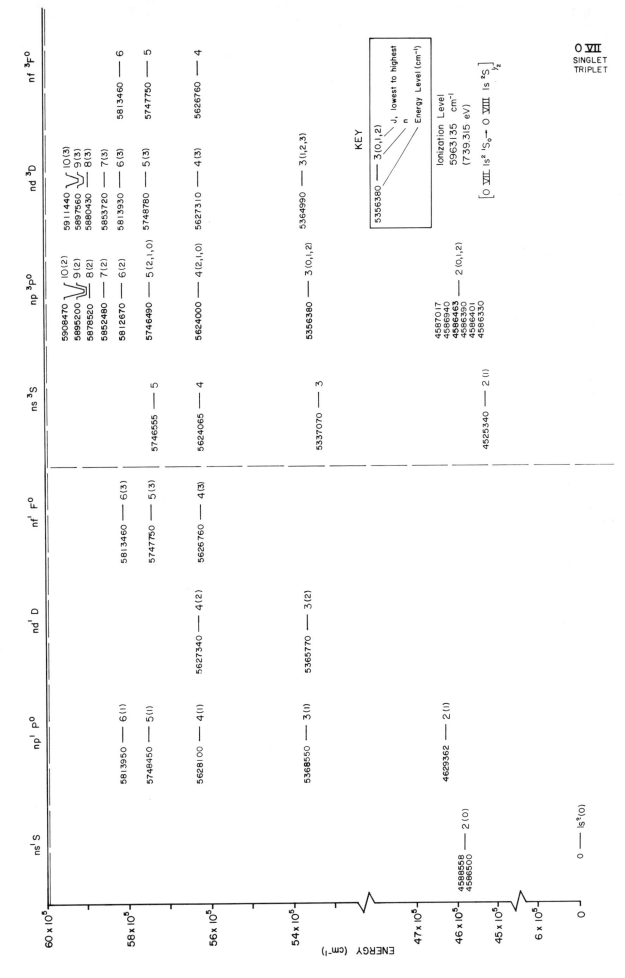

O VII ENERGY LEVELS (2 = electrons, Z = 8)
(He I Sequence, Configuration: 1s nl, Singlet and Triplet System)

205

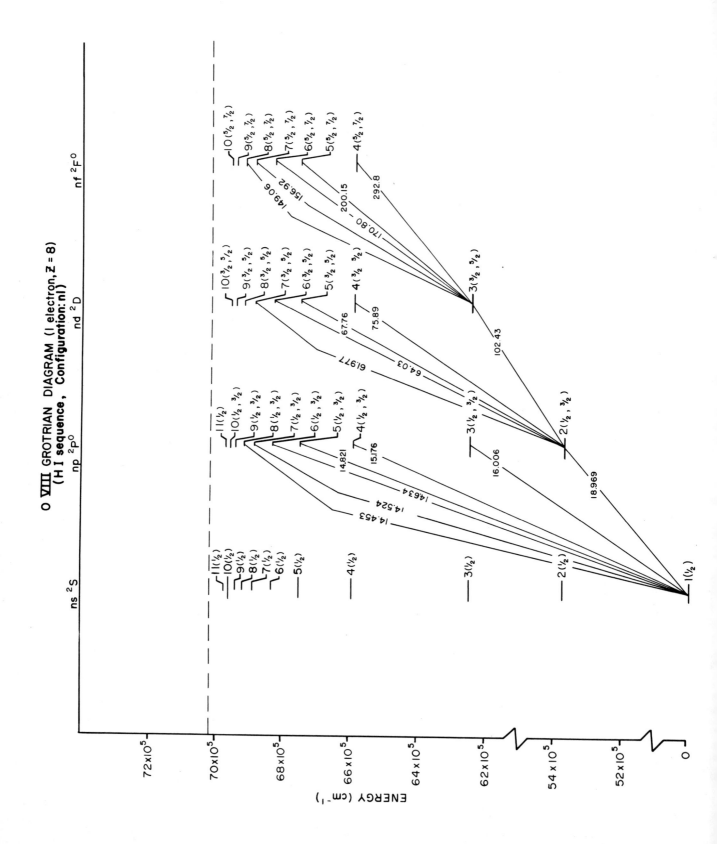

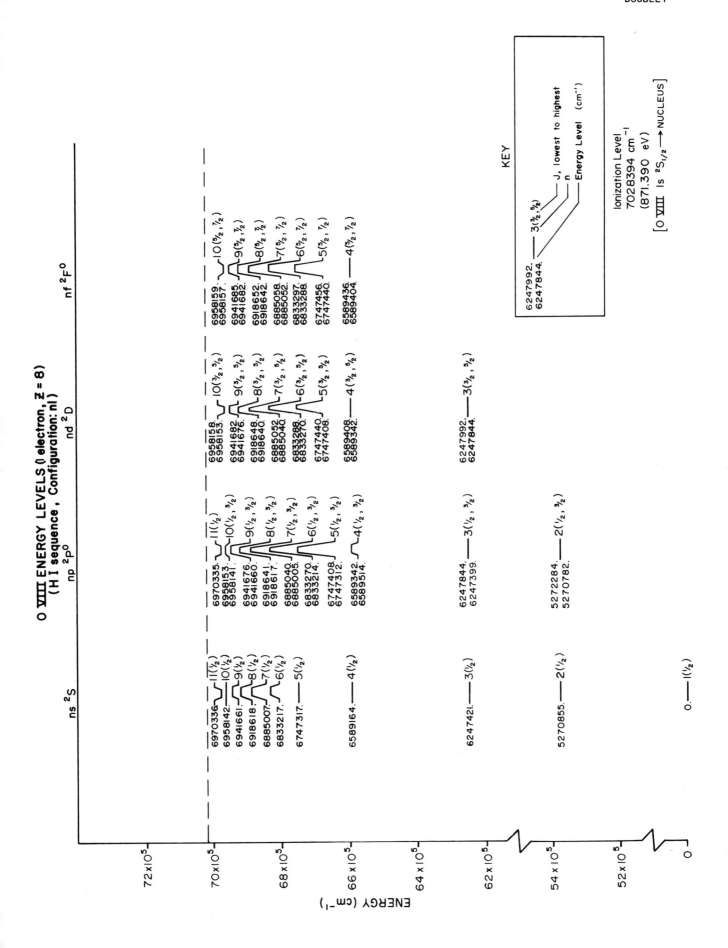

O VIII ENERGY LEVELS (1 electron, Z = 8)
(H I sequence, Configuration: nl)

O I $Z = 8$ 8 electrons

I.S. Bowen, Ap. J. **121**, 306 (1955).

   Author gives transition wavelengths from nebular observations.

I.S. Bowen, Ap. J. **132**, 1 (1960).

   Author gives transition wavelengths from nebular observations.

K.B.S. Eriksson, Ark. Fys. **30**, 199 (1965).

   Author gives observed transition wavelengths and energy levels of the ground configuration.

R.E. Huffman, J.C. Larrabee, and Y. Tanaka, J. Chem. Phys. **46**, 2213 (1967).

   Authors give line and energy level tables from observed absorption spectra in the vacuum ultraviolet. Spectrograms and an energy level diagram are included.

Comment on autoionizing triplet levels: The source is Huffman et al., (1967). Correction: Please note that the wavelengths for the transition $3s\ ^5S^\circ - 3p\ ^3P$ should be 6726.28, 6726.54.

O II $Z = 8$ 7 electrons

I.S. Bowen, Ap. J. **121**, 306 (1955).

   Author gives transition wavelengths from nebular observations.

I.S. Bowen, Ap. J. **132**, 1 (1960).

   Author gives transition wavelengths from nebular observations.

W.B. Bridges and A.N. Chester, IEEE J. Qu. Electronics **1**, 66 (1965).

   Authors give calculated and observed wavelengths of lines in ion gas lasers.

T. Sasaki, N. Kaifu, N. Itoh, K. Sakai, and I. Shimada, Sci. of Light **14**, 142 (1965).

   Authors give Grotrian diagrams.

O III $Z = 8$ 6 electrons

I.S. Bowen, Ap. J. **121**, 306 (1955).

   Author gives transition wavelengths from nebular observations.

W.B. Bridges and A.N. Chester, IEEE J. Qu. Electronics **1**, 66 (1965).

   Authors give calculated and observed wavelengths of lines in ion gas lasers.

T. Sasaki, N. Kaifu, N. Itoh, K. Sakai, and I. Shimada, Sci. of Light **14**, 142 (1965).

   Authors give Grotrian diagrams.

Comment on intermediate coupling: The system is similar to that for the isoelectronic system, N II. Please see that entry for the details.

O IV $Z = 8$ 5 electrons

W.B. Bridges and A.N. Chester, IEEE J. Qu. Electronics **1**, 66 (1965).

   Authors give calculated and observed wavelengths of lines in ion gas lasers.

O V	$Z = 8$	4 electrons

Please see the general references.

O VI	$Z = 8$	3 electrons

Please see the general references.

O VII	$Z = 8$	2 electrons

R.L. Blake, T.A. Chubb, H. Friedman, and A.E. Unzicker, Science **146**, 1037 (1964).

Authors give a table of lines observed in the solar spectrum below 25 Å.

B.C. Fawcett, A.H. Gabriel, W.G. Griffin, B.B. Jones, and R. Wilson, Nature **200**, 1303 (1963).

Authors identify spectral lines attributed to O VII.

E. Holøien and J. Midtdal, J. Phys. B: Atom. Molec. Phys. **4**, 1243 (1971).

Authors calculate energies nonrelativistically for the He I isoelectronic sequence.

G.A. Sawyer, A.J. Bearden, I. Henins, F.C. Jahoda, and F.L. Ribe, Phys. Rev. **131**, 1891 (1963).

Authors give transition wavelengths of lines in the region 15–25 Å from calculations and observations.

O VIII	$Z = 8$	1 electron

R.L. Blake, T.A. Chubb, H. Friedman, and A.E. Unzicker, Science **146**, 1037 (1964).

Authors give a table of lines observed in the solar spectrum below 25 Å.

B.C. Fawcett, A.H. Gabriel, W.G. Griffin, B.B. Jones, and R. Wilson, Nature **200**, 1303 (1963).

Authors give observed spectral lines in the range 16–400 Å.

G.A. Sawyer, A.J. Bearden, J. Henins, F.C. Jahoda, and F.L. Ribe, Phys. Rev. **131**, 1891 (1963).

Authors give observed lines of several series.

Fluor (F)

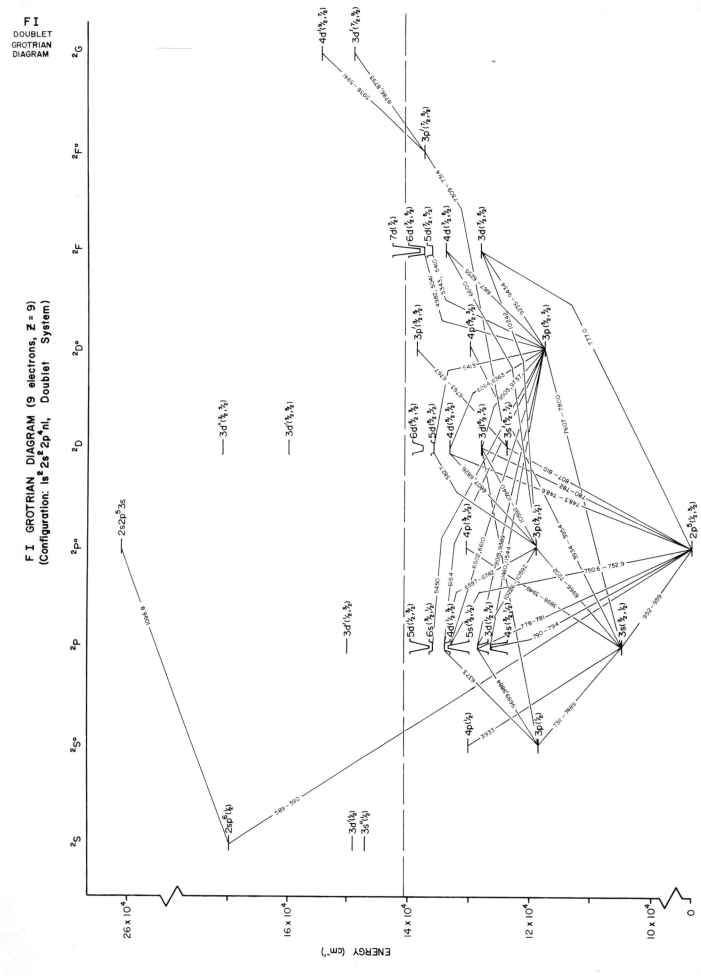

F I
DOUBLET
GROTRIAN
DIAGRAM

F I GROTRIAN DIAGRAM (9 electrons, Z = 9)
(Configuration: 1s² 2s² 2p⁴ nl, Doublet System)

ENERGY (cm⁻¹)

212

F I ENERGY LEVELS (9 electrons, Z=9)
(Configuration: $1s^2\,2s^2\,2p^4\,nl$, Doublet System)

F I
DOUBLET

ENERGY (cm⁻¹)

^{2}S

169824.5 —— $2sp^6(^1\!/_2)$

149000 —— $3d(^1\!/_2)$
146908.0 —— $3s''(^1\!/_2)$

^{2}S°

130149.1 —— $4p(^1\!/_2)$

118405.3 —— $3p(^1\!/_2)$

^{2}P

150000 —— $3d'(^1\!/_2,^3\!/_2)$

136588.0 ⎱ $5d(^1\!/_2,^3\!/_2)$
136420.1 ⎰
136151.2 —— $6s(^3\!/_2,^1\!/_2)$
135966.4
134091.3 ⎱ $4d(^1\!/_2,^3\!/_2)$
133910.4 ⎰
133223.2 —— $5s(^3\!/_2,^1\!/_2)$
132998.4
128712.3 ⎱ $3d(^1\!/_2,^3\!/_2)$
128520.2 ⎰
126581.2 ⎱ $4s(^3\!/_2,^1\!/_2)$
126282.6 ⎰

105056.3 —— $3s(^3\!/_2,^1\!/_2)$
104731.0

^{2}P°

261000? —— $2s2p^5 3s$

130489.9 —— $4p(^3\!/_2,^1\!/_2)$
130374.9

119081.8 —— $3p(^3\!/_2,^1\!/_2)$
118936.8

404.1 —— $2p^5(^1\!/_2,^3\!/_2)$
0

^{2}D

171000 —— $3d''(^3\!/_2,^5\!/_2)$

160000 —— $3d'(^3\!/_2,^5\!/_2)$

137462.7 ⎱ $6d(^5\!/_2,^3\!/_2)$
137451.3 ⎰
136113.8 —— $5d(^5\!/_2,^3\!/_2)$
136092.2
133623.8 —— $4d(^5\!/_2,^3\!/_2)$
133583.6
128219.8 —— $3d(^5\!/_2,^3\!/_2)$
128140.5
123920.0 —— $3s'(^5\!/_2,^3\!/_2)$
123921.0

^{2}D°

138703.5 —— $3p'(^3\!/_2,^5\!/_2)$
138695.6

130141.7 —— $4p(^5\!/_2,^3\!/_2)$
130016.1

117872.9 —— $3p'(^5\!/_2,^3\!/_2)$
117622.9

^{2}F

138272.0 —— $7d(^7\!/_2)$
137941.4 ⎱ $6d(^7\!/_2,^5\!/_2)$
137456.0 ⎰
136584.4 —— $5d(^7\!/_2,^5\!/_2)$
136101.6
134084.7 —— $4d(^7\!/_2,^5\!/_2)$
133606.5

128697.9 —— $3d(^7\!/_2,^5\!/_2)$
128220.4

^{2}F°

137599.0 —— $3p'(^7\!/_2,^5\!/_2)$
137590.1

^{2}G

154427.4 —— $4d'(^9\!/_2,^7\!/_2)$

148969.2 —— $3d'(^7\!/_2,^9\!/_2)$

KEY

^{2}F
128697.9 ⎱ $3d(^7\!/_2,^5\!/_2)$ ← J, lowest to highest
128220.4 ⎰ ⎫ n
⎭ Energy Level (cm⁻¹)

Ionization Level
140524.5 cm⁻¹
(17.422 eV)

nl core = $1s^2 2s^2 2p^4(^3P)$
n'l' core = $1s^2 2s^2 2p^4(^1D)$
n''l'' core = $1s^2 2s^2 2p^4(^1S)$

$[\,F\,I\ 2p^5(^2P°)\ \to\ F\,II\ 2p^4(^3P_2)\,]$

26 × 10⁴
16 × 10⁴
14 × 10⁴
12 × 10⁴
10 × 10⁴
0

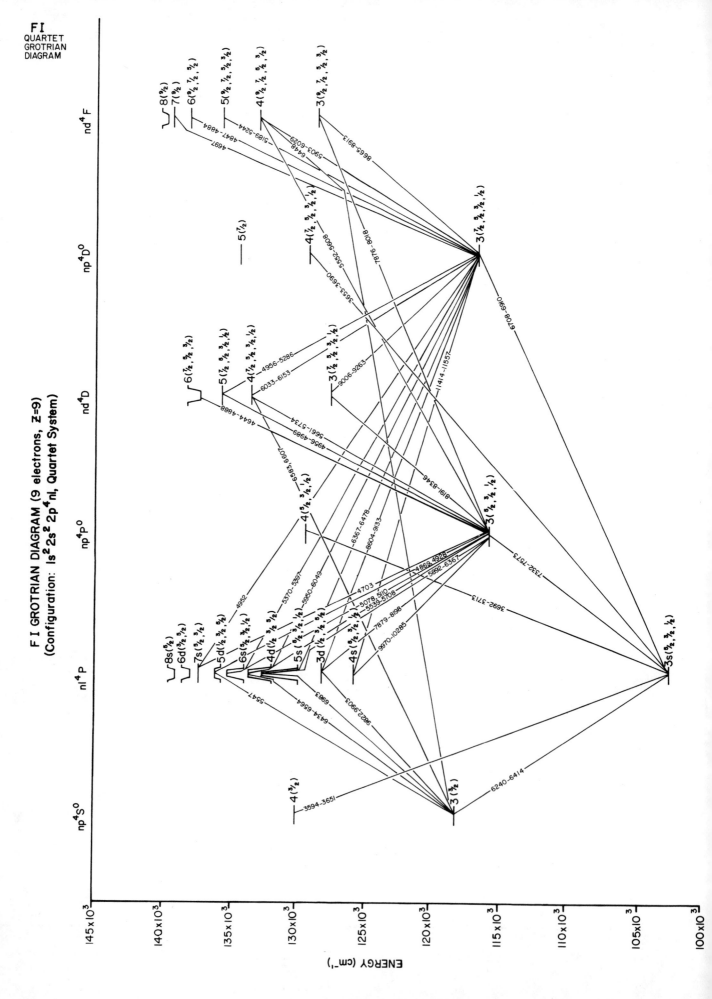

F I
QUARTET
GROTRIAN
DIAGRAM

F I GROTRIAN DIAGRAM (9 electrons, Z=9)
(Configuration: $1s^2 2s^2 2p^4 nl$, Quartet System)

214

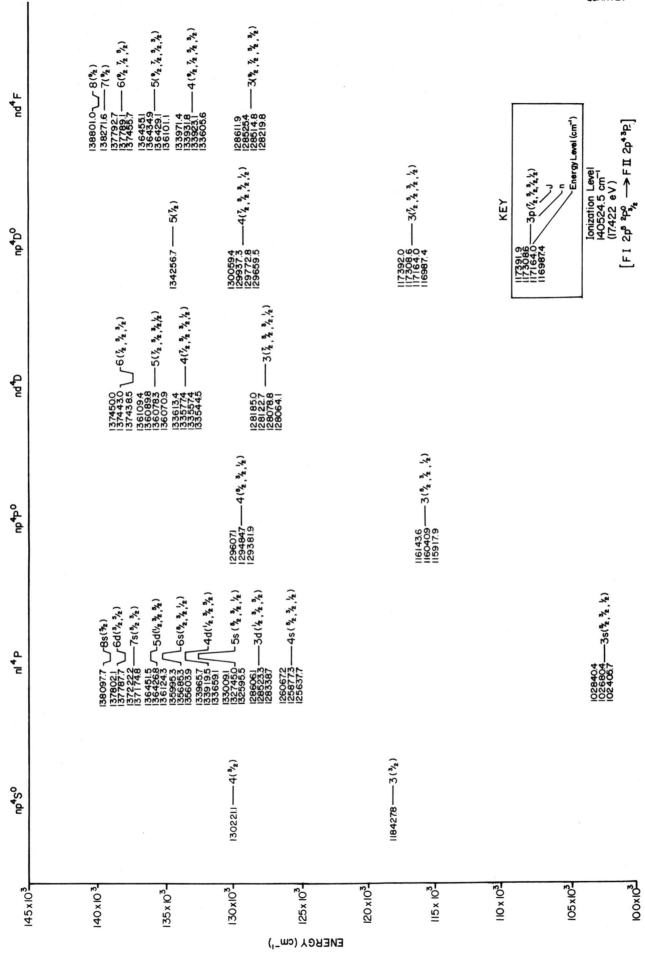

F I ENERGY LEVELS (9 electrons, Z=9)
(Configuration: 1s²2s²2p⁴nl, Quartet System)

FI
QUARTET

ENERGY (cm⁻¹)

215

F I INTERCOMBINATION GROTRIAN DIAGRAM (9 electrons, Z=9)
Configuration: $1s^2 2s^2 2p^4 nl$, Doublet to Quartet System

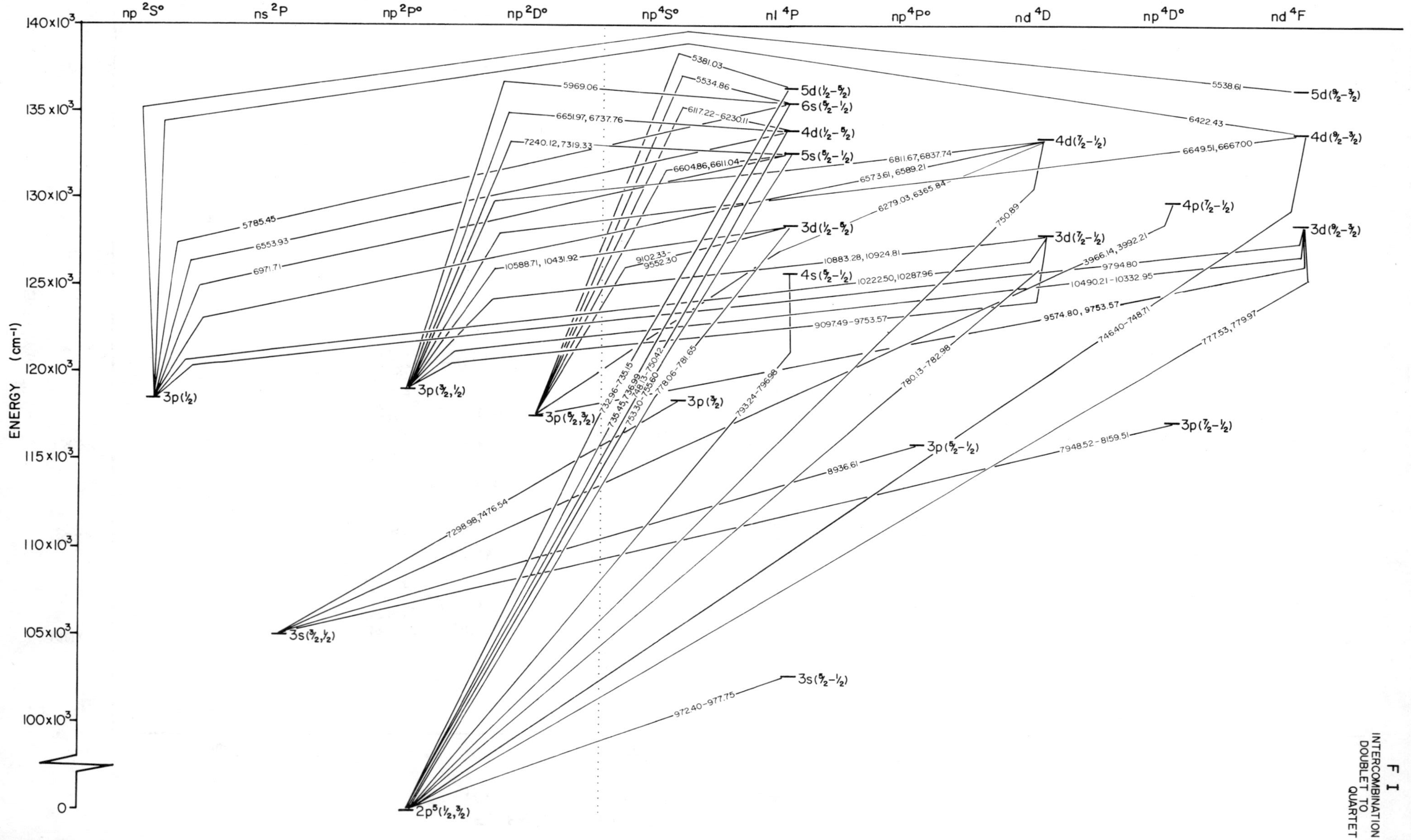

F I
INTERCOMBINATION
DOUBLET TO
QUARTET

216

F I GROTRIAN DIAGRAM (9 electrons, Z=9)
(Configuration : 1s²2s²2p⁴nl, Intercombinations, Quartet to Doublet Term System)

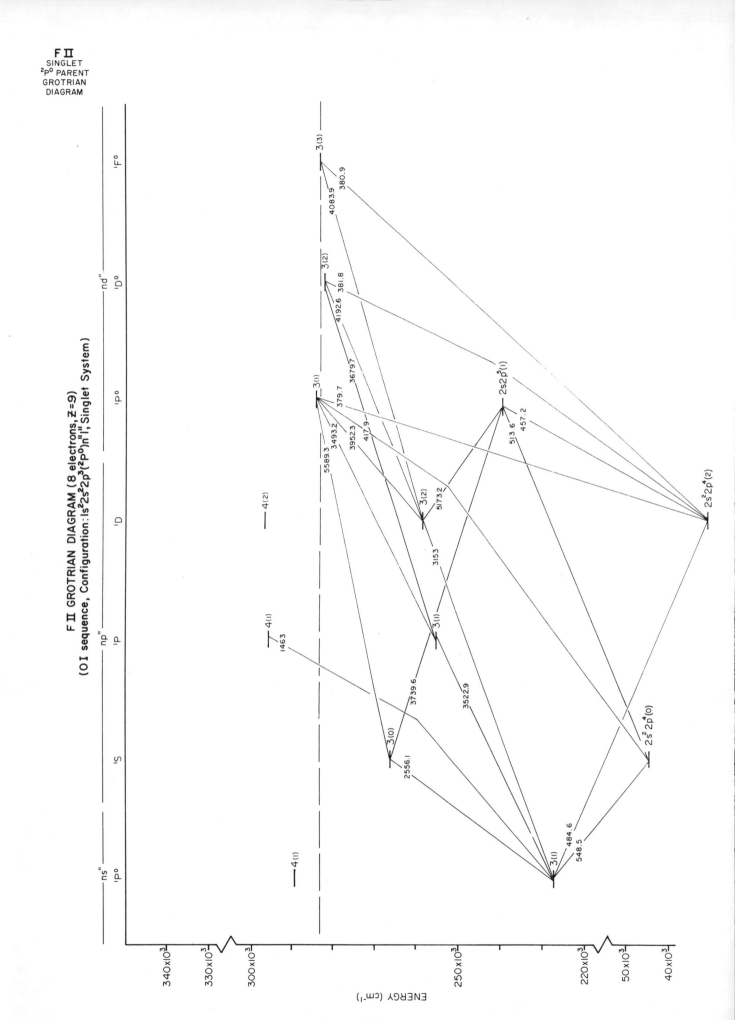

F II
SINGLET
²P⁰ PARENT
GROTRIAN
DIAGRAM

F II GROTRIAN DIAGRAM (8 electrons, Z=9)
(OI sequence, Configuration: 1s²2s²2p³(²P⁰)n"l", Singlet System)

ENERGY (cm⁻¹)

220

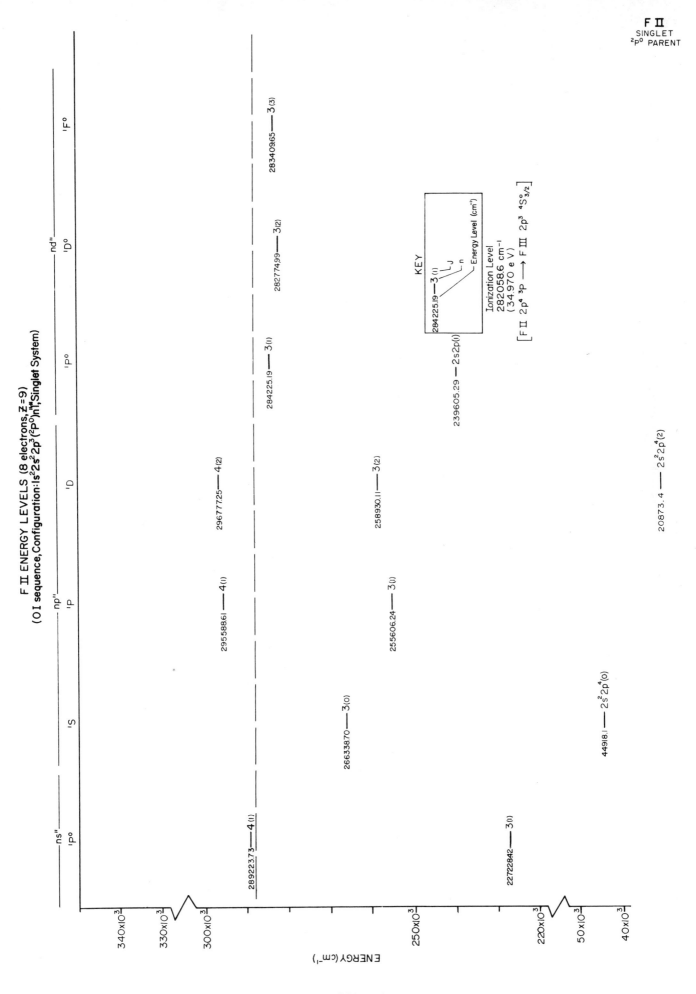

F II ENERGY LEVELS (8 electrons, Z=9)
(O I sequence, Configuration: 1s²2s²2p³(²P⁰)nl, nl",Singlet System)

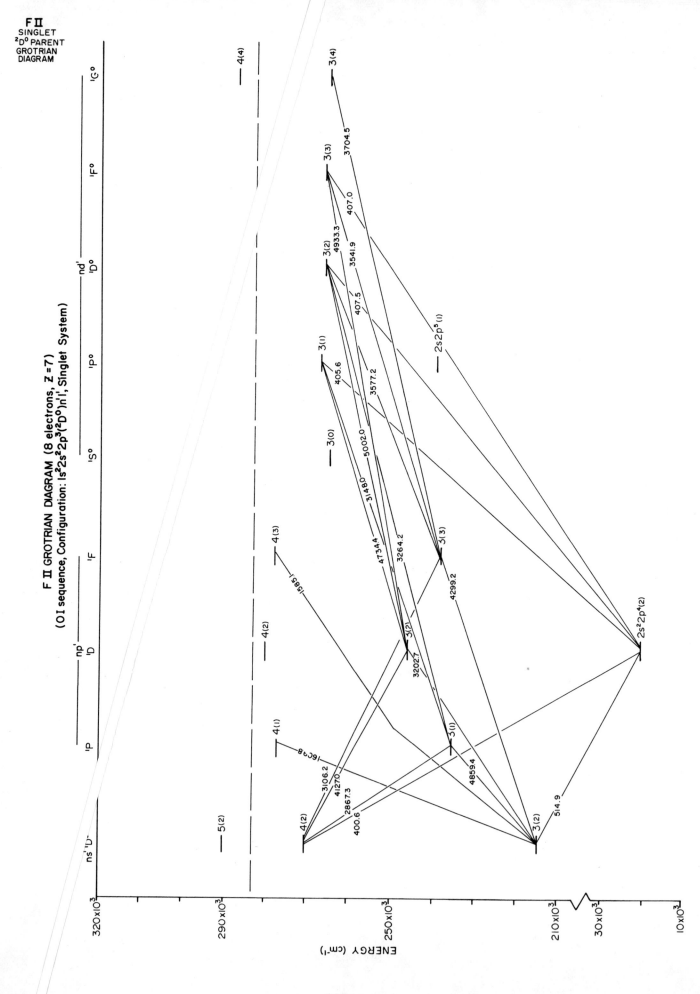

F II
SINGLET
²Dᴼ PARENT
GROTRIAN
DIAGRAM

F II GROTRIAN DIAGRAM (8 electrons, Z = 7)
(OI sequence, Configuration: 1s²2s²2p³(²Dᴼ)nl', Singlet System)

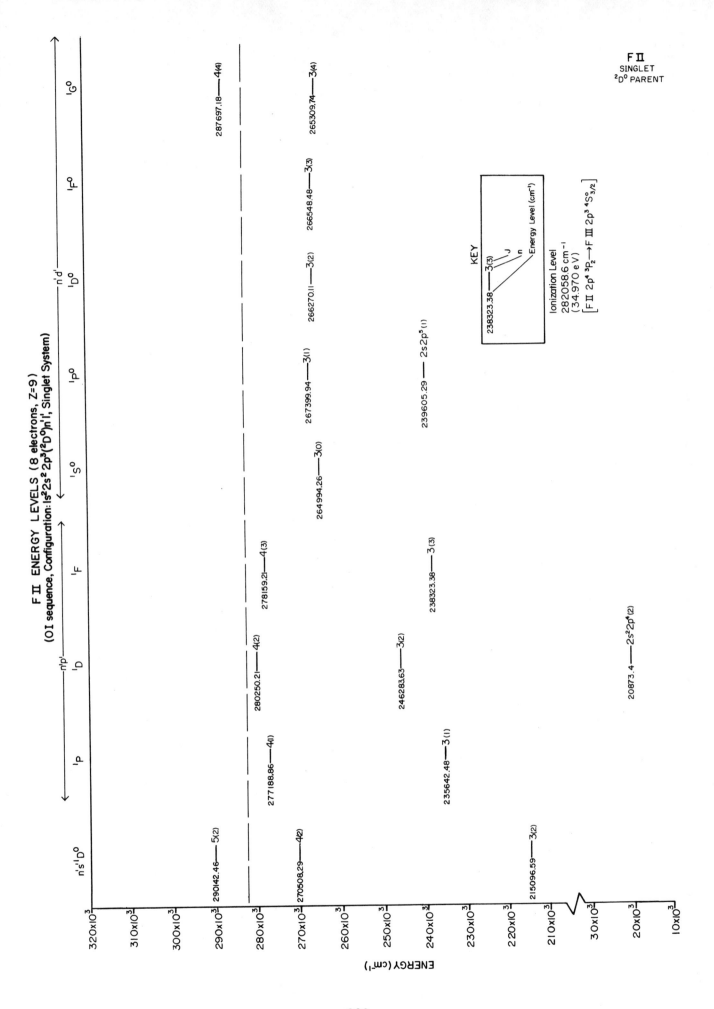

F II ENERGY LEVELS (8 electrons, Z=9)
(OI sequence, Configuration: $1s^2 2s^2 2p^3(^2D^o)n'l'$, Singlet System)

F II
SINGLET
$^2D^o$ PARENT

KEY

238323.38 ——— 3(3)
  J
  n
  Energy Level (cm⁻¹)

Ionization Level
282058.6 cm⁻¹
(34.970 eV)
$[F II 2p^4 \ ^3P_2 \rightarrow F III 2p^3 \ ^4S^o_{3/2}]$

$n's'D^o$ ^1P $n'p'D$ ^1F $^1S^o$ $^1P^o$ $n'd'D^o$ $^1F^o$ $^1G^o$

290142.46 ——— 5(2)

270508.29 ——— 4(2)

235642.48 ——— 3(1)

215096.59 ——— 3(2)

280250.21 ——— 4(2)

277188.86 ——— 4(1)

246283.63 ——— 3(2)

20873.4 ——— $2s^2 2p^4$ (2)

278159.21 ——— 4(3)

264994.26 ——— 3(0)

238323.38 ——— 3(3)

267399.94 ——— 3(1)

239605.29 ——— $2s 2p^5$ (1)

266270.11 ——— 3(2)

287697.18 ——— 4(4)

266548.48 ——— 3(3)

265309.74 ——— 3(4)

ENERGY (cm⁻¹)

320x10³
310x10³
300x10³
290x10³
280x10³
270x10³
260x10³
250x10³
240x10³
230x10³
220x10³
210x10³
30x10³
20x10³
10x10³

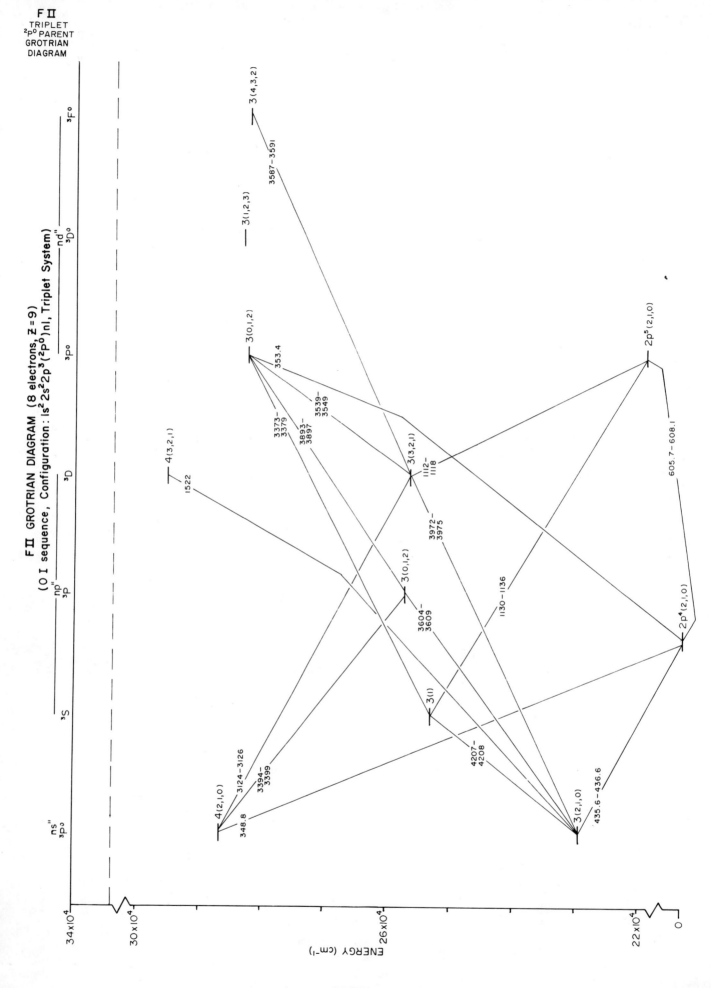

F II
TRIPLET
²Pᵒ PARENT
GROTRIAN
DIAGRAM

F II GROTRIAN DIAGRAM (8 electrons, Z = 9)
(O I sequence, Configuration: $1s^2 2s^2 2p^3(^2P^o)nl$, Triplet System)

ns"   np"   ³D   ³Pᵒ   nd"   ³Fᵒ
³Pᵒ   ³P   ³S   ³Dᵒ

ENERGY (cm⁻¹)

34×10⁴   30×10⁴   26×10⁴   22×10⁴   0

F II ENERGY LEVELS (8 electrons, Z=9)
(O I sequence, Configuration: ls² 2s² 2p³ (²P°)nl″, Triplet System)

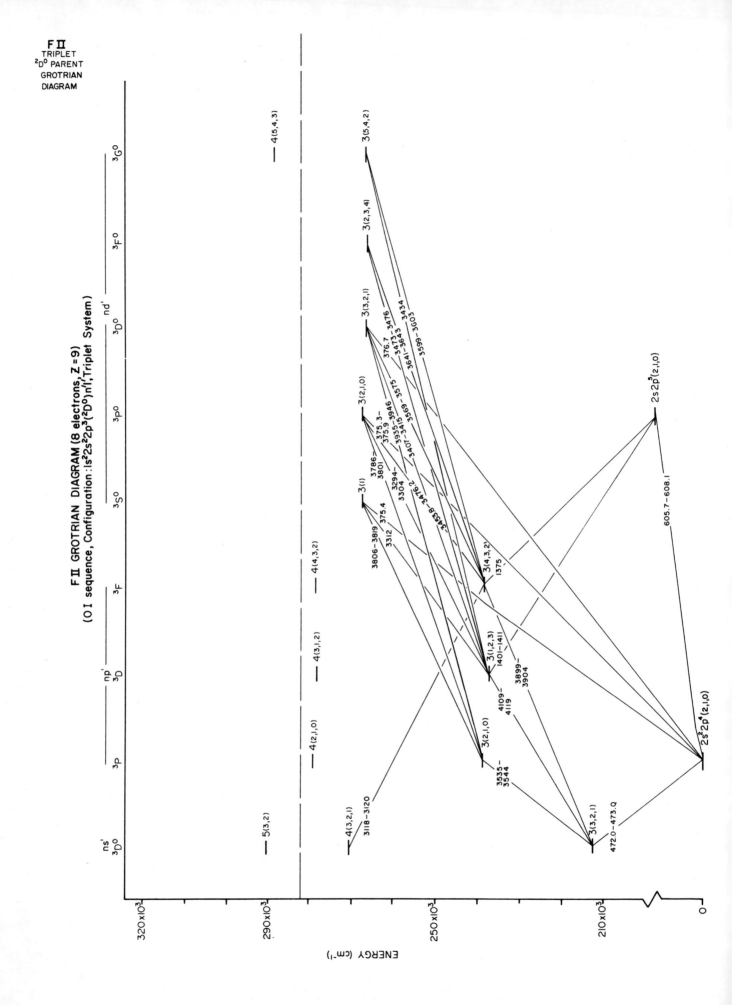

FⅡ
TRIPLET
²D⁰ PARENT
GROTRIAN
DIAGRAM

FⅡ GROTRIAN DIAGRAM (8 electrons, Z=9)
(OI sequence, Configuration :1s²2s²2p³(²D⁰)nl'Triplet System)

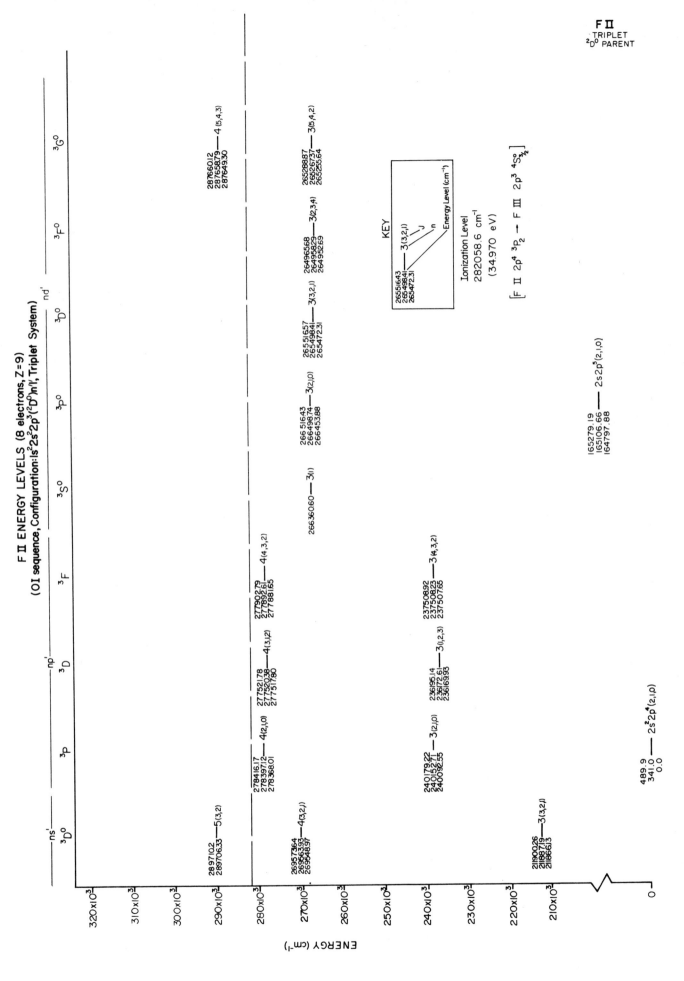

F II ENERGY LEVELS (8 electrons, Z=9)
(OI sequence, Configuration: $1s^22s^22p^3(^2D^o)n'l'$, Triplet System)

F II
TRIPLET
$^2D^o$ PARENT

KEY

Ionization Level
282058.6 cm^{-1}
(34.970 eV)

$[$ F II $2p^4$ $^3P_2$ $\rightarrow$ F III $2p^3$ $^4S^o_{3/2}]$

ENERGY (cm^{-1})

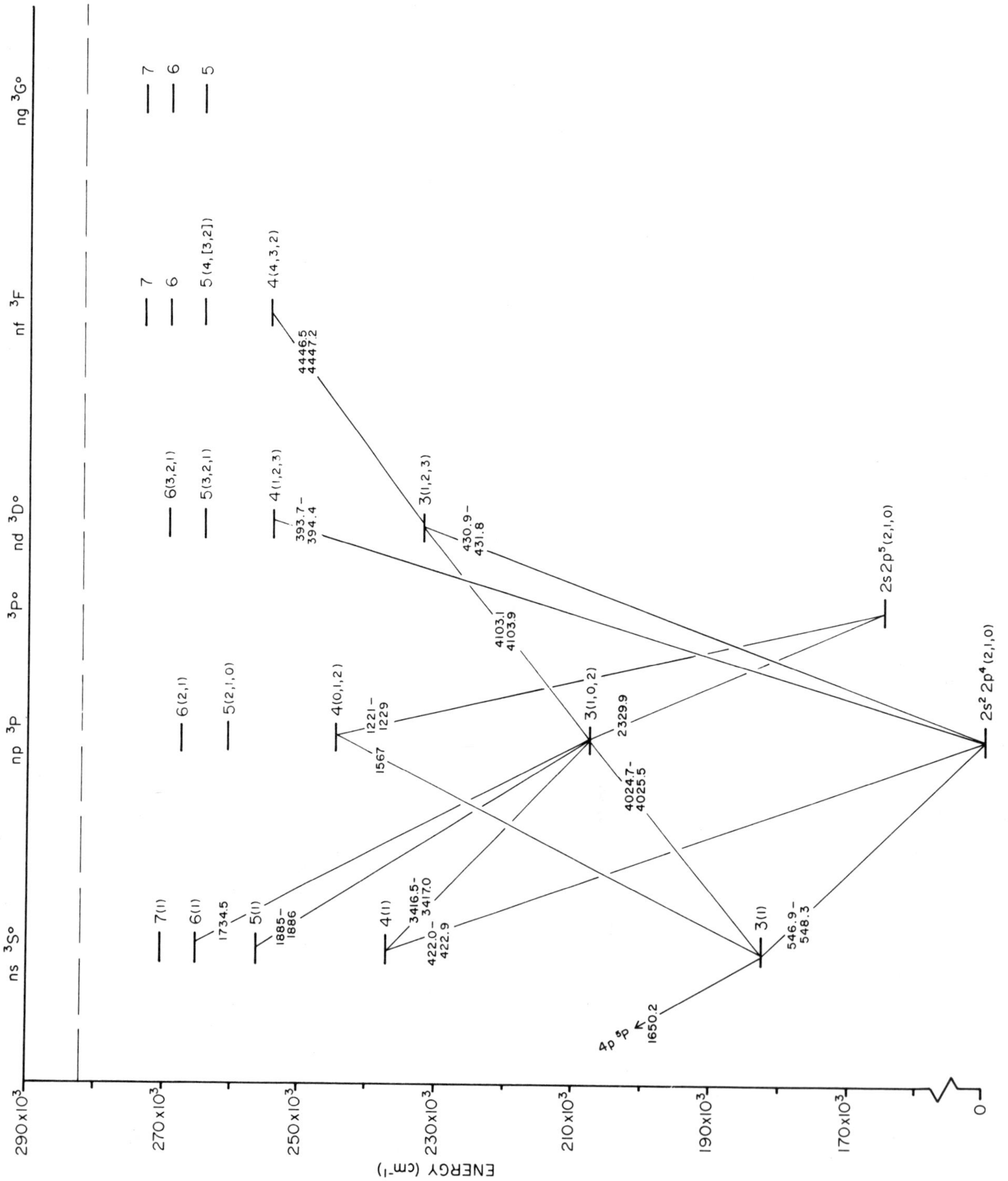

F II GROTRIAN DIAGRAM (8 electrons, Z = 9)
(O I sequence, Configuration: 1s² 2s² 2p³ (⁴S°) nl, Triplet System)

F II
TRIPLET
⁴S° PARENT
GROTRIAN
DIAGRAM

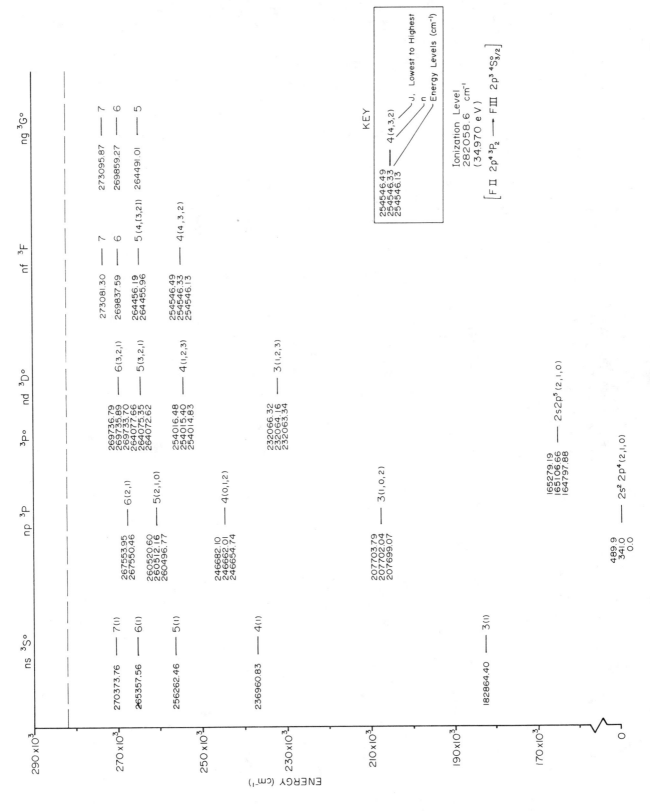

F II ENERGY LEVELS (8 electrons, Z = 9)
(O I sequence, Configuration: $1s^2 2s^2 2p^3$ ($^4S°$) nl, Triplet System)

F II
TRIPLET
$^4S°$ PARENT

229

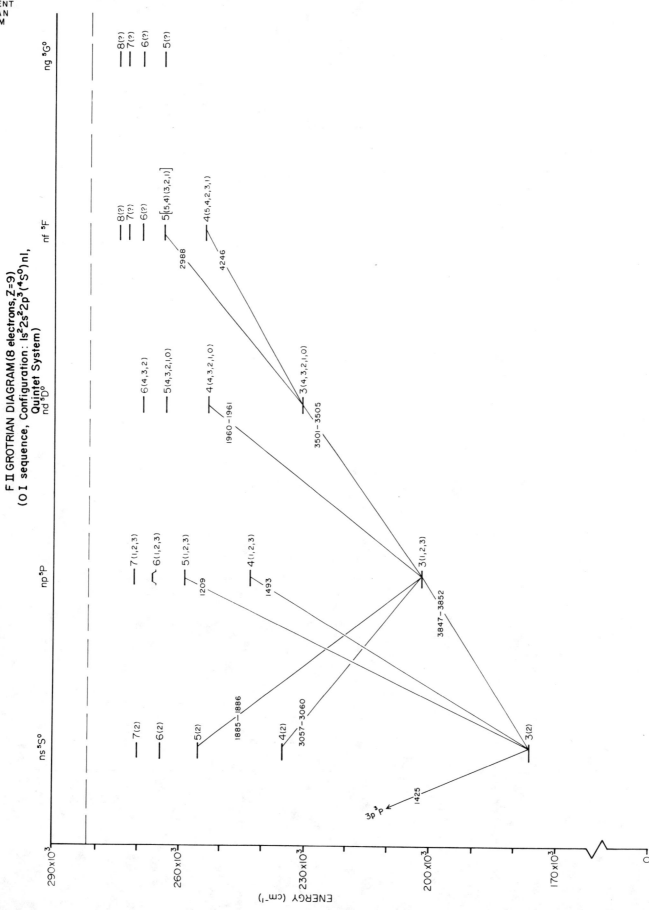

F II
QUINTET
⁴S° PARENT
GROTRIAN
DIAGRAM

F II GROTRIAN DIAGRAM (8 electrons, Z=9)
(O I sequence, Configuration: $1s^2 2s^2 2p^3 (^4S^0) nl$,
Quintet System)

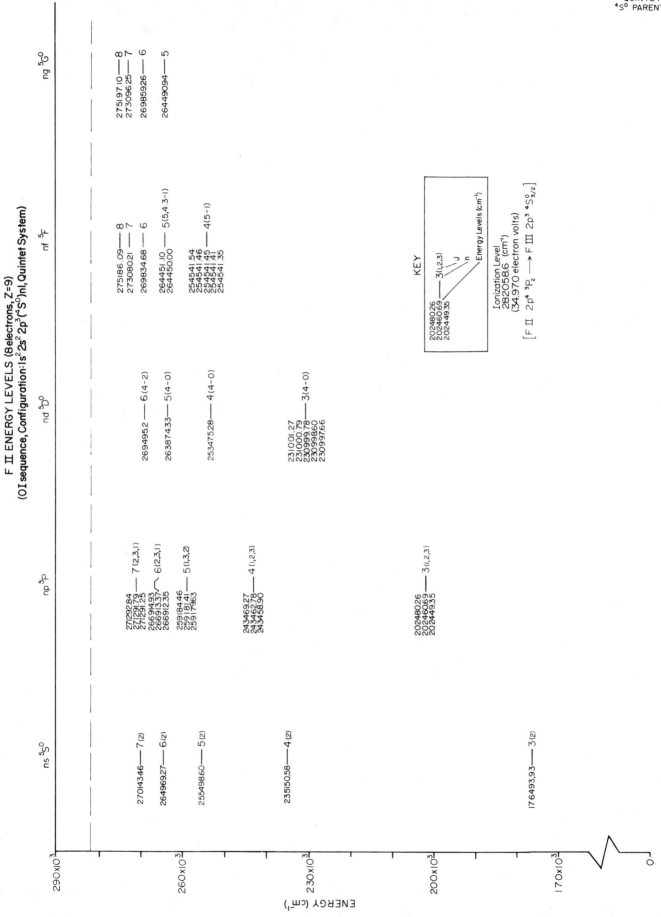

F II ENERGY LEVELS (8electrons, Z=9)
(OI sequence, Configuration: 1s² 2s² 2p³(⁴S⁰)nl, Quintet System)

F II
QUINTET
⁴S⁰ PARENT

231

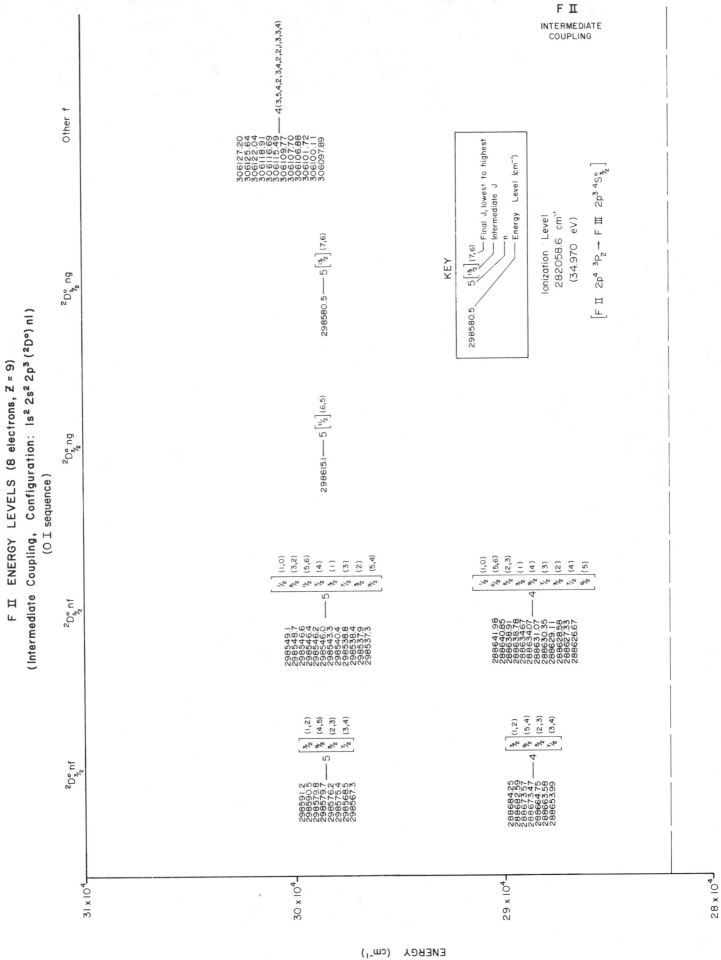

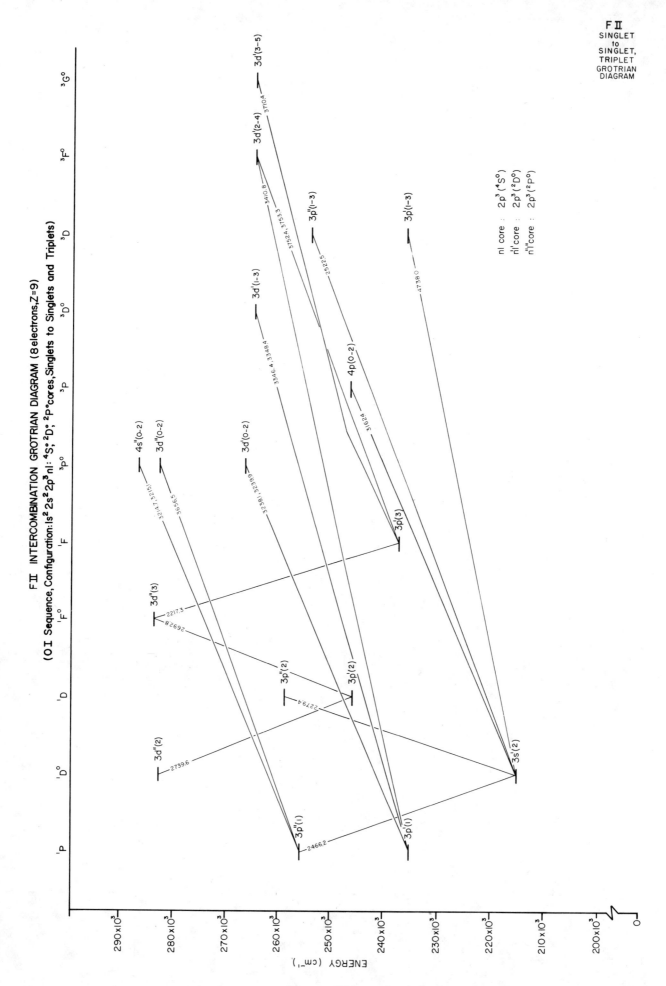

F II INTERCOMBINATION GROTRIAN DIAGRAM (8 electrons, Z=9)

(O I Sequence, Configuration: 1s² 2s² 2p³ nl: ⁴S° ²D; ²P°cores, Singlets to Singlets and Triplets)

F II
SINGLET
to
SINGLET,
TRIPLET
GROTRIAN
DIAGRAM

nl core : 2p³ (⁴S°)
nl'core : 2p³ (²D°)
nl''core : 2p³ (²P°)

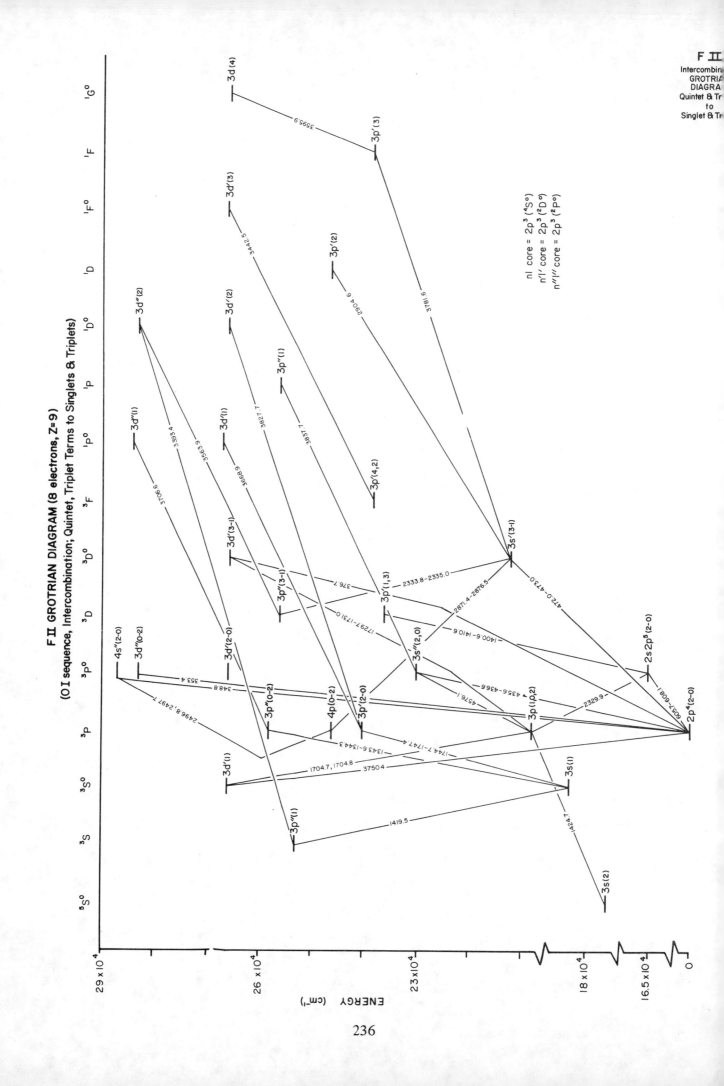

F II GROTRIAN DIAGRAM (8 electrons, Z=9)

(O I sequence, Intercombination; Quintet, Triplet Terms to Singlets & Triplets)

F II

Intercombin
GROTRIA
DIAGRA
Quintet & Tr
to
Singlet & Tr

nl core = 2p³ (⁴S°)
n'l' core = 2p³ (²D°)
n''l'' core = 2p³ (²P°)

236

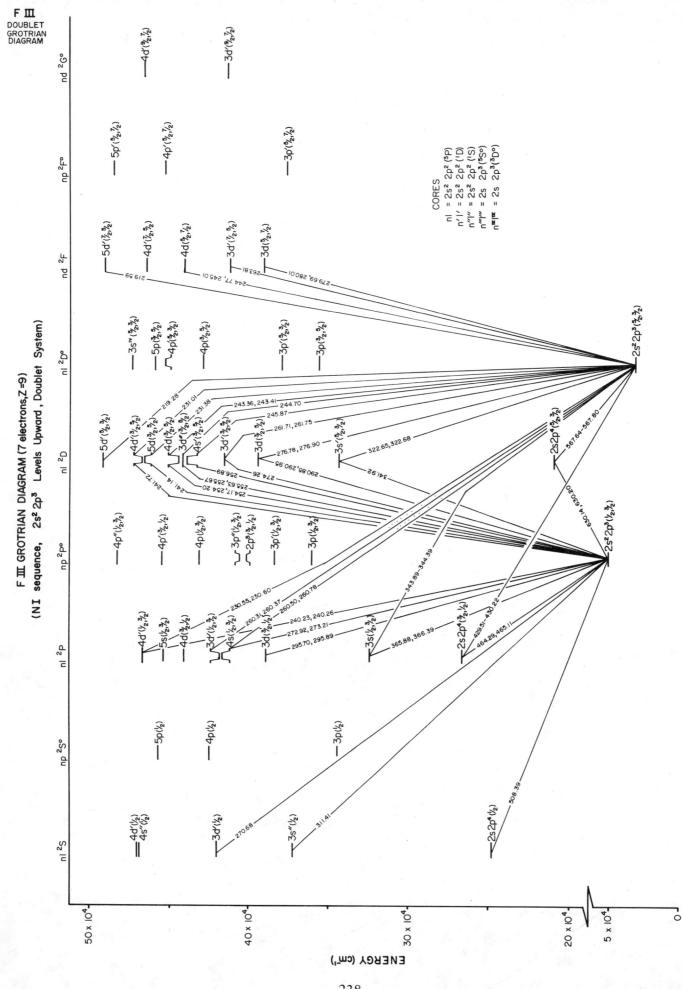

238

FIII ENERGY LEVELS (7 electrons, Z=9)

(NI sequence, Configuration 1s²2s²2p²nl, Doublet System)

ENERGY (cm⁻¹)

Column headers (left to right): nl ²S | np ²Sᵒ | nl ²P | np ²Pᵒ | nl ²D | nl ²Dᵒ | nd ²F | np²Fᵒ | nd ²Gᵒ

KEY

Core Excitation State

3d"($\frac{3}{2}$,$\frac{5}{2}$) J, Lower to Higher

n

Energy Levels (cm⁻¹)

442748.9
442694.6

Ionization Level
505 777 cm⁻¹
(62.707 eV)

[F III 2s² 2p³ ⁴S°$_{3/2}$ – F IV 2s² 2p² ³P₀]

nl core = 1s² 2s² 2p² (³P)
n'l' core = 1s² 2s² 2p² (¹D)
n"l" core = 1s² 2s² 2p² (¹S)
n'''l''' core = 1s² 2s 2p³ (⁵Sᵒ)
nᴵⱽlᴵⱽ core = 1s² 2s 2p³ (³Dᵒ)

Selected energy level values:

nl ²S: 470269.07 — 4d'(½); 469407.29 — 4s"(½); 421005.36 — 3d"(½); 372678.47 — 3s"(½); 248261.4 — 2s2p⁴(½)

np ²Sᵒ: 457376.18 — 5p(½); 425390.01 — 4p(½); 344436.96 — 3p(½)

nl ²P: 467827.11 — 4d(½,³/2); 467769.16; 454097.66 — 5s(½,³/2); 453696.10; 441167.40 — 4d(³/2,½); 441158.60; 418248.56 — 3d'(½,³/2); 418188.48; 417970.80 — 4s(½,³/2); 417585.35; 389737.63 — 3d(³/2,½); 389525.55; 324876.72 — 3s(½,³/2); 324492.55; 266944.70 — 2s2p⁴(³/2,½); 266561.85

np ²Pᵒ: 481361.17 — 4p'(½,³/2); 481343.01; 455768.72 — 4p'(³/2,½); 455648.67; 432060.64 — 4p'(½,³/2); 431967.93; 406908.95 — 3p'(½,³/2); 406504.38; 401724.33 — 2p³(³/2,½); 401205.78; 384493.06 — 3p'(½,³/2); 384358.54; 360435.08 — 3p(½,³/2); 360348.28; 51561.4 — 2s²2p³(½,³/2); 51560.6

nl ²D: 490140. — 5d'(½); 466982.86 — 4d'(³/2,⁵/2); 466978.15; 466275.50 — 5d(³/2,⁵/2); 466253.24; 445006.13 — 4d(³/2,½); 444957.11; 441748.85 — 3d'(³/2,½); 441269.88; 440828.53 — 4s(³/2,⁵/2); 440825.13; 416185.42 — 3d'(³/2,⁵/2); 416168.30; 395385.66 — 3d(³/2,⁵/2); 395268.13; 344027.58 — 3s(⁵/2,³/2); 344024.17; 210256.3 — 2s2p⁴(⁵/2,³/2); 210241.7; 341123.2 — 2s²2p³(⁵/2,³/2); 340874

nl ²Dᵒ: 474413. — 3s'ᴵⱽ(½,³/2); 474368.94; 460141.63 — 5p(³/2,⁵/2); 459804.55; 453369.77 — 4p'(³/2,½); 453343.62; 430374.75 — 4p'(³/2,⁵/2); 430012.48; 380307.13 — 3p'(⁵/2,³/2); 380250.52; 356371.82 — 3p(³/2,⁵/2); 355981.90

nd ²F: 489494.53 — 5d'(⁷/2,⁵/2); 464939.50 — 4d'(⁷/2,⁵/2); 464921.74; 442636.27 — 4d(⁵/2,⁷/2); 442278.93; 413195.00 — 3d'(⁷/2,⁵/2); 413143.86; 391626.91 — 3d(⁵/2,⁷/2); 391257.42

np²Fᵒ: 483521.14 — 5p'(⁵/2,⁷/2); 483510.17; 452770.85 — 4p'(⁵/2,⁷/2); 452748.54; 376879.17 — 3p'(⁵/2,⁷/2); 376814.02

nd ²Gᵒ: 466190.97 — 4d'(⁹/2,⁷/2); 466190.47; 414897.72 — 3d'(⁹/2,⁷/2); 414894.89

239

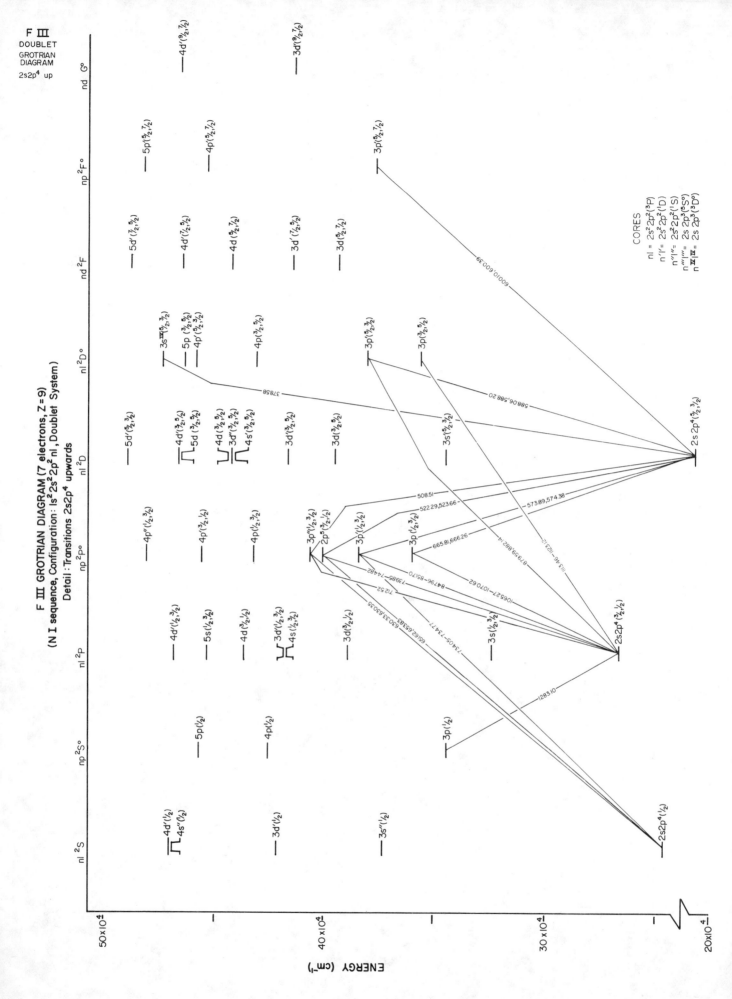

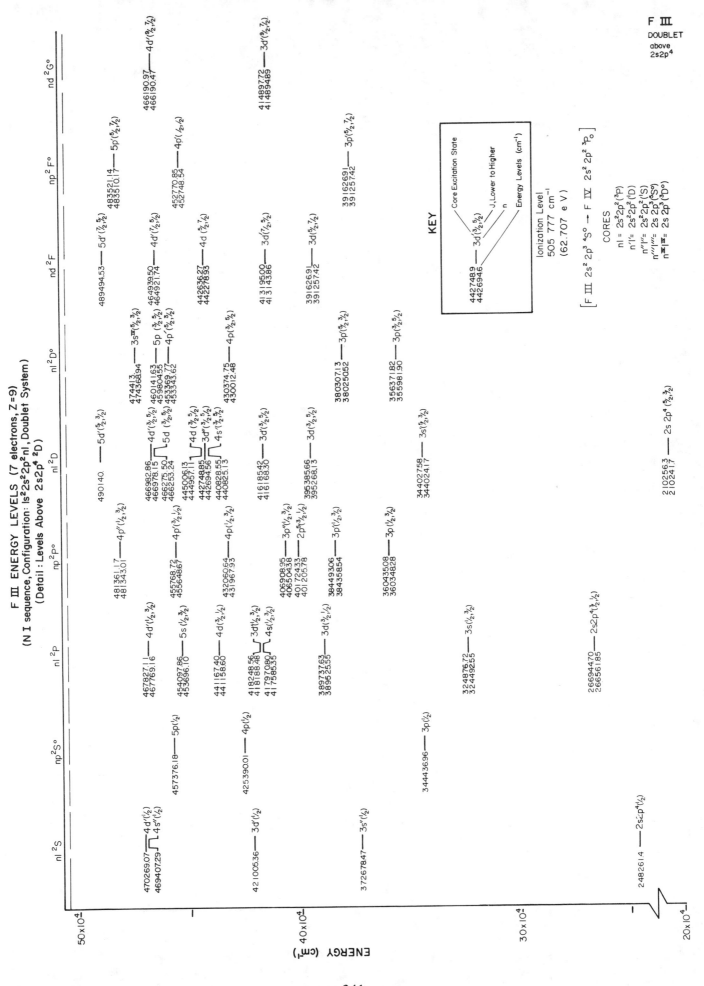

F III ENERGY LEVELS (7 electrons, Z=9)
(N I sequence, Configuration: 1s²2s²2p²nl, Doublet System)
(Detail: Levels Above 2s2p⁴ ²D)

F III
DOUBLET
above
2s2p⁴

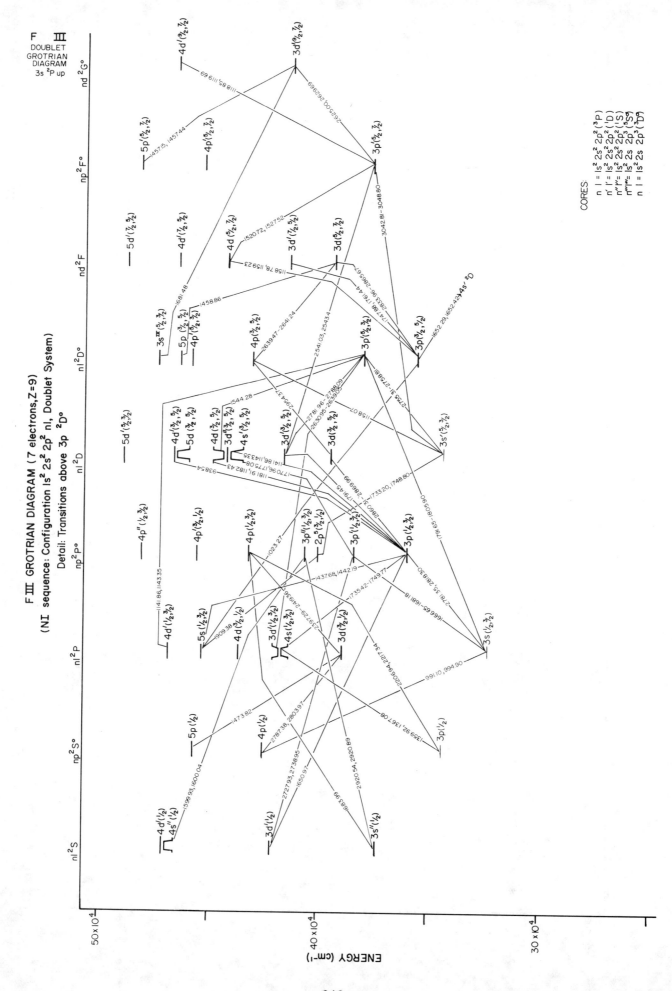

F III GROTRIAN DIAGRAM (7 electrons, Z=9)
(NI sequence: Configuration 1s² 2s² 2p² nl, Doublet System)
Detail: Transitions above 3p ²D°

F III
DOUBLET
GROTRIAN
DIAGRAM
3s ²P up

CORES:
n l = 1s² 2s² 2p² (³P)
n' l' = 1s² 2s² 2p² (¹D)
n" l" = 1s² 2s² 2p² (¹S)
n'" l'" = 1s² 2s 2p³ (⁵S°)
n l = 1s² 2s 2p³ (³D°)

ENERGY (cm⁻¹)

242

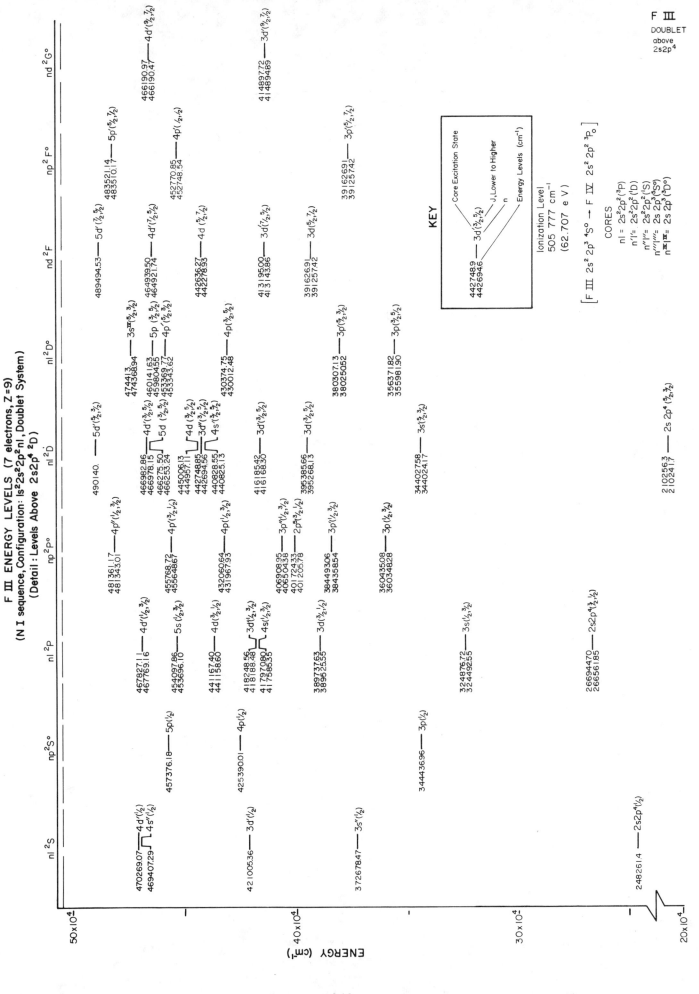

F III ENERGY LEVELS (7 electrons, Z=9)
(N I sequence, Configuration: 1s²2s²2p²nl, Doublet System)
(Detail : Levels Above 2s2p⁴ ²D)

F III
DOUBLET
above
2s2p⁴

243

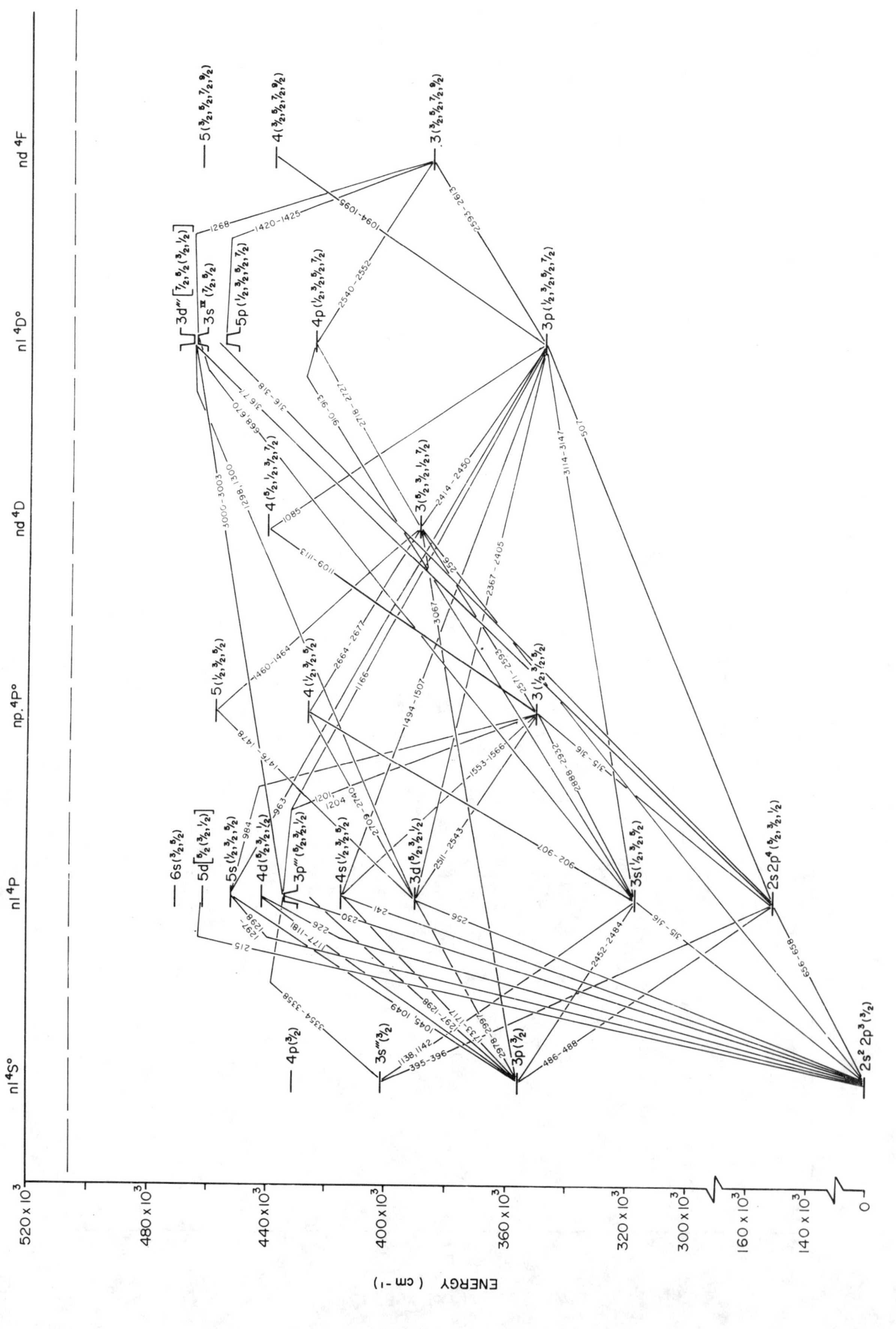

F III
QUARTET
GROTRIAN
DIAGRAM

F III GROTRIAN DIAGRAM (7 electrons, Z = 9)
(N I sequence, Configuration: 1s² 2s² 2p² nl, Quartet System)

244

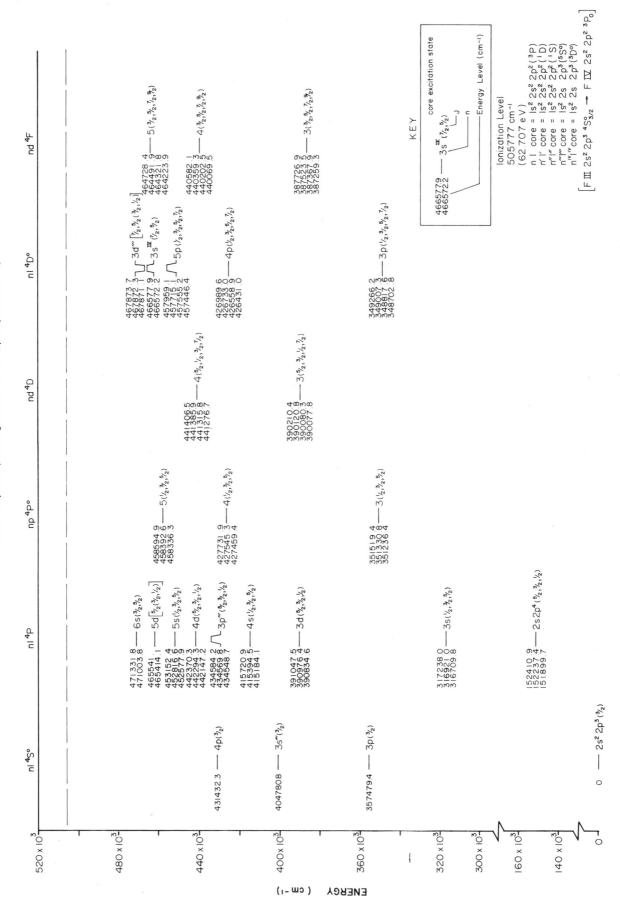

F III ENERGY LEVELS (7 electrons, Z = 9)

(N I sequence, Configuration: 1s² 2s² 2p² nl, Quartet System)

F III
QUARTET

245

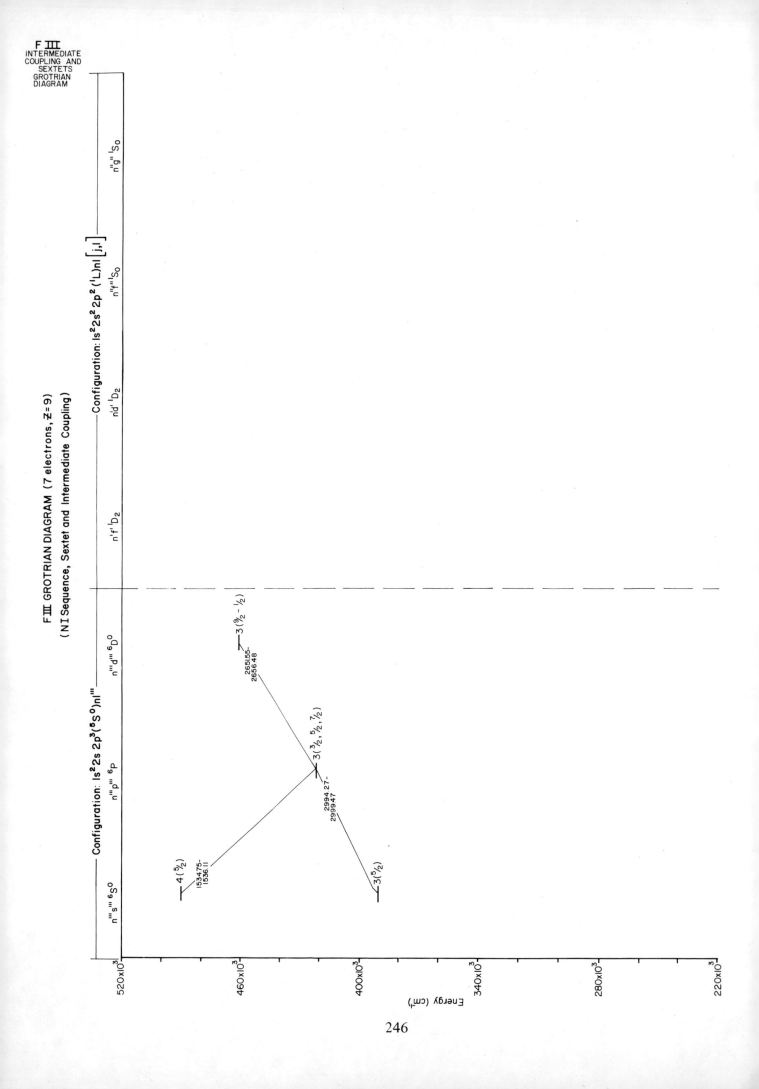

F III
INTERMEDIATE
COUPLING AND
SEXTETS
GROTRIAN
DIAGRAM

F III GROTRIAN DIAGRAM (7 electrons, Z = 9)
(NI Sequence, Sextet and Intermediate Coupling)

Configuration: $1s^2 2s 2p^3 (^5S^o) nl'''$

Configuration: $1s^2 2s^2 2p^2 (^1L) nl [j,l]$

$n'''s''' \ ^6S^o$ $n'''p''' \ ^6P$ $n'''d''' \ ^6D^o$ $n'f' \ ^1D_2$ $nd' \ ^1D_2$ $n''f''' \ ^1S_0$ $n''g'' \ ^1S_0$

$4(^5/_2)$
153475–
153611

$3(^3/_2, \ ^5/_2, \ ^7/_2)$
299427–
299947

$3(^9/_2 - ^1/_2)$
265155–
265648

$3(^5/_2)$

Energy (cm⁻¹)

520x10³   460x10³   400x10³   340x10³   280x10³   220x10³

FIII ENERGY LEVELS (7 electrons, Z=9)
(NI Sequence, Sextet and Intermediate Coupling)

— Configuration: 1s² 2s 2p³(⁵Sº)nl''' —

—————— Configuration: 1s² 2s² 2p²(¹L)nl [j,l] ——————

n'''s'''⁶Sº   n'''p'''⁶P   n'''d'''⁶Dº   n'f'¹D₂   n'd'¹D₂   n''f'''¹Sº   n''g'''¹Sº

490418.04 — 4 (⁵/₂)

462964.67
462961.82
462958.24 — 3 (⁹/₂ – ¹/₂)
462954.27
462951.93

425318.80
425282.70 — 3 (³/₂, ⁵/₂, ⁷/₂)
425261.06

39193.58 — 3(⁵/₂)

491461.65
491460.25
491168.25 —— 5
491132.38
491131.37

469564.01
469561.58
469243.66
469241.41
469120.35
469119.52
468683.43 —— 4
468682.91
468606.05
468605.94

$\left[\begin{array}{l}5 \\ 3 \\ 4\end{array}\right.$ ⁰ (⁹/₂, ¹¹/₂)
5 (¹/₂)
3 (⁷/₂)
4 (⁷/₂, ⁹/₂)

$\left[\begin{array}{l}1 \\ 5 \\ 2 \\ 3 \\ 4\end{array}\right.$ ⁰ (¹/₂, ³/₂)
5 (⁹/₂, ¹¹/₂)
2 (³/₂, ⁵/₂)
3 (⁵/₂, ⁷/₂)
4 (⁷/₂, ⁹/₂)

491592.47
491539.20 —— 5
491492.10
491408.79
491407.75
491395.56

497323.25 —— 4 [3] ⁰ (⁵/₂, ⁷/₂)
497322.46

$\left[\begin{array}{l}2 \\ 6 \\ 3 \\ 4 \\ 5\end{array}\right.$ (⁵/₂, ³/₂)
(¹³/₂, ¹¹/₂)
(⁷/₂, ⁵/₂)
(⁹/₂, ⁷/₂)
(¹¹/₂, ⁹/₂)

519773.70 — 5 [4] ⁰ (⁹/₂, ⁷/₂)

KEY

Ionization Level
505777 cm⁻¹
(62.707 eV)

$\left[\text{FIII } 2s^2\,2p^3\,^4S^0 \rightarrow \text{FIV } 2s^2\,2p^2\,^3P_0\right]$

39193.58 — 3(⁵/₂)
parity: ⁰
J: (⁵/₂)
n: 3
Energy Level (cm⁻¹)

n''''''' core: 2s 2p³ (⁵Sº)

KEY

497323.25 — 4 [3] ⁰ (⁵/₂, ⁷/₂)
497322.46 — 4 [3] (³/₂, ¹/₂)

parity
Final J, lower to higher
j,l coupling
n
Energy Level (cm⁻¹)

n'l' core: 2p² (¹D)
n''l'' core: 2p² (¹S)

Energy (cm⁻¹)

520x10³
460x10³
400x10³
340x10³
280x10³
220x10³

247

F III ENERGY LEVELS (7 electrons, Z = 9)

(N I sequence, Configuration: $1s^2 2s^2 2p^2(^3P)nl[j,l](J)$, Intermediate Coupling)

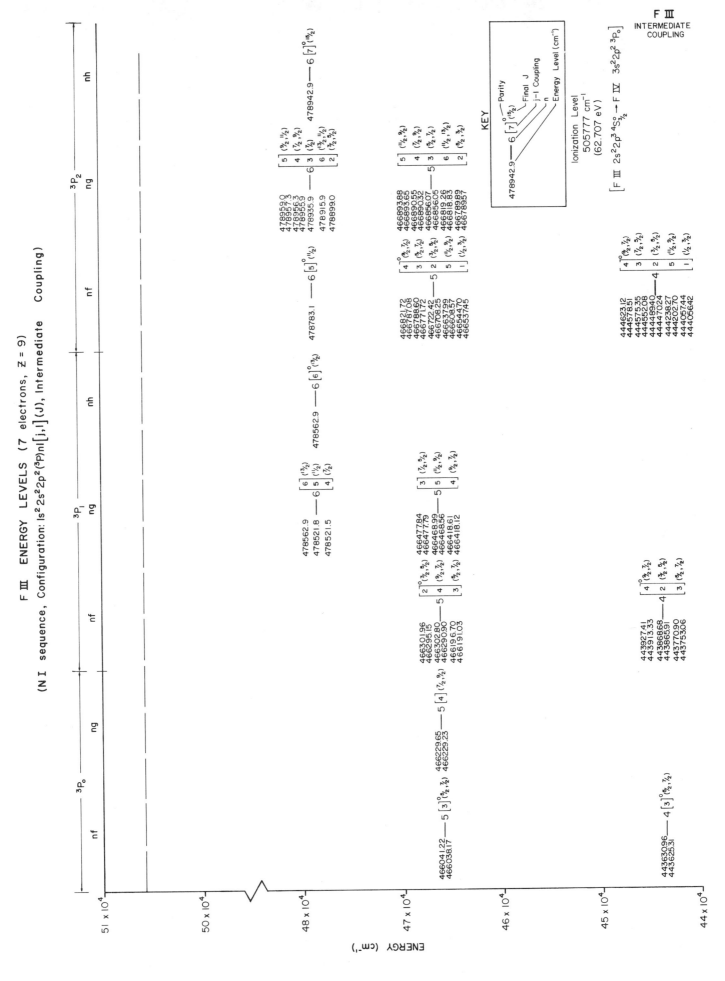

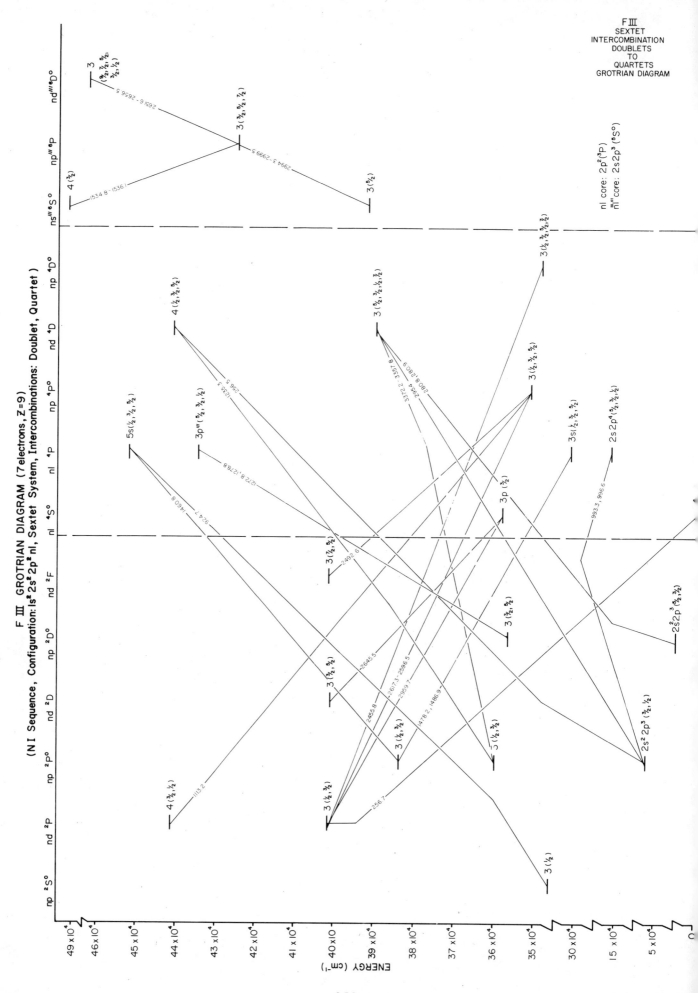

F III GROTRIAN DIAGRAM (7 electrons, Z=9)

(N I Sequence, Configuration: 1s² 2s² 2p² nl, Sextet System, Intercombinations: Doublet, Quartet)

F III
SEXTET
INTERCOMBINATION
DOUBLETS
TO
QUARTETS
GROTRIAN DIAGRAM

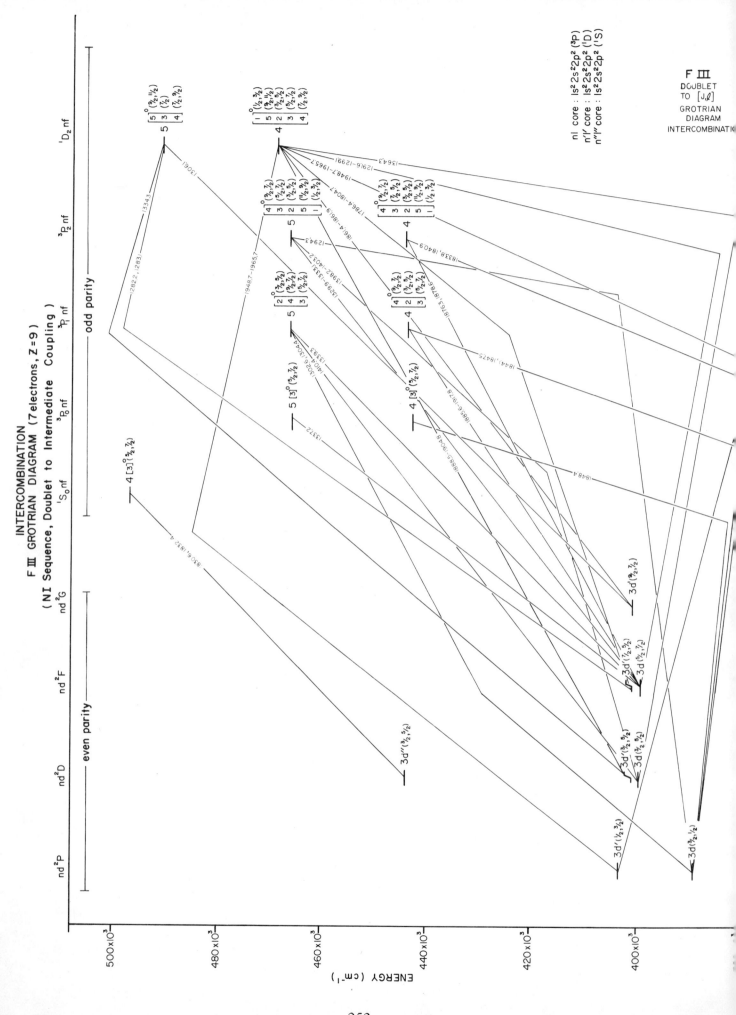

INTERCOMBINATION
F III GROTRIAN DIAGRAM (7 electrons, Z = 9)
(NI Sequence, Doublet to Intermediate Coupling)

F III
DOUBLET
TO [J,ℓ]
GROTRIAN
DIAGRAM
INTERCOMBINATION

nl core : 1s²2s²2p² (³P)
n'l' core : 1s²2s²2p² (¹D)
n''l'' core : 1s²2s²2p² (¹S)

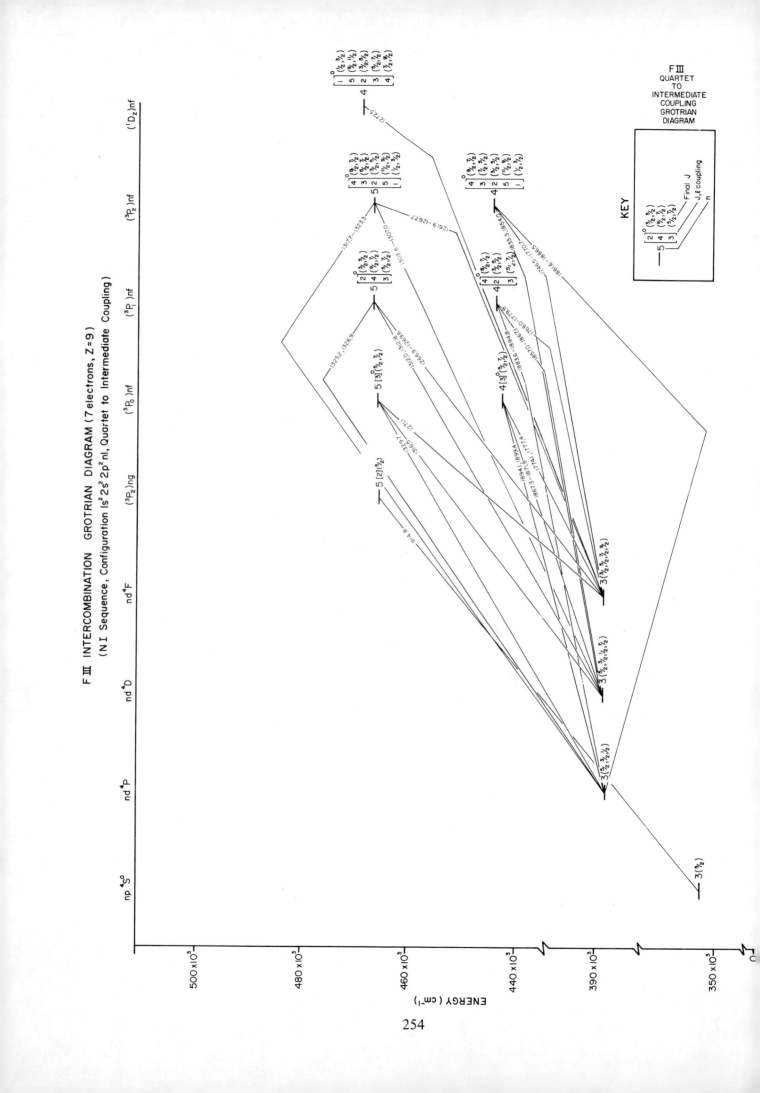

F III INTERCOMBINATION GROTRIAN DIAGRAM (7 electrons, Z = 9)

(N I Sequence, Configuration 1s² 2s² 2p² nl, Quartet to Intermediate Coupling)

254

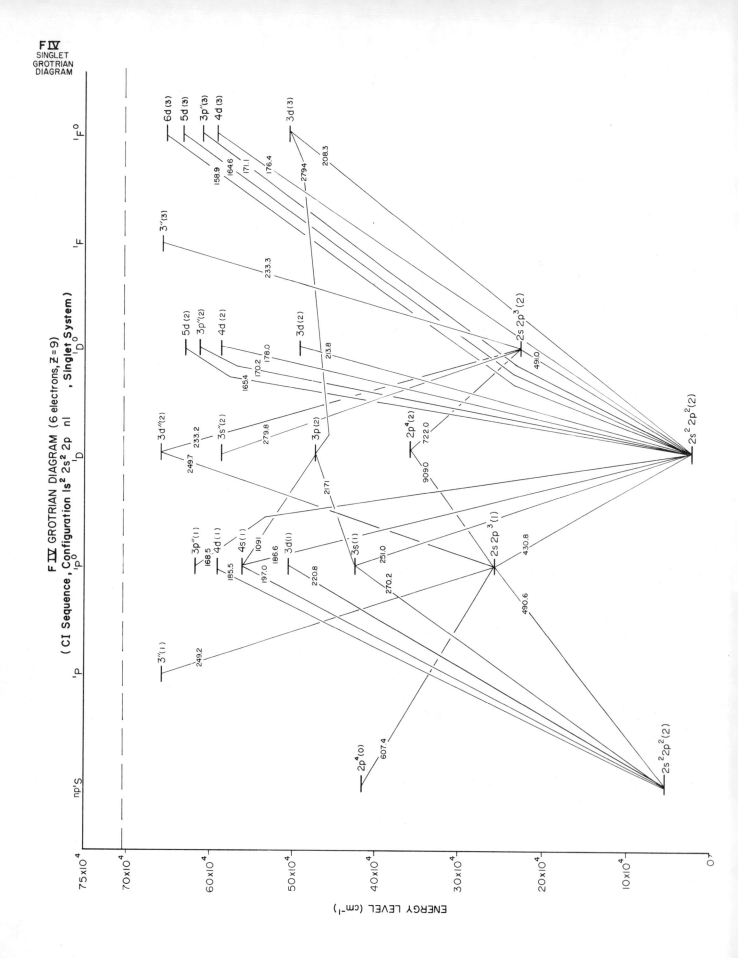

F IV
SINGLET
GROTRIAN
DIAGRAM

F IV GROTRIAN DIAGRAM (6 electrons, Z = 9)
(CI Sequence, Configuration $1s^2 2s^2 2p$ nl , Singlet System)

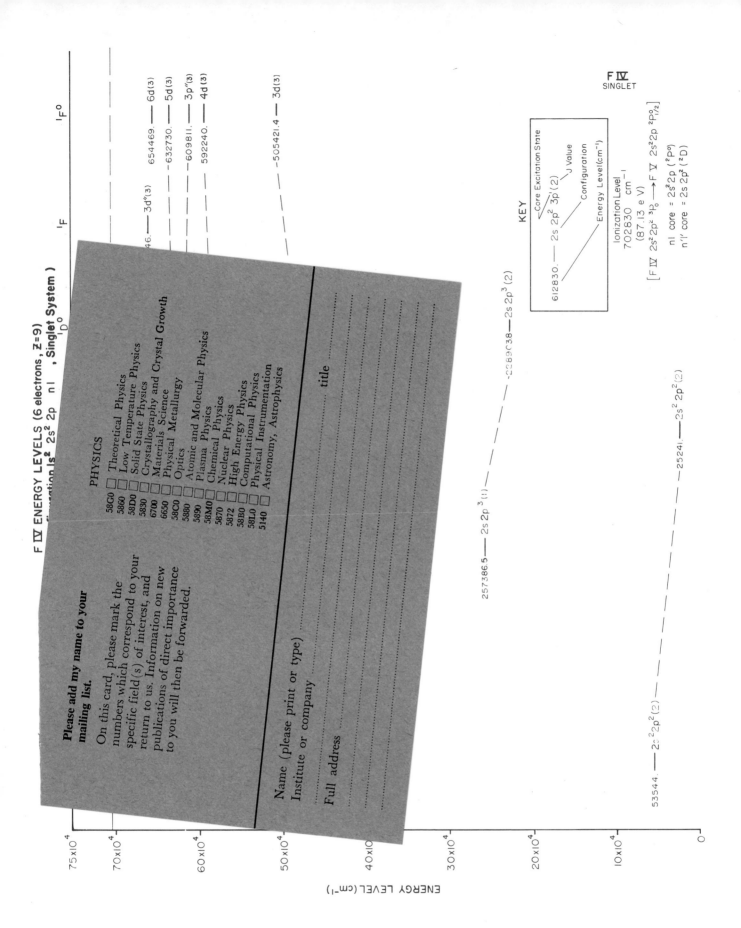

F IV ENERGY LEVELS (6 electrons, Z=9)
Configuration 1s² 2s² 2p nl , Singlet System

F IV
SINGLET

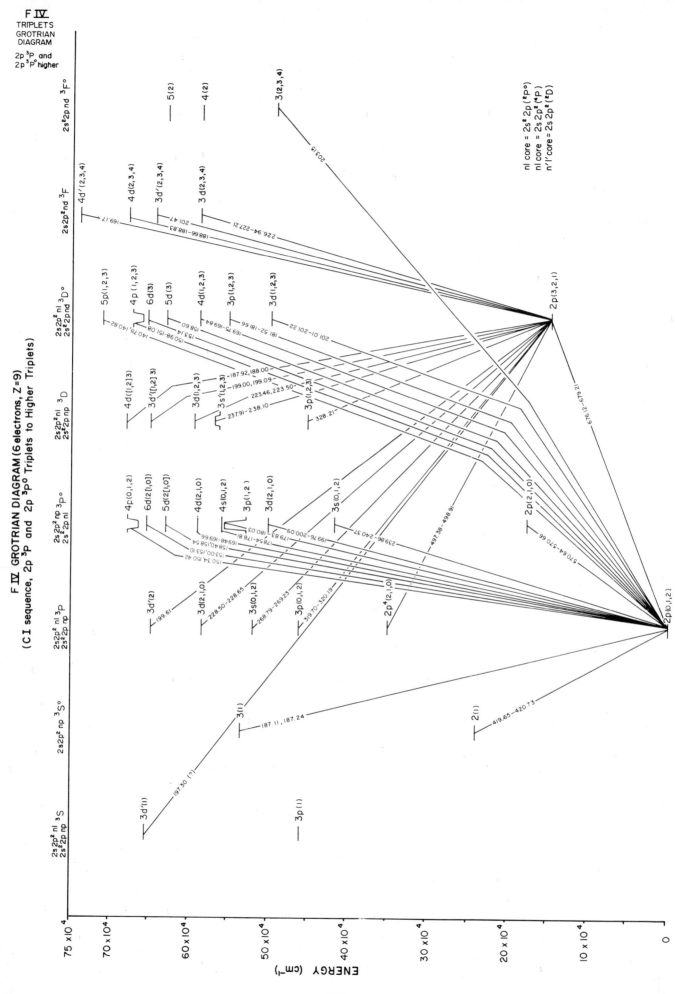

F IV
TRIPLETS
GROTRIAN
DIAGRAM

2p³P and
2p³P° higher

F IV GROTRIAN DIAGRAM (6 electrons, Z=9)
(C I sequence, 2p³P and 2p³P° Triplets to Higher Triplets)

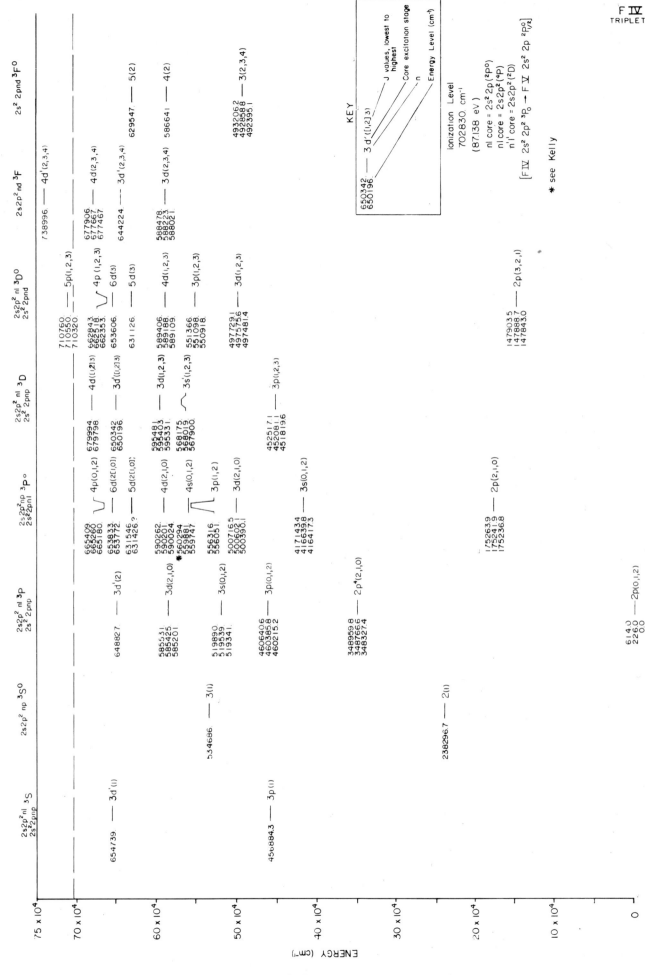

Energy Level Diagram of Energy Levels (6 Electrons, 2-9)

(CI sequence, Configuration: $1s^2\,2s^2\,2p\,nl$ and $1s^2\,2s2p^2\,nl$, Triplet System)

259

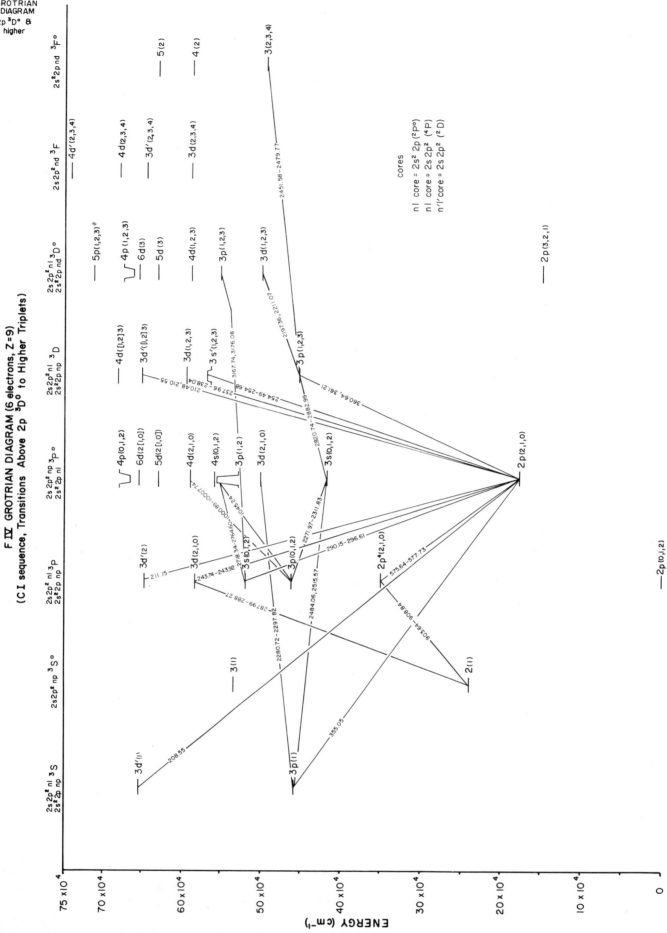

F IV
TRIPLET
GROTRIAN
DIAGRAM
2p³D° & higher

F IV GROTRIAN DIAGRAM (6 electrons, Z=9)
(C I sequence, Transitions Above 2p ³D° to Higher Triplets)

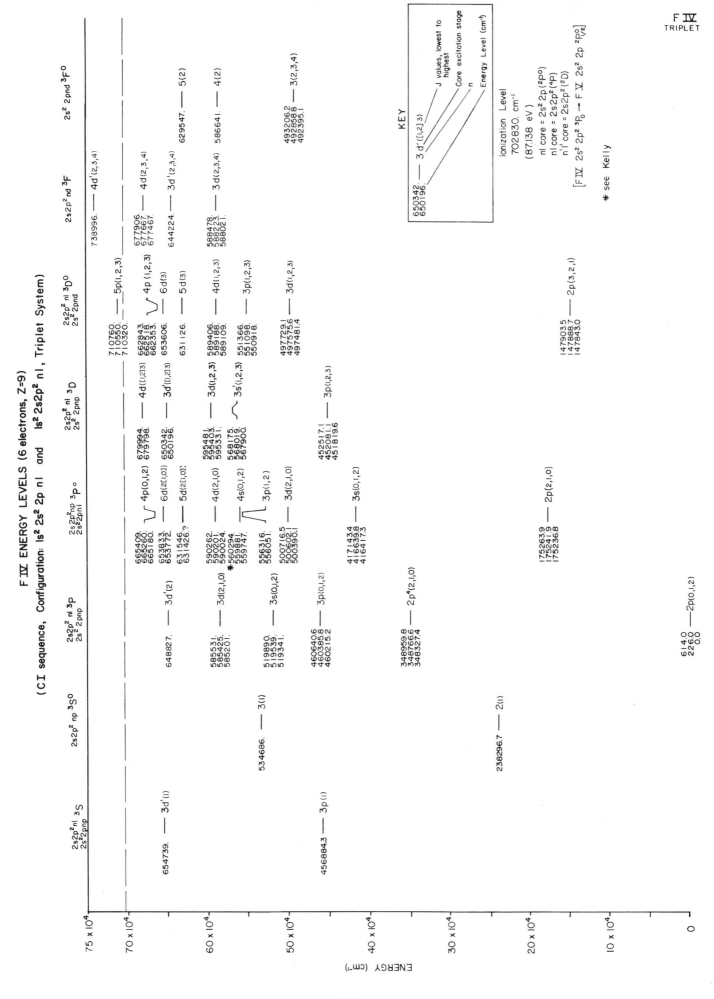

F Ⅳ ENERGY LEVELS (6 electrons, Z=9)

(CI sequence, Configuration: 1s² 2s² 2p nl   and   1s² 2s2p² nl, Triplet System)

F Ⅳ
TRIPLET

261

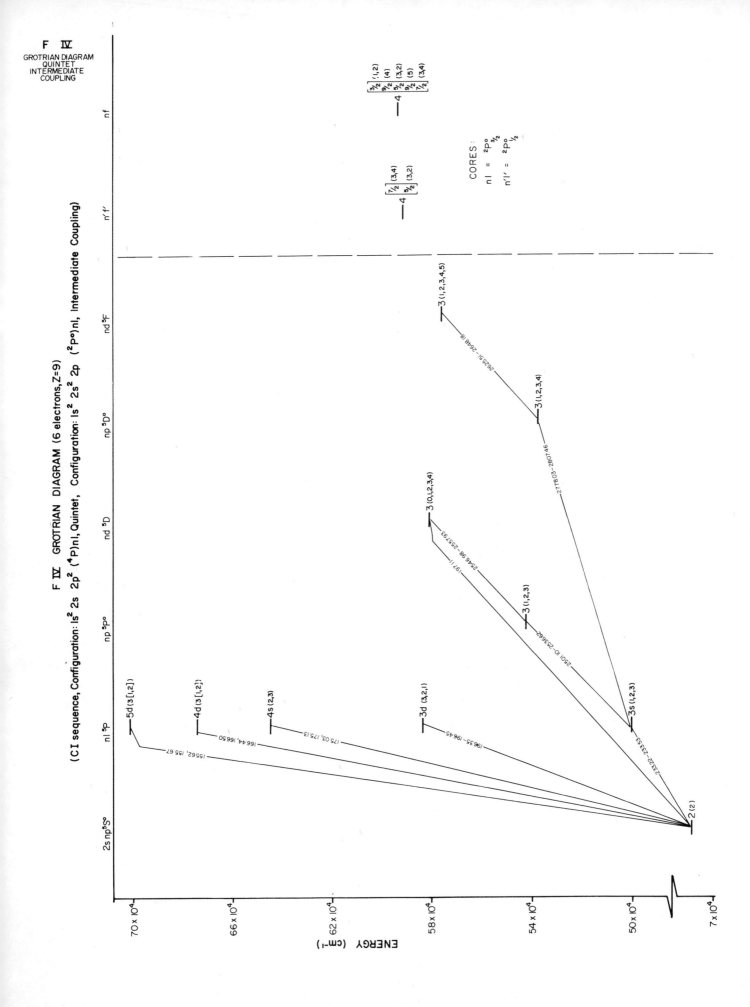

F IV

GROTRIAN DIAGRAM
QUINTET
INTERMEDIATE
COUPLING

F IV GROTRIAN DIAGRAM (6 electrons, Z=9)

(CI sequence, Configuration: 1s² 2s 2p² (⁴P)nl, Quintet, Configuration: 1s² 2s² 2p (²P°)nl, Intermediate Coupling)

CORES:
nl = ²P°₃/₂
n′l′ = ²P°₁/₂

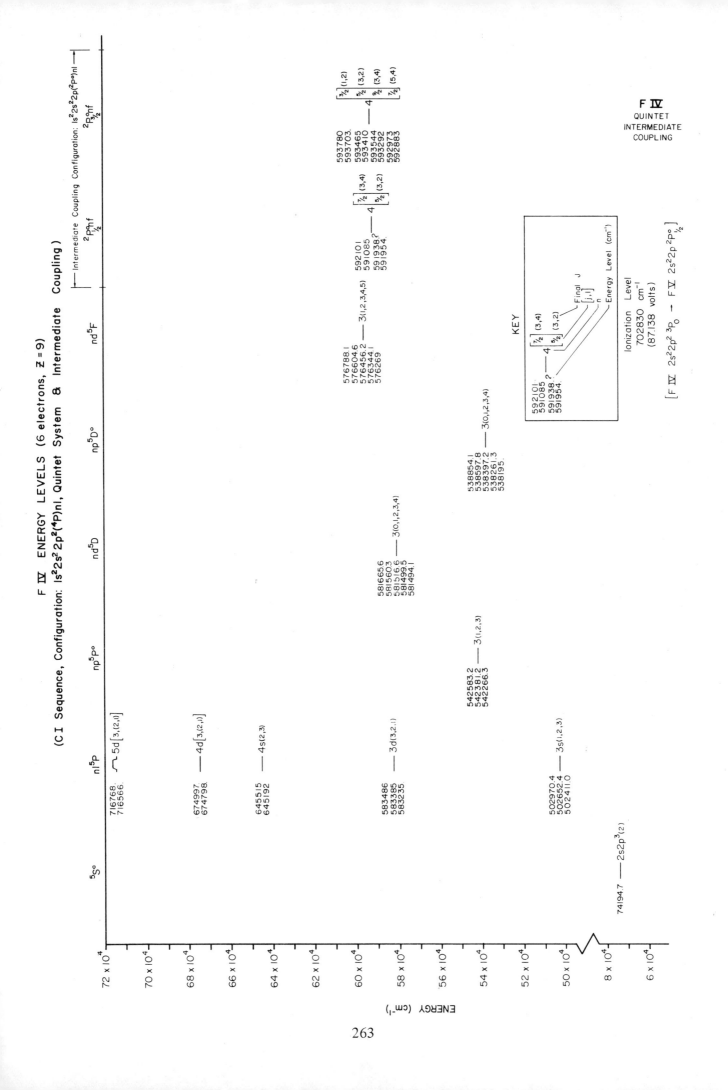

F IV ENERGY LEVELS (6 electrons, Z = 9)

(C I Sequence, Configuration: 1s²2s²2p²(⁴P)nl, Quintet System & Intermediate Coupling)

F IV
QUINTET
INTERMEDIATE
COUPLING

263

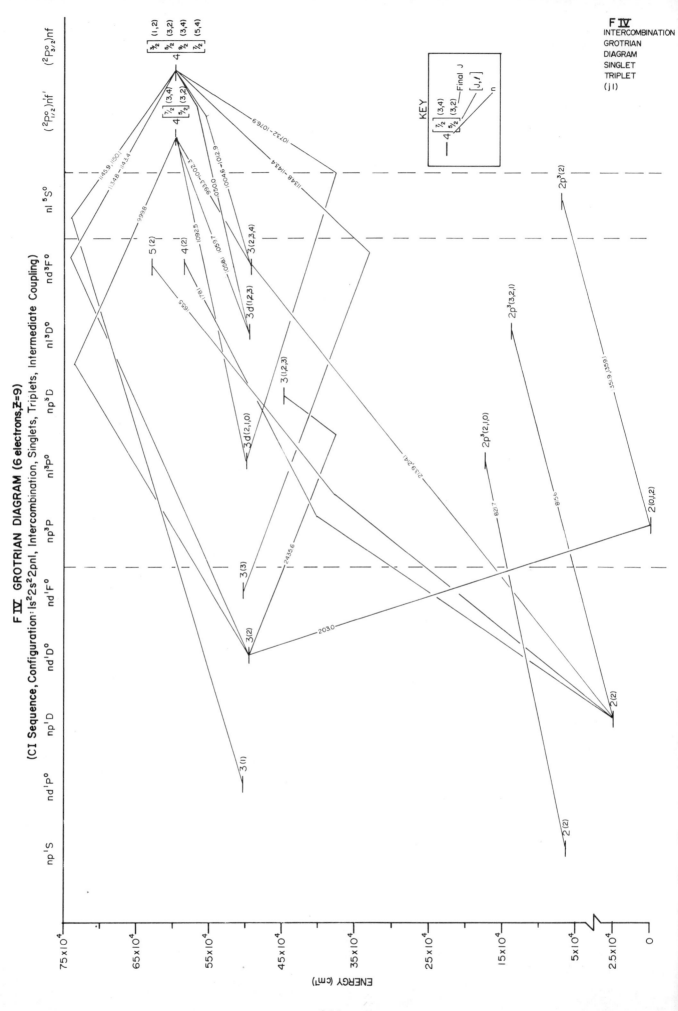

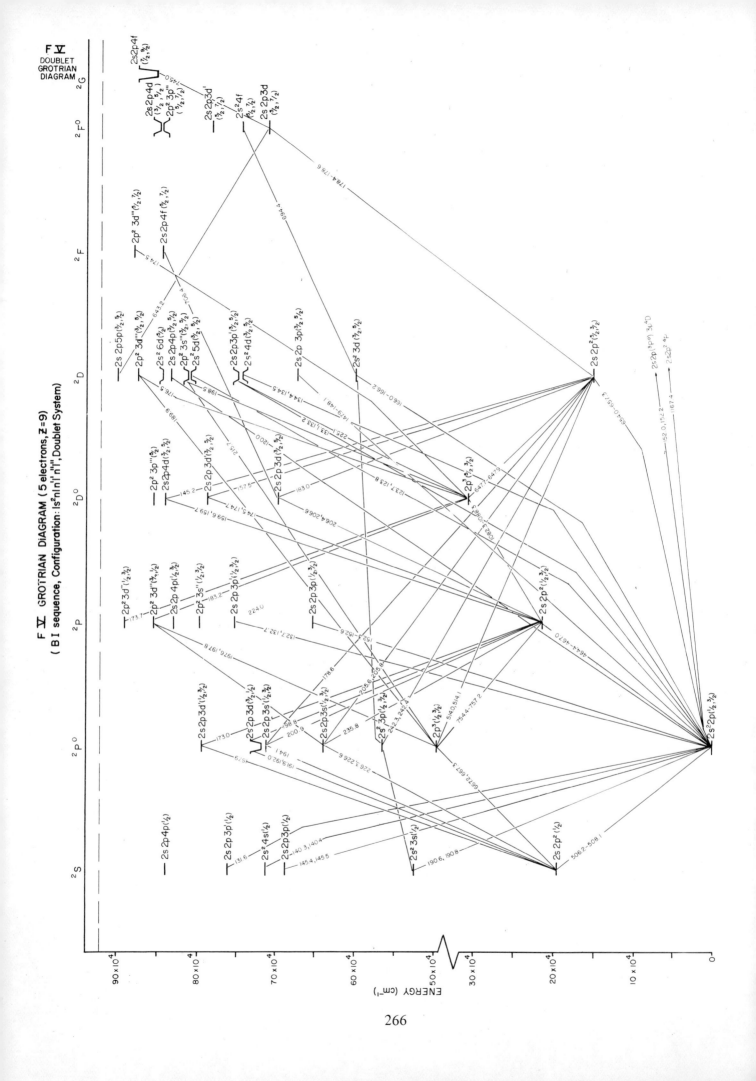

F V GROTRIAN DIAGRAM (5 electrons, Z=9)
(B I sequence, Configuration: ls²nl n'l' n''l'', Doublet System)

F V
DOUBLET
GROTRIAN
DIAGRAM

266

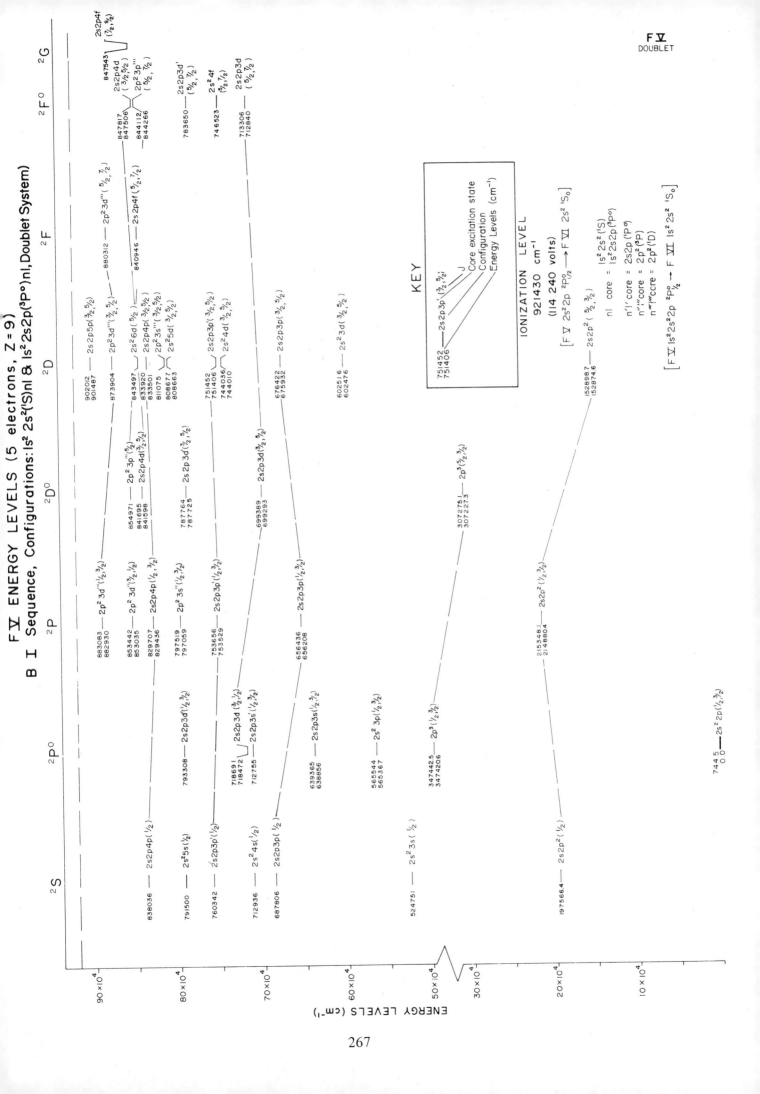

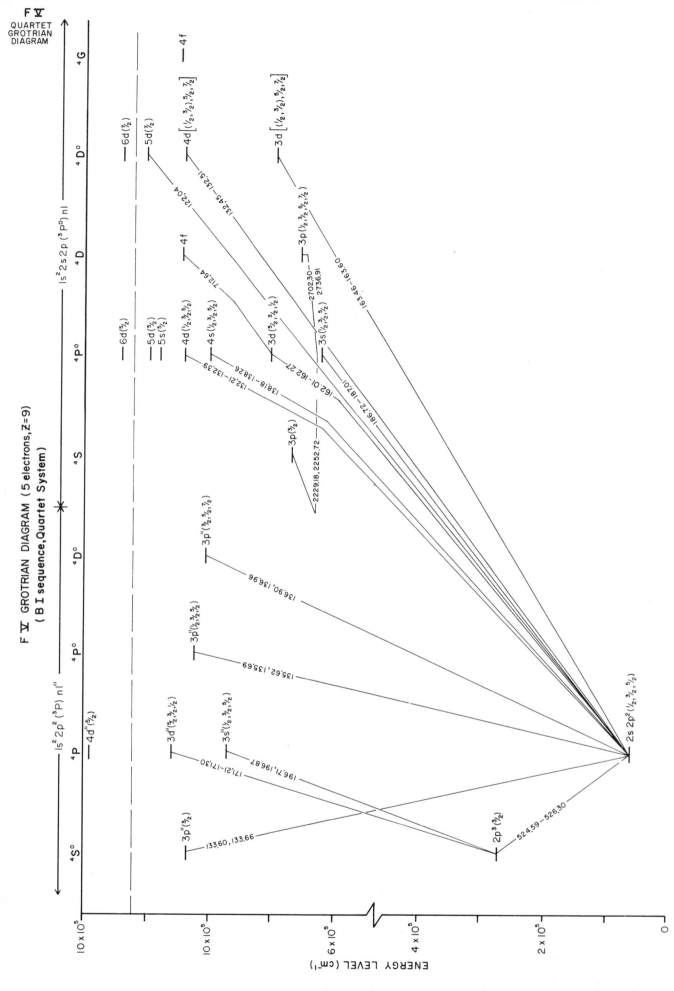

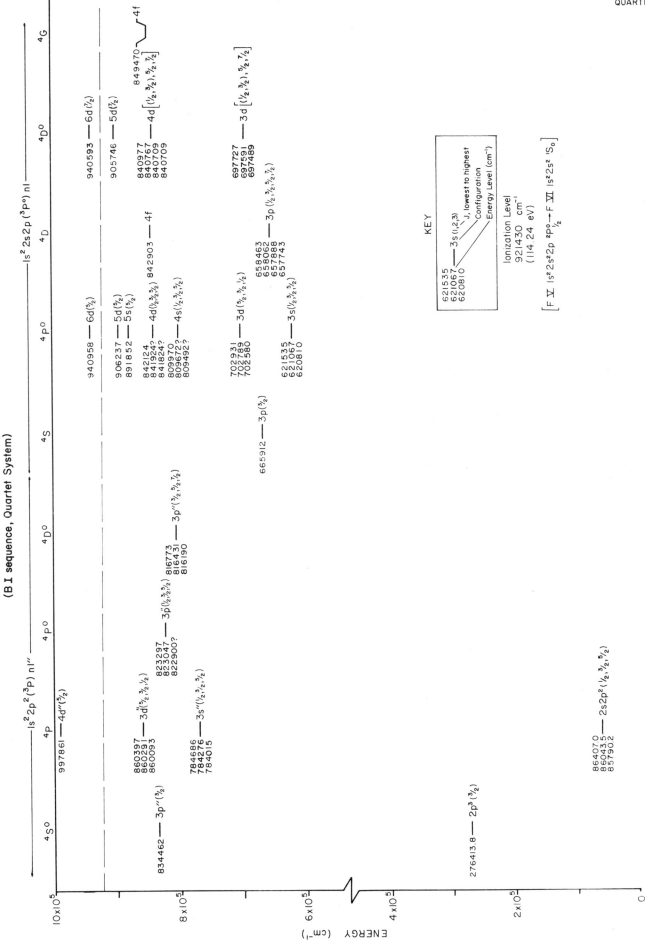

F V ENERGY LEVELS (5 electrons, Z=9)
(B I sequence, Quartet System)

F V
QUARTET

269

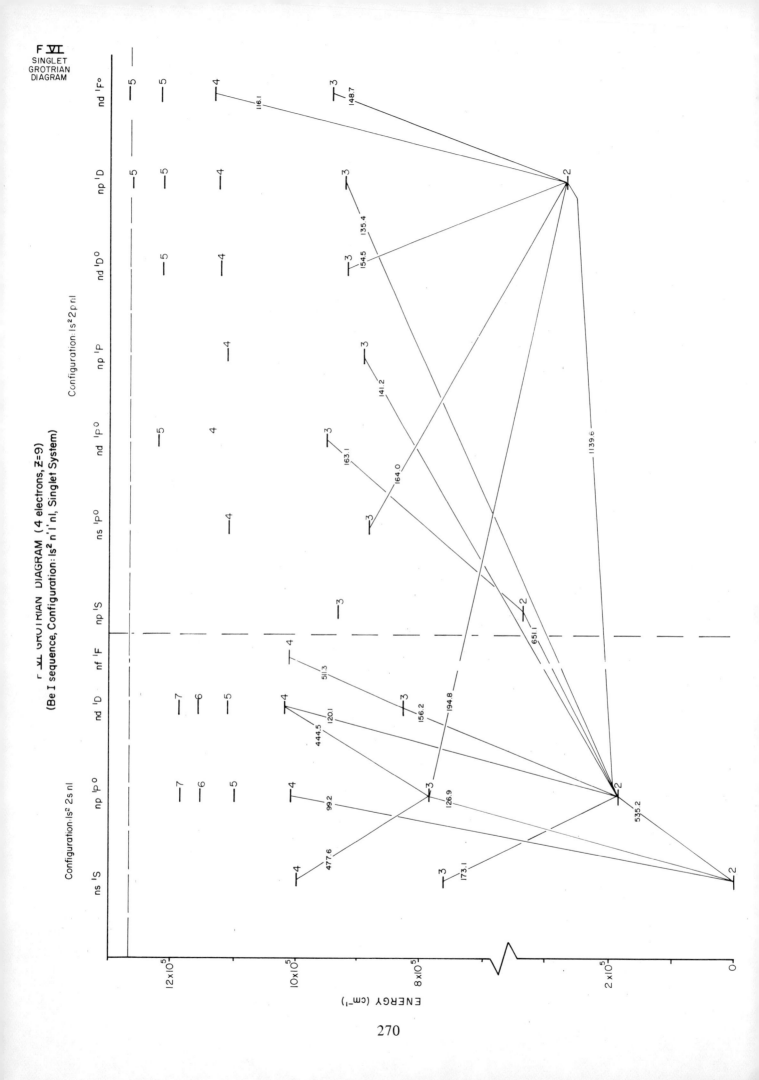

F VI
SINGLET
GROTRIAN
DIAGRAM

F VI GROTRIAN DIAGRAM (4 electrons, Z=9)
(Be I sequence, Configuration: 1s² n′l′nl, Singlet System)

Configuration: 1s² 2s nl

Configuration: 1s²2pnl

ENERGY (cm⁻¹)

270

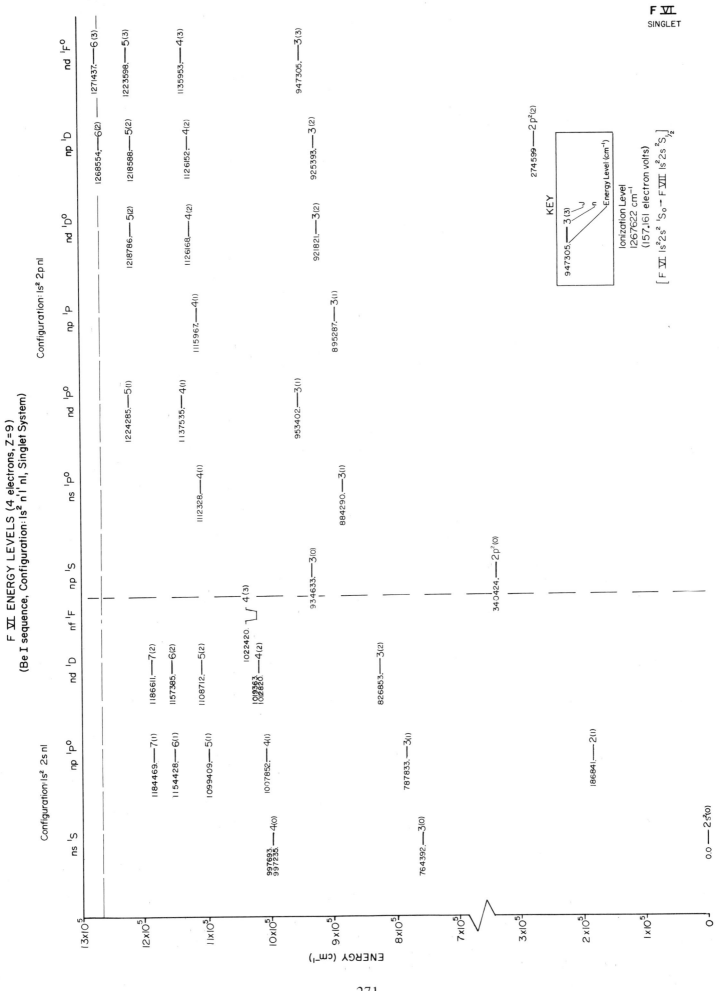

F VI ENERGY LEVELS (4 electrons, Z=9)
(Be I sequence, Configuration: 1s² nl'nl, Singlet System)

F VI
SINGLET

271

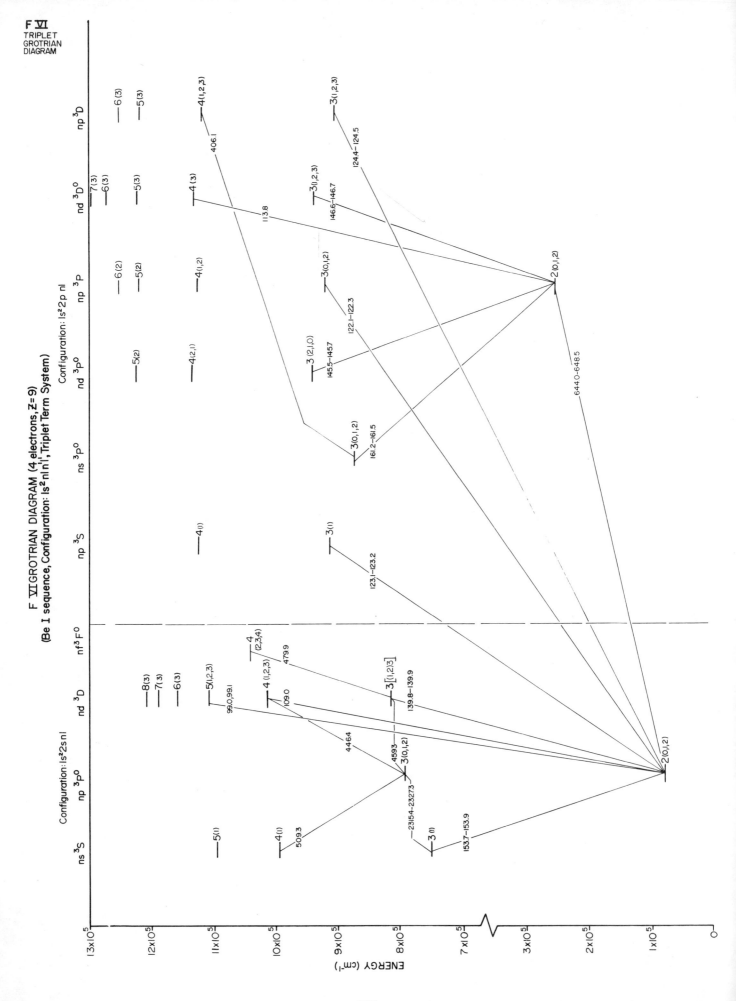

F VI
TRIPLET
GROTRIAN
DIAGRAM

F VI GROTRIAN DIAGRAM (4 electrons, Z=9)
(Be I sequence, Configuration: 1s²nl n'l', Triplet Term System)

272

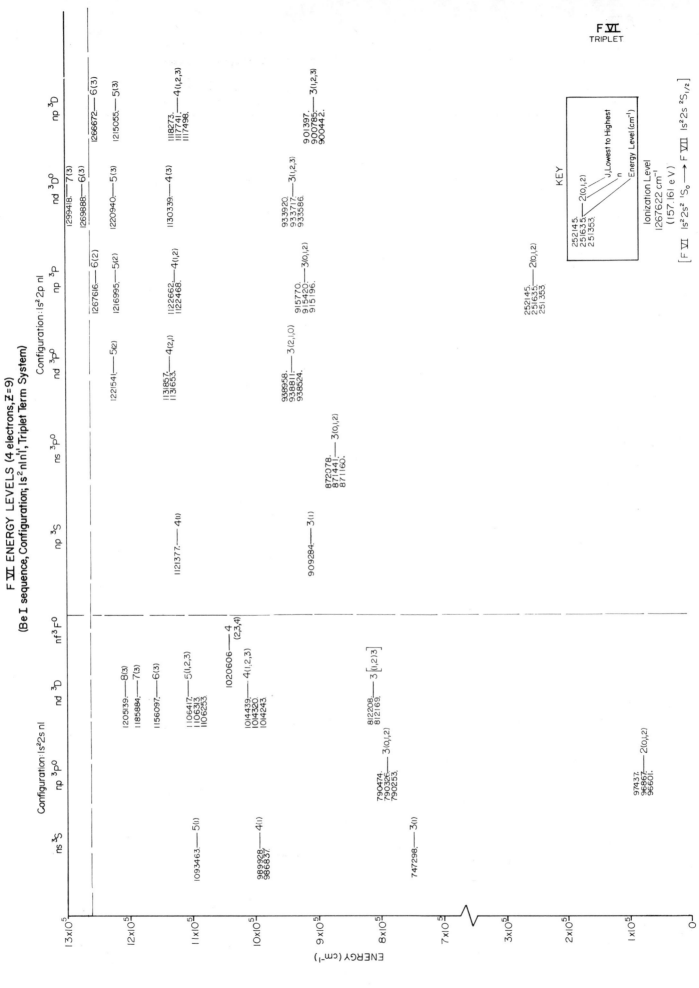

F VI ENERGY LEVELS (4 electrons, Z=9)
(Be I sequence, Configuration; 1s² nl n'l', Triplet Term System)

273

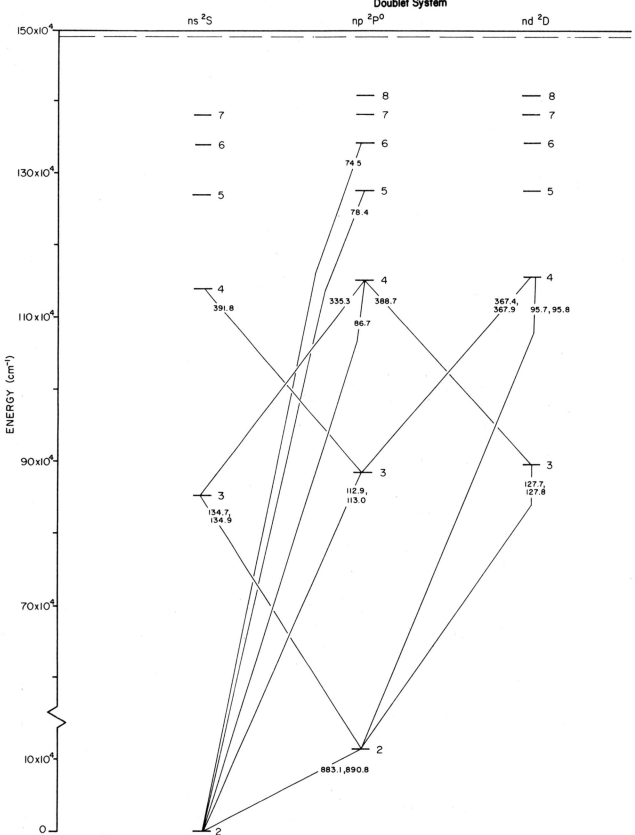

F VII GROTRIAN DIAGRAM (3 electrons, Z = 9)
Li I sequence
Doublet System

F VII ENERGY LEVELS (3 electrons, Z = 9)
(Li I sequence, Configuration : Is² nl)
Doublet System

ns ²S   np ²Pᴼ   nd ²D

TERM VALUE (cm⁻¹)

150 × 10⁴

140 × 10⁴ — 1408848. —— 8(½,³⁄₂)   1409538. —— 8(³⁄₂,⁵⁄₂)
1380775. —— 7(½)   1382858. —— 7(½,³⁄₂)   1383841. —— 7(³⁄₂,⁵⁄₂)

1339216. —— 6(½)   1342877. —— 6(½,³⁄₂)   1344141. —— 6(³⁄₂,⁵⁄₂)

130 × 10⁴
1269826. —— 5(½)   1276194. —— 5(½,³⁄₂)   1278404. —— 5(³⁄₂,⁵⁄₂)

120 × 10⁴
1157255. 1157223. —— 4(³⁄₂,⁵⁄₂)
1140416. —— 4(½)   1152977. —— 4(½,³⁄₂)

110 × 10⁴

100 × 10⁴

90 × 10⁴
895722. 895632. —— 3(³⁄₂,⁵⁄₂)
885418. 885136. —— 3(½,³⁄₂)
854625. —— 3(½)

80 × 10⁴

KEY

70 × 10⁴
113235. 112258. —— 2(½,³⁄₂)

60 × 10⁴

10 × 10⁴
113238. 112263. —— 2(½,³⁄₂)

0 ——— 0.0 —— 2(½)

KEY:
113235. 112258. —— 2(½,³⁄₂)
J, lowest to highest
n
Energy Level (cm⁻¹)

Ionization Level
1493629 cm⁻¹
(185.182 eV)
[F VII Is²2s ²S₁/₂ → F VIII Is² ¹S₀]

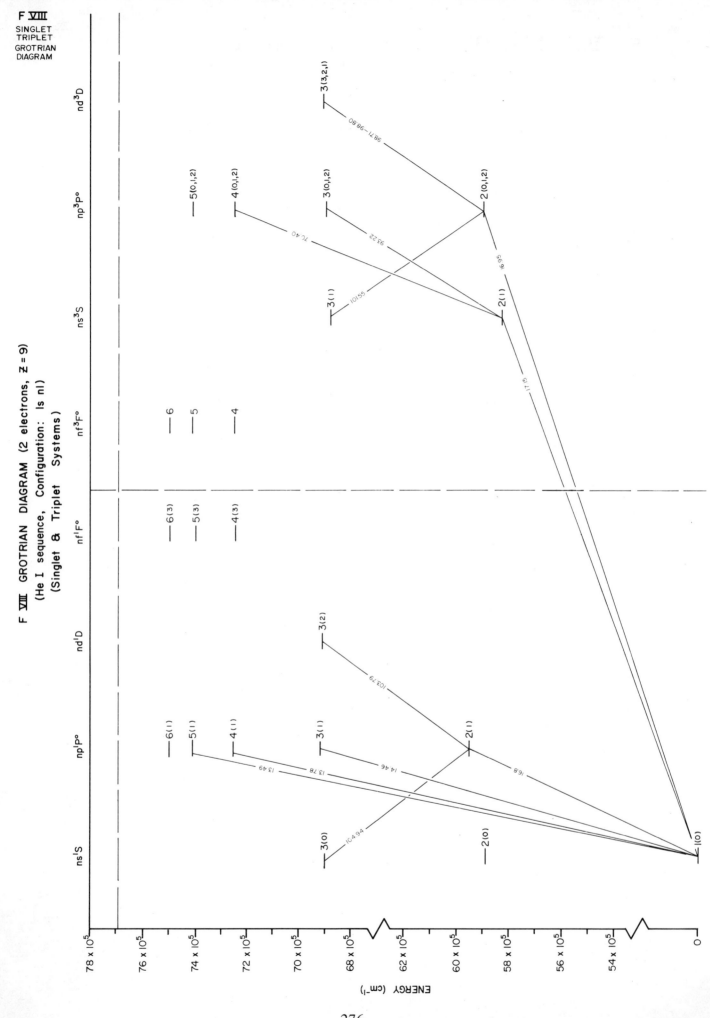

F VIII
SINGLET
TRIPLET
GROTRIAN
DIAGRAM

F VIII GROTRIAN DIAGRAM (2 electrons, Z = 9)
(He I sequence, Configuration: 1s nl)
(Singlet & Triplet Systems)

276

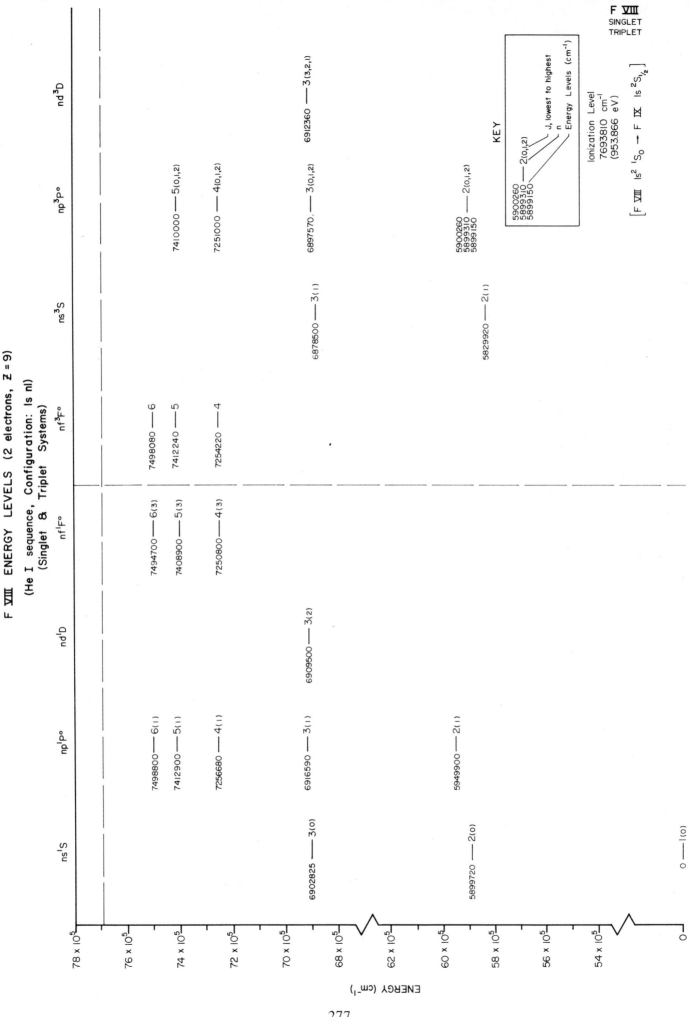

F VIII ENERGY LEVELS (2 electrons, Z = 9)

(He I sequence, Configuration: 1s nl)
(Singlet & Triplet Systems)

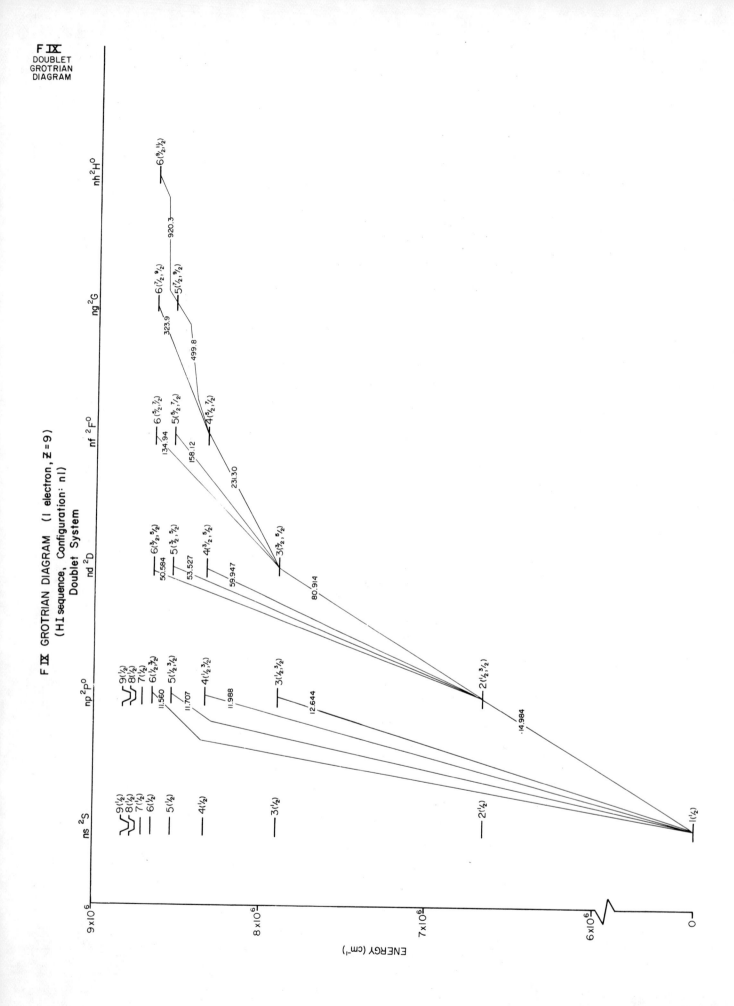

F IX
DOUBLET
GROTRIAN
DIAGRAM

F IX GROTRIAN DIAGRAM (1 electron, Z = 9)
(H I sequence, Configuration: nl)
Doublet System

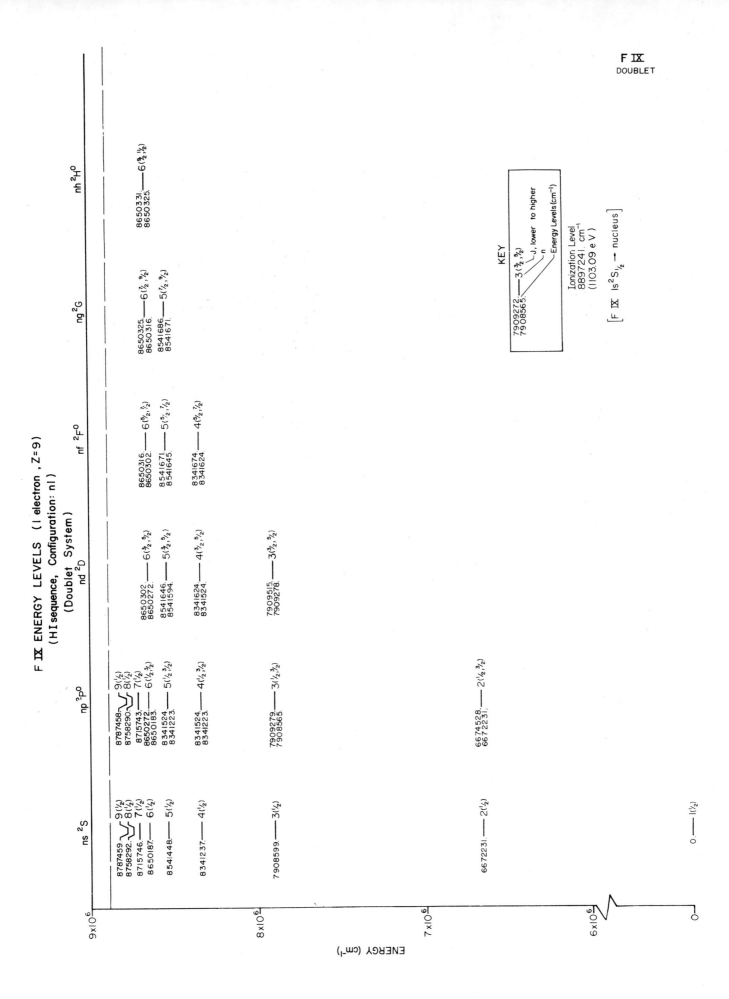

F IX ENERGY LEVELS (1 electron, Z=9)
(H I sequence, Configuration: nl)
(Doublet System)

F IX
DOUBLET

F I $\hspace{8em}$ $Z = 9$ $\hspace{8em}$ 9 electrons

J.E. Hansen and W. Persson, Physica Scripta **8**, 197 (1973).

   Authors give $2p^4$ ns and $2p^4$ nd configurations in F I and Ne II.

F II $\hspace{8em}$ $Z = 9$ $\hspace{8em}$ 8 electrons

I.S. Bowen, Ap. J. **132**, 1 (1960).

   Author gives observed wavelengths.

H. Palenius, J. Opt. Soc. Amer. **56**, 828 (1966).

   Author gives transition wavelengths.

Comment on intermediate coupling: According to H. Palenius, Ark. Fys. **39**, 15 (1968), coupling with f-electrons does not produce clear configurations. The pair-coupling approximation (please see N II) is used to describe some of the spectroscopic terms, but terms we list under "other f" cannot be so identified.

F III $\hspace{8em}$ $Z = 9$ $\hspace{8em}$ 7 electrons

I.S. Bowen, Ap. J. **132**, 1 (1960).

   Author gives observed wavelengths.

For the intermediate-coupling designation, please see H. Palenius, Physica Scripta **1**, 113 (1970).

F IV $\hspace{8em}$ $Z = 9$ $\hspace{8em}$ 6 electrons

I.S. Bowen, Ap. J. **121**, 306 (1955).

   Author gives transition wavelengths from nebular observations.

I.S. Bowen, Ap. J. **132**, 1 (1960).

   Author gives wavelengths from nebular observations.

For the intermediate-coupling designation, please see H. Palenius, University of Lund (Sweden) Report (May, 1971).
Note: The first triplet Grotrian diagram is labeled $2p$ ^{3}P and $2p$ ^{3}P°. The correct designation is $2p$ ^{3}P and $2p$ ^{3}D°. The corner designation should read $2p$ ^{3}P and $2p$ ^{3}D° to higher triplets.

F V $\hspace{8em}$ $Z = 9$ $\hspace{8em}$ 5 electrons

   Please see the general references.

F VI $\hspace{8em}$ $Z = 9$ $\hspace{8em}$ 4 electrons

Please see the general references. Note, also, that a line has been omitted: $2s^2$ ^{1}S–$2s$ $2p$ ^{3}P°: 1032.34.

F VII $Z = 9$ 3 electrons

   Please see the general references.

F VIII $Z = 9$ 2 electrons

E. Holøien and J. Midtdal, J. Phys. B: Atom. Molec. Phys. **4**, 1243 (1971).

   Authors calculate energies nonrelativistically in treatment of He I isoelectronic sequence.

F IX $Z = 9$ 1 electron

   Please see the general references.

F VII $Z = 9$ 3 electrons

Neon (Ne)

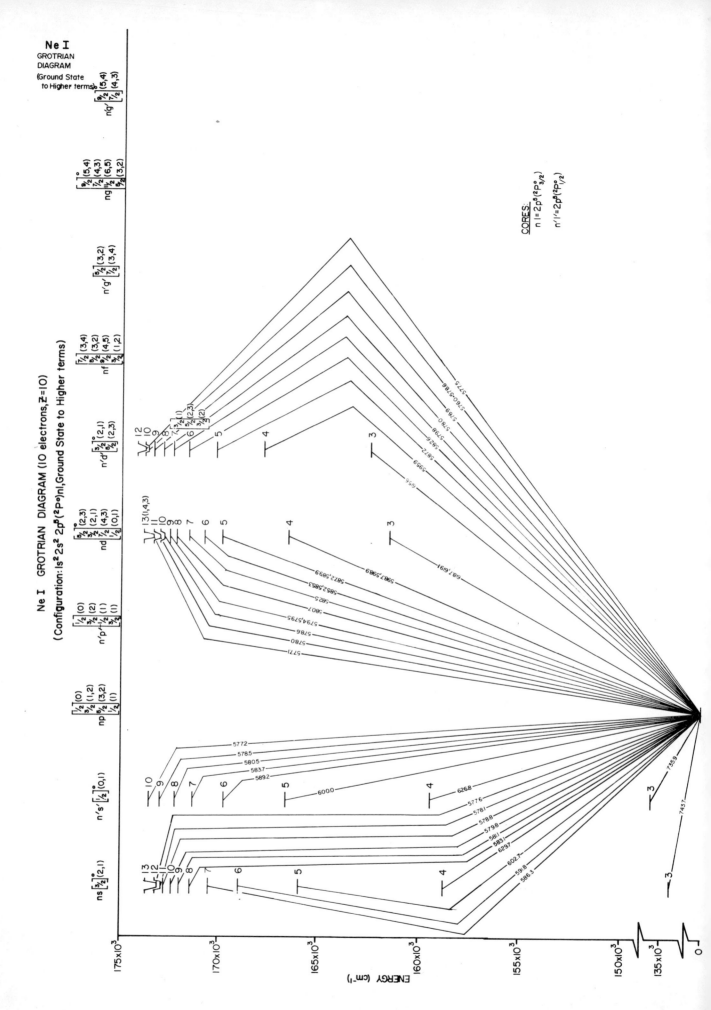

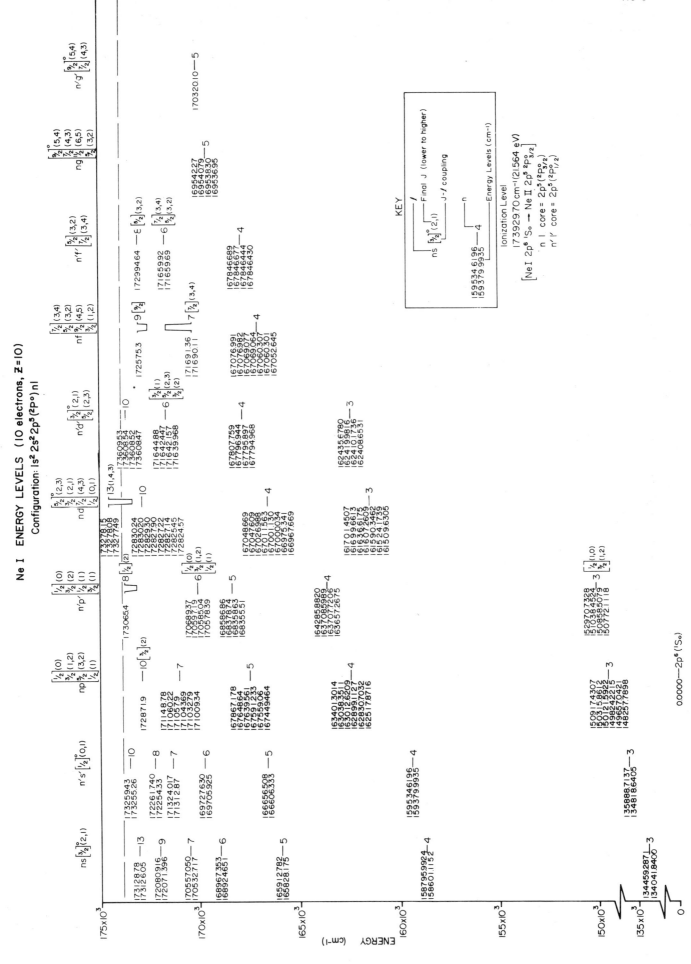

Ne I   ENERGY LEVELS   (10 electrons, Z=10)

Configuration: 1s²2s²2p⁵(²P°)nl

Ne I

285

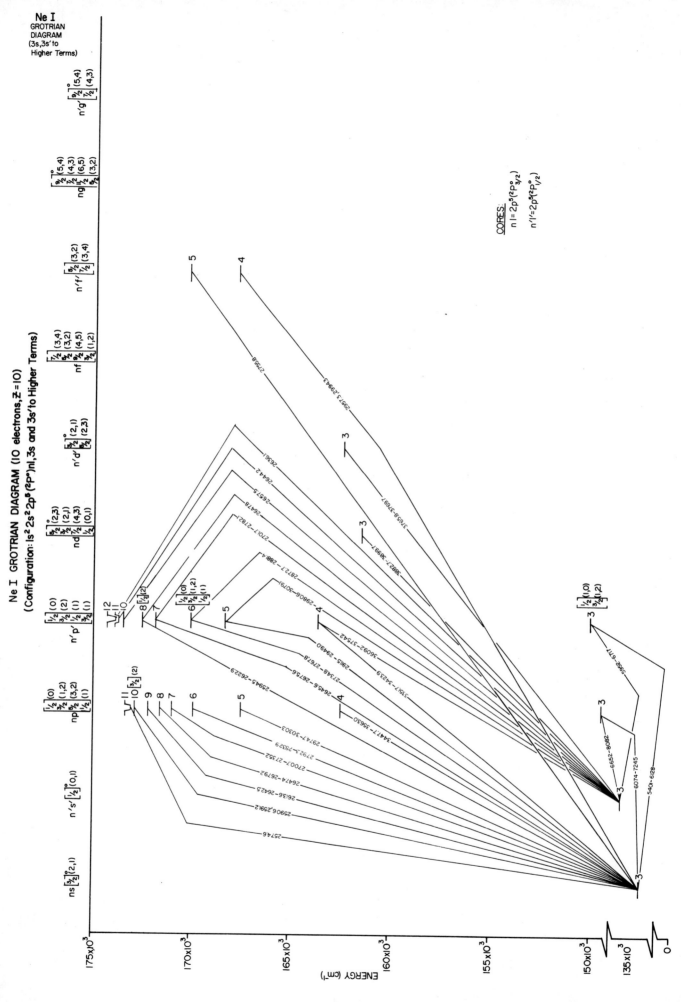

Ne I ENERGY LEVELS (10 electrons, Z=10)
Configuration: 1s² 2s² 2p⁵(²P°)nl

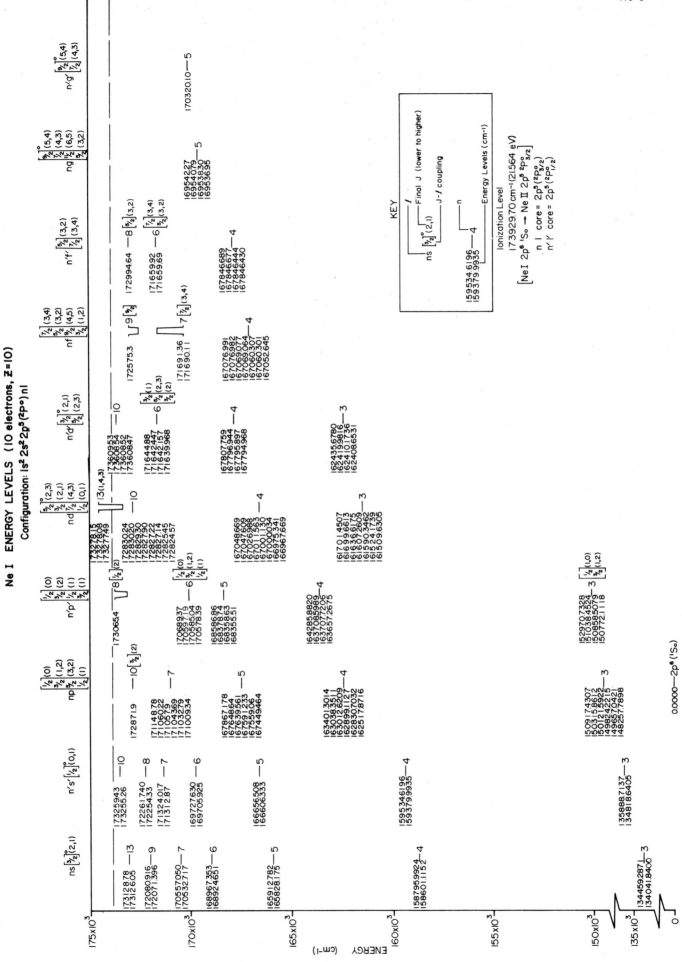

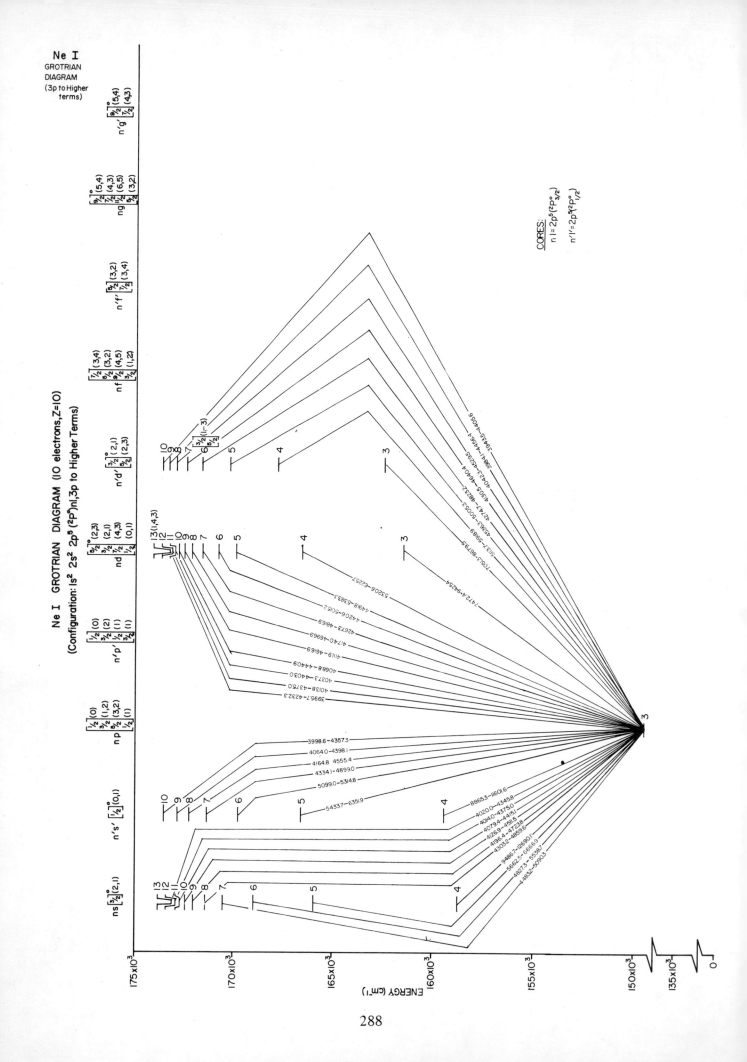

288

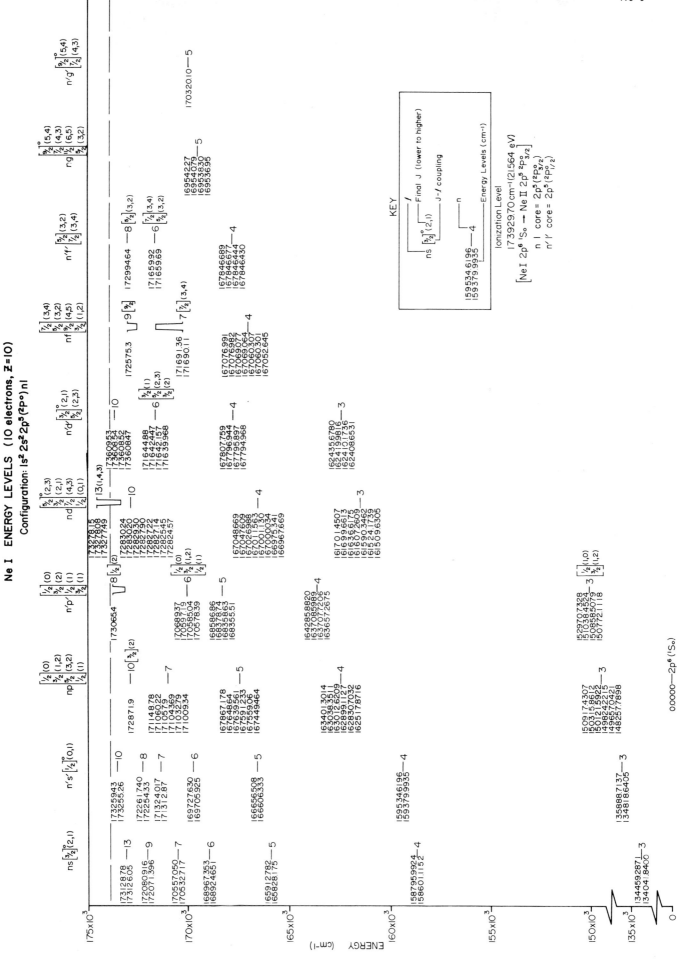

Ne I ENERGY LEVELS (IO electrons, Z=IO)
Configuration: 1s² 2s²2p⁵(²P°)nl

Ne I

289

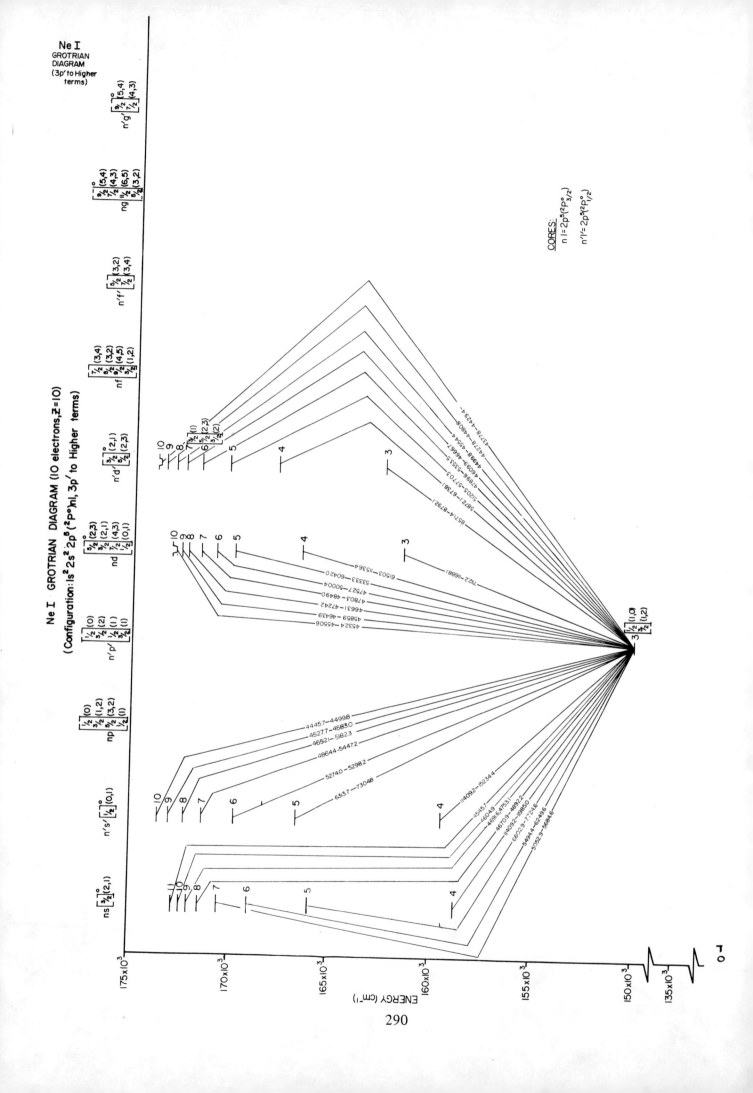

Ne I
GROTRIAN
DIAGRAM
(3p' to Higher
terms)

Ne I GROTRIAN DIAGRAM (10 electrons, Z=10)
(Configuration: 1s²2s²2p⁵(²P°)nl, 3p' to Higher terms)

CORES:
n l = 2p⁵(²P°₃/₂)
n'l'= 2p⁵(²P°₁/₂)

ENERGY (cm⁻¹)

290

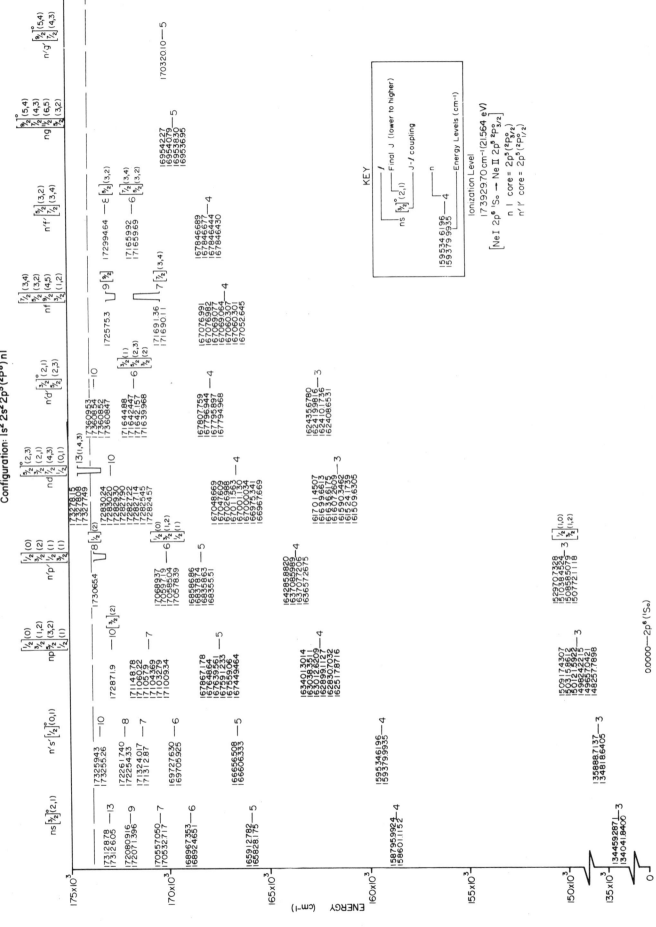

Ne I

Ne I ENERGY LEVELS (10 electrons, Z=10)

Configuration: 1s² 2s² 2p⁵(²P°)nl

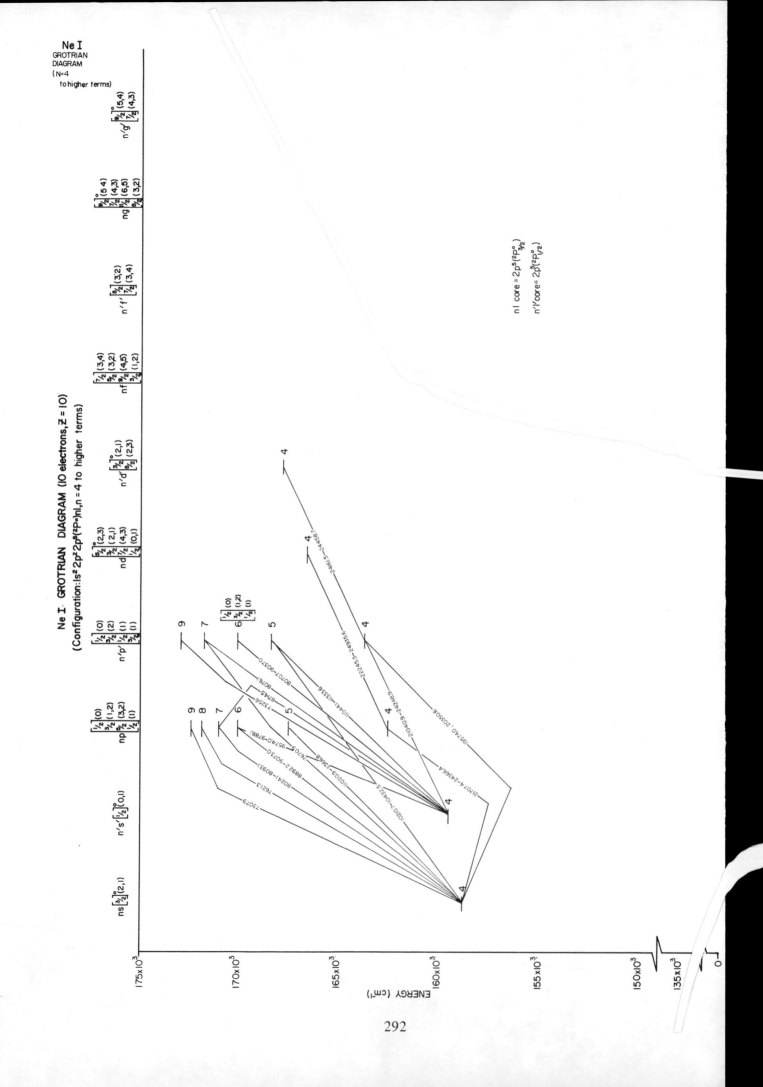

Ne I
GROTRIAN
DIAGRAM
(N=4
to higher terms)

Ne I GROTRIAN DIAGRAM (10 electrons, Z = 10)
(Configuration: 1s² 2p² 2p⁶(²Pᵒ)nl, n = 4 to higher terms)

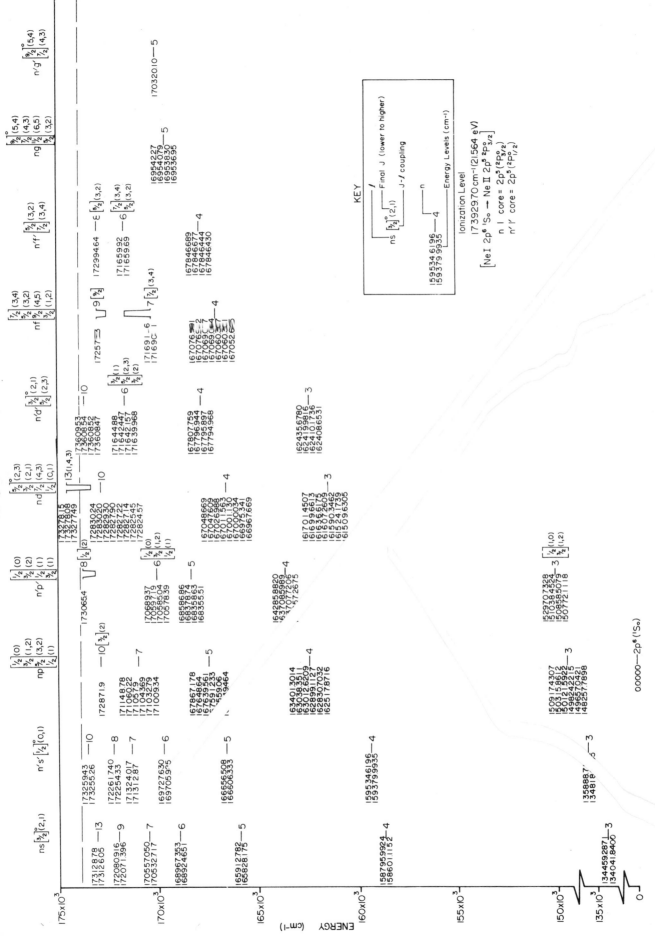

Ne I

Ne I ENERGY LEVELS (10 electrons, Z=0)

Configuration: 1s² 2s²2p⁵(²P°)nl

293

Ne I
2p⁶ ¹S₀ to High
TERMEDIATE COUPLING
GROTRIAN DIAGRAM

Ne I GROTRIAN DIAGRAM (10 electrons, Z = 10)
(Configuration: 1s² 2s² 2p⁵(ᵐL_J)nl, High Intermediate Coupling)

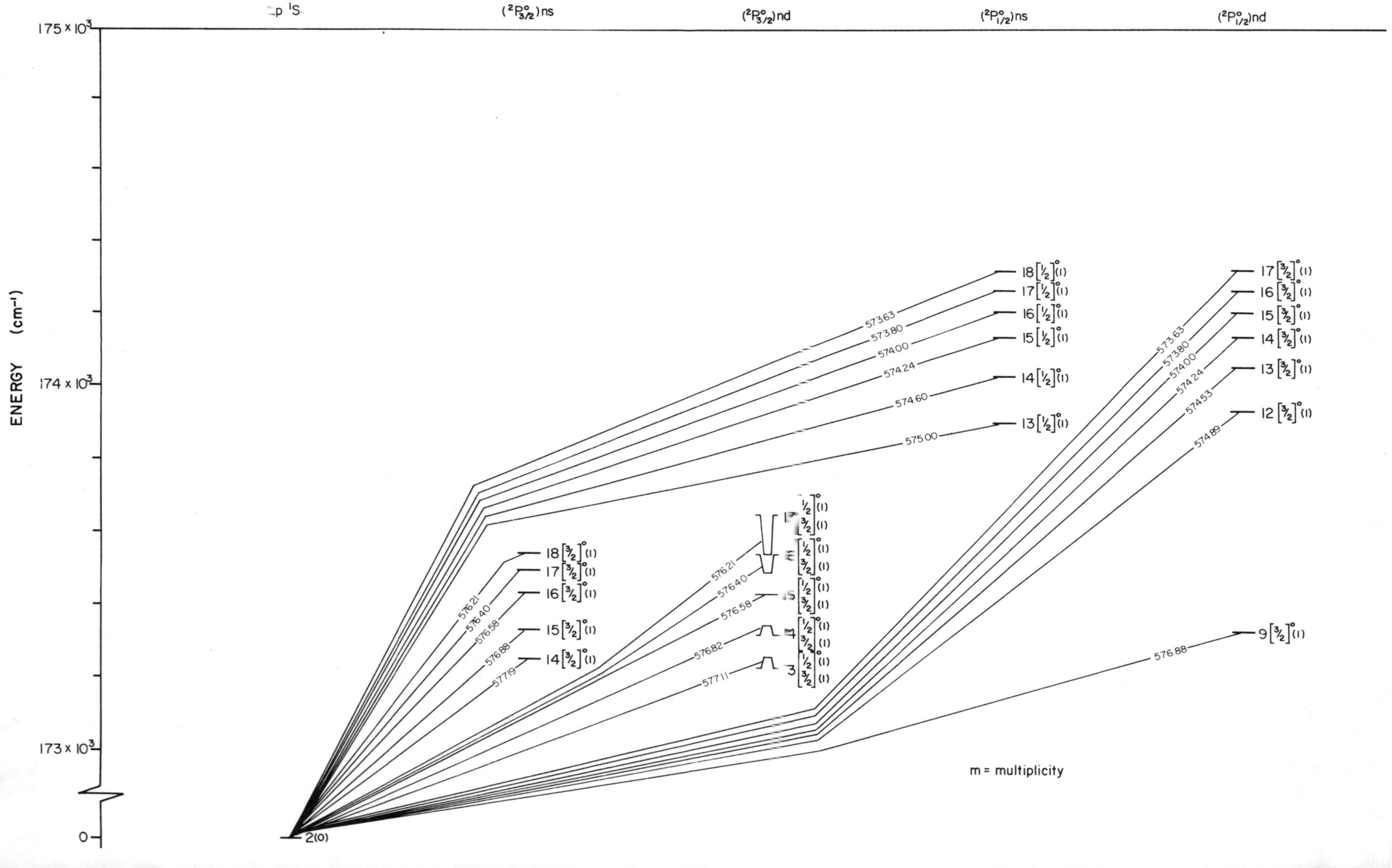

Ne I ENERGY LEVELS   (10 electrons, Z =10)

(Configuration: $1s^2 2s^2 2p^5(^m L_j)nl$, High Intermediate Coupling)

Ne I
HIGH
INTERMEDIATE
COUPLING

ENERGY (cm^{-1})

175×10^3

174×10^3

173×10^3

$(^2P^\circ_{3/2})$ns

173548 — $18[\tfrac{3}{2}]^\circ(1)$
173491 — $17[\tfrac{3}{2}]^\circ(1)$
173436 — $16[\tfrac{3}{2}]^\circ(1)$
173346 — $15[\tfrac{3}{2}]^\circ(1)$
173253 — $14[\tfrac{3}{2}]^\circ(1)$

$(^2P^\circ_{3/2})$nd

173548 — $17 \begin{cases} [\tfrac{1}{2}]^\circ(1) \\ [\tfrac{3}{2}]^\circ(1) \end{cases}$
173491 — $16 \begin{cases} [\tfrac{1}{2}]^\circ(1) \\ [\tfrac{3}{2}]^\circ(1) \end{cases}$
173436 — $15 \begin{cases} [\tfrac{1}{2}]^\circ(1) \\ [\tfrac{3}{2}]^\circ(1) \end{cases}$
173364 — $14 \begin{cases} [\tfrac{1}{2}]^\circ(1) \\ [\tfrac{3}{2}]^\circ(1) \end{cases}$
173277 — $13 \begin{cases} [\tfrac{1}{2}]^\circ(1) \\ [\tfrac{3}{2}]^\circ(1) \end{cases}$

$(^2P^\circ_{1/2})$ns

174328 — $18[\tfrac{1}{2}]^\circ(1)$
174277 — $17[\tfrac{1}{2}]^\circ(1)$
174216 — $16[\tfrac{1}{2}]^\circ(1)$
174143 — $15[\tfrac{1}{2}]^\circ(1)$
174034 — $14[\tfrac{1}{2}]^\circ(1)$
173913 — $13[\tfrac{3}{2}]^\circ(1)$

$(^2P^\circ_{1/2})$nd

174328 — $17[\tfrac{3}{2}]^\circ(1)$
174277 — $16[\tfrac{1}{2}]^\circ(1)$
174216 — $15[\tfrac{3}{2}]^\circ(1)$
174143 — $14[\tfrac{3}{2}]^\circ(1)$
174055 — $13[\tfrac{3}{2}]^\circ(1)$
173946 — $12[\tfrac{3}{2}]^\circ(1)$

173346 — $9[\tfrac{3}{2}]^\circ(1)$

m = multiplicity

KEY

174216 — $16[\tfrac{1}{2}]^\circ(1)$

odd parity
Final j
j-ℓ coupling
n
Energy Level (cm^{-1})

Ionization Level
173 939 cm^{-1}
(21.564 eV)

[Ne I $2p^6$ $^1S_0$ — Ne II $2p^5$ $^2P^\circ_{3/2}$]

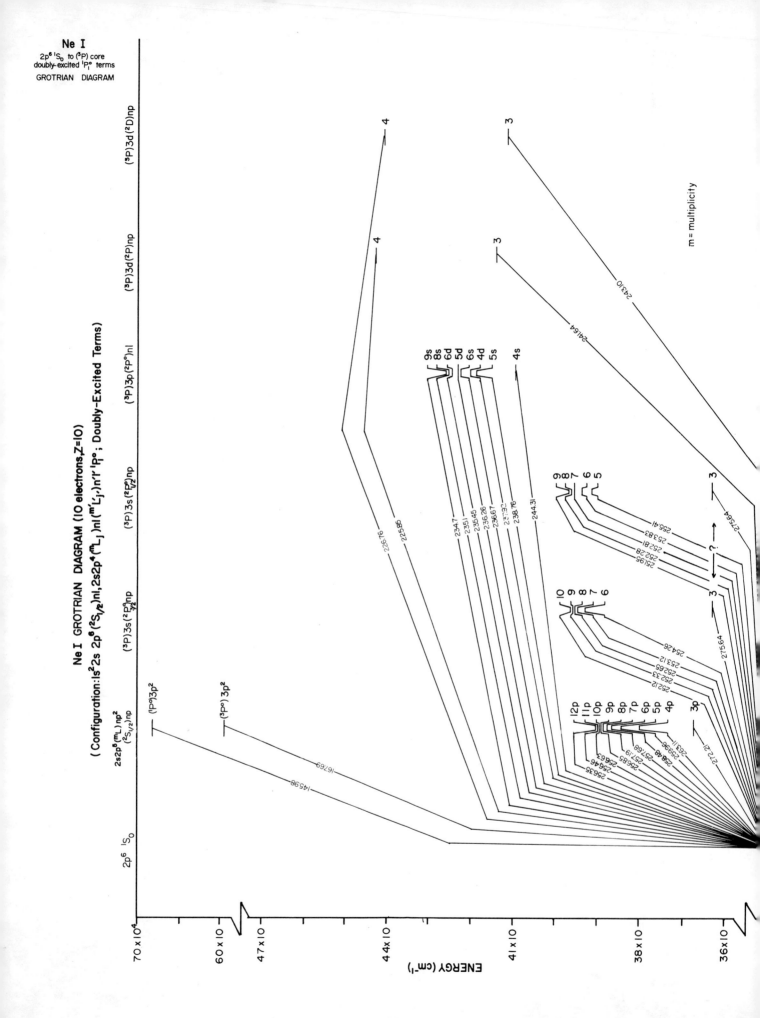

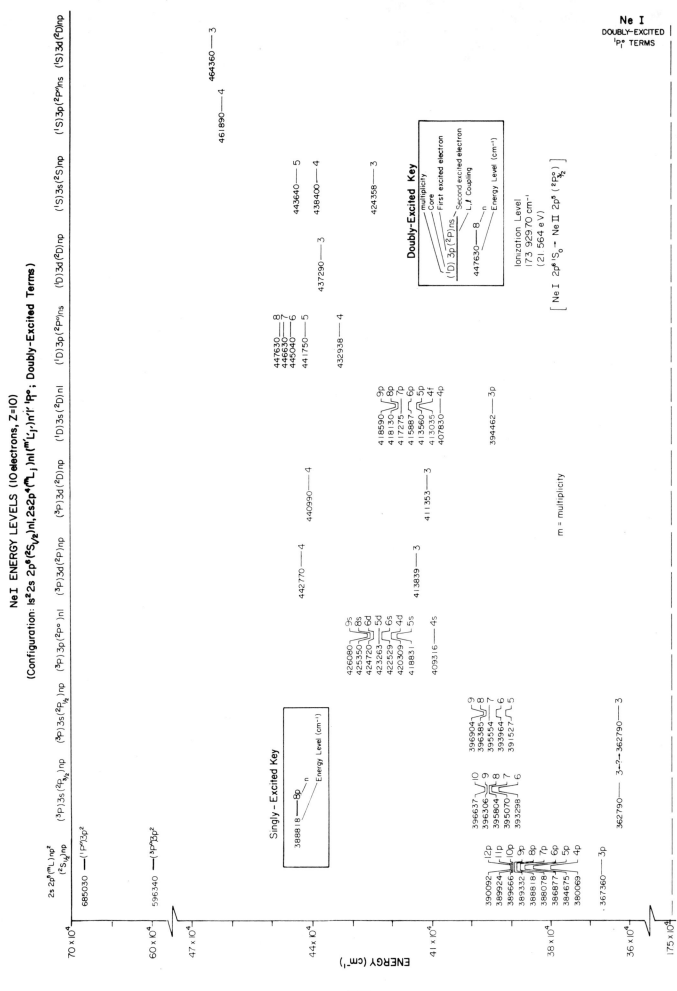

Ne I ENERGY LEVELS (10 electrons, Z=10)

(Configuration: $1s^2 2s\, 2p^6(^2S_{1/2})nl,\ 2s2p^6(^{m}L_J)nl(^{m'}L'_{J'})nl'\,^1P^o;$ Doubly-Excited Terms)

Ne I
DOUBLY-EXCITED
$^1P^o$ TERMS

Doubly-Excited Key

multiplicity
Core
First excited electron
Second excited electron
L,l Coupling
n
Energy Level (cm⁻¹)

$(^1D)\,3p\,(^2P)ns$
$447630 \relbar 8$

Ionization Level
173 92970 cm⁻¹
(21.564 eV)

$[\ \text{Ne I } 2p^6\,^1S_0 \rightarrow \text{Ne II } 2p^5\,(^2P^o)_{3/2}\ ]$

Column headers (left to right):

$2s\,2p^6(^mL)np^2$ $(^2S_{1/2})np$ | $(^3P)3s(^2P_{3/2})np$ | $(^3P)3s(^2P_{1/2})np$ | $(^3P)3p(^2P^o)nl$ | $(^3P)3d(^2P)np$ | $(^3P)3d(^2D)np$ | $(^1D)3s(^2D)nl$ | $(^1D)3s(^2D)nl$ | $(^1D)3p(^2P^o)ns$ | $(^1D)3d(^2D)np$ | $(^1S)3s(^2S)np$ | $(^1S)3p(^2P^o)ns$ | $(^1S)3d(^2D)np$

$685030 \relbar (^1P^o)3p^2$

$596340 \relbar (^3P^o)3p^2$

$461890 \relbar 4$
$464360 \relbar 3$

$447630 \relbar 8$
$446630 \relbar 7$
$445040 \relbar 6$
$441750 \relbar 5$
$432938 \relbar 4$

$443640 \relbar 5$
$438400 \relbar 4$

$437290 \relbar 3$

$424358 \relbar 3$

$442770 \relbar 4$

$440990 \relbar 4$

$418590 \relbar 9p$
$418130 \relbar 8p$
$417275 \relbar 7p$
$415887 \relbar 6p$
$413560 \relbar 5p$
$413035 \relbar 4f$
$407830 \relbar 4p$

$426080 \relbar 9s$
$425350 \relbar 8s$
$424720 \relbar 6d$
$423263 \relbar 5d$
$422529 \relbar 6s$
$420309 \relbar 4d$
$418831 \relbar 5s$

$409316 \relbar 4s$

$413839 \relbar 3$

$411353 \relbar 3$

$394462 \relbar 3p$

Singly-Excited Key

$388818 \relbar 8p$
n
Energy Level (cm⁻¹)

$396904 \relbar 9$
$396385 \relbar 8$
$395554 \relbar 7$
$393964 \relbar 6$
$391527 \relbar 5$

$396637 \relbar 10$
$396306 \relbar 9$
$395804 \relbar 8$
$395070 \relbar 7$
$393298 \relbar 6$

m = multiplicity

$362790 \relbar 3-?-362790 \relbar 3$

$390092 \relbar 12p$
$389924 \relbar 11p$
$389666 \relbar 10p$
$389332 \relbar 9p$
$388818 \relbar 8p$
$388078 \relbar 7p$
$386877 \relbar 6p$
$384675 \relbar 5p$
$380069 \relbar 4p$

$367360 \relbar 3p$

ENERGY (cm⁻¹) axis (vertical):
70×10^4
60×10^4
47×10^4
44×10^4
41×10^4
38×10^4
36×10^4
175×10^4

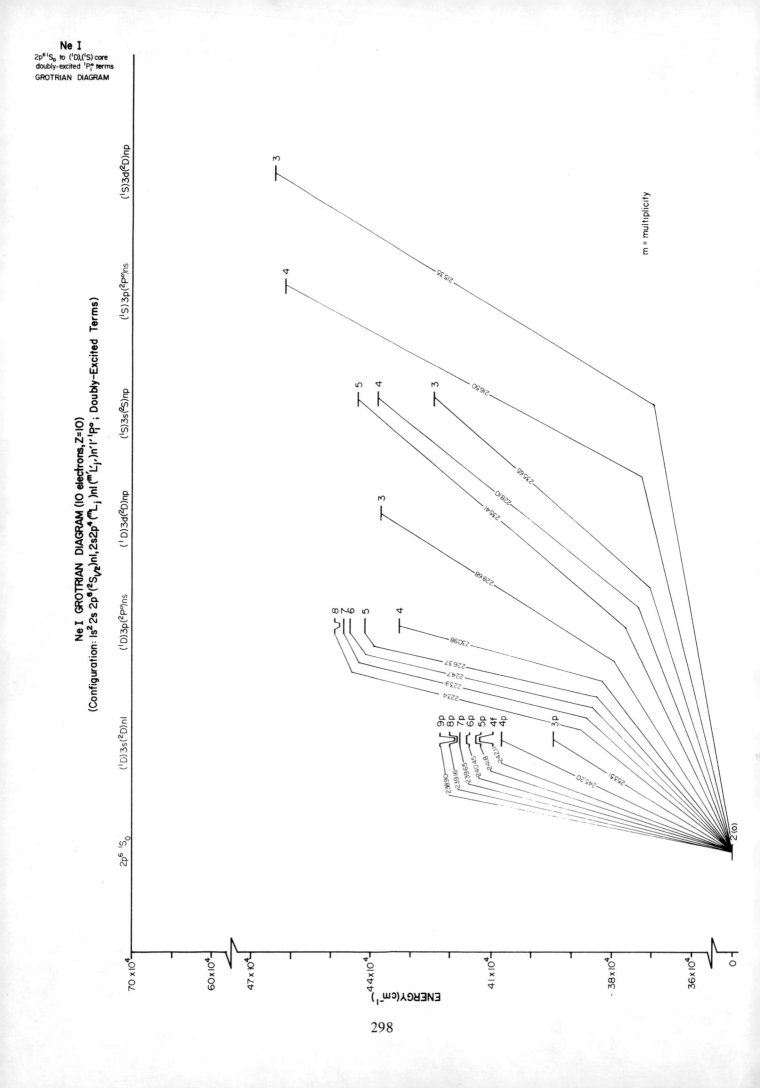

NeI ENERGY LEVELS (10 electrons, Z=10)

(Configuration: $1s^2 2s\, 2p^6(^2S_{1/2})nl, 2s2p^4(^{m'}L'_{J'})nl\,(^mL_J)nl(^{m'}L'_{J'})nl'l'\, ^1P^\circ$; Doubly-Excited Terms)

Ne I
DOUBLY–EXCITED
$^1P^\circ_1$ TERMS

Column headings (left to right):

$2s\,2p^6(^mL_J)np^2$ / $(^2S_{1/2})np$ | $(^3P)3s(^2P_{3/2})np$ | $(^3P)3s(^2P_{1/2})np$ | $(^3P)3p(^2P^\circ)nl$ | $(^3P)3d(^2P)np$ | $(^3P)3d(^2D)np$ | $(^1D)3s(^2D)nl$ | $(^1D)3p(^2P^\circ)ns$ | $(^1D)3d(^2D)np$ | $(^1S)3s(^2S)np$ | $(^1S)3p(^2P^\circ)ns$ | $(^1S)3d(^2D)np$

Singly – Excited Key

388818 —— 8p
n
Energy Level (cm⁻¹)

Doubly-Excited Key

multiplicity
Core
First excited electron
Second excited electron
L, ℓ Coupling
$(^1D)\,3p\,(^2P^\circ)ns$
447630 —— 8
n
Energy Level (cm⁻¹)

Ionization Level
173 929 70 cm⁻¹
(21.564 eV)

[Ne I $2p^6\ ^1S_0 \rightarrow$ Ne II $2p^5\ (^2P^\circ_{3/2})$]

m = multiplicity

Data values:

685030 —— $(^1P^\circ)3p^2$

596340 —— $(^3P)3p^2$

442770 —— 4

440990 —— 4

464360 —— 3
461890 —— 4

443640 —— 5
438400 —— 4

424358 —— 3

437290 —— 3

447630 —— 8
446630 —— 7
445040 —— 6
441750 —— 5

432938 —— 4

418590 —— 9p
418130 —— 8p
417275 —— 7p
415887 —— 6p
413560 —— 5p
413035 —— 4f
407830 —— 4p

394462 —— 3p

413839 —— 3

411353 —— 3

426080 —— 9s
425350 —— 8s
424720 —— 6d
423263 —— 5d
422529 —— 6s
420309 —— 4d
418831 —— 5s

409316 —— 4s

396904 —— 9
396385 —— 8
395554 —— 7
393964 —— 6
391527 —— 5

396637 —— 10
396306 —— 9
395804 —— 8
395070 —— 7
393298 —— 6

362790 —— 3 —?—> 362790 —— 3

390092 —— 12p
389924 —— 11p
389666 —— 10p
389332 —— 9p
388818 —— 8p
388078 —— 7p
386877 —— 6p
384675 —— 5p
380069 —— 4p

367360 —— 3p

ENERGY (cm⁻¹)

70 × 10⁴
60 × 10⁴
47 × 10⁴
44 × 10⁴
41 × 10⁴
38 × 10⁴
36 × 10⁴
175 × 10⁴

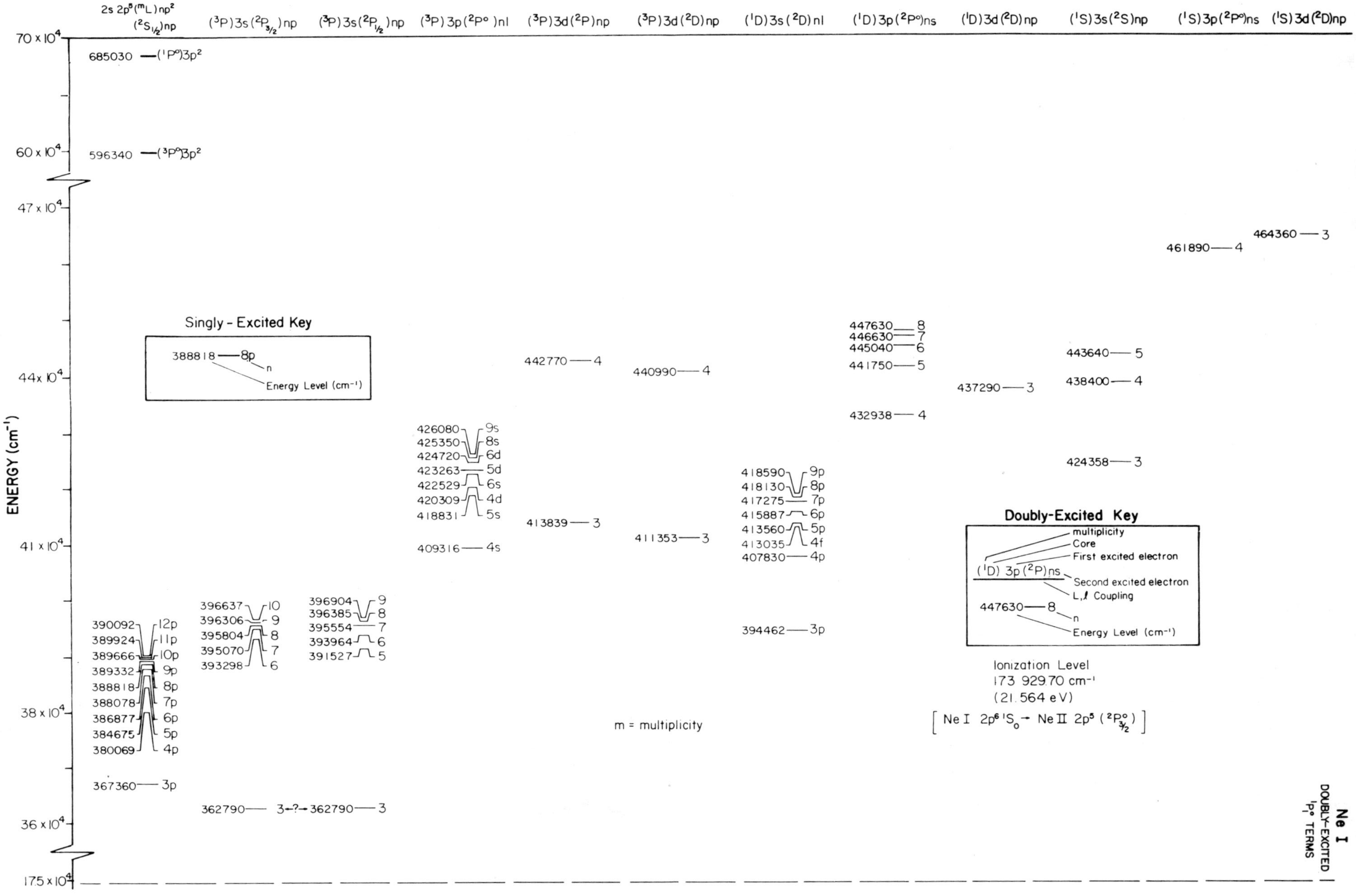

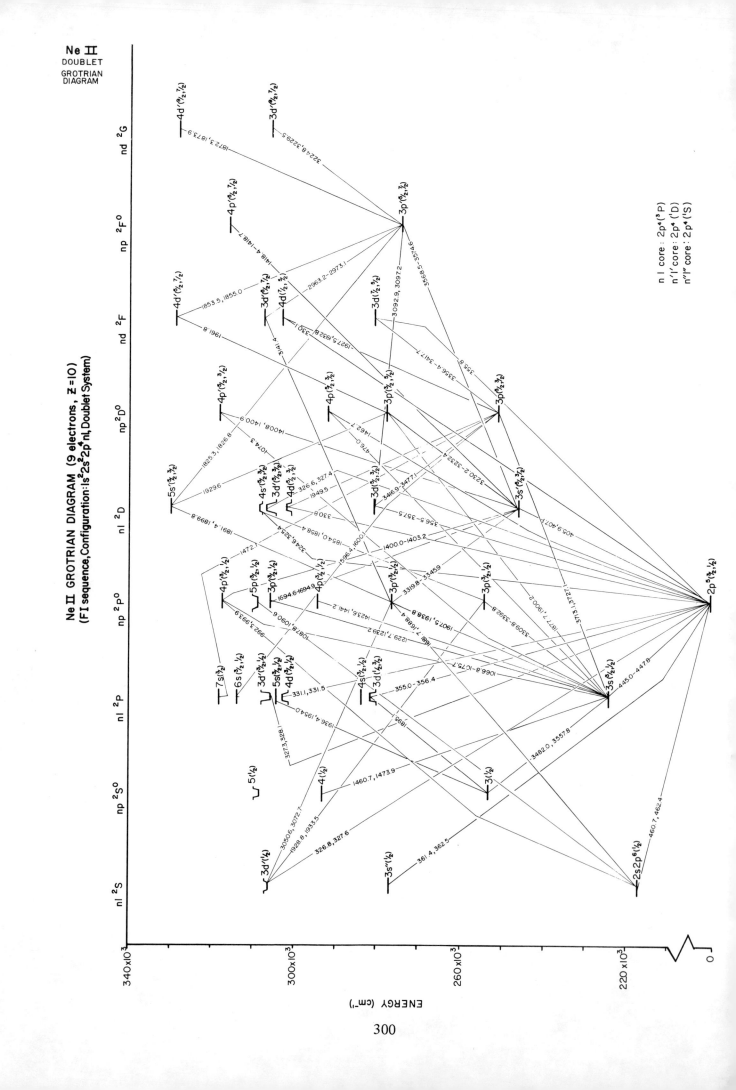

Ne II
DOUBLET
GROTRIAN
DIAGRAM

Ne II GROTRIAN DIAGRAM (9 electrons, Z = IO)
(FI sequence, Configuration: 1s²2s²2p⁴nl, Doublet System)

n l core : 2p⁴(³P)
n'l' core: 2p⁴(¹D)
n"l" core : 2p⁴(¹S)

ENERGY (cm⁻¹)

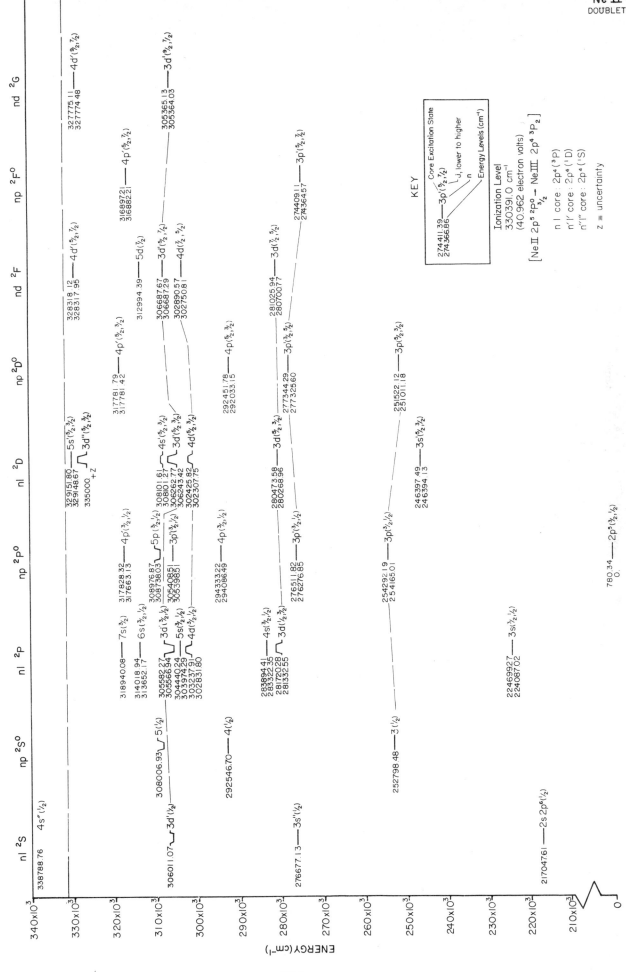

Ne II
DOUBLET

Ne II ENERGY LEVELS (9 electrons, Z=10)
(F I sequence, Configuration: 1s² 2s² 2p⁴ nl, Doublet System)

301

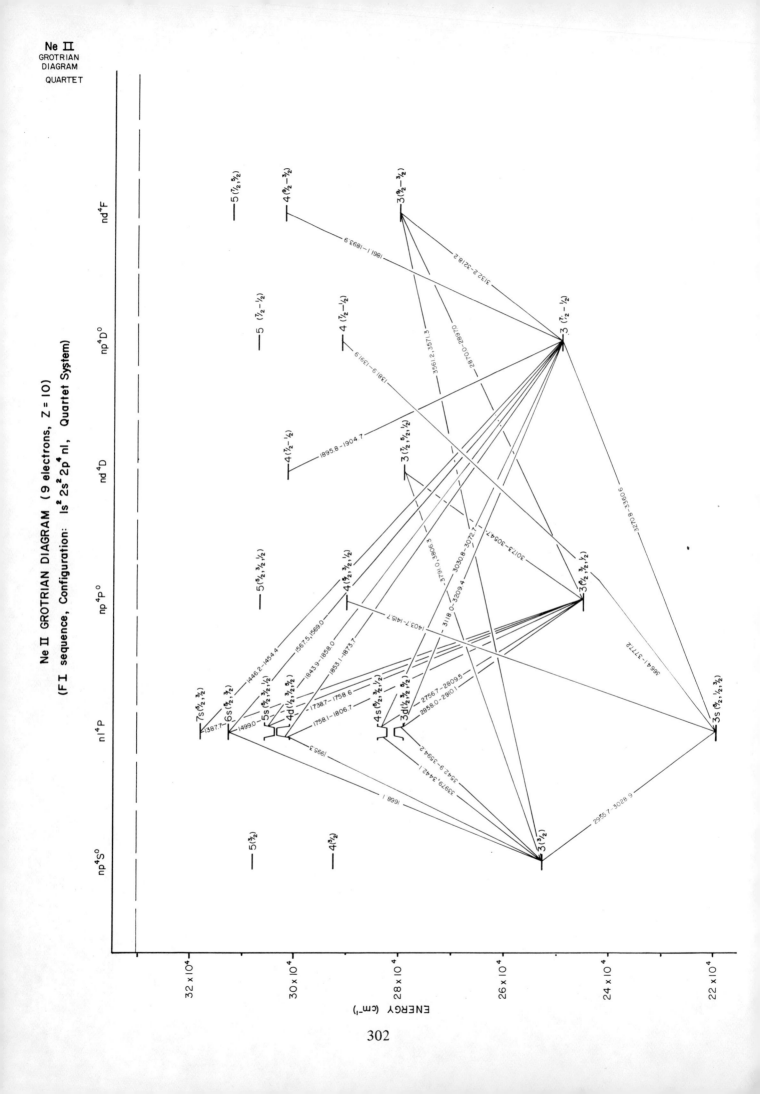

Ne II
GROTRIAN
DIAGRAM
QUARTET

Ne II GROTRIAN DIAGRAM (9 electrons, Z = 10)
(FI sequence, Configuration: ls² 2s² 2p⁴ nl, Quartet System)

Ne II ENERGY LEVELS (9 electrons, Z = 10)

(F I sequence, Configuration: 1s²2s²2p⁴nl, Quartet System)

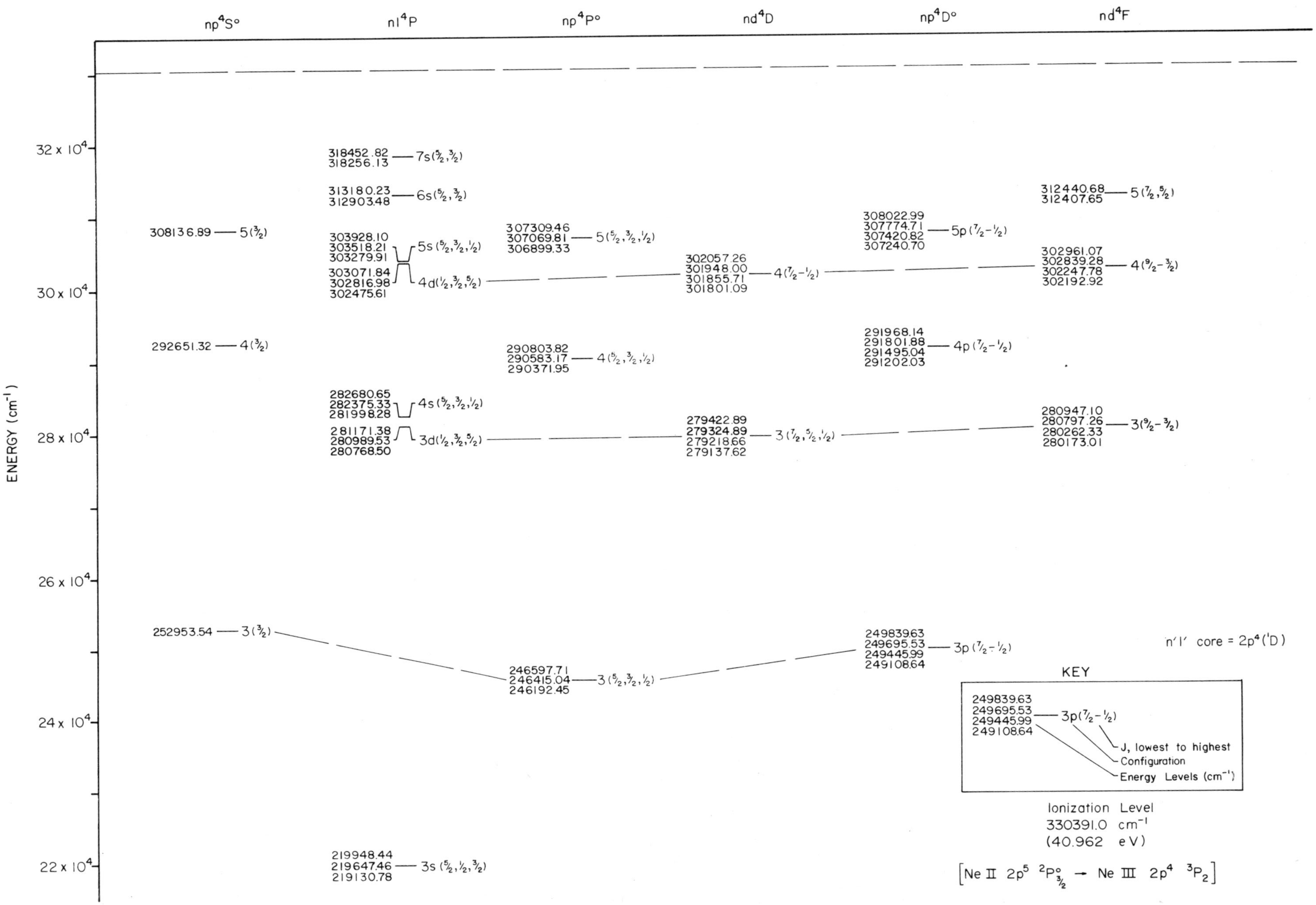

Ne II

QUARTET

ENERGY (cm⁻¹)

303

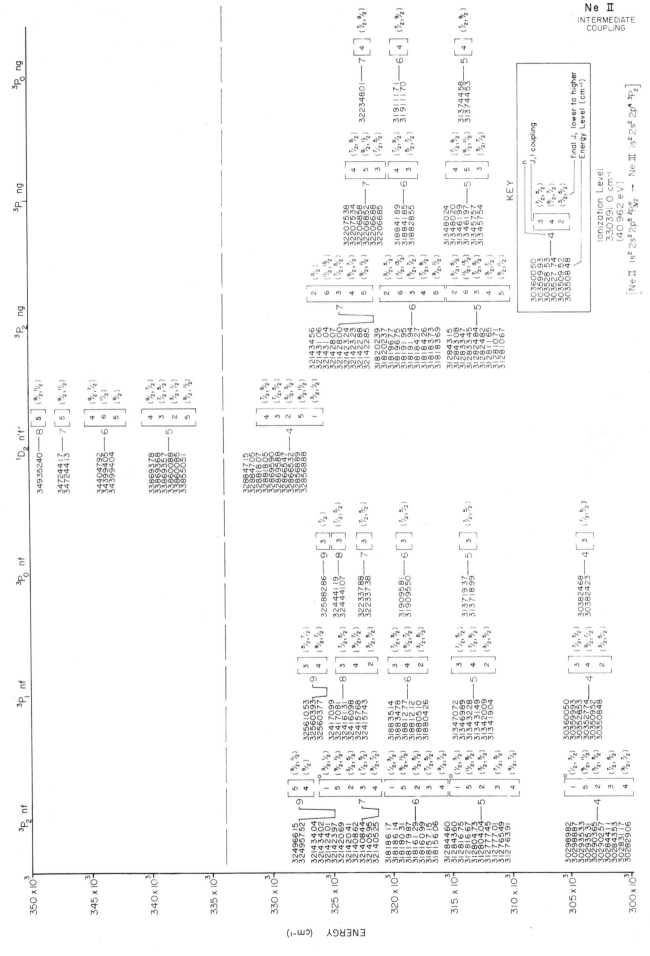

Ne II INTERMEDIATE COUPLING (9 electrons, Z=10)

F I sequence, Configuration $1s^2 2s^2 2p^4$ $(^3P)nl$, $1s^2 2s^2 2p^4$ $(^1D)n'l'$ $[J,l](J)$

Ne II
INTERMEDIATE
COUPLING

305

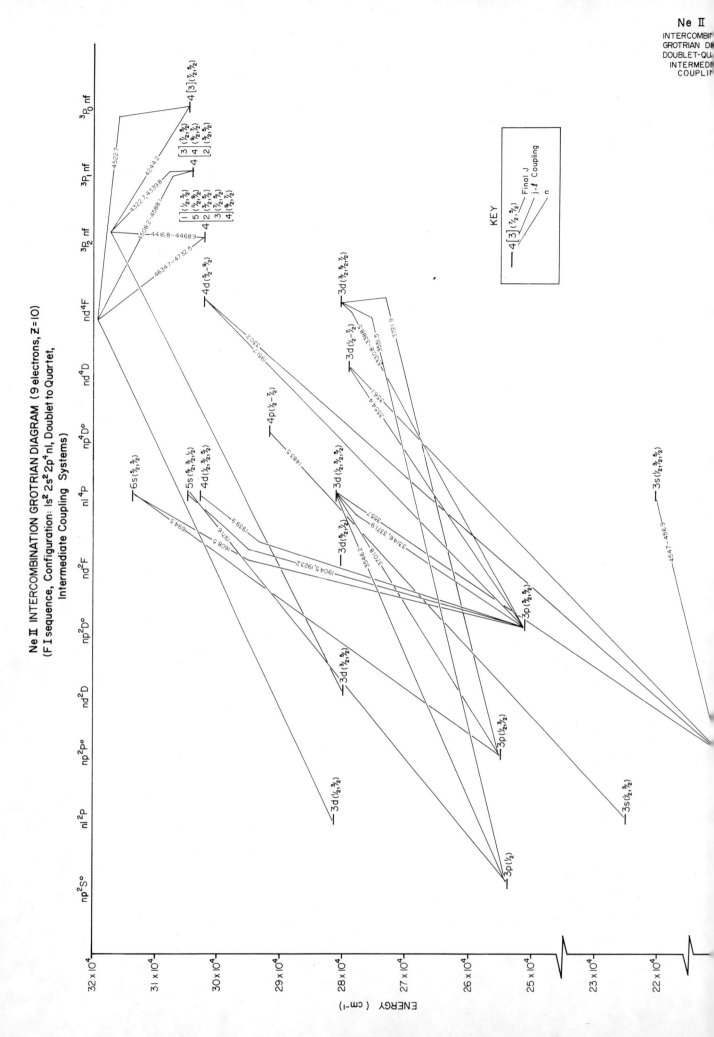

Ne II INTERCOMBINATION GROTRIAN DIAGRAM (9 electrons, Z̄=10)
(F I sequence, Configuration: Is² 2s² 2p⁴ nl, Doublet to Quartet, Intermediate Coupling Systems)

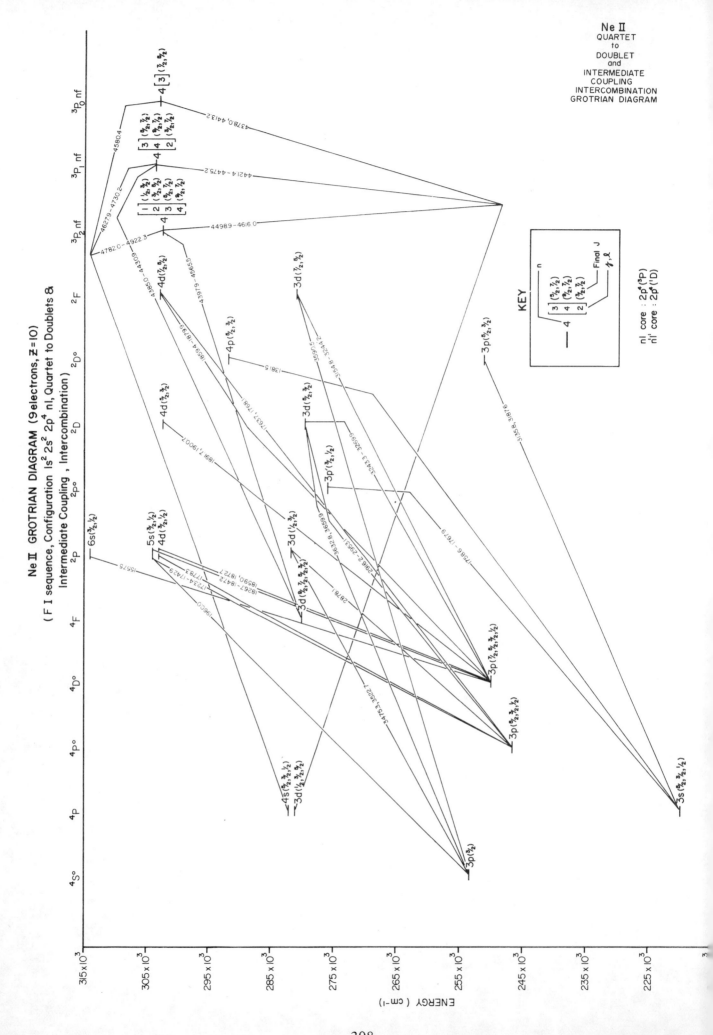

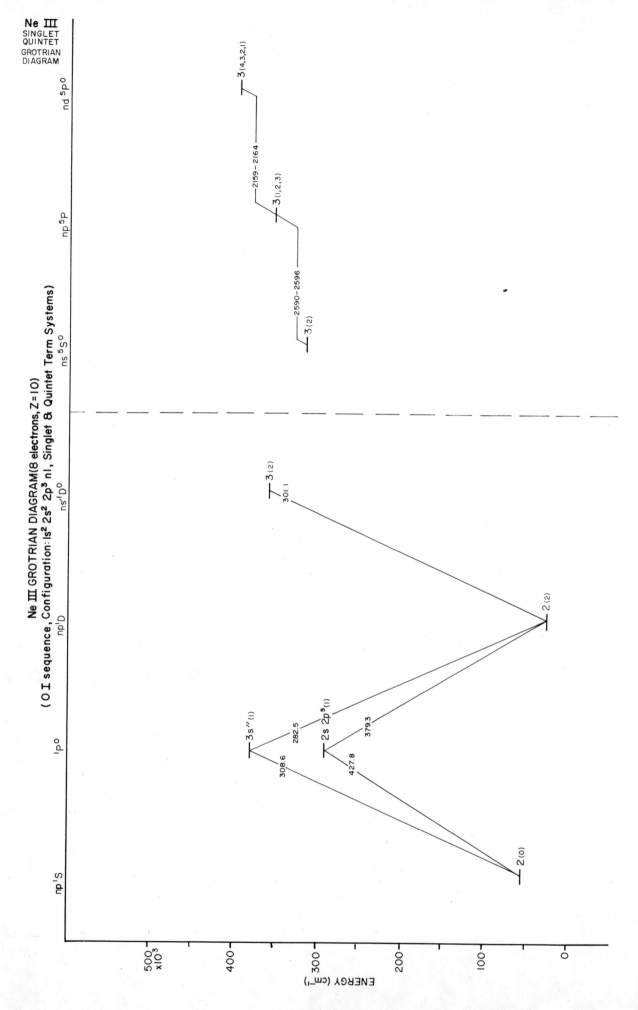

Ne III
SINGLET
QUINTET
GROTRIAN
DIAGRAM

Ne III GROTRIAN DIAGRAM (8 electrons, Z=10)

(O I sequence, Configuration: 1s² 2s² 2p³ nl, Singlet & Quintet Term Systems)

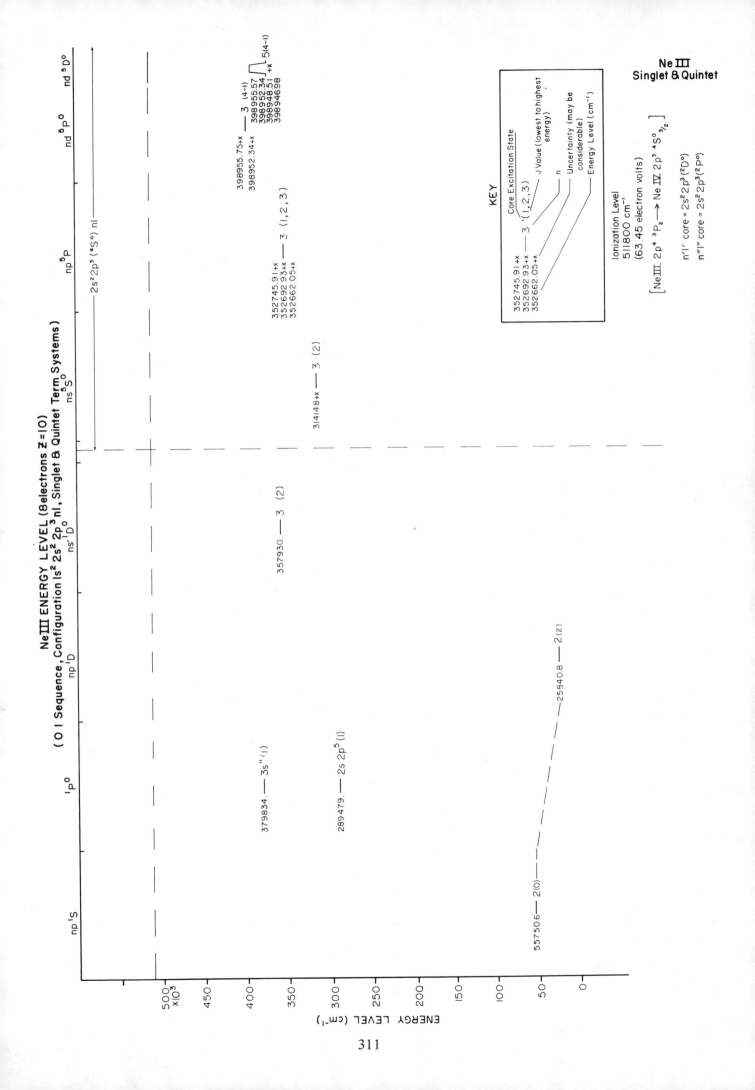

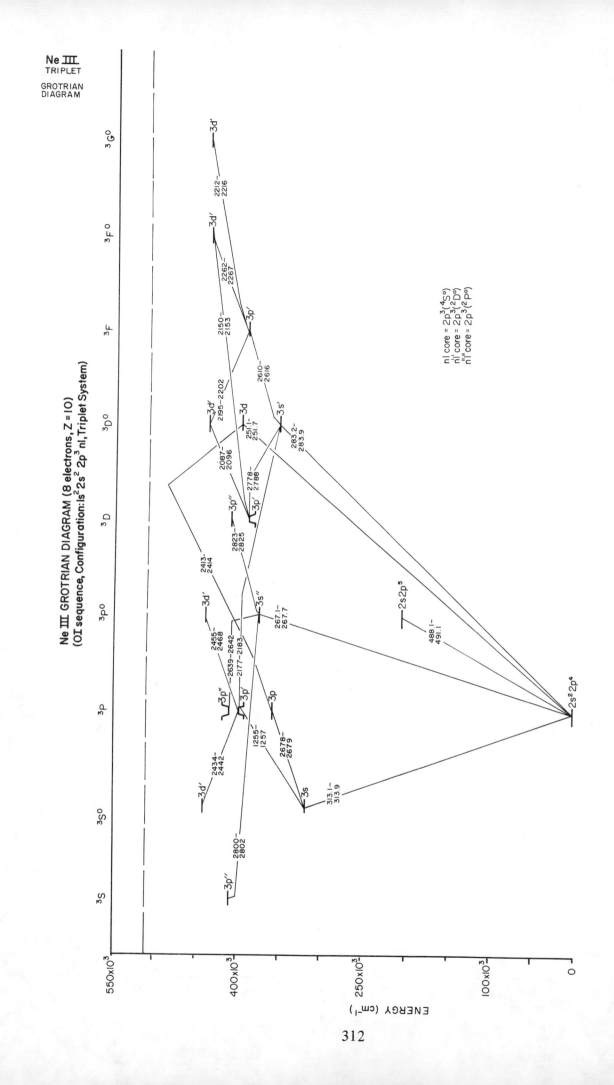

Ne III
TRIPLET

GROTRIAN
DIAGRAM

Ne III GROTRIAN DIAGRAM (8 electrons, Z =10)
(OI sequence, Configuration: $1s^2 2s^2 2p^3 nl$, Triplet System)

nl core = $2p^3(^4S°)$
n'l' core = $2p^3(^2D°)$
n''l'' core = $2p^3(^2P°)$

ENERGY (cm^{-1})

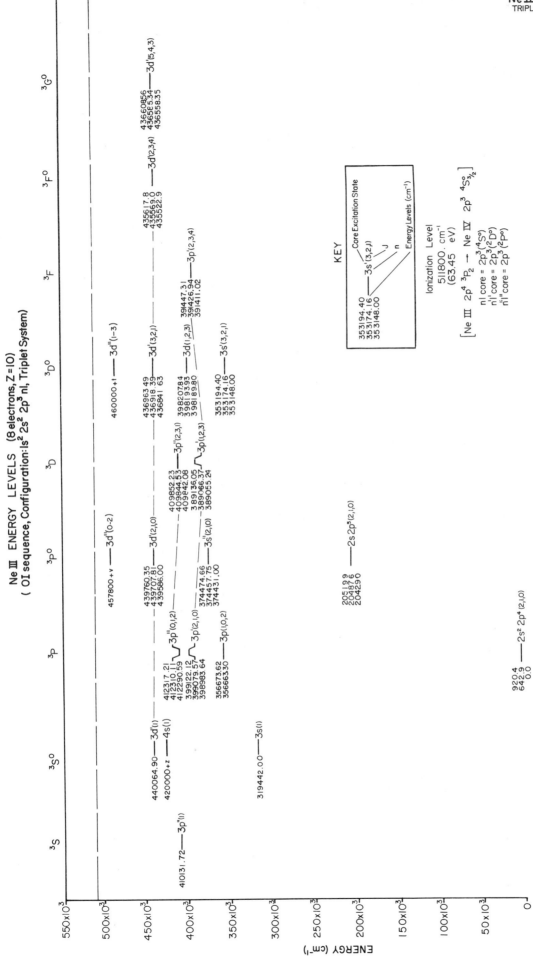

Ne III ENERGY LEVELS (8 electrons, Z=10)
(OI sequence, Configuration: 1s² 2s² 2p³ nl, Triplet System)

Ne III
TRIPLET

313

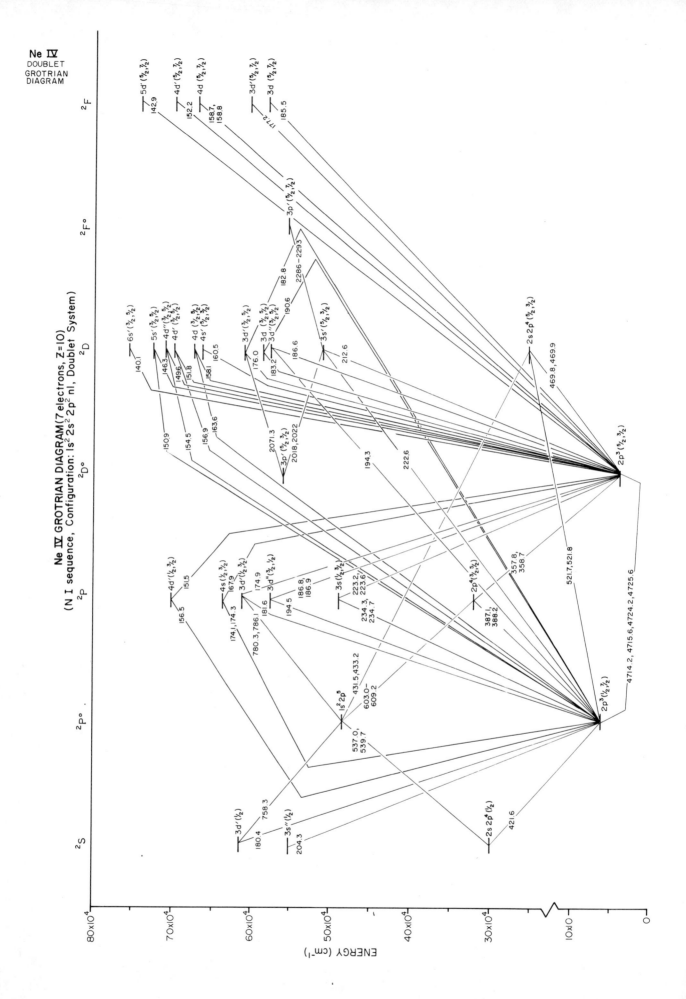

Ne IV
DOUBLET
GROTRIAN
DIAGRAM

Ne IV GROTRIAN DIAGRAM (7 electrons, Z=10)
(N I sequence, Configuration: 1s² 2s² 2p² nl, Doublet System)

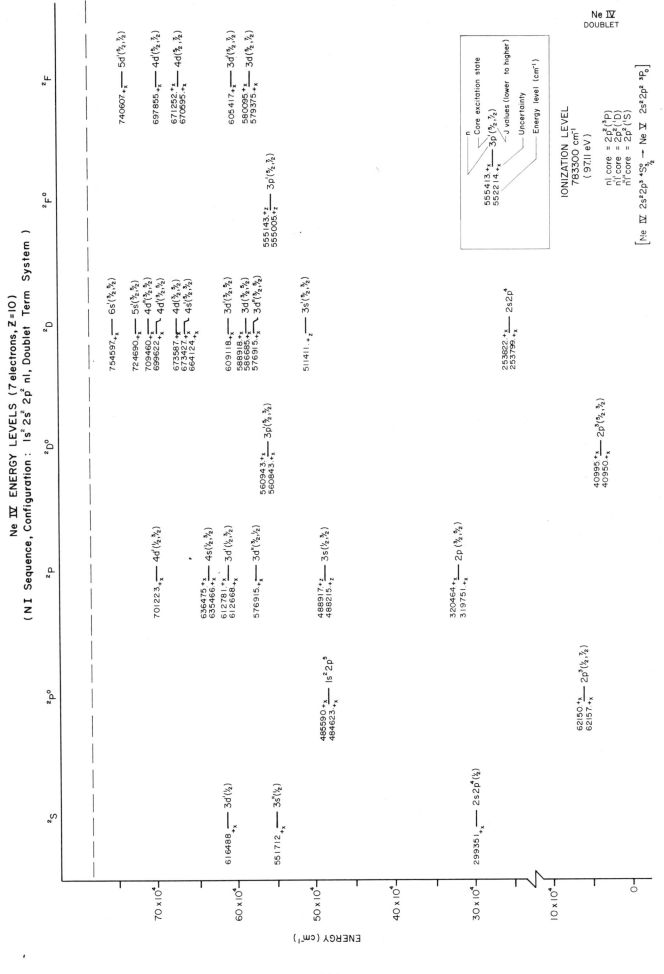

Ne IV ENERGY LEVELS (7 electrons, Z=10)
(N I Sequence, Configuration : 1s² 2s² 2p² nl, Doublet Term System)

ENERGY (cm⁻¹)

315

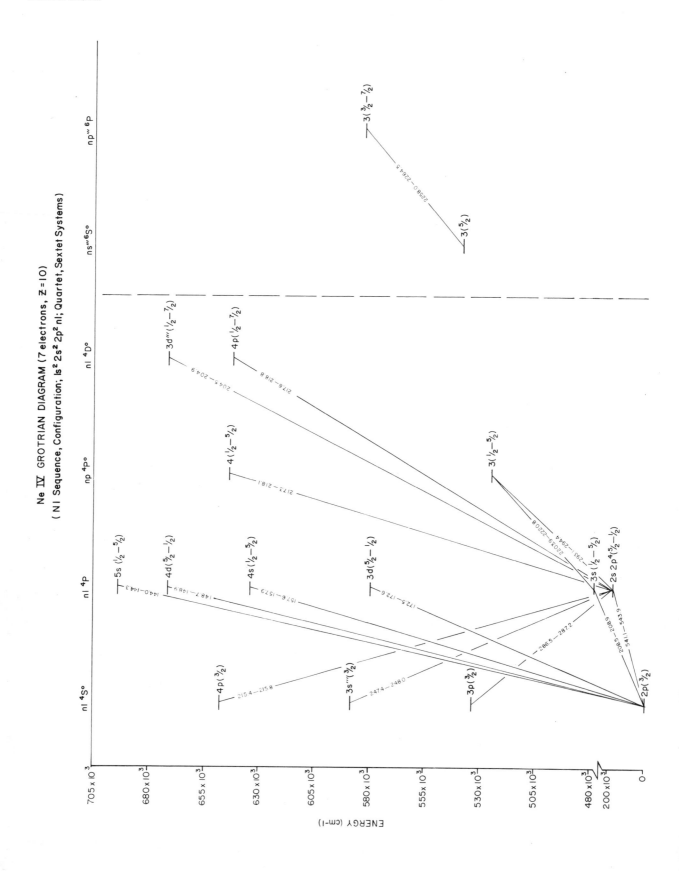

Ne IV
QUARTET
SEXTET
GROTRIAN DIAGRAM

Ne IV GROTRIAN DIAGRAM (7 electrons, Z=10)
(NI Sequence, Configuration; 1s² 2s² 2p² nl; Quartet, Sextet Systems)

316

Ne IV ENERGY LEVELS (7 electrons, Z = 10)

(NI sequence, Configuration; 1s² 2s² 2p² nl; Quartet, Sextet Systems)

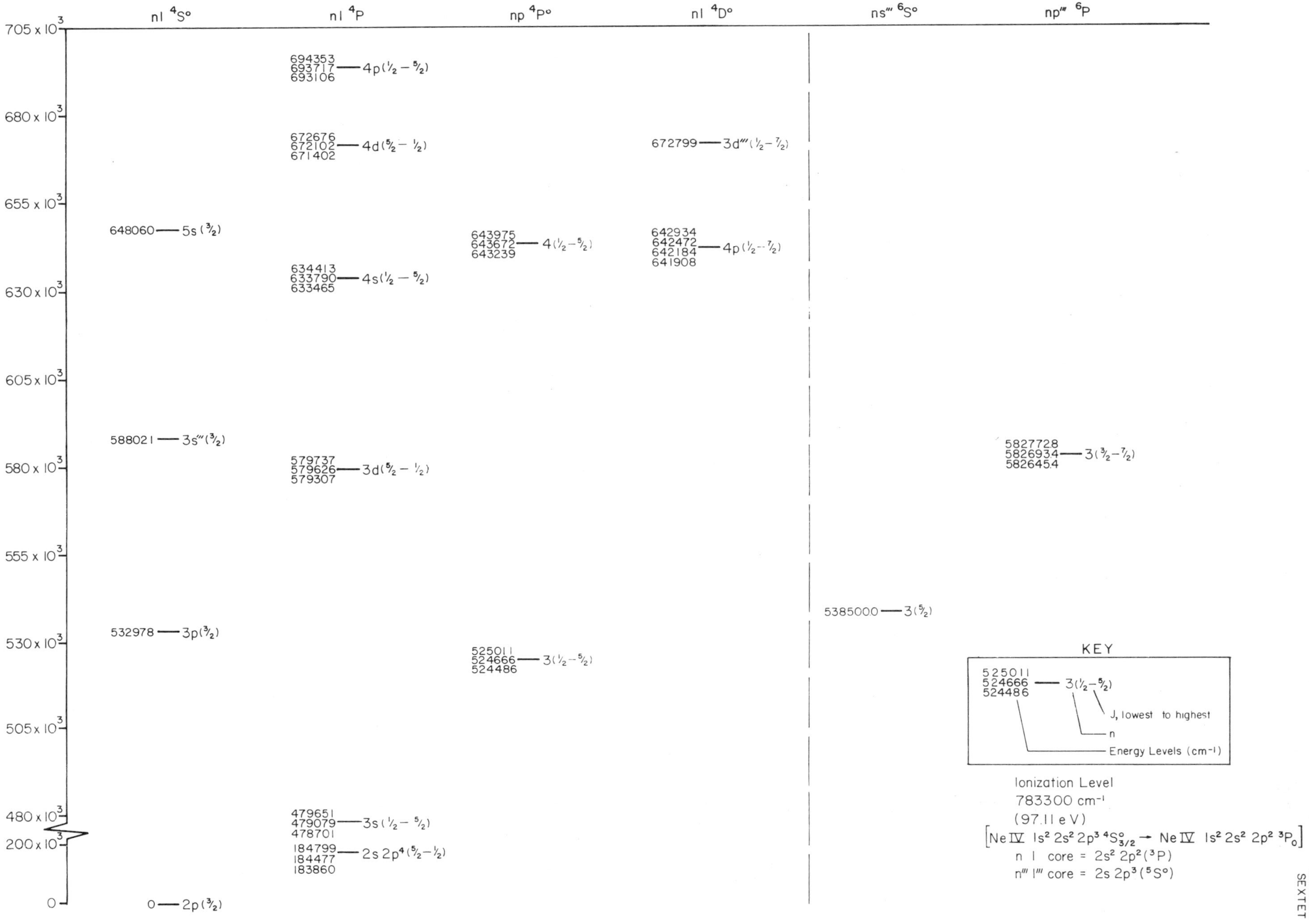

ENERGY (cm⁻¹)

317

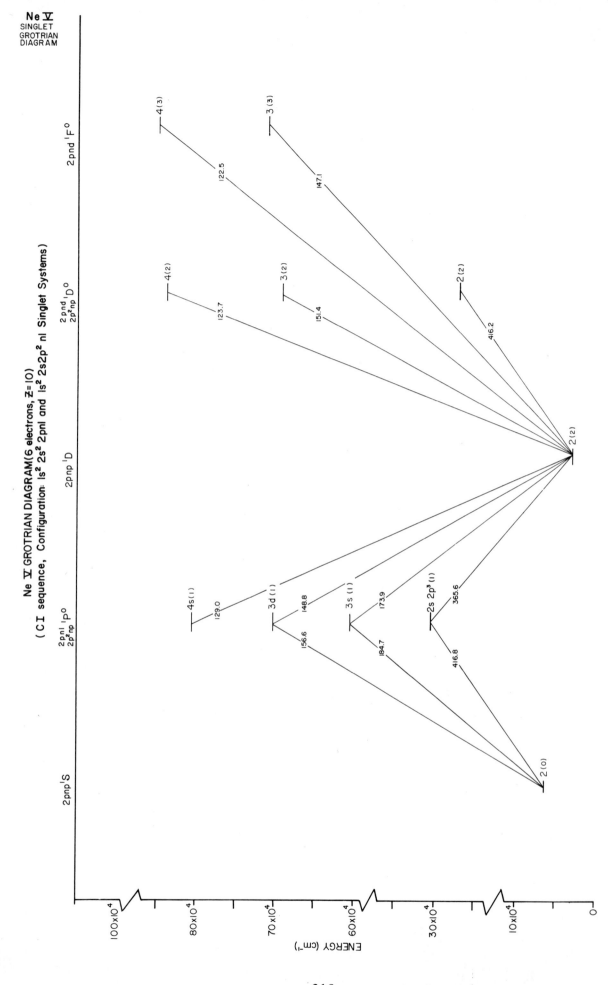

Ne V
SINGLET
GROTRIAN
DIAGRAM

Ne V GROTRIAN DIAGRAM (6 electrons, Z=10)
(CI sequence, Configuration: $1s^2\,2s^2\,2pnl$ and $1s^2\,2s2p^2\,nl$ Singlet Systems)

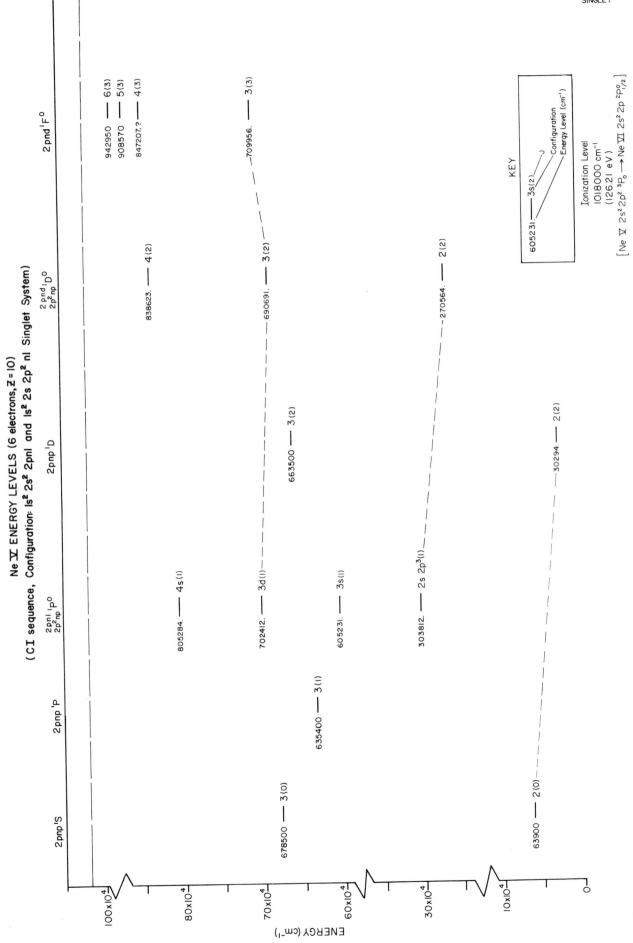

Ne V ENERGY LEVELS (6 electrons, Z = 10)

(CI sequence, Configuration: 1s² 2s² 2pnl and 1s² 2s 2p² nl Singlet System)

ENERGY (cm⁻¹)

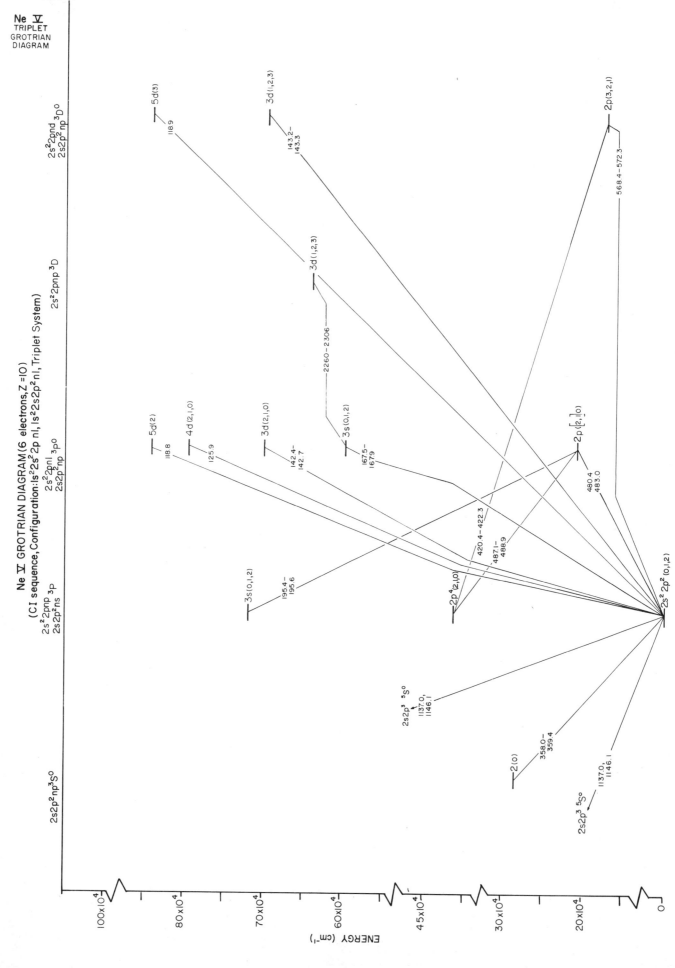

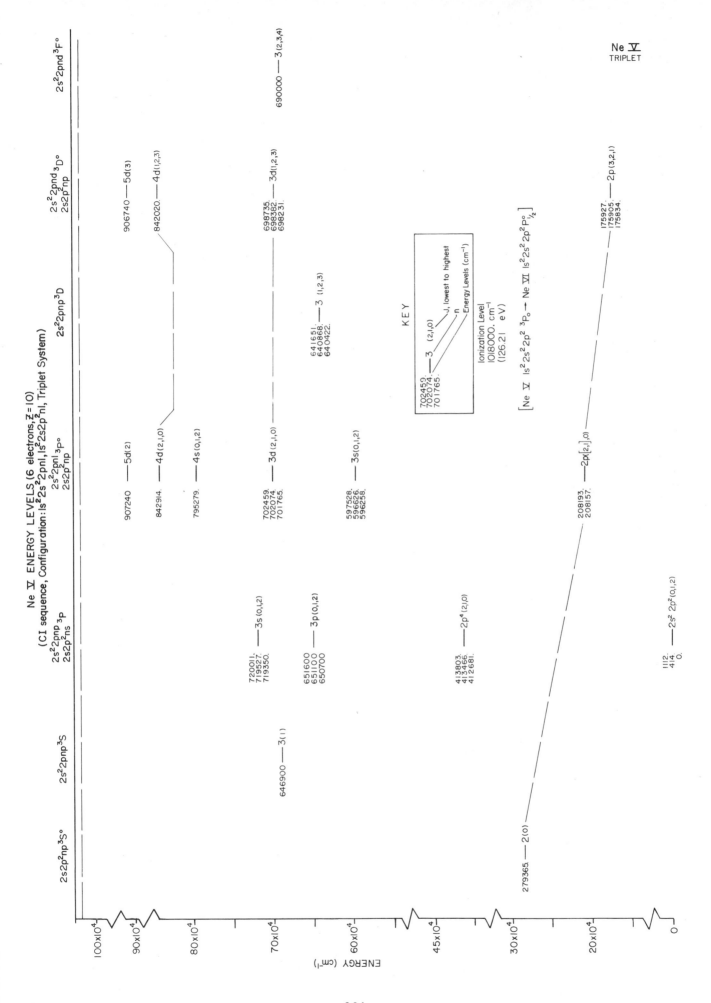

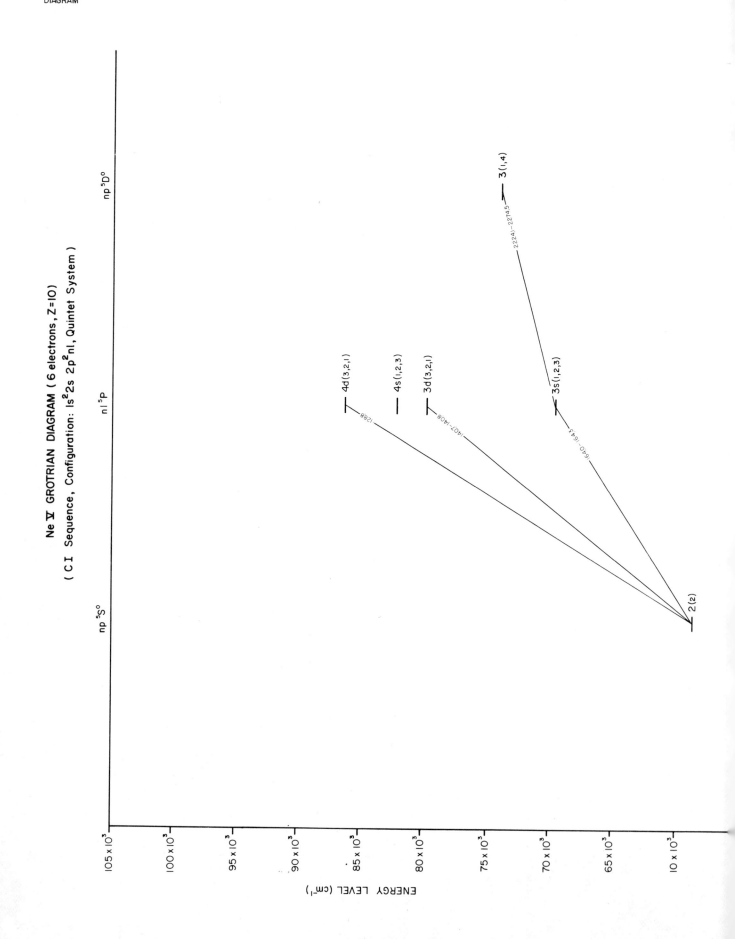

Ne Ⅴ
QUINTET
GROTRIAN
DIAGRAM

Ne Ⅴ GROTRIAN DIAGRAM (6 electrons, Z=10)
(C I Sequence, Configuration: 1s²2s 2p²nl, Quintet System)

np ⁵S°   nl ⁵P   np ⁵D°

4d(3,2,1)
4s(1,2,3)
3d(3,2,1)
3s(1,2,3)
3(1,4)

2(2)

ENERGY LEVEL (cm⁻¹)

105 x 10³
100 x 10³
95 x 10³
90 x 10³
85 x 10³
80 x 10³
75 x 10³
70 x 10³
65 x 10³
10 x 10³

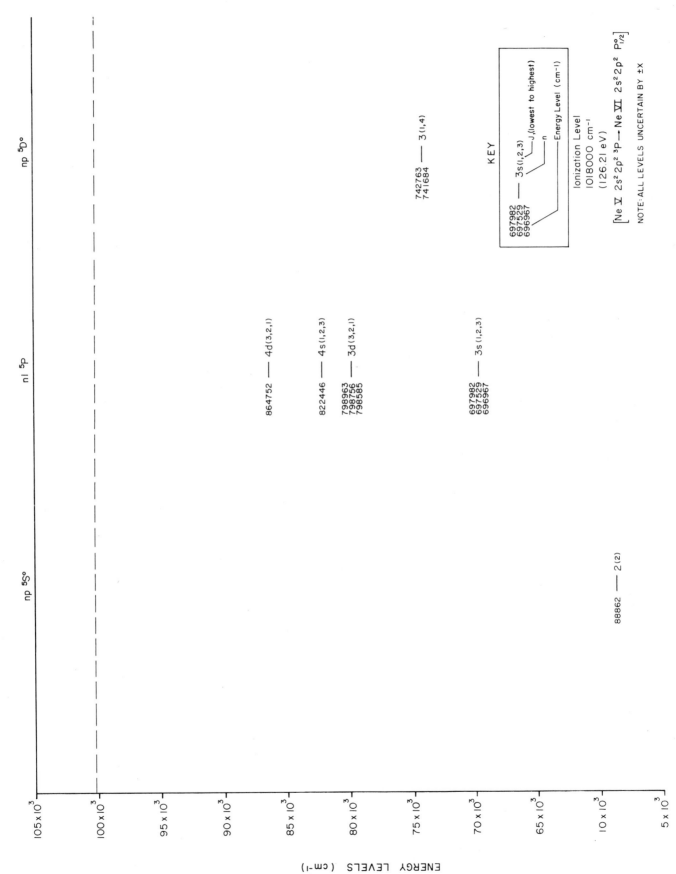

Ne V
QUINTET

Ne V   ENERGY LEVEL   (6 electrons, Z=10)
(CI sequence, Configuration: $1s^2\,2s\,2p^2\,nl$, Quintet System)

np $^5S°$ nl ^5P np $^5D°$

88862 —— 2(2)

697982
697529 —— 3s(1,2,3)
696967

798963
798756 —— 3d(3,2,1)
798585

822446 —— 4s(1,2,3)

864752 —— 4d(3,2,1)

742763 —— 3(1,4)
741684

KEY

697982
697529 —— 3s(1,2,3)
696967
          └ J,(lowest to highest)
        └ n
        └ 3s(1,2,3)
                └ Energy Level (cm⁻¹)

Ionization Level
1018000 cm⁻¹
(126.21 eV)
[Ne V $2s^2\,2p^2$ ^3P → Ne VI $2s^2\,2p^2$ $P°_{1/2}$]

NOTE: ALL LEVELS UNCERTAIN BY ±X

ENERGY LEVELS (cm⁻¹)

105 × 10³
100 × 10³
95 × 10³
90 × 10³
85 × 10³
80 × 10³
75 × 10³
70 × 10³
65 × 10³
10 × 10³
5 × 10³

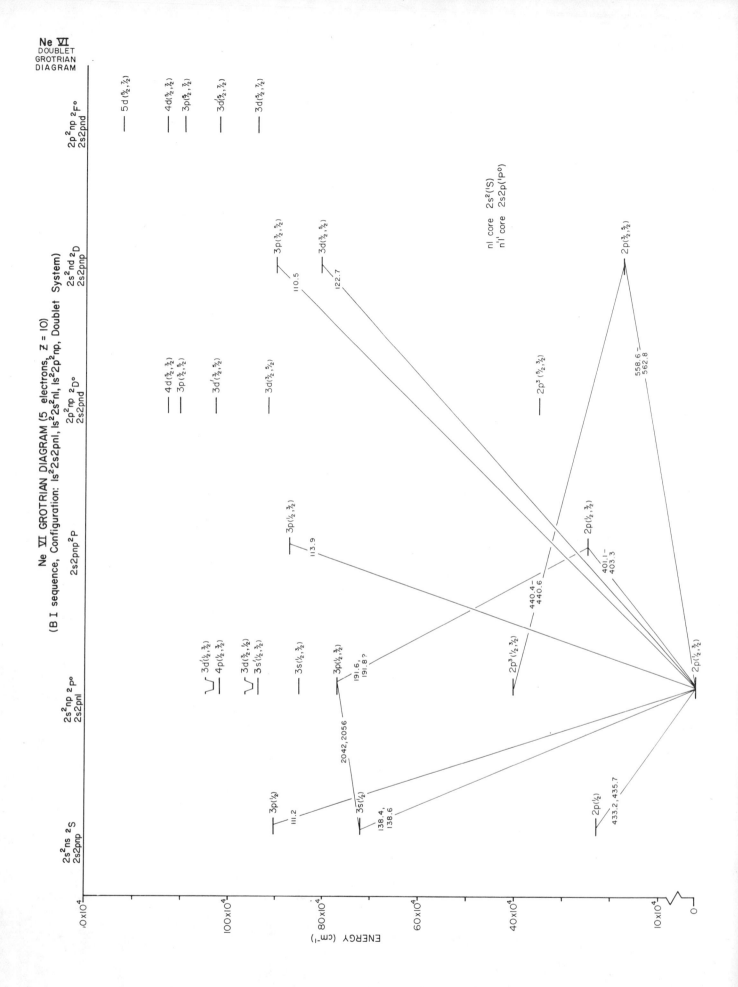

Ne VI GROTRIAN DIAGRAM (5 electrons, Z = 10)
(B I sequence, Configuration: 1s²2s²2pnl, 1s²2s²nl, 1s²2p²np, Doublet System)

Ne VI
DOUBLET
GROTRIAN
DIAGRAM

ENERGY (cm⁻¹)

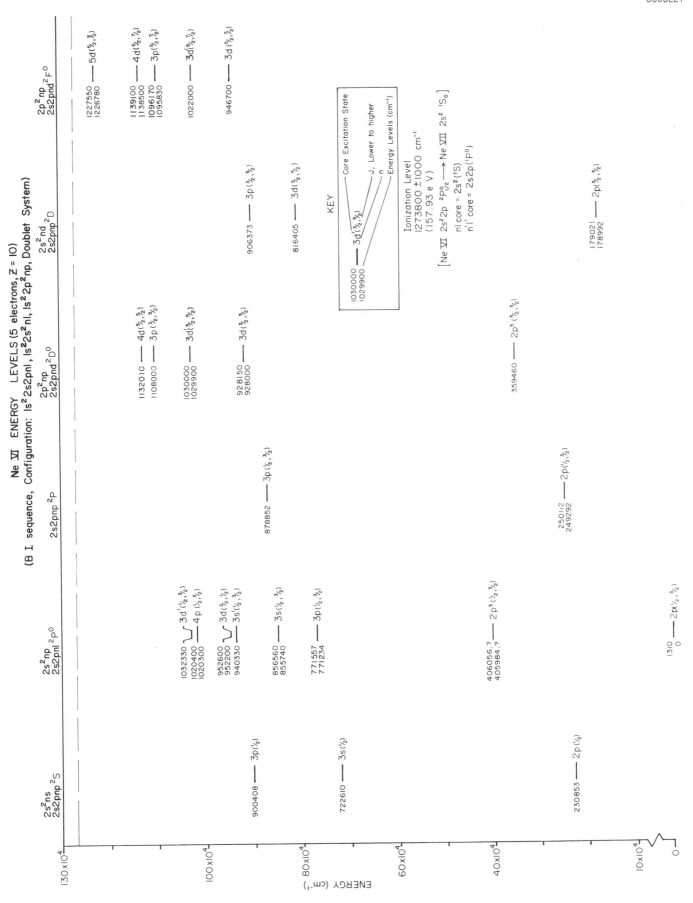

325

Ne VI
QUARTET
GROTRIAN
DIAGRAM

Ne VI GROTRIAN DIAGRAM (5 electrons, Z=10)
(B I Sequence Configurations: $1s^2 2s 2p(^3P^o)nl$, $1s^2 2p^2(^3P)nl$, Quartet System)

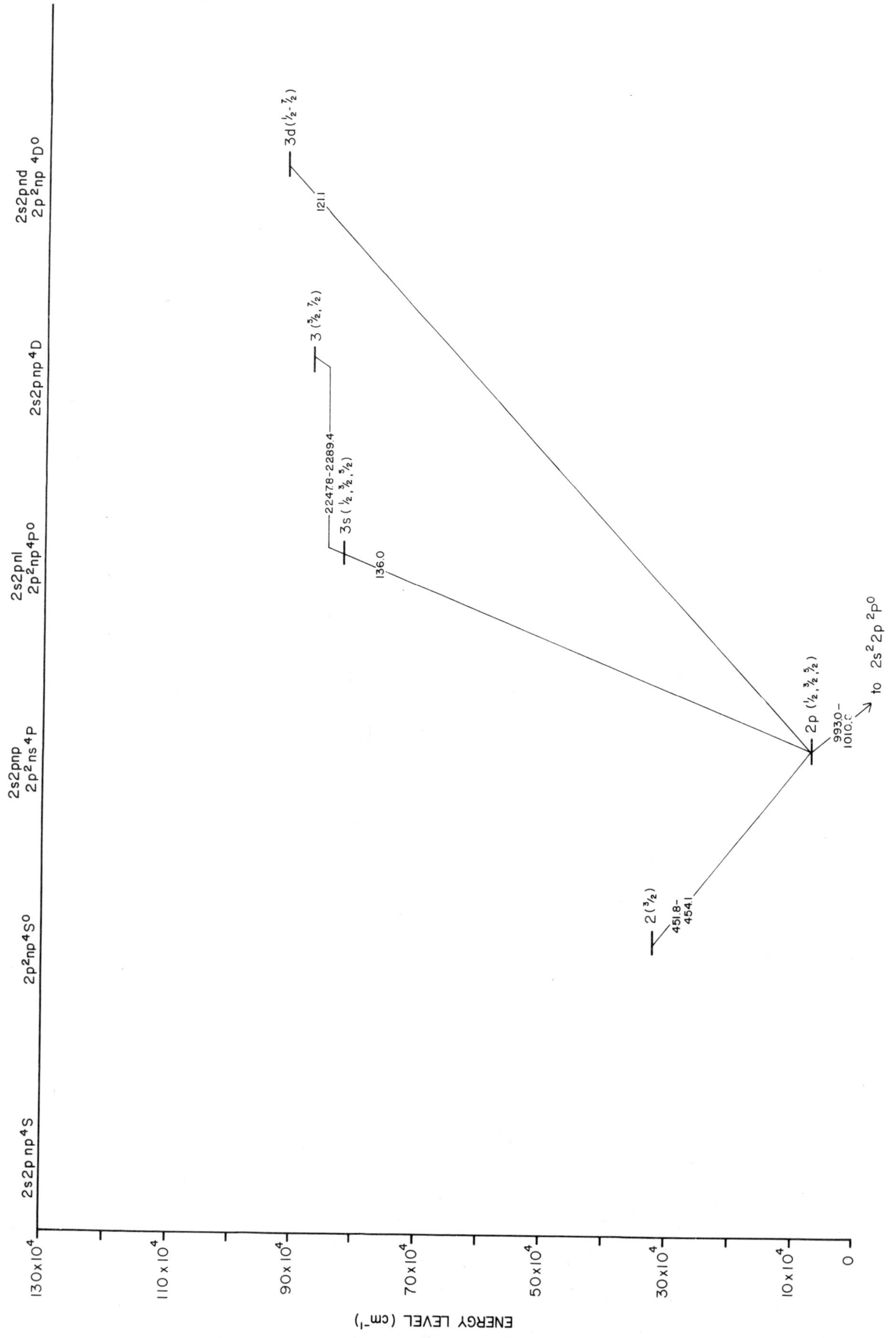

Ne VI ENERGY LEVELS (5 electrons, Z=10)
(B I Sequence Configurations: 1s²2s2p(³P°)nl, 1s²2p²(³P)nl, Quartet System)

2s2pnd
2p²np ⁴D°

1224180 —— 5d (½-⁷/₂)

1131100
1130900 —— 4d (½-⁷/₂)
1130650
1130600

1062050 —— 3p (½-⁷/₂)

926577 —— 3d (½-⁷/₂)

2s2pnp ⁴D

880740+x
880070+x —— 3 (⁵/₂, ⁷/₂)

2s2pnl
2p²np ⁴P°

1132800
1132700 —— 4d (⁵/₂, ³/₂, ½)
1132600

1066140
1065570 —— 3p (½, ³/₂, ⁵/₂)
1065200

932150+u
931900+u —— 3d (⁵/₂, ½, ½)
931600+u

834835+y
834050+y —— 3s (½, ³/₂, ⁵/₂)
833573+y

2s2pnp
2p²ns ⁴P

1023230 —— 3s (½, ³/₂, ⁵/₂)

101347
100704 —— 2p (½, ³/₂, ⁵/₂)
100261

2p²np ⁴S°

1083040 —— 3 (³/₂)

321579 —— 2 (³/₂)

2s2pnp ⁴S

888000 —— 3p (³/₂)
+x

KEY

Configuration
J values, lowest to highest energy
834835+y ——— 3s (½, ³/₂, ⁵/₂)
834050+y
833573+y
Uncertainty
x = 1000 (cm⁻¹)
Energy Level (cm⁻¹)

Ionization level
1273800 cm⁻¹
157.93 eV

[Ne VI 2s²2p ²P°½ → Ne VII 2s² ¹S₀]

ENERGY LEVEL (cm⁻¹)

130×10⁴
110×10⁴
90×10⁴
70×10⁴
50×10⁴
30×10⁴
10×10⁴
0

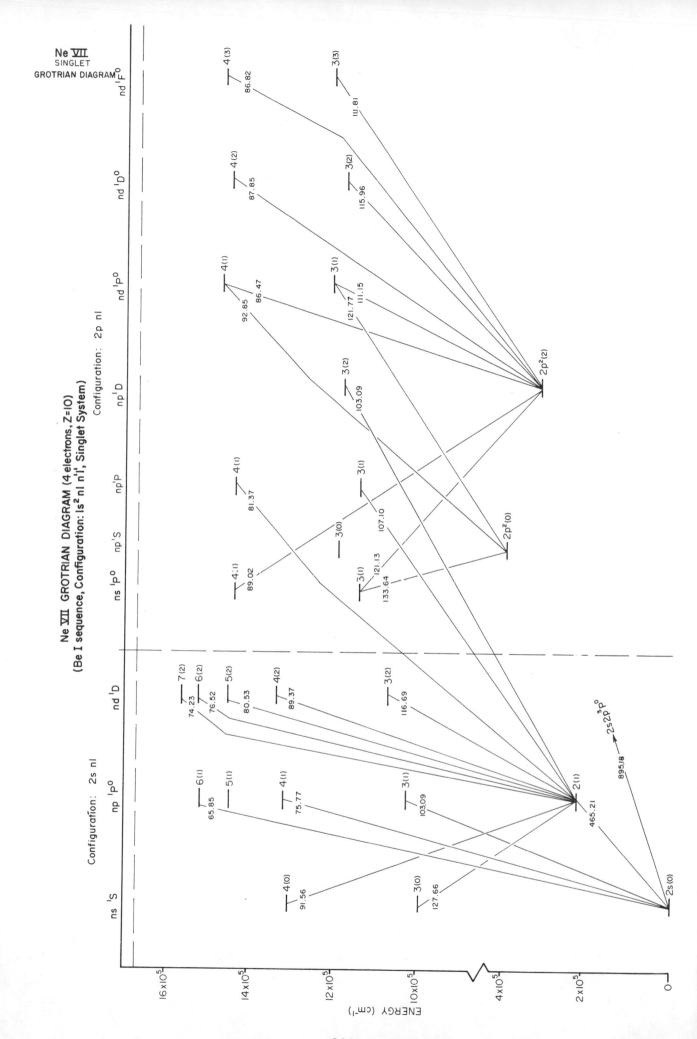

Ne VII
SINGLET
GROTRIAN DIAGRAM

Ne VII GROTRIAN DIAGRAM (4 electrons, Z=10)
(Be I sequence, Configuration: 1s² nl n'l', Singlet System)

Configuration: 2p nl

Configuration: 2s nl

ENERGY (cm⁻¹)

328

Ne VII ENERGY LEVELS (4 electrons, Z=10)
(Be I sequence, Configuration: 1s² nln'l', Triplet System)

Configuration: 2p nl

nd ³Dº

1584380 —— 5(1,2,3)

1461270 —— 4(1,2,3)

1194640
1194310 3(1,2,3)
1194060

nd ³Pº

1463190 —— 4(2,1,0)

1200750
1200460 3(2,1,0)
1200000

np ³D

1576760 —— 5(1,2,3)

1446220 —— 4(1,2,3)

1157080
1155400 3(1,2,3)
1154350

np ³P

1172470
1172140 3(0,1,2)
1172000

290722
289839 2p²(0,1,2)
289328

ns ³Pº

1121780
1120765 3(0,1,2)
1120270

np ³S

1166060 —— 3(1)

Configuration: 2s nl

³D

1587130 —— 8(1,2,3)
1563160 —— 7(1,2,3)
1520400 —— 6(1,2,3)
1456300 —— 5(1,2,3)

1328240 —— 4(1,2,3)

1054410 —— 3(1,2,3)

np ³Pº

1028754. 7+y
1028499. 3+y 3(0,1,2)
1028366. 5+y

1127C0
111706 2(0,1,2)
111251

ns ³S

1436360 —— 5(1)

1299250 —— 4(1)

978320 —— 3(1)

KEY

1576760 +x —— 5(1,2,3)
             J, Lowest to Highest
             n
             Uncertainty
             Energy Level (cm⁻¹)

Ionization Level
1671792 cm⁻¹
(207.27 electron volts)

[Ne VII 1s² 2s² ¹S₀ → Ne VIII 1s² 2s ²S₍½₎]

NOTE: All levels ±x except for 2s3p³Pº

ENERGY (cm⁻¹)

17x10⁵
14x10⁵
11x10⁵
4x10⁵
1x10⁵
0

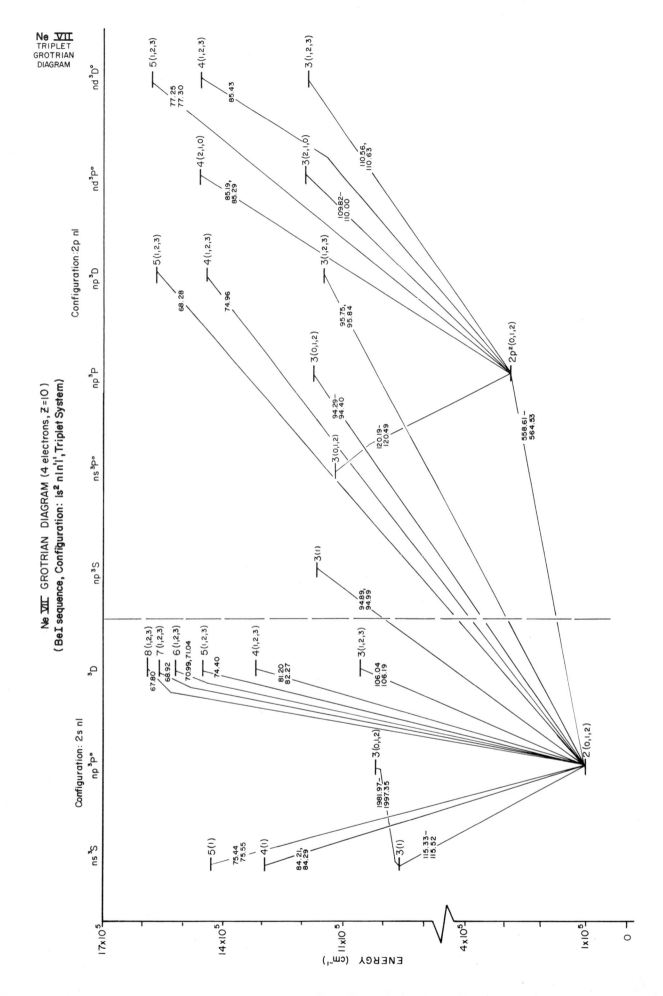

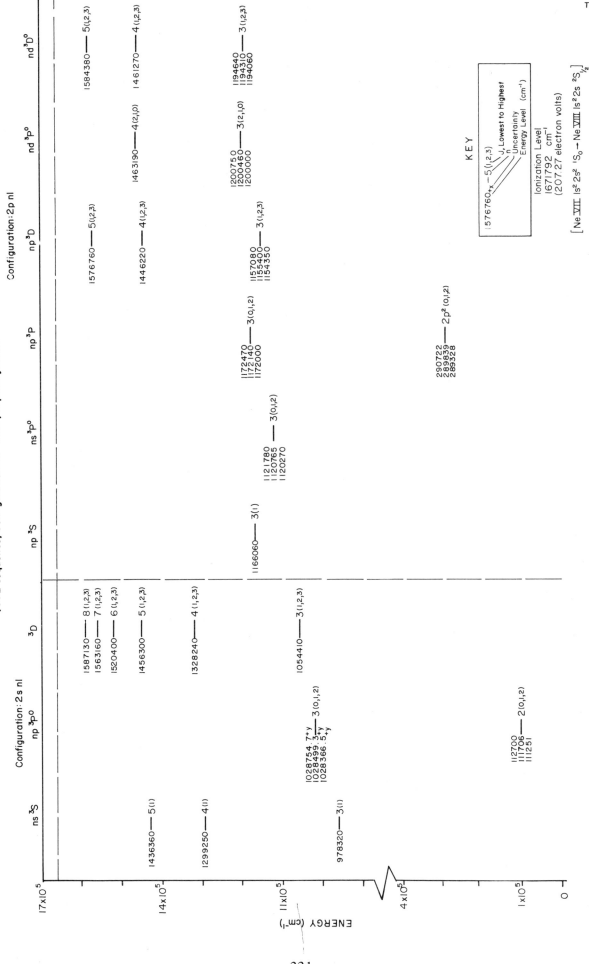

Ne VII ENERGY LEVELS (4 electrons, Z=10)
(Be I sequence, Configuration: 1s² nln'l', Triplet System)

Ne VII
TRIPLET

[Ne VII 1s² 2s² ¹S₀ → Ne VIII 1s² 2s ²S₁/₂]

NOTE: All levels ± x except for 2s3p ³P°

331

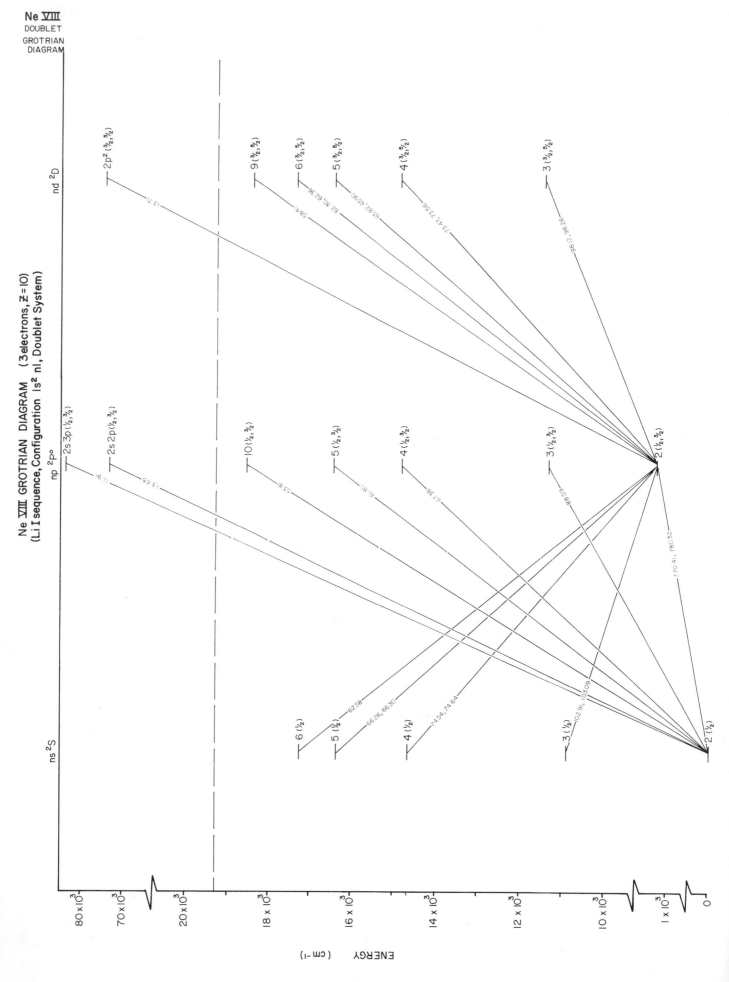

Ne VIII
DOUBLET
GROTRIAN
DIAGRAM

Ne VIII GROTRIAN DIAGRAM (3 electrons, Z = 10)
(Li I sequence, Configuration 1s² nl, Doublet System)

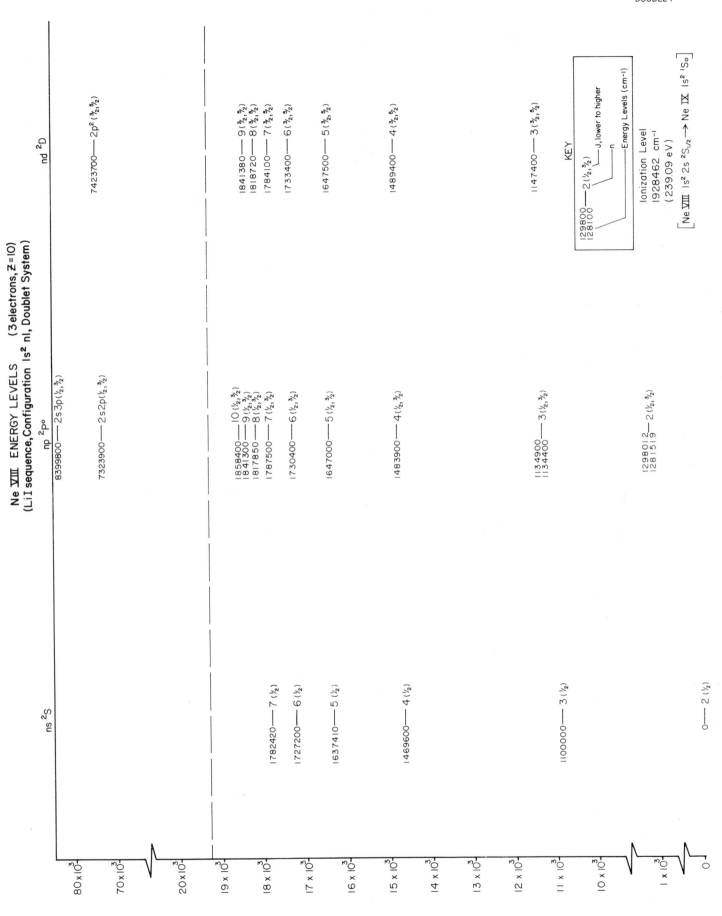

Ne VIII ENERGY LEVELS (3 electrons, Z=10)
(Li I sequence, Configuration 1s² nl, Doublet System)

Ne VIII
DOUBLET

333

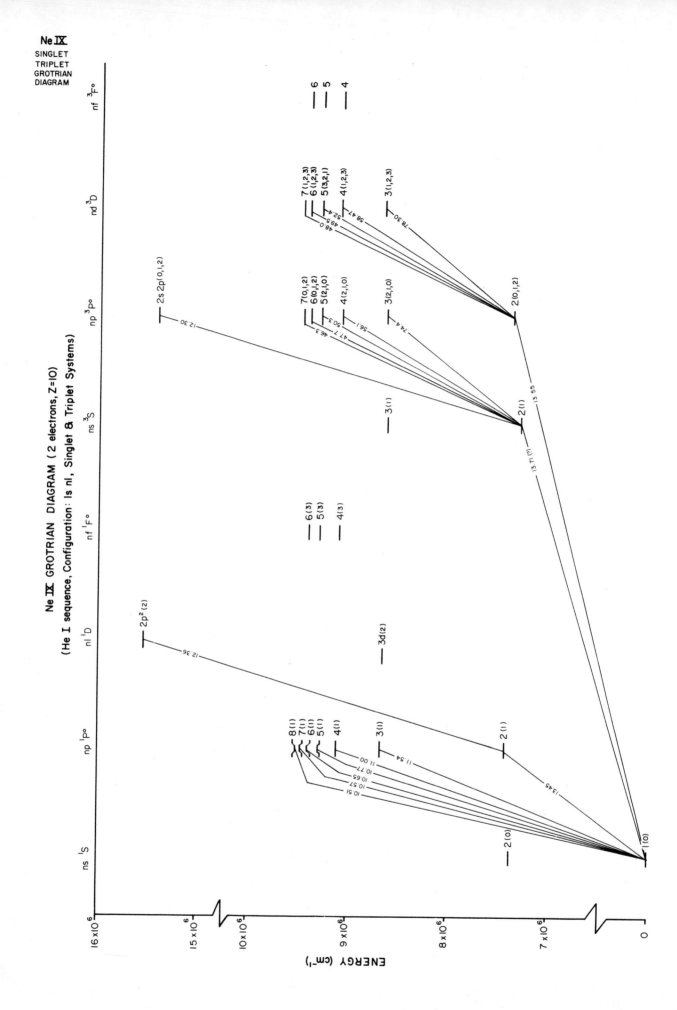

Ne IX
SINGLET
TRIPLET
GROTRIAN
DIAGRAM

Ne IX GROTRIAN DIAGRAM (2 electrons, Z=10)
(He I sequence, Configuration: Is nl, Singlet & Triplet Systems)

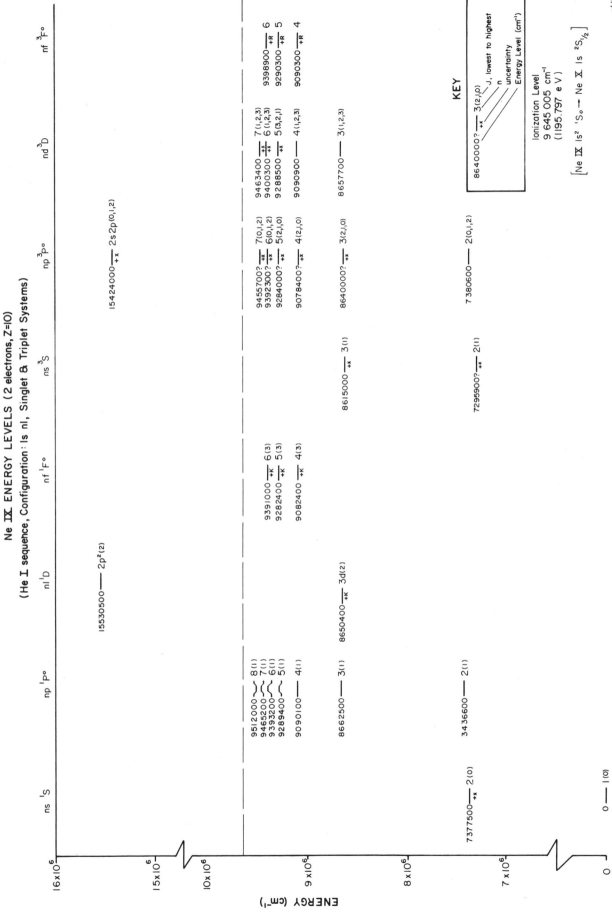

Ne IX ENERGY LEVELS (2 electrons, Z=10)

(He I sequence, Configuration: 1s nl, Singlet & Triplet Systems)

Ne IX
SINGLET
TRIPLET

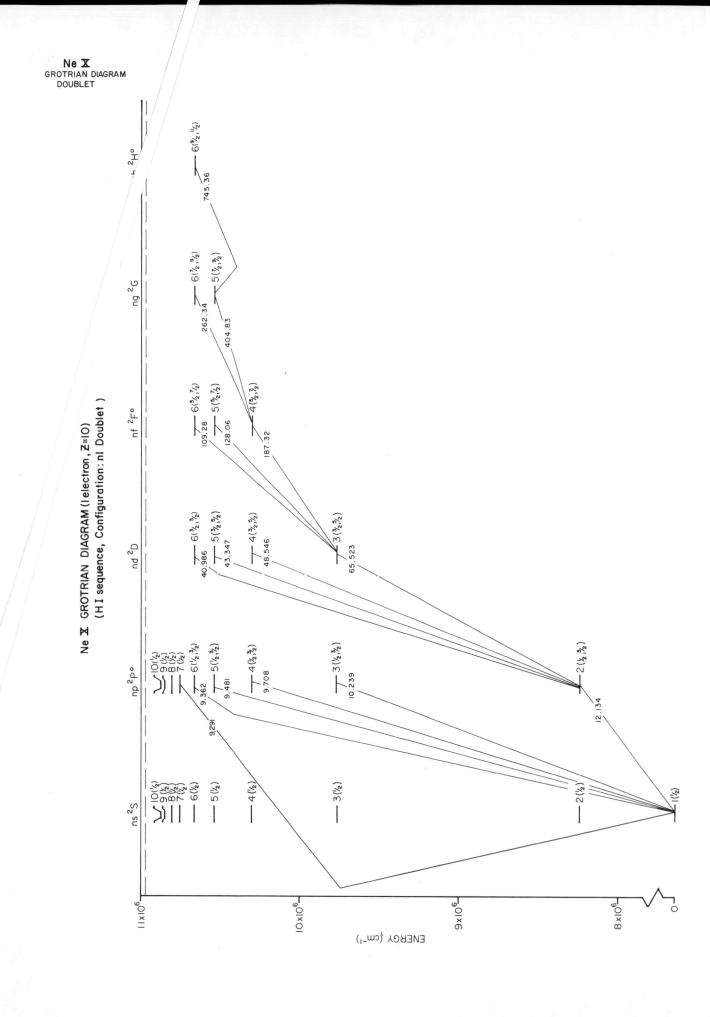

Ne X
GROTRIAN DIAGRAM
DOUBLET

Ne X GROTRIAN DIAGRAM (1 electron, Z=10)
(H I sequence, Configuration: nl Doublet)

ENERGY (cm⁻¹)

336

Ne X ENERGY LEVELS (1 electron, Z =10)
(H I sequence, Configuration: nl, Doublet)

ENERGY (cm⁻¹) axis: 11×10^6, 10×10^6, 9×10^6, 8×10^6, 0

ns ²S
10877088. — 10(½)
10851328. — 9(½)
10815314. — 8(½)
10762778. — 7(½)
10681826. — 6(½)
10547550. — 5(½)
10300312. — 4(½)
9766025. — 3(½)
8239098. — 2(½)
0. — 1(½)

np ²P°
10877086. — 10(½)
10851327. — 9(½)
10815311. — 8(½)
10762775. — 7(½)
10681956. / 10681820. — 6(½, 3/2)
10547774. / 10547539. — 5(½, 3/2)
10300750. / 10300291. — 4(½, 3/2)
9767065. / 9765976. — 3(½, 3/2)
8242608. / 8238936. — 2(½, 3/2)

nd ²D
10682001. / 10681953. — 6(3/2, 5/2)
10547852. / 10547774. — 5(3/2, 5/2)
10300902. / 10300750. — 4(3/2, 5/2)
9767425. / 9767063. — 3(3/2, 5/2)

nf ²F°
10682023. / 10682001. — 6(5/2, 7/2)
10547891. / 10547852. — 5(5/2, 7/2)
10300978. / 10300902. — 4(5/2, 7/2)

ng ²G
10682037. / 10682023. — 6(7/2, 9/2)
10547914. / 10547891. — 5(7/2, 9/2)

nh ²H°
10682046. / 10682037. — 6(9/2, 11/2)

KEY
9767425. — 3(3/2, 5/2)
9767063. — J, Lowest to Highest
n
Energy Levels (cm⁻¹)

Ionization Level
10986875. cm⁻¹
(1362.16 electron volts)
[Ne X 1s ²S₁/₂ → NUCLEUS]

337

Ne I $Z = 10$ 10 electrons

O. Andrade, M. Gallardo, and K. Bockasten, Appl. Phys. Letters **11**, 99 (1967).

Authors give a table of lines observed in noble-gas lasers.

K. Codling, R.P. Madden, and D.L. Ederer, Phys. Rev. **155**, 26 (1967).

Authors give a line table of resonance lines in the absorption spectrum in the range 80–570 Å.

K.G. Ericsson and L.R. Lidholt, IEEE J. Qu. Electronics **3**, 94 (1967).

Authors give wavelengths of superradiant transitions, observed and calculated.

W.L. Faust, R.A. McFarlane, C.K.N. Patel, and C.G.B. Garrett, Phys. Rev. **133A**, 1476 (1964).

Authors give a line table, a term table, and a partial Grotrian diagram based on observations in the region 20 000–350 000 Å.

C.J. Humphreys, E. Paul, Jr., R.D. Cowan, and K.L. Andrew, J. Opt. Soc. Amer. **57**, 855 (1967).

Authors give a line table and an energy level array from observations in the region 39 000–40 160 Å.

U. Litzén, Ark. Fys. **38**, 317 (1968).

Author gives a line table based on observations and an energy level table of values observed and calculated.

Not all levels are shown, but the most energetic in each column is given. The coupling scheme is taken from C.E. Moore, *Atomic Energy Levels*, N.B.S. Circular 467, Vol. 1 (1965).

All spectral lines have been shown, but at the expense of dividing the Grotrian Diagram into eight parts. The information for high intermediate coupling and doubly-excited terms comes from Codling, Madden, and Ederer. Note that the origin of λ 275.64 Å is uncertain; two possible parent terms are shown, linked by a question mark. The designation, including the j-values, at a given level is either shown explicitly or, if not shown, is the same as that at the column heading.

Ne II $Z = 10$ 9 electrons

W.B. Bridges and A.N. Chester, IEEE J. Qu. Electronics **1**, 66 (1965).

Authors give a table of lines observed and calculated in ion gas lasers.

K.G. Ericsson and L.R. Lidholt, IEEE J. Qu. Electronics **3**, 94 (1967).

Authors give wavelengths of superradiant transitions, observed and calculated.

U. Fink, S. Bashkin, and W.S. Bickel, J. Quant. Spectrosc. Radiat. Transfer **10**, 1241 (1970).

Authors give a line table of calculated and observed values in the region 3195–4515 Å.

J.E. Hansen and W. Persson, Physica Scripta **8**, 197 (1973).

Authors give $2p^4\,ns$ and $2p^4\,nd$ configurations in F I and Ne II.

K.W. Meissner, R.D. Van Veld, and P.G. Wilkinson, J. Opt. Soc. Amer. **48**, 1001 (1958).

Authors give line tables in the vacuum ultraviolet region.

W. Persson and L. Minnhagen, Ark. Fys. **37**, 273 (1967).

Authors give line tables, energy level tables, and a partial energy level diagram for lines observed in the region 10 200–2500 Å.

For intermediate-coupling terms, please see W. Persson, Physica Scripta **3**, 133 (1971).

Ne III $Z = 10$ 8 electrons

I.S. Bowen, Ap. J. **121**, 306 (1955).

   Author gives wavelengths from nebular observations.

I.S. Bowen, Ap. J. **132**, 1 (1960).

   Author gives wavelengths from nebular observations.

W.B. Bridges and A.N. Chester, IEEE J. Qu. Electronics **1**, 66 (1965).

   Authors give a line table from observations and calculations for ion gas lasers.

U. Fink, S. Bashkin, and W.S. Bickel, J. Quant. Spectrosc. Radiat. Transfer **10**, 1241 (1970).

   Authors give a line table of calculated and observed values in the region 2590–2785 Å.

Ne IV $Z = 10$ 7 electrons

I.S. Bowen, Ap. J. **121**, 306 (1955).

   Author gives wavelengths from nebular observations.

I.S. Bowen, Ap. J. **132**, 1 (1960).

   Author gives line table of values predicted and observed in nebulae.

W.B. Bridges and A.N. Chester, IEEE J. Qu. Electronics **1**, 66 (1965).

   Authors give a line table from observations and calculations for ion gas lasers.

Ne V $Z = 10$ 6 electrons

I.S. Bowen, Ap. J. **121**, 306 (1955).

   Author gives wavelengths from nebular observations.

I.S. Bowen, Ap. J. **132**, 1 (1960).

   Author gives line table based on laboratory and nebular observations.

Ne VI $Z = 10$ 5 electrons

L.J. Shamey, J. Opt. Soc. Amer. **61**, 942 (1971).

   Author gives a line table and a table of calculated energy levels.

Ne VII $Z = 10$ 4 electrons

L.L. House and G.A. Sawyer, Ap. J. **139**, 775 (1964).

   Authors give a line table of lines observed in the region 128–75 Å.

Ne VIII $Z = 10$ 3 electrons

L.L. House and G.A. Sawyer, Ap. J. **139**, 775 (1964).

   Authors give a line table for lines observed in the region 60–104 Å.

Ne IX $Z = 10$ 2 electrons

E. Holøien and J. Midtdal, J. Phys. B: Atom. Molec. Phys. **4**, 1243 (1971).

Authors calculate nonrelativistically energies for the He I isoelectronic sequence.

Ne X $Z = 10$ 1 electron

Please see the general references.

Sodium (Na)

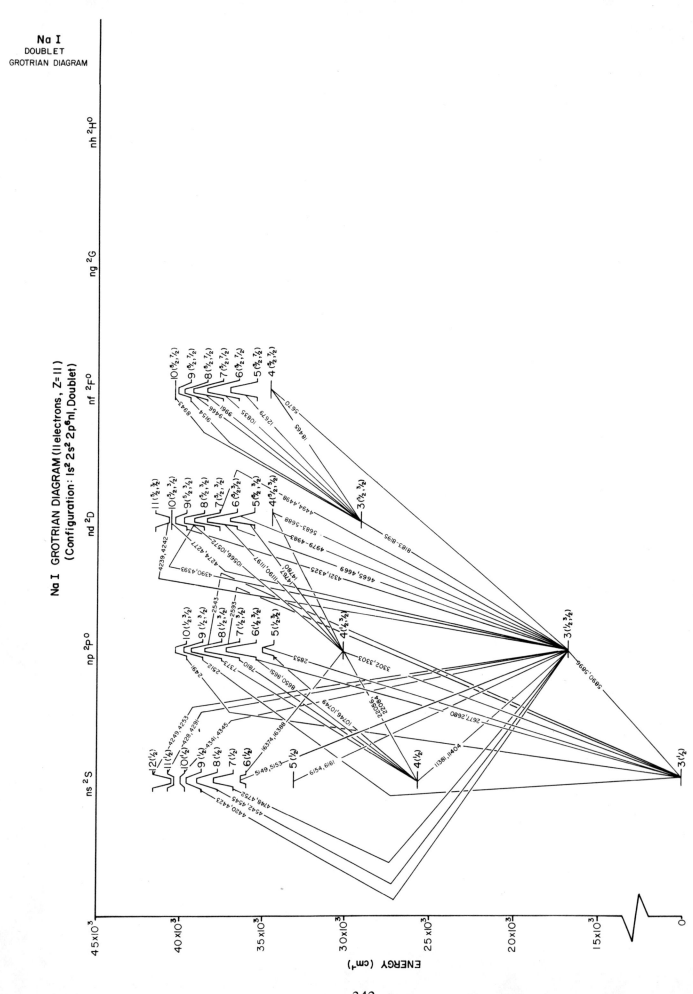

Na I
DOUBLET
GROTRIAN DIAGRAM

Na I GROTRIAN DIAGRAM (11 electrons, Z=11)
(Configuration: 1s² 2s² 2p⁶nl, Doublet)

ENERGY (cm⁻¹)

342

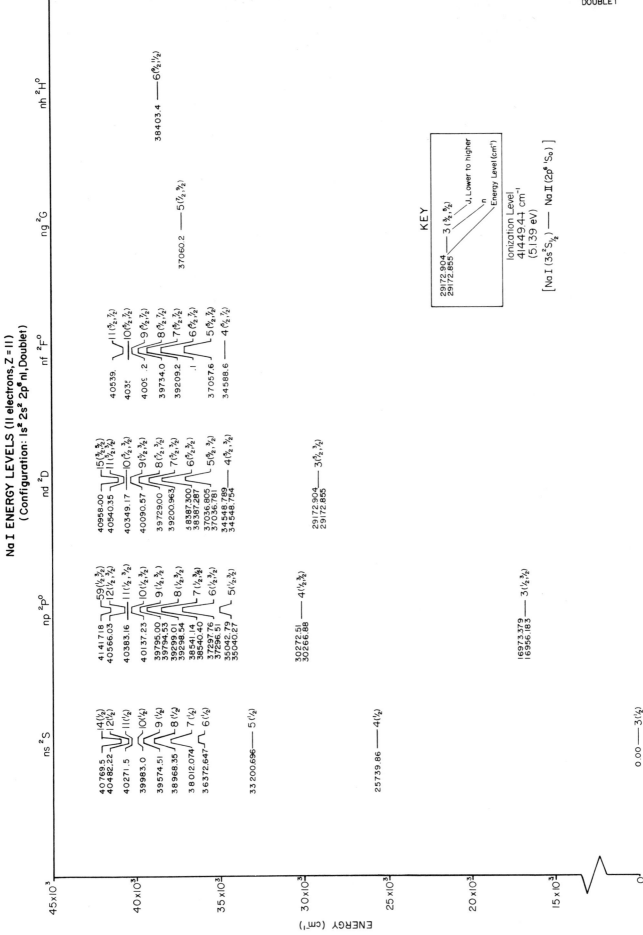

Na I
DOUBLET

Na I ENERGY LEVELS (11 electrons, Z = 11)
(Configuration: 1s² 2s² 2p⁶ nl, Doublet)

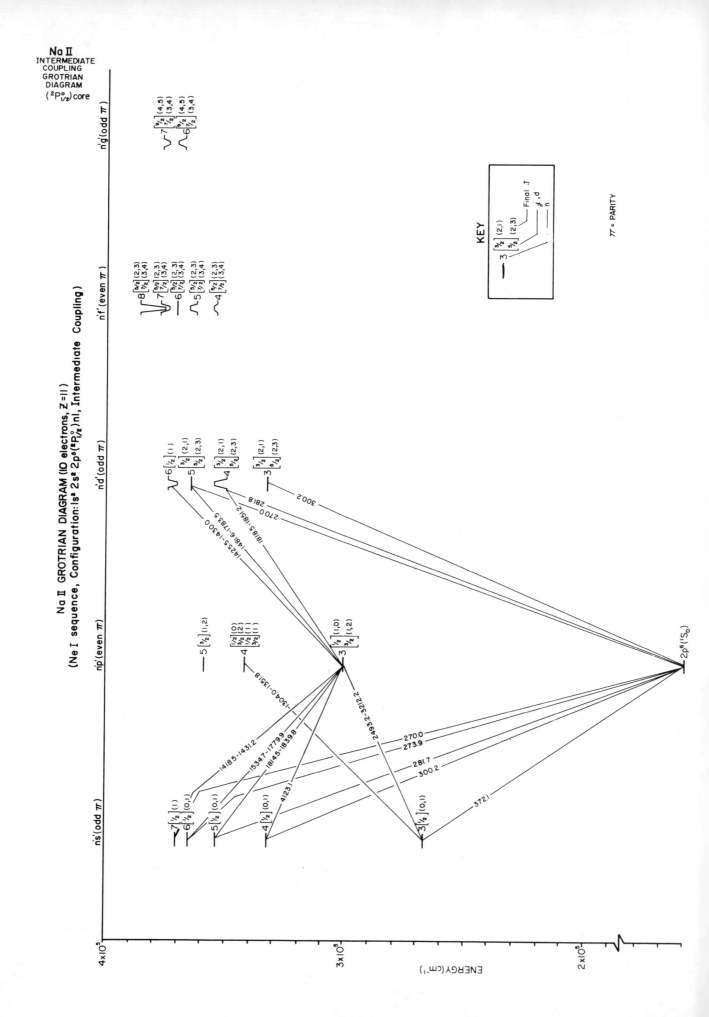

Na II GROTRIAN DIAGRAM (10 electrons, Z=11)
(Ne I sequence, Configuration: 1s² 2s² 2p⁵(²P°₁/₂)nl, Intermediate Coupling)

Na II
INTERMEDIATE
COUPLING
GROTRIAN
DIAGRAM
(²P°₁/₂) core

Na II ENERGY LEVELS (10 electrons, Z=11)
(Ne I sequence, Configuration: 1s² 2s² 2p⁵($^2P^o_{1/2}$)nl, Intermediate Coupling)

ENERGY (cm⁻¹) — scale: 4×10^5, 3×10^5, 2×10^5

n'g'(odd π)
373800.1 — 7 $9/2$ [4,5] $7/2$ [3,4]
370564.6 — 6 $9/2$ [4,5] $7/2$ [3,4]

nf'(even π)
375893.8 / 375893.7 — 8 $5/2$ [2,3] $7/2$ [3,4]
373791.8 / 373791.7 — 7 $9/2$ [2,3] $7/2$ [3,4]
370552.2 / 370551.5 — 6 $9/2$ [2,3] $7/2$ [3,4]
365178.4 / 365177.6 — 5 $9/2$ [2,3] $7/2$ [3,4]
352883 / 352886.2 — 4 $5/2$ [2,3] $7/2$ [3,4]

nd'(odd π)
370039.3 — 6 [$1/2$] (1)
364935.2
364747.2 — 5 $3/2$ [2,1] $9/2$ [2,3]
364607.8
364542.1
354880.2
354710.8 — 4 $3/2$ [2,1] $5/2$ [2,3]
354562.9
354530.1
333166.8
332966.5 — 3 $3/2$ [2,1] $5/2$ [2,3]
332845.7
332806.0

np'(even π)
359737.9 — 5 $3/2$ [1,2]
359425.0
343473.1 — 4 $1/2$ [0]
343260.5 — $3/2$ [2]
342974.8 — $1/2$ [1]
342742.4 — $3/2$ [1]
308864.3 — 3 $1/2$ [1,0]
300511.1 — $3/2$ [1,2]
300107.7
299889.2

ns'(odd π)
370382.5 — 7 [$1/2$] (1)
365047.2 — 6 [$1/2$] (0,1)
365013.7
355001.5 — 5 [$1/2$] (0,1)
354863.6
333111.6 — 4 [$1/2$] (0,1)
332713.8
268766.8 — 3 [$1/2$] (0,1)
266285.4

0 — 2p⁶ (¹S₀)

KEY

268766.8
266285.4 — 3 [$1/2$] (0,1)
Final J (lower to higher)
J-ℓ Coupling
n
Energy Levels (cm⁻¹)

Ionization Level
381395 cm⁻¹
(47.286 eV)
[Na II 2p⁶ ¹S₀ → Na III 2p⁵ $^2P^o_{3/2}$]

π = PARITY

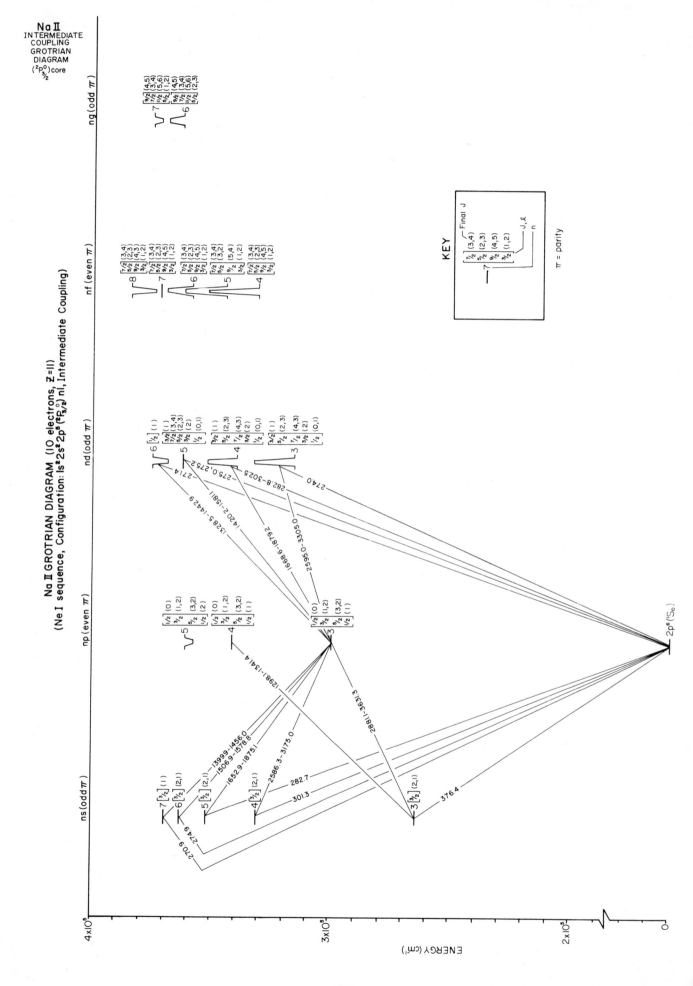

Na II ENERGY LEVELS (10 electrons, Z=11)
(Ne I sequence, Configuration: 1s²2s²2p⁵(²P°₃/₂)nl, Intermediate Coupling)

$$\text{Na II INTERMEDIATE COUPLING } (^2P^o_{3/2})\text{core}$$

ns (odd π)

369074.0 — 7 [³⁄₂](1)
363730.0 — 6 [³⁄₂](2,1)
363614.0
353723.1 — 5 [³⁄₂](2,1)
353540.6
331877.6 — 4 [³⁄₂](2,1)
331500.3
265693.2 — 3 [³⁄₂](2,1)
264928.2

np (even π)

358774.6 [¹⁄₂](0)
358549.5 ³⁄₂(1,2)
358512.8 — 5
358157.1 ⁵⁄₂(3,2)
358064.2 ¹⁄₂(2)
357684.3
342347.7 [¹⁄₂](0)
341961.1 ³⁄₂(1,2)
341910.7 — 4
341461.5 ⁵⁄₂(3,2)
341136 ¹⁄₂(1)
340243.5
300391.6 [¹⁄₂](0)
299193.7 ³⁄₂(1,2)
298169.1 — 3
297639.3 ⁵⁄₂(3,2)
297252.5 ¹⁄₂(1)
293224.2

nd (odd π)

368497.8 — 6 [¹⁄₂](1)
363636.0 ³⁄₂(1)
363445.5 ⁷⁄₂(3,4)
363445.5 ⁵⁄₂(2,3) — 5
363350.6 ³⁄₂(2)
363336.3 ¹⁄₂(0,1)
353605.0 ³⁄₂(1)
353487.6 ⁵⁄₂(2,3)
353466.9 — 4 ⁷⁄₂(4,3)
353243.9 ³⁄₂(2)
353206.3 ¹⁄₂(0,1)
353155.6
352973.7
331748.7 ³⁄₂(1)
331716.6 ⁵⁄₂(2,3)
331169.4 — 3 ⁷⁄₂(4,3)
331128.0 ³⁄₂(2)
331126.7 ¹⁄₂(0,1)
330792.9
330640.6
330553.2

nf (even π)

374536.4 ⁷⁄₂(3,4)
374520.6 ⁵⁄₂(2,3) — 8
374524.7 ⁹⁄₂(4,3)
374518.7 ³⁄₂(1,2)
372436.6 ⁷⁄₂(3,4)
372429.4 ⁵⁄₂(2,3) — 7
372419.4 ⁹⁄₂(4,5)
372411.4 ³⁄₂(1,2)
369203.3 ⁷⁄₂(3,4)
369190.7 ⁵⁄₂(2,3) — 6
369175.5 ⁹⁄₂(4,5)
369162.6 ³⁄₂(1,2)
363841.6 ⁷⁄₂(3,4)
363819.0 ⁵⁄₂(3,2)
363794.2 — 5 ⁹⁄₂(5,4)
363772.9 ³⁄₂(1,2)
353976.5 ⁷⁄₂(3,4)
353930.2 ⁵⁄₂(2,3) — 4
353885.2 ⁹⁄₂(4,5)
353842.7 ³⁄₂(1,2)

ng (odd π)

372437.9 ⁹⁄₂(4,5)
372436.5 ⁷⁄₂(3,4) — 7
372430.8 ¹¹⁄₂(5,6)
372428.3 ⁹⁄₂(1,2)
369207.4 ⁹⁄₂(4,5)
369193.6 ⁷⁄₂(3,4) — 6
369188.7 ¹¹⁄₂(5,6)
        ⁹⁄₂(2,3)

KEY

265693.2 — 3 [³⁄₂](2,1)
264928.2

— Final J, lower to higher
— J-ℓ Coupling
— n
— Energy Levels (cm⁻¹)

Ionization Level
381395 cm⁻¹
(47.286 eV)
[Na II 2p⁶ ¹S₀ → Na III 2p⁵ ²P°₃/₂]

π = parity

0 — 2p⁶(¹S₀)

ENERGY (cm⁻¹)
4×10⁵
3×10⁵
2×10⁵
0

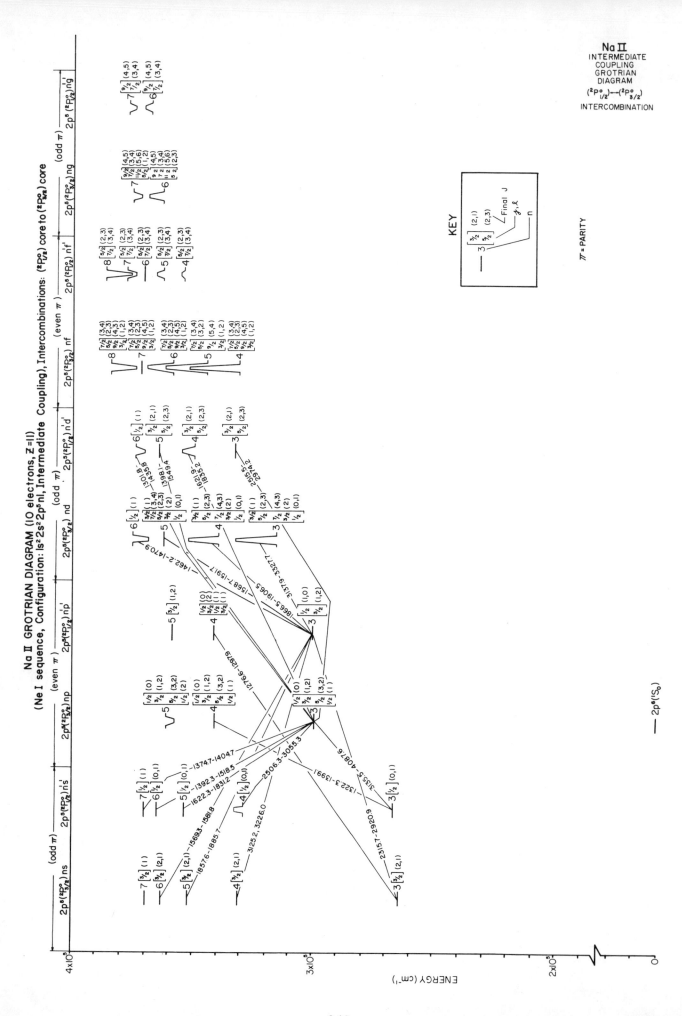

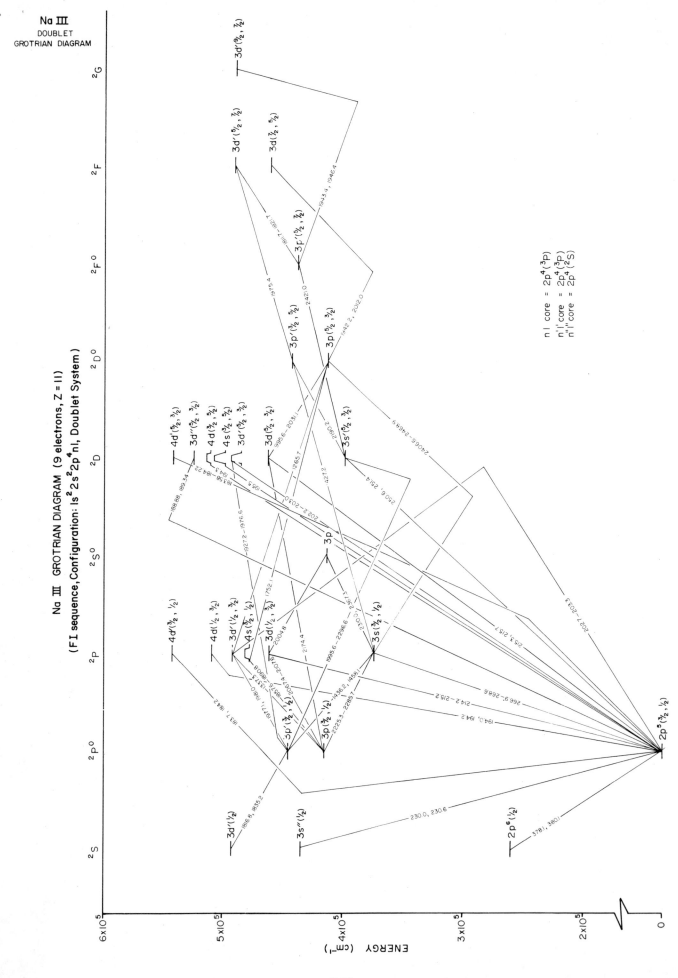

Na III
DOUBLET
GROTRIAN DIAGRAM

Na III GROTRIAN DIAGRAM (9 electrons, Z = 11)

(FI sequence, Configuration: 1s² 2s² 2p⁴ nl, Doublet System)

nl core = 2p⁴(³P)
n'l' core = 2p⁴(³P)
n''l'' core = 2p⁴(²S)

ENERGY (cm⁻¹)

350

Na III
DOUBLET

Na III ENERGY LEVELS (9 electrons, Z = 11)

(F I sequence, Configuration: 1s² 2s² 2p⁴ nl, Doublet System)

ENERGY (cm⁻¹)

Columns: ²S ²Pᵒ ²S ²Pᵒ ²Sᵒ ²D ²Dᵒ ²Fᵒ ²F ²G

²S
- 547910.7 —— 4d'(¹/₂)
- 543638. —— 4s''(¹/₂)
- 493849.24 —— 3d'(¹/₂)
- 435028.00 —— 3s''(¹/₂)
- 264455.1 —— 2p⁶(¹/₂)

²Pᵒ
- 478884.07 } 3p''
- 478842.99 (¹/₂,³/₂)
- 448107.31 } 3p'(³/₂,¹/₂)
- 447547.96
- 418556.54 } 3p(³/₂,¹/₂)
- 418417.50
- 1366.4 —— 2p⁵(³/₂,¹/₂)
- 0.0

²P
- 544280. —— 4d'(¹/₂)
- 544226. —— 5s(¹/₂)
- 523560. —— 4d(³/₂,¹/₂)
- 522720.
- 515984.
- 515142.
- 493293.98 —— 3d'(¹/₂,³/₂)
- 493192.06
- 482388.55 } 4s'(³/₂,¹/₂)
- 482402.20
- 466788.03 } 3d(¹/₂,³/₂)
- 466011.91
- 374679.91 } 3s(³/₂,¹/₂)
- 373632.32

²Sᵒ
- 416909.31 —— 3p(¹/₂)

²D
- 522416. —— 5s(³/₂,⁵/₂)
- 544793. } 4d'(³/₂,³/₂)
- 544749.
- 529497.70 —— 3d''(⁵/₂,³/₂)
- 529461.64
- 515365. } 4d(³/₂,⁵/₂)
- 515016.
- 511434.3 } 4s(³/₂,⁵/₂)
- 511433.8
- 494685.86 } 3d'(⁵/₂,³/₂)
- 494602.73
- 465017.83 } 3d(³/₂,³/₂)
- 464390.17
- 399182.3 } 3s(³/₂,¹/₂)
- 399174.7

²Dᵒ
- 445873.20 } 3p' (³/₂,⁵/₂)
- 445797.52
- 415172.28 } 3p(⁵/₂,³/₂)
- 414281.85

²Fᵒ
- 441055.69 —— 3p'(³/₂,⁷/₂)
- 440940.20

²F
- 544914. —— 4d'(⁷/₂)
- 514700. } 4d(⁵/₂,⁷/₂)
- 514688.
- 495435.20 } 3d' (⁵/₂,⁷/₂)
- 495429.75
- 465398.6 } 3d(⁷/₂,⁵/₂)
- 463970.9

²G
- 492316.41 —— 3d'(⁹/₂,⁷/₂)
- 492313.91

KEY

399182.3 ——— 3s'(⁵/₂,³/₂) ⎤ Core Excitation State
399174.7 ———   ⎦ J, lower to higher
                          n
                          Energy Level (cm⁻¹)

Ionization Level
577800 cm⁻¹
(71.64 eV)

[Na III 1s²2s²2p⁵3p⁰ → Na IV 1s²2s²2p⁴ ³P₂]

nl core = 2p⁴(³P)
n'l' core = 2p⁴(¹D)
n''l''core = 2p⁴(¹S)

351

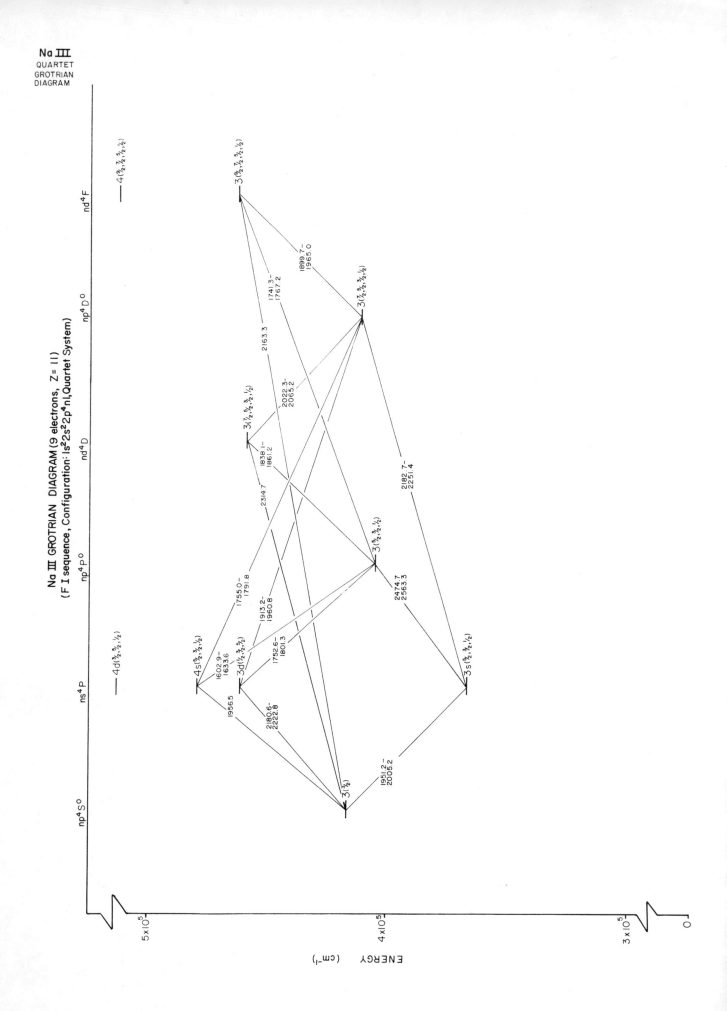

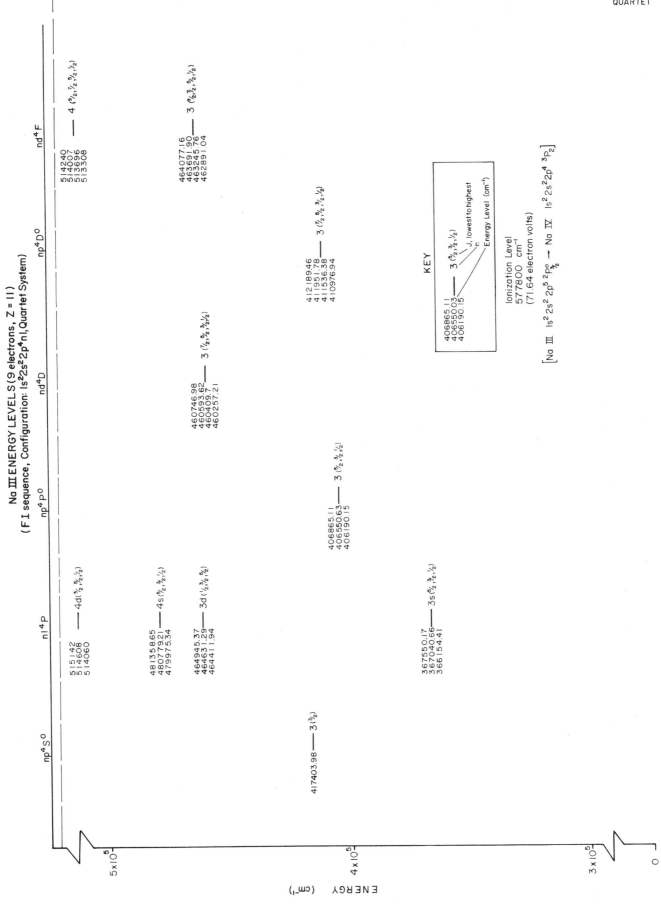

Na III ENERGY LEVELS (9 electrons, Z = 11)
(F I sequence, Configuration: $1s^2 2s^2 2p^4 nl$, Quartet System)

KEY

406865.11
406550.03 ——— 3 $(^5/_2, ^3/_2, ^1/_2)$
406190.15

J, lowest to highest
n
Energy Level (cm⁻¹)

Ionization Level
577800 cm⁻¹
(71.64 electron volts)

[Na III $1s^2 2s^2 2p^5 {}^2P^o_{3/2} \rightarrow$ Na IV $1s^2 2s^2 2p^4 {}^3P_2$]

ENERGY (cm⁻¹)

353

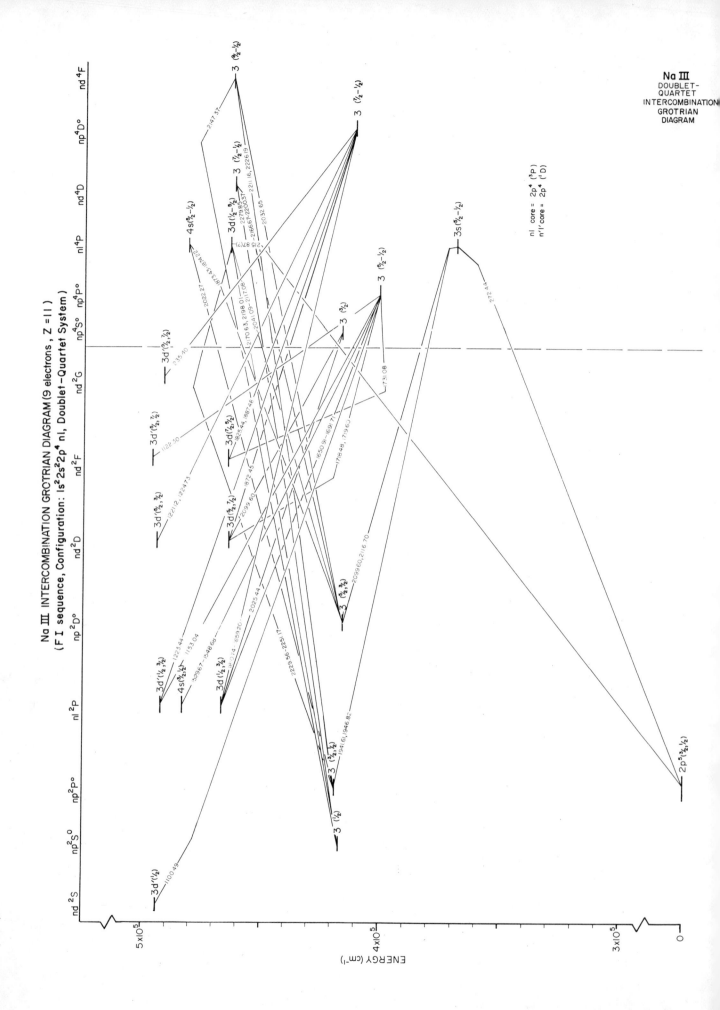

Na III INTERCOMBINATION GROTRIAN DIAGRAM (9 electrons, Z = 11)
(F I sequence, Configuration: 1s²2s²2p⁴ nl, Doublet-Quartet System)

Na III
DOUBLET-
QUARTET
INTERCOMBINATION
GROTRIAN
DIAGRAM

nl core = 2p⁴ (³P)
n'l' core = 2p⁴ (¹D)

ENERGY (cm⁻¹)

354

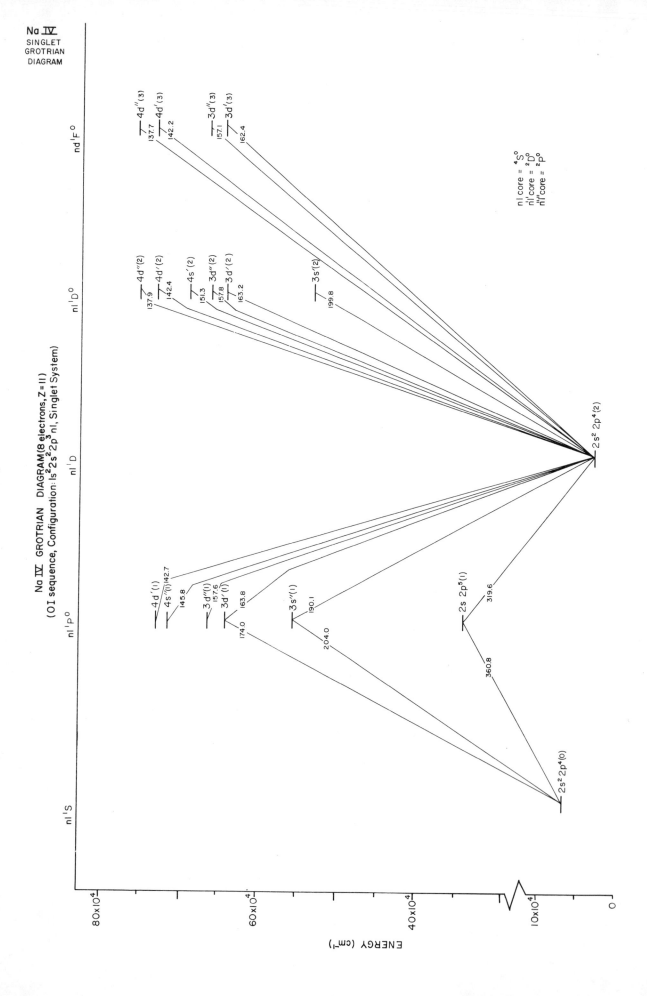

Na IV
SINGLET
GROTRIAN
DIAGRAM

Na IV GROTRIAN DIAGRAM (8 electrons, Z=11)
(OI sequence, Configuration: 1s²2s²2p³nl, Singlet System)

nl core = ⁴S°
nl' core = ²D°
nl" core = ²P°

ENERGY (cm⁻¹)

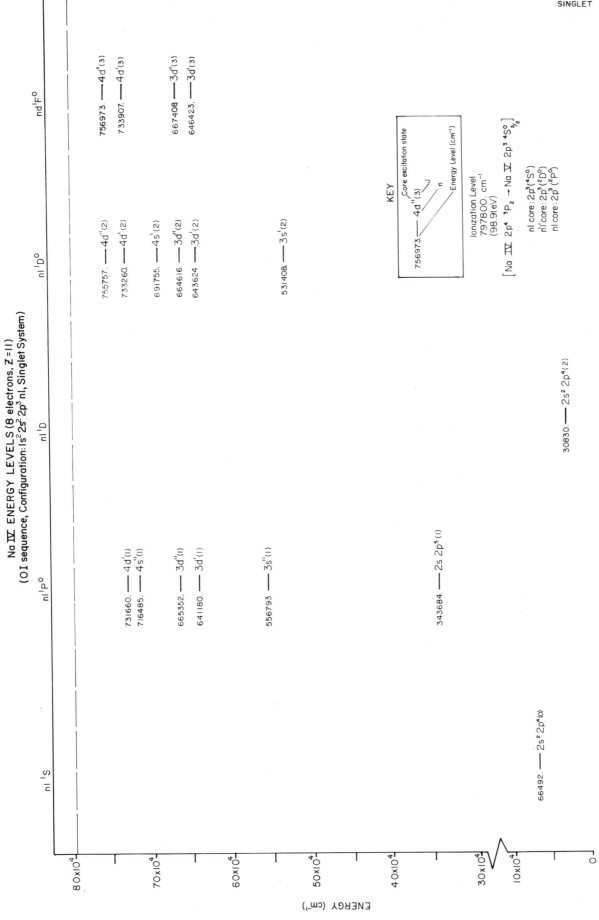

Na IV
SINGLET

Na IV ENERGY LEVELS (8 electrons, Z = II)
(OI sequence, Configuration: $1s^2 2s^2 2p^3$ nl, Singlet System)

KEY

756973. —— $4d''(3)$

Core excitation state
J
n
Energy Level (cm⁻¹)

Ionization Level
797800. cm⁻¹
(98.9leV)

$\left[\text{Na IV } 2p^4 \, {}^3P_2 \rightarrow \text{Na V } 2p^3 \, {}^4S^0 \right]_{3/2}$

nl core: $2p^3(^4S^0)$
n'l' core: $2p^3(^2D^0)$
nl core: $2p^3(^2P^0)$

nd ¹F⁰

756973. —— $4d''(3)$
733907. —— $4d'(3)$
667408 —— $3d''(3)$
646423. —— $3d'(3)$

nl ¹D⁰

755757. —— $4d''(2)$
733260. —— $4d'(2)$
691755. —— $4s'(2)$
664616 —— $3d''(2)$
643624 —— $3d'(2)$
531408 —— $3s'(2)$

nl ¹D

30830.—— $2s^2 2p^4(2)$

nl ¹P⁰

731660. —— $4d'(1)$
716485. —— $4s''(1)$
665352. —— $3d''(1)$
641180. —— $3d'(1)$
556793 —— $3s''(1)$
343684 —— $2s\,2p^5(1)$

nl ¹S

66492.—— $2s^2 2p^4(0)$

ENERGY (cm⁻¹)

80×10^4
70×10^4
60×10^4
50×10^4
40×10^4
30×10^4
10×10^4
0

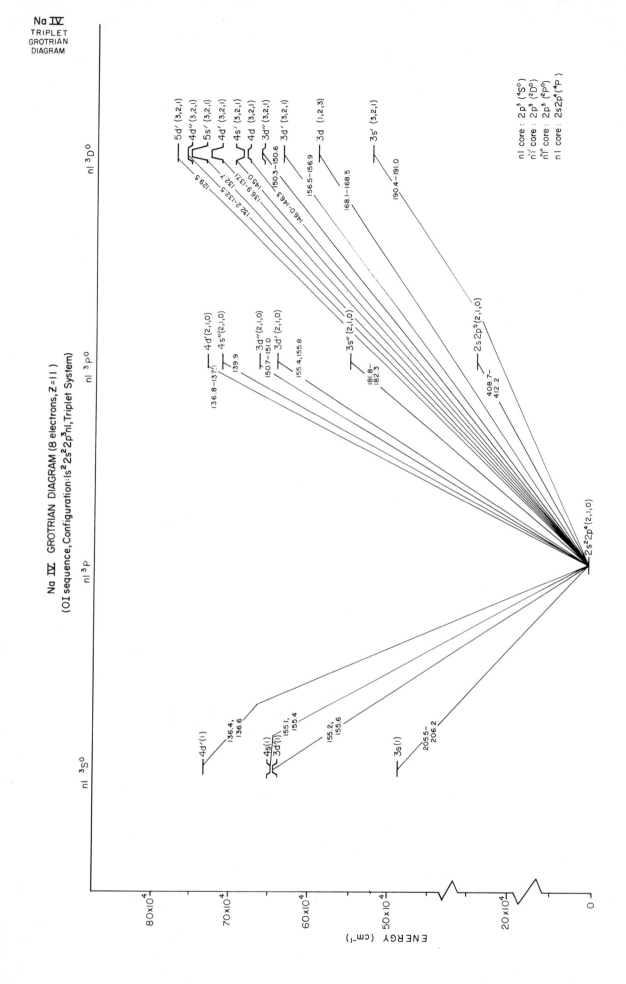

Na IV
TRIPLET
GROTRIAN
DIAGRAM

Na IV GROTRIAN DIAGRAM (8 electrons, Z=11)
(OI sequence, Configuration: 1s²2s²2p³nl, Triplet System)

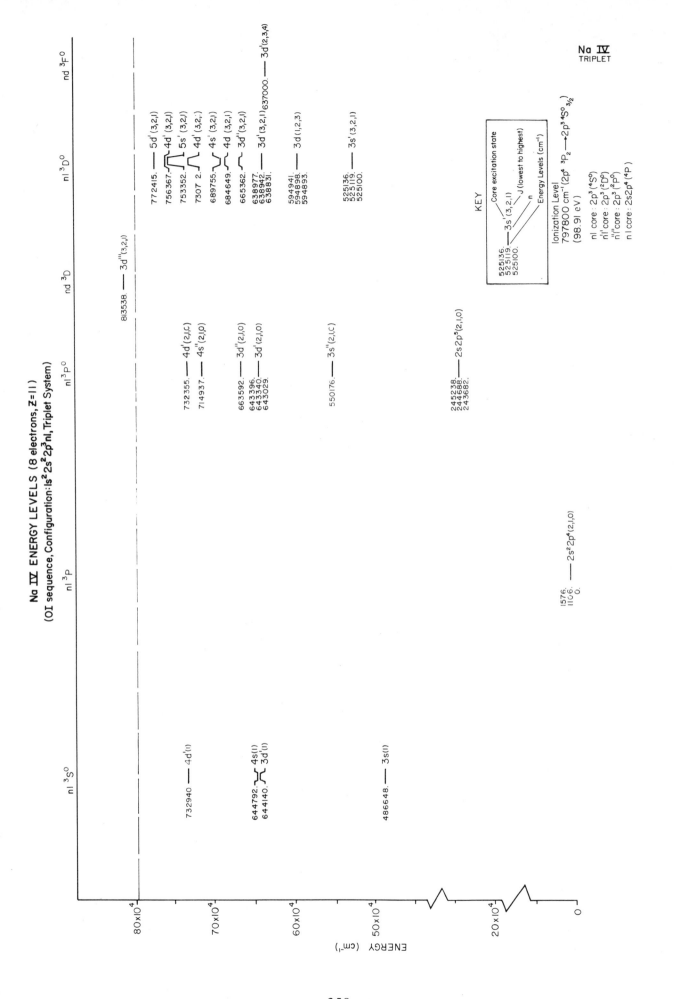

Na IV ENERGY LEVELS (8 electrons, Z=11)
(OI sequence, Configuration: 1s²2s²2p³nl, Triplet System)

Na IV
TRIPLET

ENERGY (cm⁻¹)

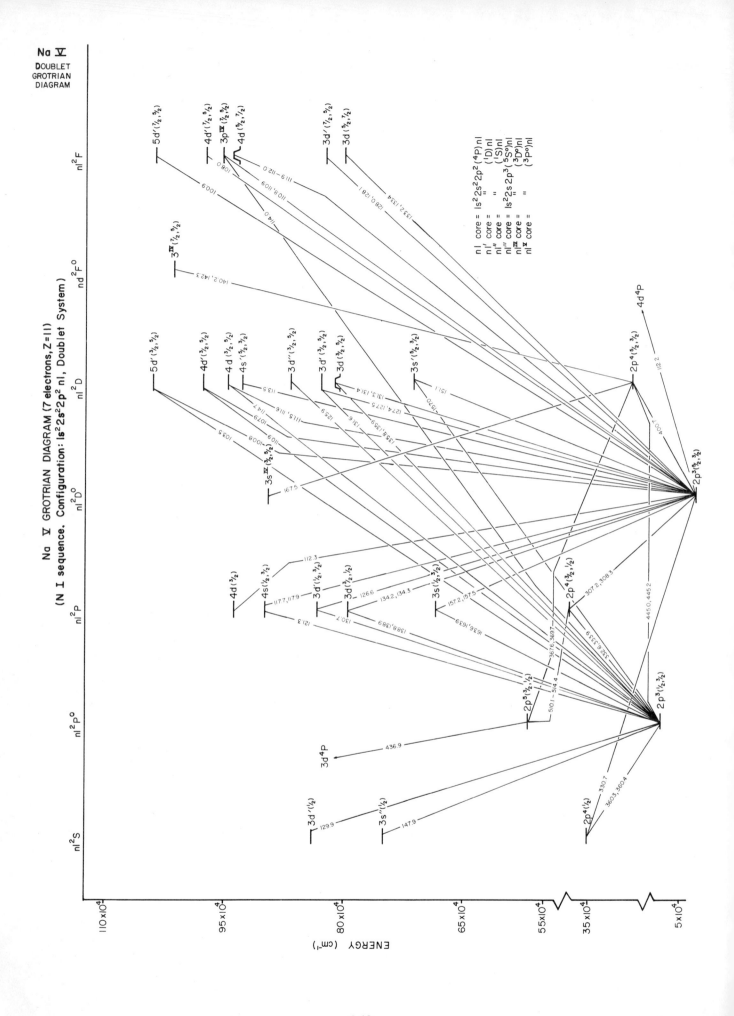

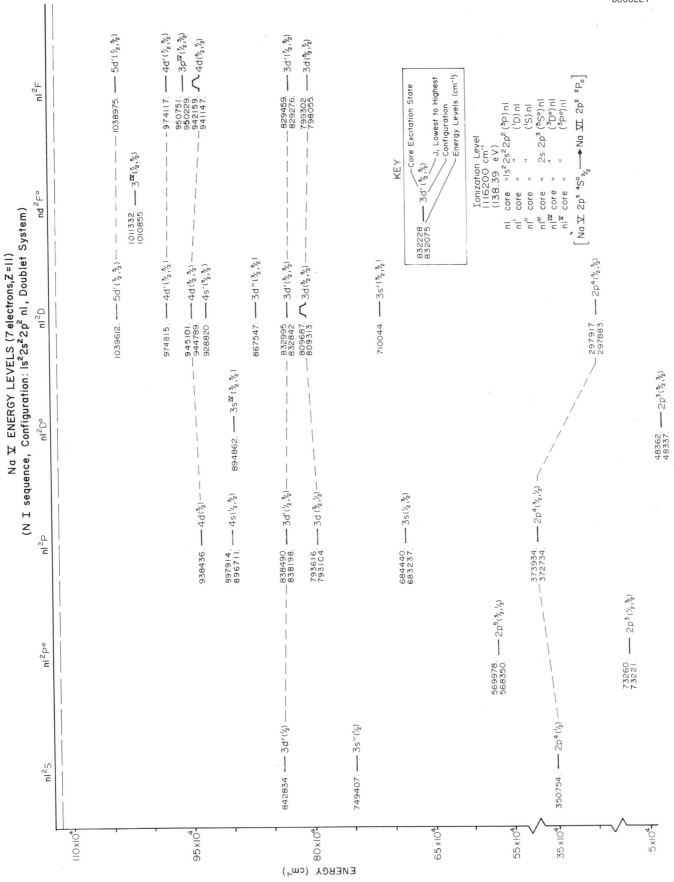

Na V ENERGY LEVELS (7 electrons, Z=11)
(N I sequence, Configuration: 1s²2s²2p² nl, Doublet System)

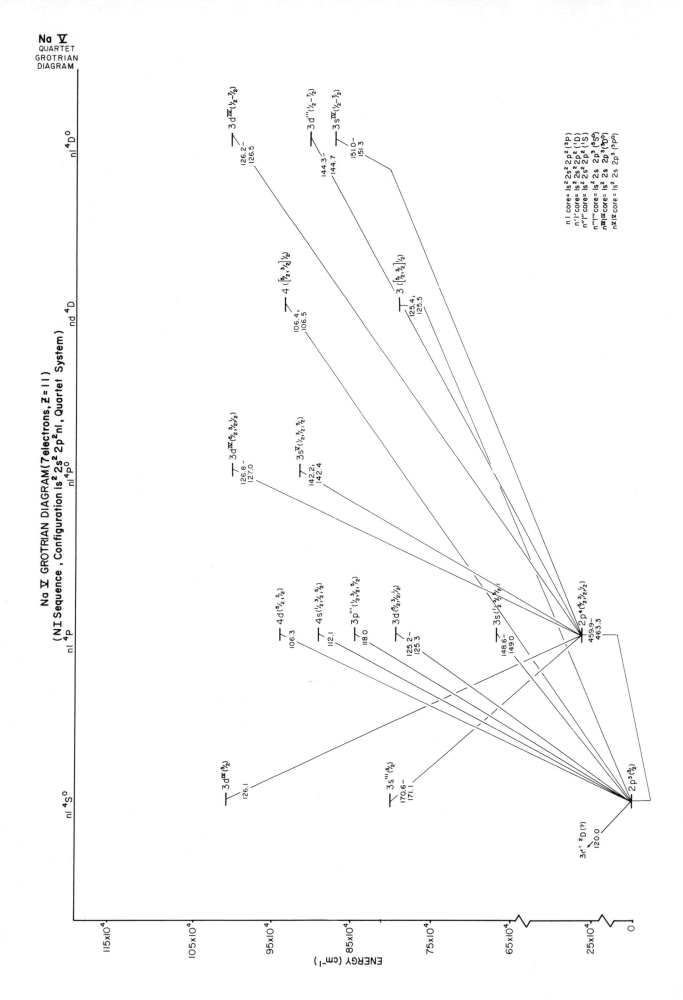

Na V
QUARTET
GROTRIAN
DIAGRAM

Na V GROTRIAN DIAGRAM (7 electrons, Z=11)
(NI Sequence, Configuration 1s² 2s² 2p²nl, Quartet System)

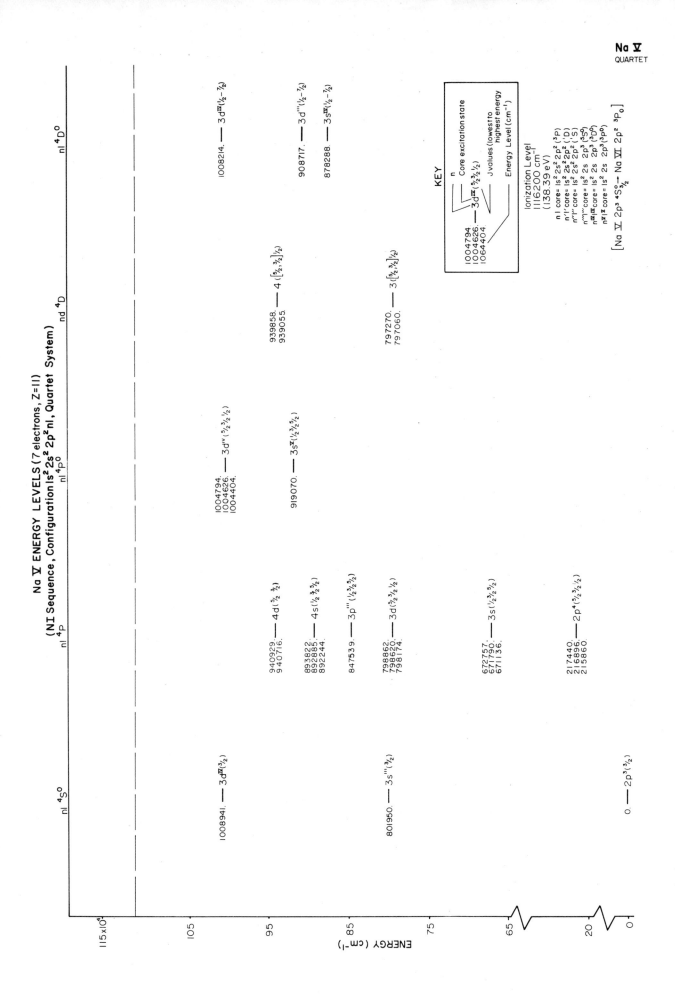

Na V ENERGY LEVELS (7 electrons, Z=11)
(N I Sequence, Configuration 1s² 2s² 2p² nl, Quartet System)

Na Ⅴ
QUARTET

363

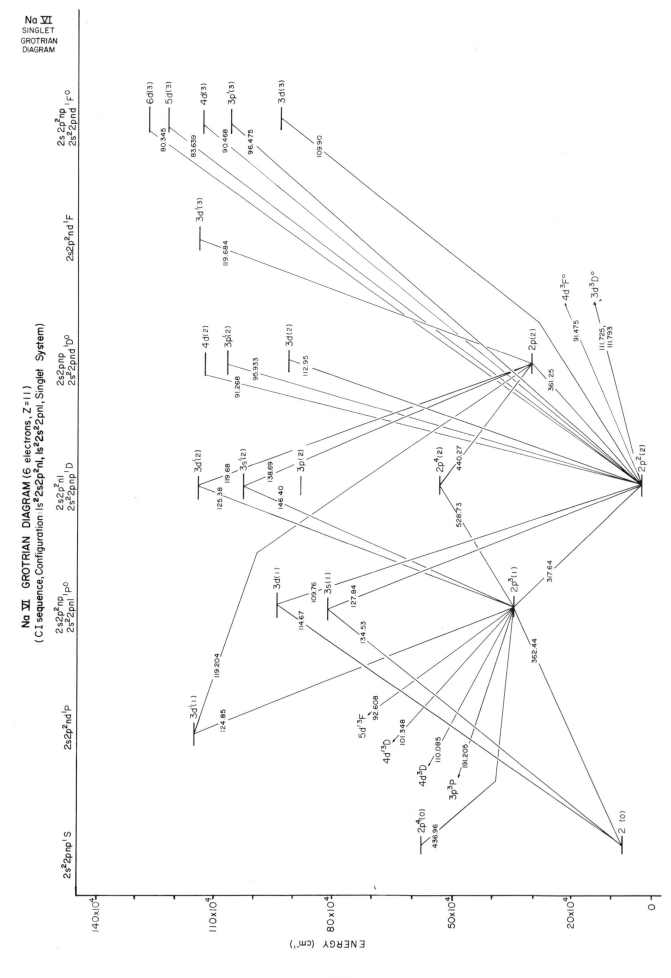

Na VI GROTRIAN DIAGRAM (6 electrons, Z=11)
(CI sequence, Configuration: 1s²2s2p²nl, 1s²2s²2pnl, Singlet System)

Na VI
SINGLET
GROTRIAN
DIAGRAM

Na VI   ENERGY LEVELS (6 electrons, Z = 11)

(C I sequence, Configuration: $1s^2 2s2p^3nl, 1s^2 2s^2 2pnl$, Singlet System)

Na VI
SINGLET

KEY

920826.———3d'(2)

3d'(2)———J value
n
Energy Level (cm⁻¹)

Ionization Level
1388500 cm⁻¹
(172.15 eV)

[Na VI $2s^2 2p^2$ $^3P_0$ → Na VII $2s^2 2p$ $^2P^o_{1/2}$]

nl core = $2s^2 2p(^2P^o)$,
nl core = $2s2p^2(^4P)$,
n'l' core = $2s2p^2(^2D)$

$2s2p^2np\ ^1F^o$
$2s^2 2pnd$

1280111.———6d(3)
1231092.———5d(3)
1140841.———4d(3)
1072016.———3p'(3)

945429.———3d(3)

$2s2p^2nd\ ^1F$

1147828.———3d'(3)

$2s2pnp\ ^1D^o$
$2s^2 2pnd$

1131152.———4d(2)
1077872.———3p'(2)

920826.———3d(2)

312295.———2p(2)

$2s2p^2nl\ ^1D$
$2s^2 2pnp$

1147855.———3d'(2)

1033341.———3s'(2)

888000.———3p(2)

539430.———2p⁴(2)

35478.———2p²(2)

$2s2p^2np\ ^1P^o$
$2s^2 2pnl$

946512.———3d(1)

817718.———3s(1)

350299.———2p³(1)

$2s2p^2nd\ ^1P$

1151260.———3d'(1)

$2s^2 2pnp\ ^1S$

579153.———2p⁴(0)

74394.———2 (0)

140x10⁴

110x10⁴

80x10⁴

50x10⁴

20x10⁴

0

ENERGY (cm⁻¹)

365

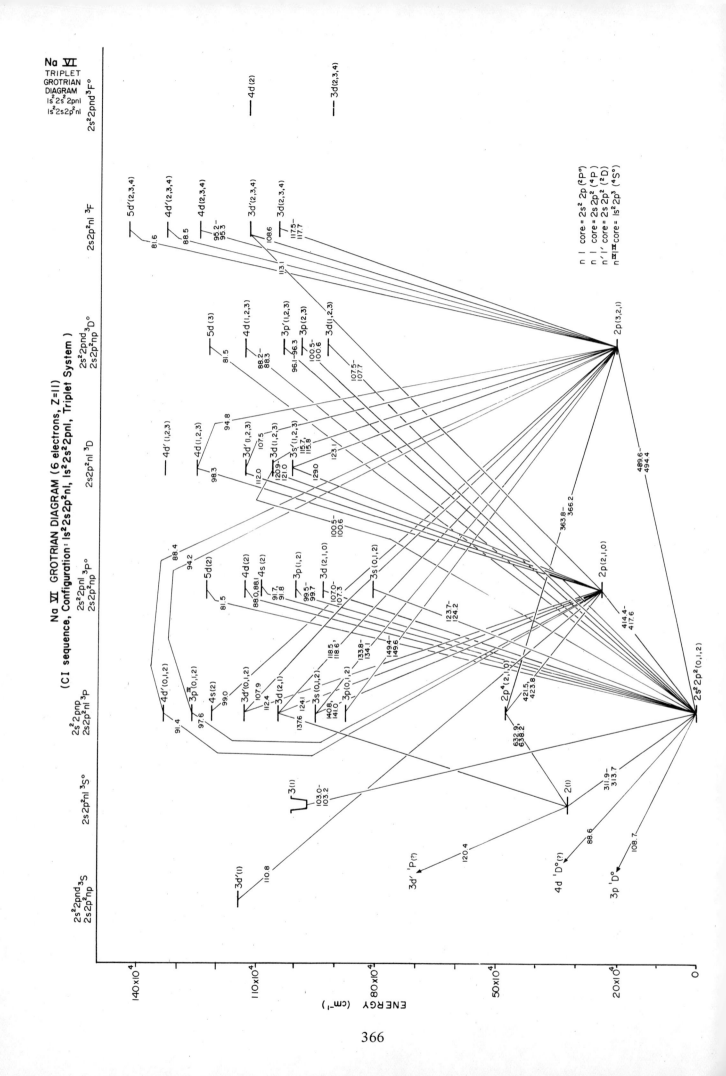

Na VI
TRIPLET
GROTRIAN
DIAGRAM
1s² 2s² 2pnl
1s² 2s2p² nl

Na VI GROTRIAN DIAGRAM (6 electrons, Z=11)
(CI sequence, Configuration: 1s²2s²2pnl, 1s²2s2p²nl, Triplet System)

366

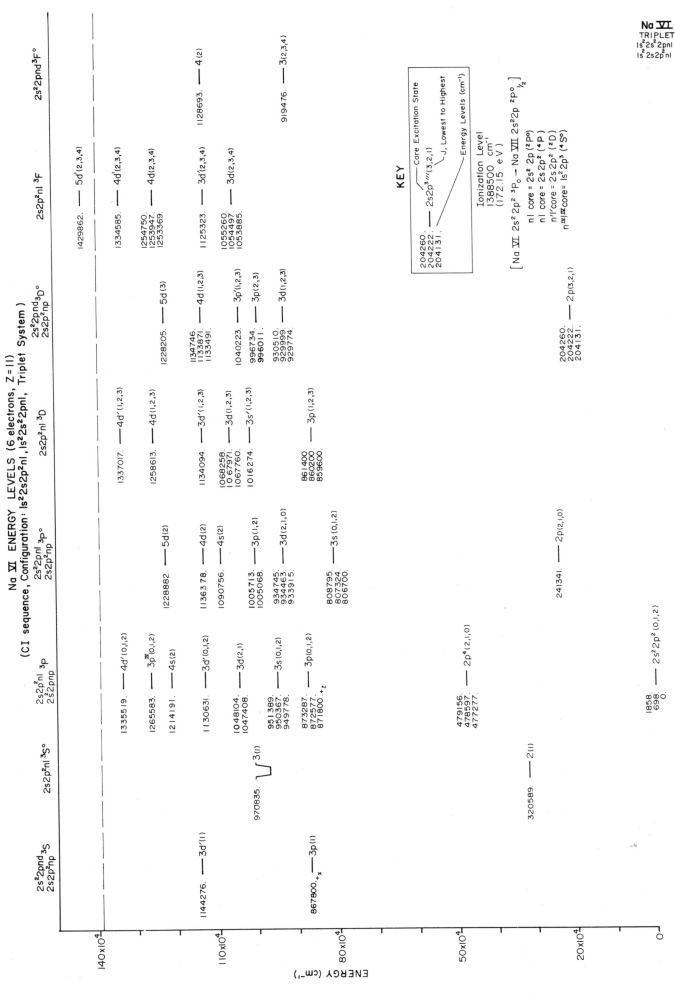

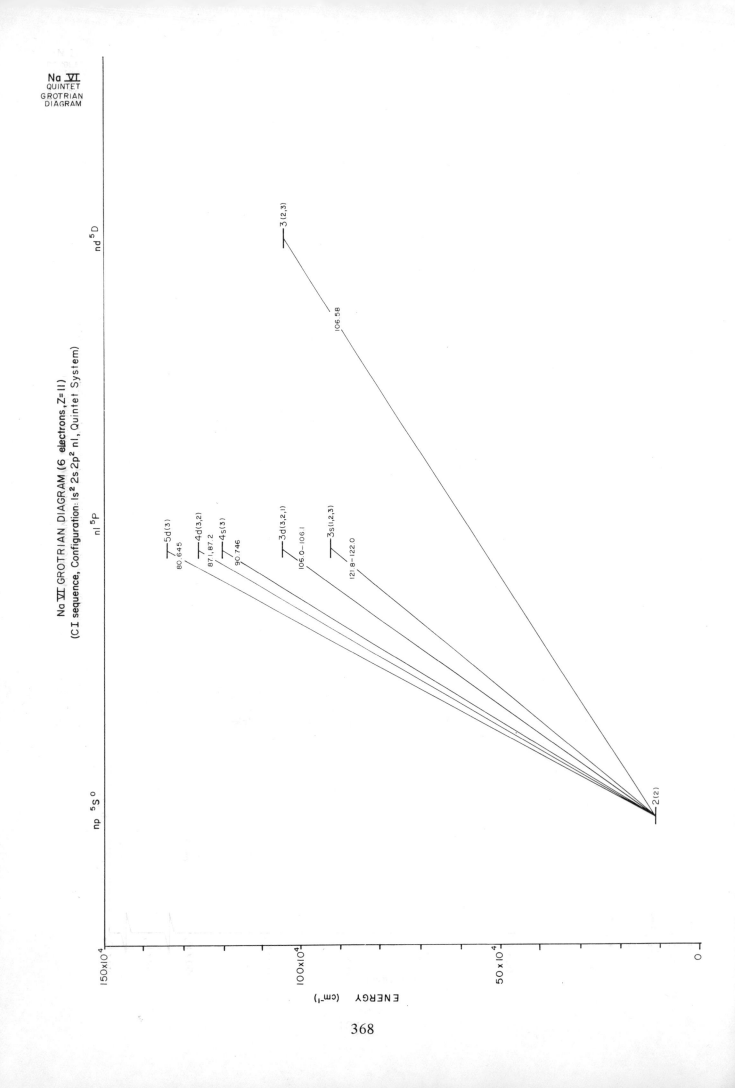

Na $\overline{\text{VI}}$
QUINTET
GROTRIAN
DIAGRAM

Na $\overline{\text{VI}}$ GROTRIAN DIAGRAM (6 electrons, Z=11)
(C I sequence, Configuration: ls² 2s 2p² nl, Quintet System)

nd ⁵D

nl ⁵P

np ⁵Sº

3(2,3)

106.58

5d(3)
80.645
4d(3,2)
87.1,87.2
4s(3)
90.746
3d(3,2,1)
106.0—106.1
3s(1,2,3)
121.8—122.0

2(2)

ENERGY (cm⁻¹)

150x10⁴

100x10⁴

50 x10⁴

0

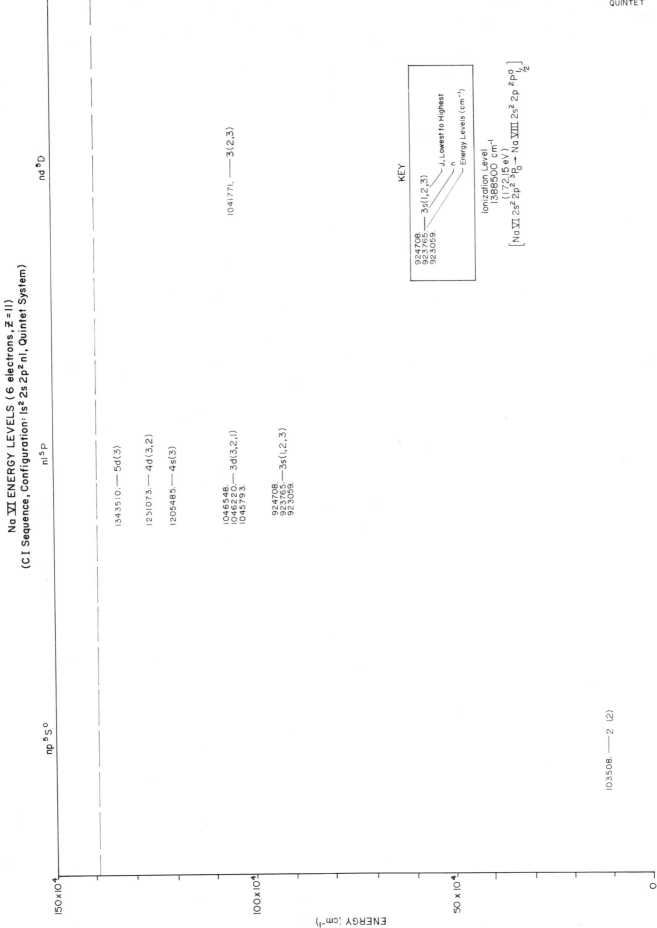

Na VI ENERGY LEVELS (6 electrons, Z =11)
(C I Sequence, Configuration: 1s² 2s 2p² nl, Quintet System)

Na VI
QUINTET

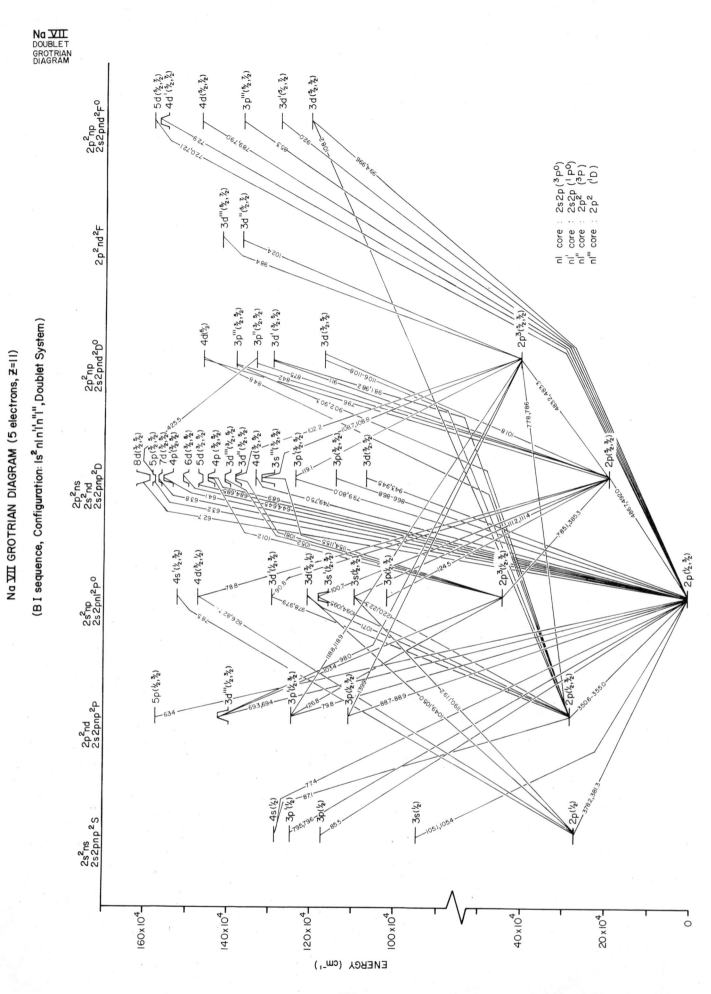

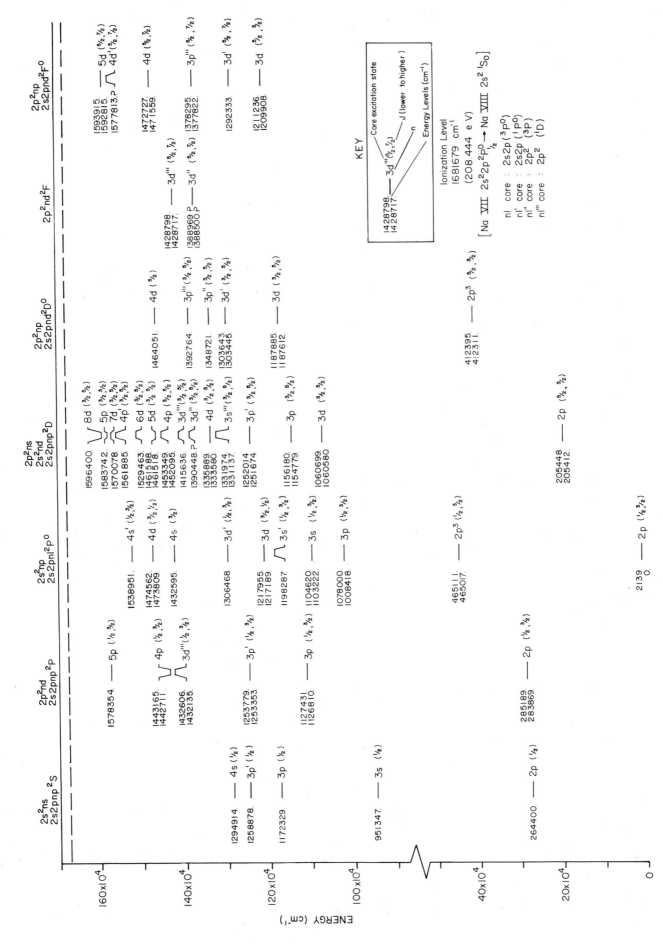

Na VII ENERGY LEVELS (5 electrons, Z = 11)

(B I sequence, Configuration: 1s²nln'l'n''l'', Doublet System)

371

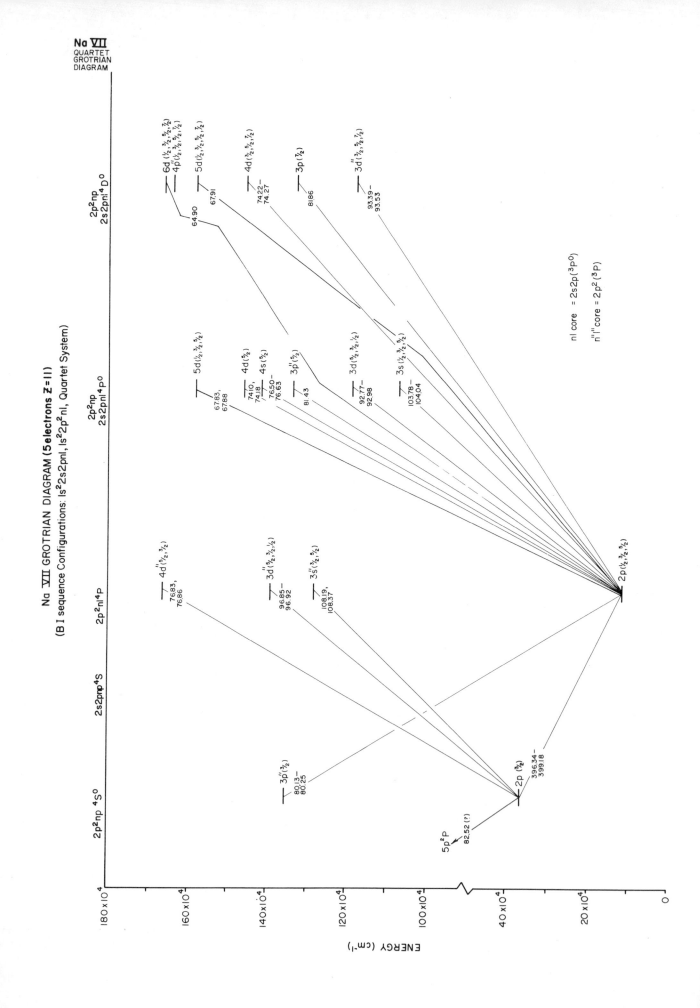

Na VII GROTRIAN DIAGRAM (5 electrons Z=11)

(B I sequence Configurrations: Is²2s2pnl, Is²2p²nl, Quartet System)

Na VII
QUARTET
GROTRIAN
DIAGRAM

nl core = 2s2p(³P⁰)
n''l''core = 2p²(³P)

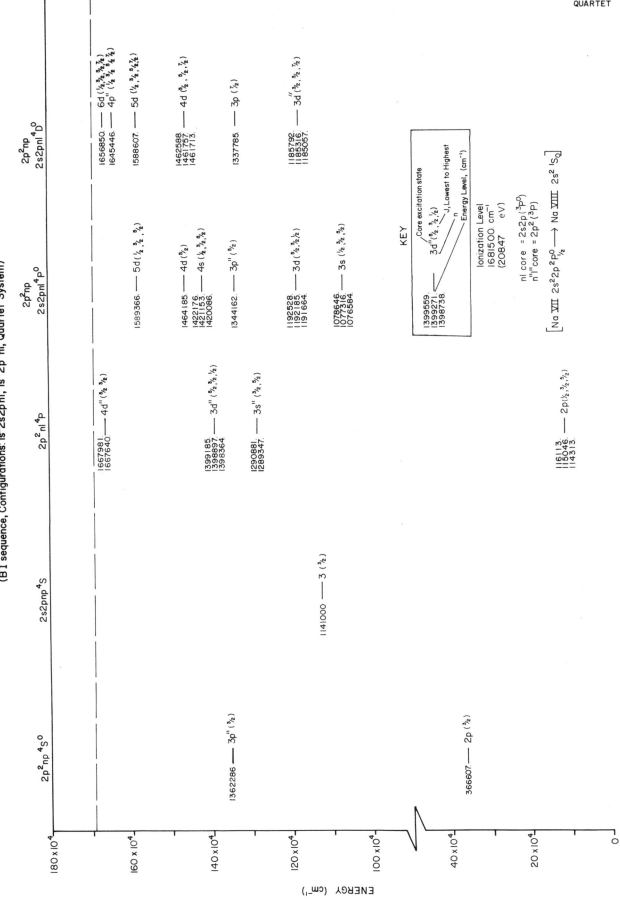

Na VII ENERGY LEVELS (5 electrons, $\bar{Z}$=11)
(B I sequence, Configurations: 1s²2s2pnl, 1s²2p²nl, Quartet System)

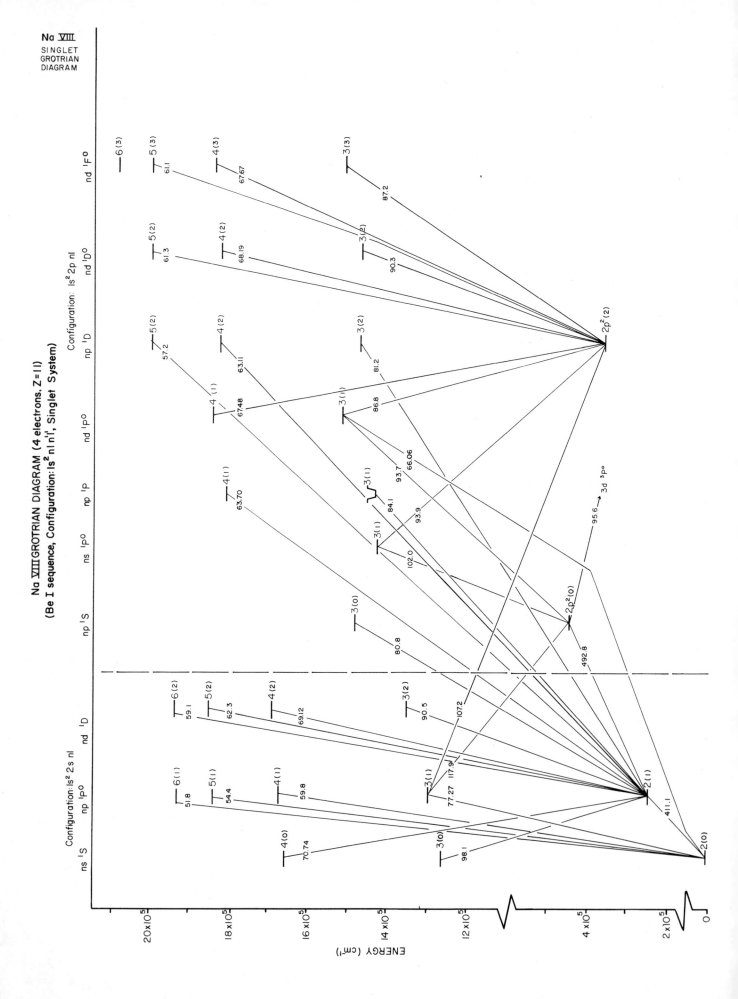

Na VIII
SINGLET
GROTRIAN
DIAGRAM

Na VIII GROTRIAN DIAGRAM (4 electrons, Z=11)
(Be I sequence, Configuration: ls² nl n'l', Singlet System)

Configuration: ls² 2p nl

nd ¹F°   nd ¹D°

Configuration: ls² 2s nl

nd ¹D   np ¹P°   ns ¹S

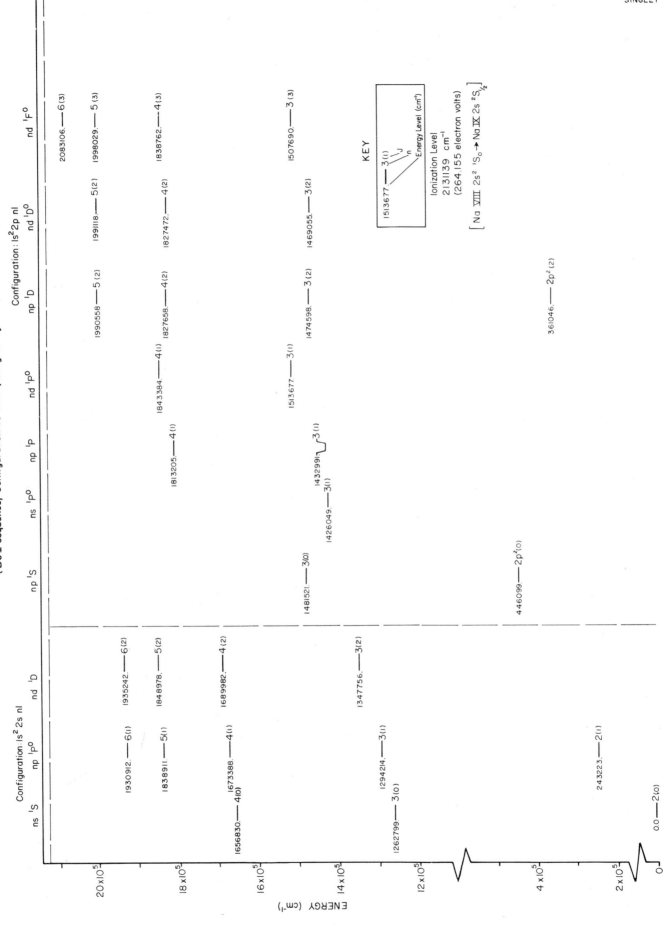

Na VIII ENERGY LEVELS (4 electrons, Z=11)
(Be I sequence, Configuration: 1s²nl n'l', Singlet System)

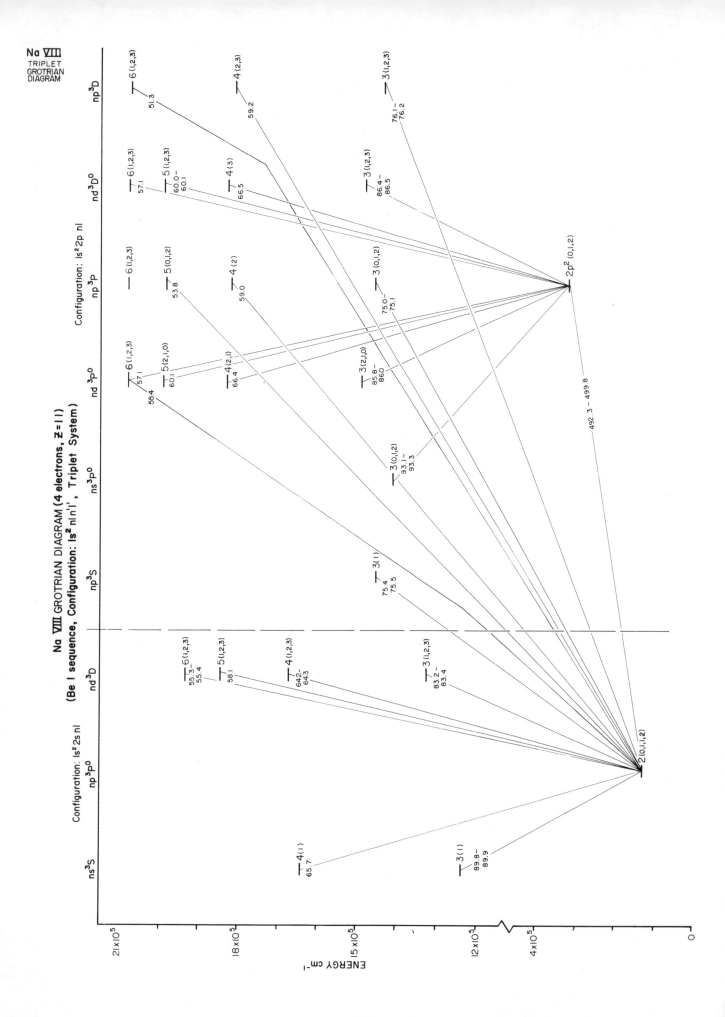

Na VIII
TRIPLET
GROTRIAN
DIAGRAM

Na VIII GROTRIAN DIAGRAM (4 electrons, Z = 11)
(Be I sequence, Configuration: 1s²nln'l', Triplet System)

Configuration: 1s²2p nl

Configuration: 1s²2s nl

ENERGY cm⁻¹

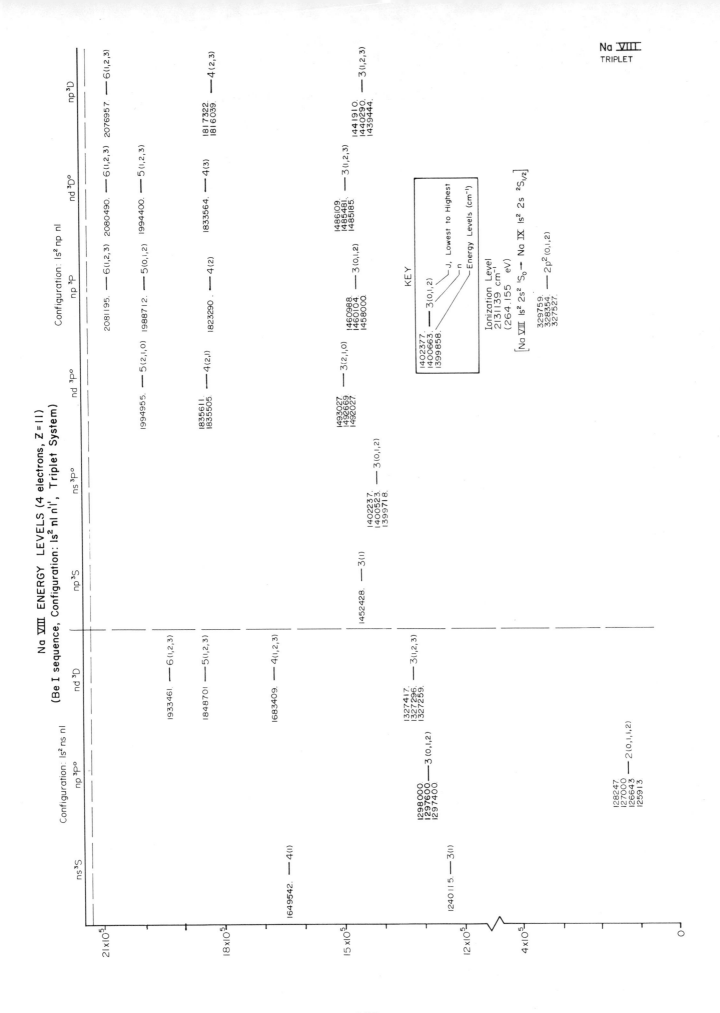

Na VIII ENERGY LEVELS (4 electrons, Z = 11)
(Be I sequence, Configuration: 1s² nl nl'l', Triplet System)

Na VIII
TRIPLET

Na IX GROTRIAN DIAGRAM (3 electrons, Z = 11)
(Li I sequence, Configuration: ls² nl)

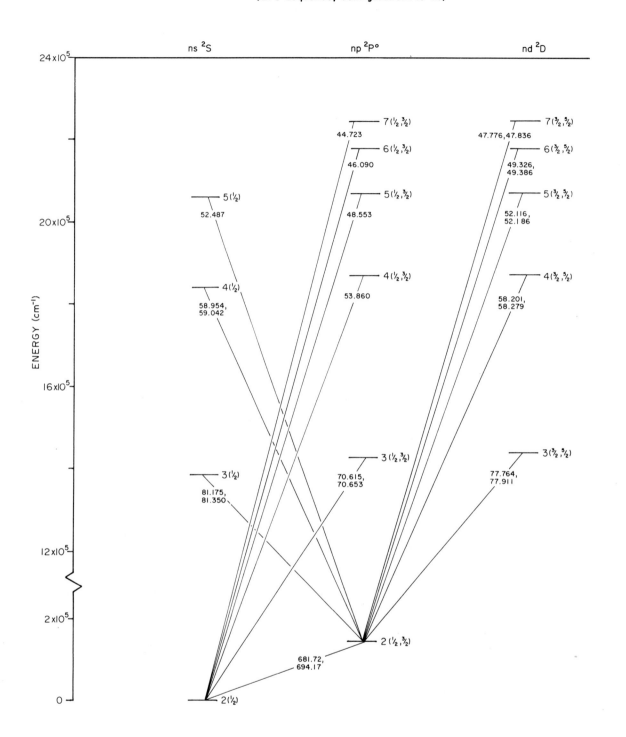

Na IX ENERGY LEVELS   (3 electrons, Z = 11)
(Li I sequence,   Configuration : 1s² nl)

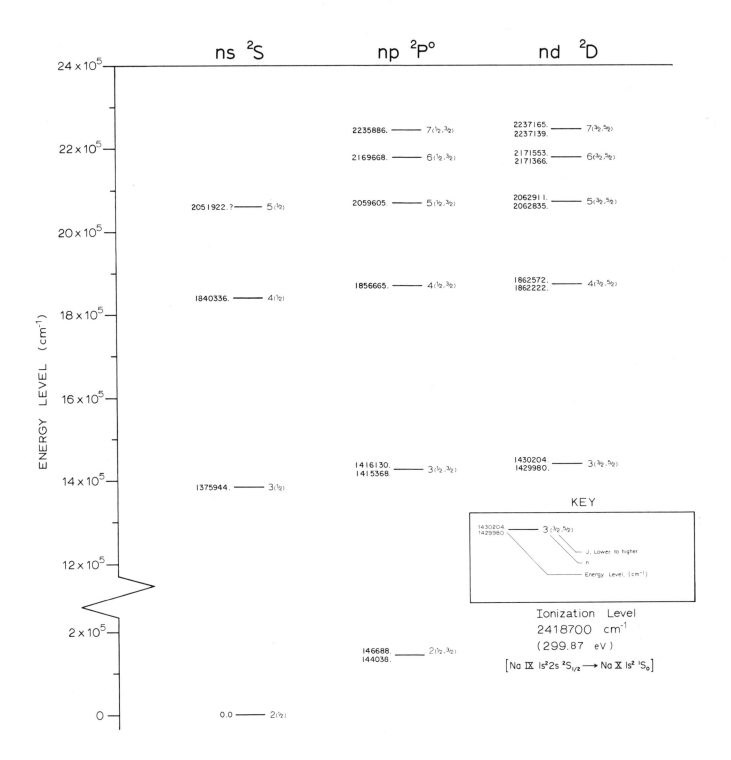

ns ²S   np ²Pº   nd ²D

ENERGY LEVEL (cm⁻¹)

24 × 10⁵

2235886. ——— 7(½,³⁄₂)   2237165.
2237139. ——— 7(³⁄₂,⁵⁄₂)

22 × 10⁵   2169668. ——— 6(½,³⁄₂)   2171553.
2171366. ——— 6(³⁄₂,⁵⁄₂)

2051922.? ——— 5(½)   2059605. ——— 5(½,³⁄₂)   2062911.
2062835. ——— 5(³⁄₂,⁵⁄₂)

20 × 10⁵

1856665. ——— 4(½,³⁄₂)   1862572.
1862222. ——— 4(³⁄₂,⁵⁄₂)

1840336. ——— 4(½)

18 × 10⁵

16 × 10⁵

1416130.
1415368. ——— 3(½,³⁄₂)   1430204.
1429980. ——— 3(³⁄₂,⁵⁄₂)

14 × 10⁵   1375944. ——— 3(½)

KEY

1430204
1429980 ——— 3(³⁄₂,⁵⁄₂)

J, Lower to higher
n
Energy Level, (cm⁻¹)

12 × 10⁵

Ionization Level
2418700 cm⁻¹
(299.87 eV)

[Na IX 1s²2s ²S₁/₂ ⟶ Na X 1s² ¹S₀]

2 × 10⁵

146688.
144038. ——— 2(½,³⁄₂)

0   0.0 ——— 2(½)

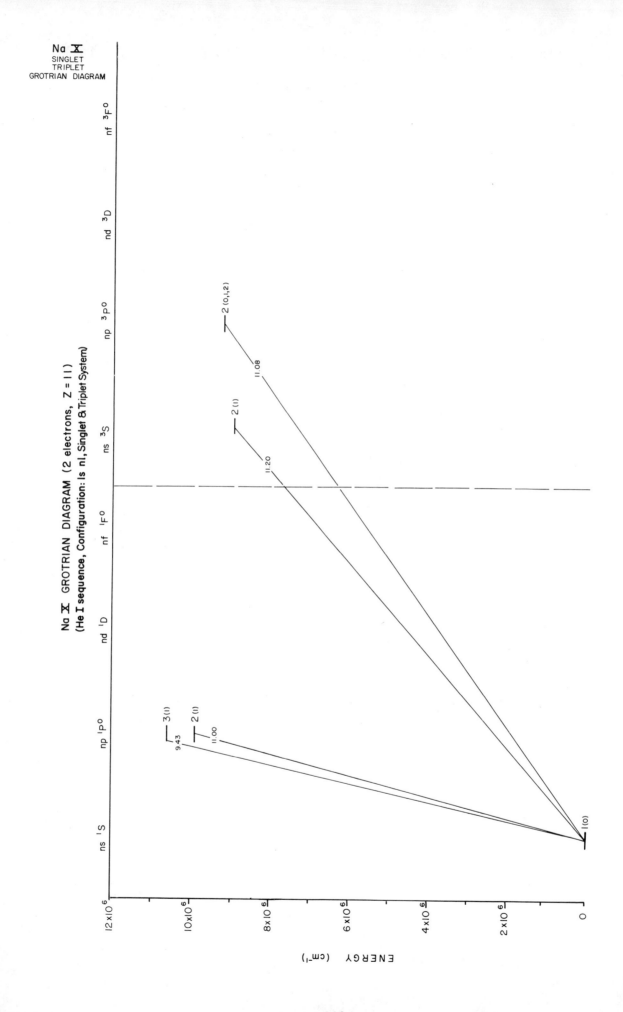

Na X GROTRIAN DIAGRAM (2 electrons, Z = 11)
(He I sequence, Configuration: ls nl, Singlet & Triplet System)

Na X
SINGLET
TRIPLET
GROTRIAN DIAGRAM

ENERGY (cm⁻¹)

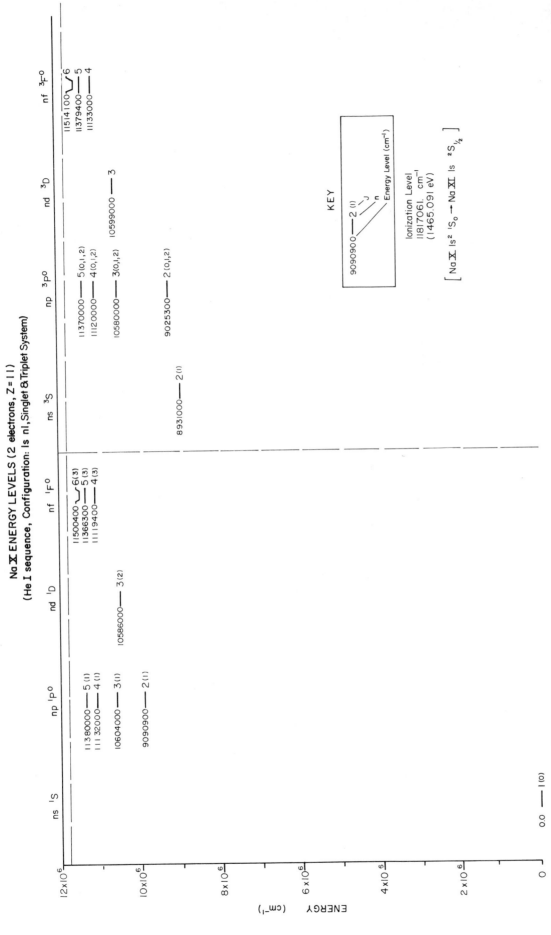

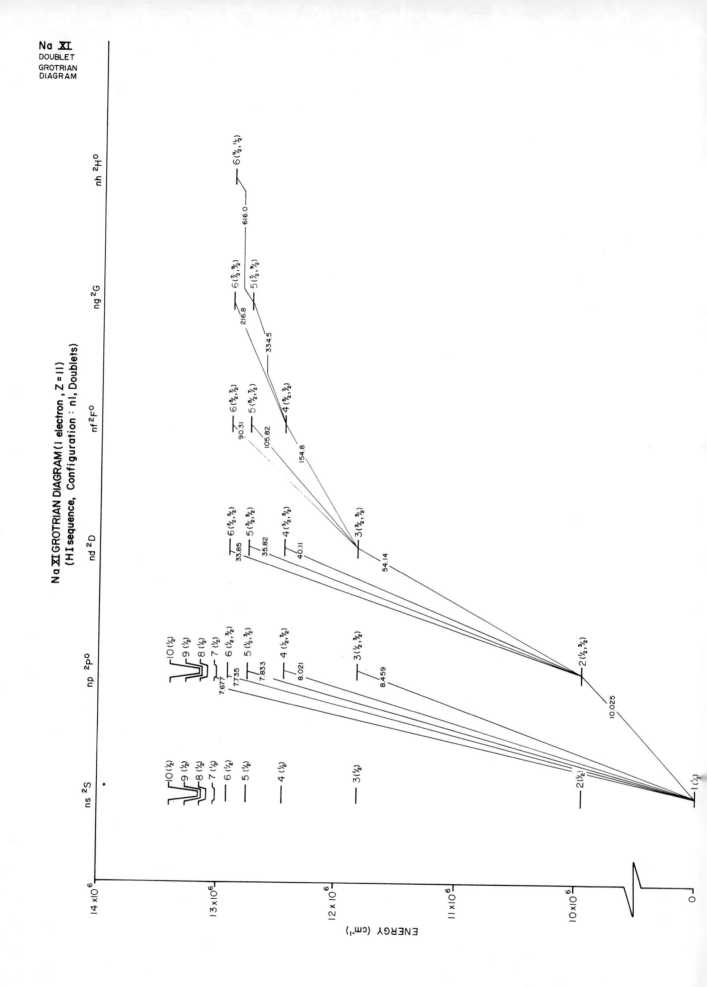

Na XI
DOUBLET
GROTRIAN
DIAGRAM

Na XI GROTRIAN DIAGRAM (I electron , Z = II)
(H I sequence, Configuration : nl, Doublets)

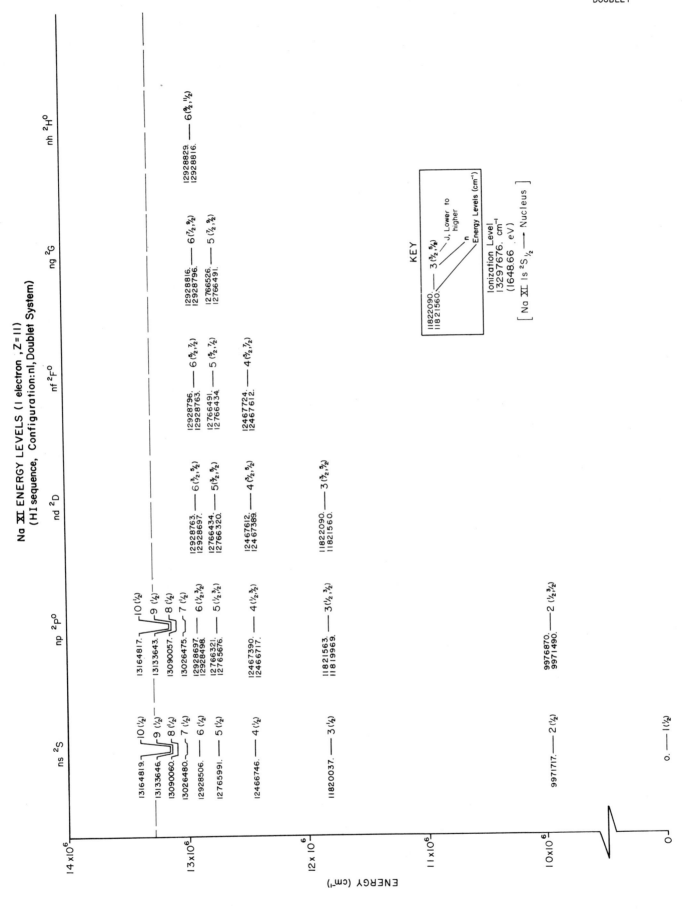

Na XI ENERGY LEVELS (1 electron, Z=11)
(HI sequence, Configuration: nl, Doublet System)

Na XI
DOUBLET

Na I $Z = 11$ 11 electrons

L. Johansson, Ark. Fys. **20**, 135 (1961).

    Author gives a line table of lines observed in the region 2380–12 680 Å.

Na II $Z = 11$ 10 electrons

Please see the general references.
The intermediate-coupling designation is taken from Kelly's unpublished tabulation of energy levels. He quotes A.M. Crooker, unpublished data (1966).

Na III $Z = 11$ 9 electrons

    Please see the general references.

Na IV $Z = 11$ 8 electrons

I.S. Bowen, Ap. J. **132**, 1 (1960).

    Author gives transition wavelengths from nebular and laboratory observations.

B. Edlén, Handbuch der Phys. **27**, 172 (1964).

    Author gives corrections to values in C.E. Moore's *Atomic Energy Levels.*

Na V $Z = 11$ 7 electrons

I.S. Bowen, Ap. J. **132**, 1 (1960).

    Author gives transition wavelengths from laboratory observations.

B. Edlén, Handbuch der Phys. **27**, 172 (1964).

    Author gives corrections to values in C.E. Moore's *Atomic Energy Levels.*

Na VI $Z = 11$ 6 electrons

    Please see the general references.

Na VII $Z = 11$ 5 electrons

B. Edlén, Handbuch der Phys. **27**, 172 (1964).

    Author gives corrections to values in C.E. Moore's *Atomic Energy Levels.*

    Please note that the following lines were omitted:

$2s^2\,2p\;^2P^\circ - 2p^2\,(^3P)\,3s\;^4P$	77.56 (?)	
$2s\,2p^2\;^2D\;-$	$4d\;^4D$	79.62 (?)
$2p^3\;^2P^\circ\;-$	$4d\;^4P$	83.18 (?)
$-$	$3d\;^4P$	107.09

Na VIII $Z = 11$ 4 electrons

Please see the general references.
Note that the following line was omitted: $2s^2\ {}^1S-2s\ 2p\ {}^3P°$: 789.6.

Na IX $Z = 11$ 3 electrons

Please see the general references.

Na X $Z = 11$ 2 electrons

Please see the general references.

Na XI $Z = 11$ 1 electron

Please see the general references.

Na VIII $Z = 11$ 4 electrons

$2s^2\ {}^1S-2s\ 2p\ {}^3P°$: 789.6.

Magnesium (Mg)

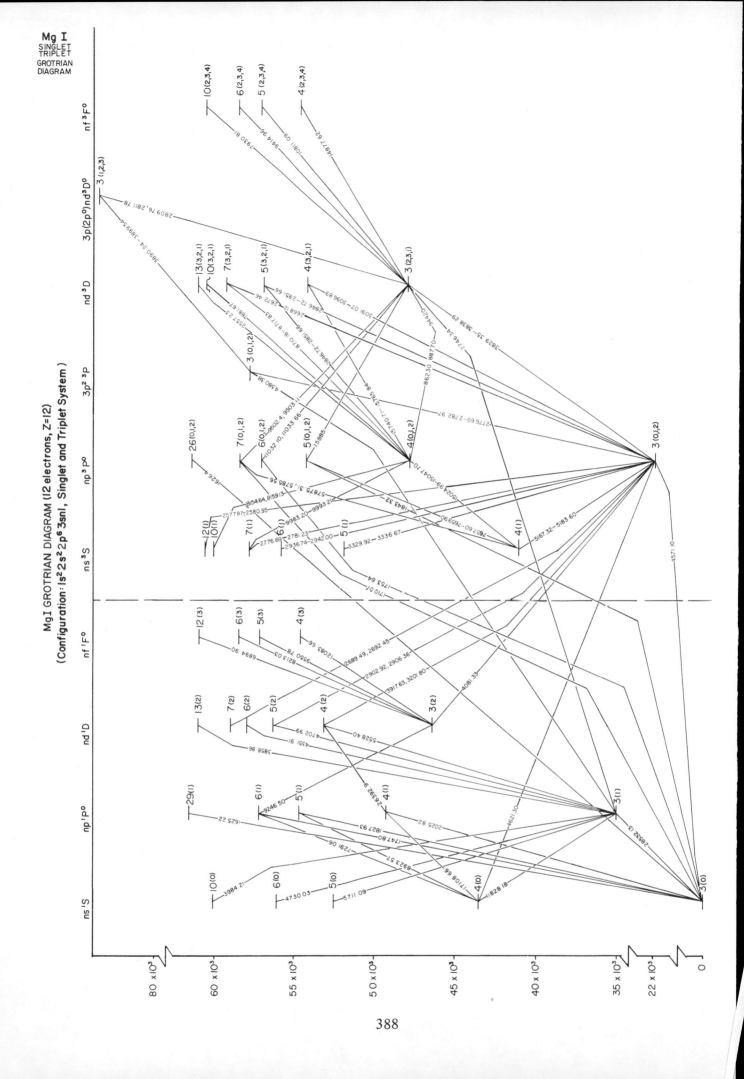

MgI GROTRIAN DIAGRAM (12 electrons, Z=12)
(Configuration: 1s²2s²2p⁶3snl, Singlet and Triplet System)

Mg I
SINGLET
TRIPLET
GROTRIAN
DIAGRAM

388

Mg I ENERGY LEVELS (12 electrons, Z=12)
(Configuration: 1s² 2s² 2p⁶ 3snl, Singlet and Triplet System)

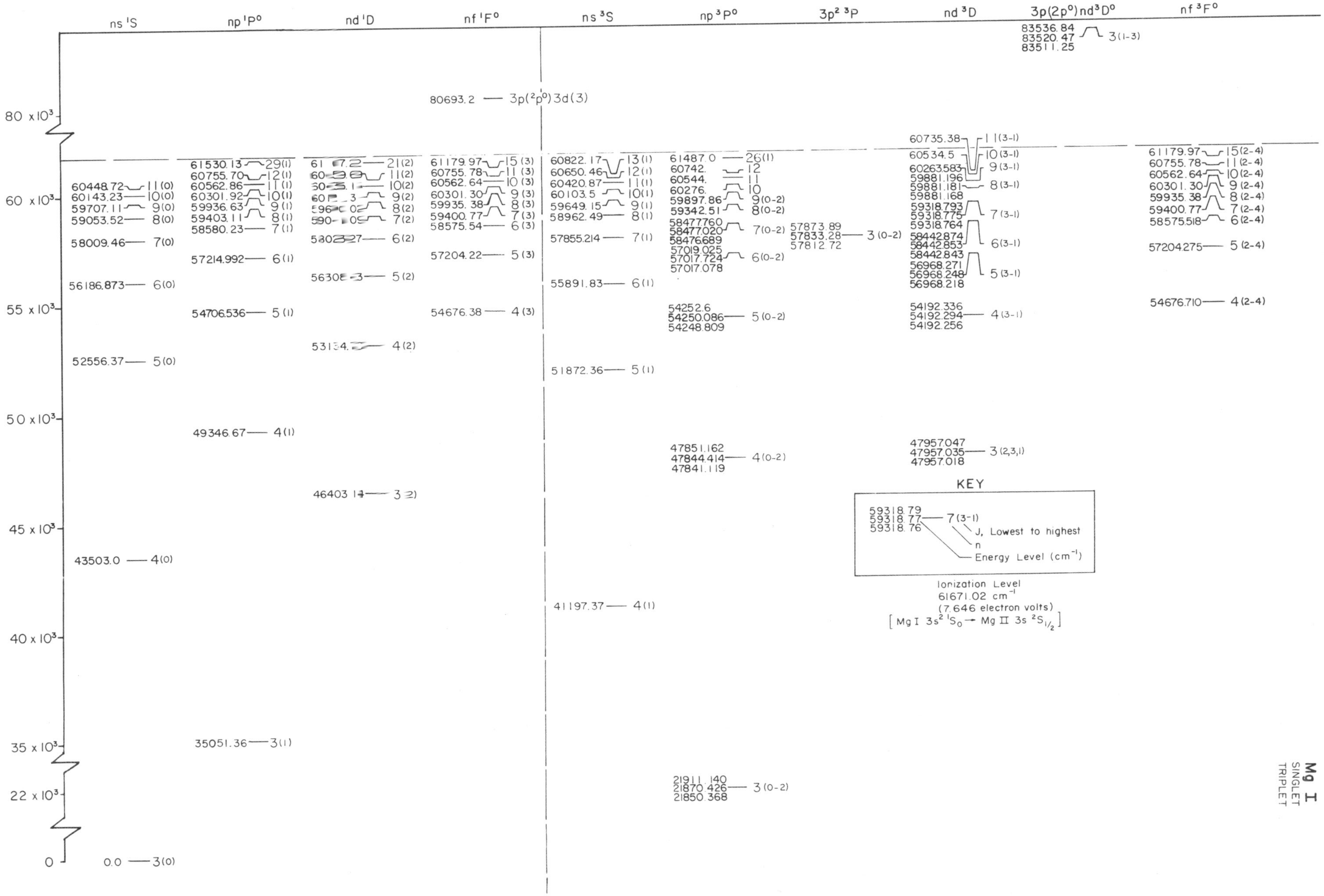

Column headings (left to right):

ns ¹S   np ¹P°   nd ¹D   nf ¹F°   ns ³S   np ³P°   nd ³D   3p² ³P   3p(2p°)nd³D°   nf ³F°

ns ¹S
- 60448.72 — 11(0)
- 60143.23 — 10(0)
- 59707.11 — 9(0)
- 59053.52 — 8(0)
- 58009.46 — 7(0)
- 56186.873 — 6(0)
- 52556.37 — 5(0)
- 43503.0 — 4(0)
- 0.0 — 3(0)

np ¹P°
- 61530.13 — 29(1)
- 60755.70 — 12(1)
- 60562.86 — 11(1)
- 60301.92 — 10(1)
- 59936.63 — 9(1)
- 59403.11 — 8(1)
- 58580.23 — 7(1)
- 57214.992 — 6(1)
- 54706.536 — 5(1)
- 49346.67 — 4(1)
- 35051.36 — 3(1)

nf ¹F°
- 61179.97 — 15(3)
- 60755.78 — 11(3)
- 60562.64 — 10(3)
- 60301.30 — 9(3)
- 59935.38 — 8(3)
- 59400.77 — 7(3)
- 58575.54 — 6(3)
- 57204.22 — 5(3)
- 54676.38 — 4(3)
- 80693.2 — 3p(2p°)3d(3)

ns ³S
- 60822.17 — 13(1)
- 60650.46 — 12(1)
- 60420.87 — 11(1)
- 60103.5 — 10(1)
- 59649.15 — 9(1)
- 58962.49 — 8(1)
- 57855.214 — 7(1)
- 55891.83 — 6(1)
- 51872.36 — 5(1)
- 41197.37 — 4(1)

np ³P°
- 61487.0 — 26(1)
- 60742. — 12
- 60544. — 10
- 60276. — 9(0-2)
- 59897.86 — 8(0-2)
- 59342.51 —
- 58477.760 / 58477.009 / 58476.825 — 7(0-2)
- 57019.025 / 57017.724 / 57017.078 — 6(0-2)
- 54252.6 / 54250.086 / 54248.809 — 5(0-2)
- 47851.162 / 47844.414 / 47841.19 — 4(0-2)
- 21911.440 / 21870.426 / 21850.368 — 3(0-2)

nd ³D
- 60735.38 — 11(3-1)
- 60534.5 — 10(3-1)
- 60263.583 — 9(3-1)
- 59881.196 / 59881.181 / 59881.168 — 8(3-1)
- 59318.793 / 59318.775 / 59318.764 — 7(3-1)
- 58442.874 / 58442.853 / 58442.843 — 6(3-1)
- 56968.271 / 56968.248 / 56968.218 — 5(3-1)
- 54192.336 / 54192.294 / 54192.256 — 4(3-1)
- 47957.047 / 47957.035 / 47957.018 — 3(2,3,1)

3p² ³P
- 57873.89
- 57833.28
- 57812.72

3p(2p°)nd³D°
- 83536.84 / 83520.47 / 83511.25 — 3(1-3)

nf ³F°
- 61179.97 — 15(2-4)
- 60755.78 — 11(2-4)
- 60562.64 — 10(2-4)
- 60301.30 — 9(2-4)
- 59935.38 — 8(2-4)
- 59400.77 — 7(2-4)
- 58575.518 — 6(2-4)
- 57204.275 — 5(2-4)
- 54676.710 — 4(2-4)

KEY
- 59318.79
- 59318.79 — 7(3-1)
- 59318.76 — J, Lowest to highest
- n
- Energy Level (cm⁻¹)

Ionization Level
61671.02 cm⁻¹
(7.646 electron volts)
[Mg I 3s² ¹S₀ → Mg II 3s ²S₁/₂]

Y-axis: 80 × 10³, 60 × 10³, 55 × 10³, 50 × 10³, 45 × 10³, 40 × 10³, 35 × 10³, 22 × 10³, 0

389

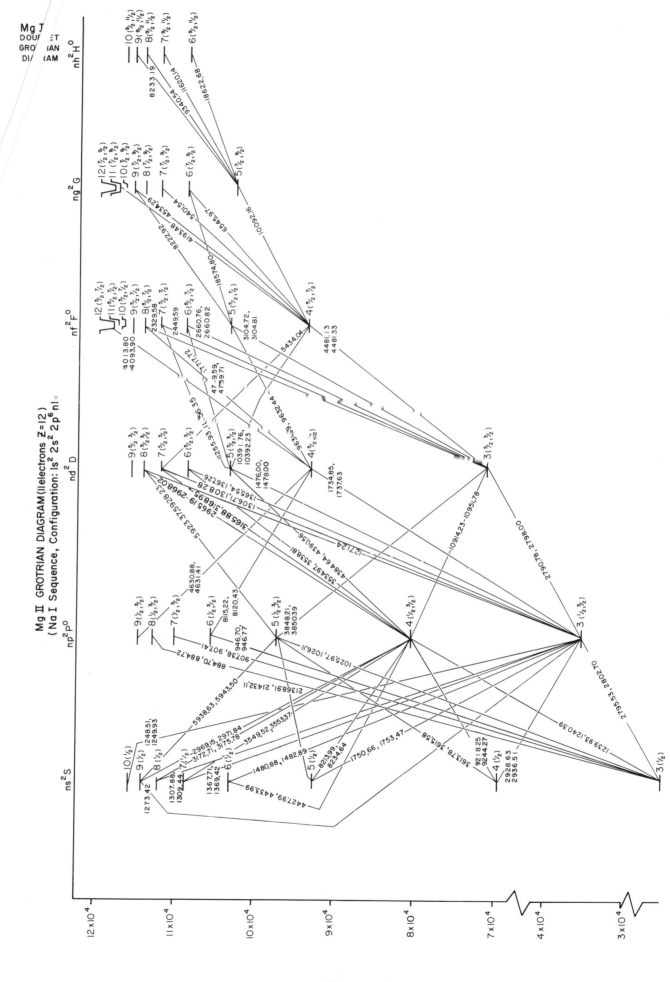

Mg II GROTRIAN DIAGRAM(11electrons Z̄=12)
(Na I Sequence, Configuration: 1s² 2s² 2p⁶ nl

Mg II ENERGY LEVELS (11 electrons, Z=12)
(Na I Sequence, Configuration: 1s² 2s² 2p⁶ nl Doublet)

Mg II
DOUBLET

ns ²S

115764.99 — 10 (½)
114289.36 — 9 (½)
112129.20 — 8 (½)

108784.33 — 7 (½)

103196.75 — 6 (½)

92790.51 — 5 (½)

69804.95 — 4 (½)

0.00 — 3 (½)

np ²P⁰

114898.72 — 9 (½, ³⁄₂)
114896.79
113033.09 — 8 (½, ³⁄₂)
113030.25
110207.99 — 7 (½, ³⁄₂)
110203.58

105629.72 — 6 (½, ³⁄₂)
105622.34

97468.92 — 5 (½, ³⁄₂)
97455.12

80650.02 — 4 (½, ³⁄₂)
80613.50

35760.88 — 3 (½, ³⁄₂)
35669.31

nd ²D

115794.41 — 9 (⁵⁄₂, ³⁄₂)
114332.74 — 8 (⁵⁄₂, ³⁄₂)
114332.68
112197.16 — 7 (⁵⁄₂, ³⁄₂)
112197.06

108900.19 — 6 (⁵⁄₂, ³⁄₂)
108900.03

103419.70 — 5 (⁵⁄₂, ³⁄₂)
103420.00

93311.11 — 4 (⁵⁄₂, ³⁄₂)
93310.59

71491.06 — 3 (⁵⁄₂, ³⁄₂)
71490.19

nf ²F⁰

118218.5 — 12 (⁵⁄₂, ⁷⁄₂)
117638.31 — 11 (⁵⁄₂, ⁷⁄₂)
116875.25 — 10 (⁵⁄₂, ⁷⁄₂)
115844.60 — 9 (⁵⁄₂, ⁷⁄₂)
114403.55 — 8 (⁵⁄₂, ⁷⁄₂)
112301.47 — 7 (⁵⁄₂, ⁷⁄₂)

109062.34 — 6 (⁵⁄₂, ⁷⁄₂)

103689.89 — 5 (⁵⁄₂, ⁷⁄₂)

93799.70 — 4 (⁵⁄₂, ⁷⁄₂)

ng ²G

118220.2 — 12 (⁷⁄₂, ⁹⁄₂)
117639.51 — 11 (⁷⁄₂, ⁹⁄₂)
116877.54 — 10 (⁷⁄₂, ⁹⁄₂)
115847.67 — 9 (⁷⁄₂, ⁹⁄₂)
114407.88 — 8 (⁷⁄₂, ⁹⁄₂)
112307.79 — 7 (⁷⁄₂, ⁹⁄₂)

109072.05 — 6 (⁷⁄₂, ⁹⁄₂)

103705.66 — 5 (⁷⁄₂, ⁹⁄₂)

nh ²H⁰

116878.04 — 10 (⁹⁄₂, ¹¹⁄₂)
115848.28 — 9 (⁹⁄₂, ¹¹⁄₂)
114408.74 — 8 (⁹⁄₂, ¹¹⁄₂)
112309.06 — 7 (⁹⁄₂, ¹¹⁄₂)

109074.0 — 6 (⁹⁄₂, ¹¹⁄₂)

KEY

71491.32 ──── 3 (³⁄₂, ⁵⁄₂)
71490.41

                L values (lower to higher)
                n
                Energy Level (cm⁻¹)

Ionization Level
121267.61 cm⁻¹
15.035 eV

[Mg II 2p⁶ 3s ²S₁/₂ ⟶ Mg III 2p⁶ ¹S₀]

ENERGY (cm⁻¹)

12×10⁴
11×10⁴
10×10⁴
9×10⁴
8×10⁴
7×10⁴
4×10⁴
3×10⁴
0

391

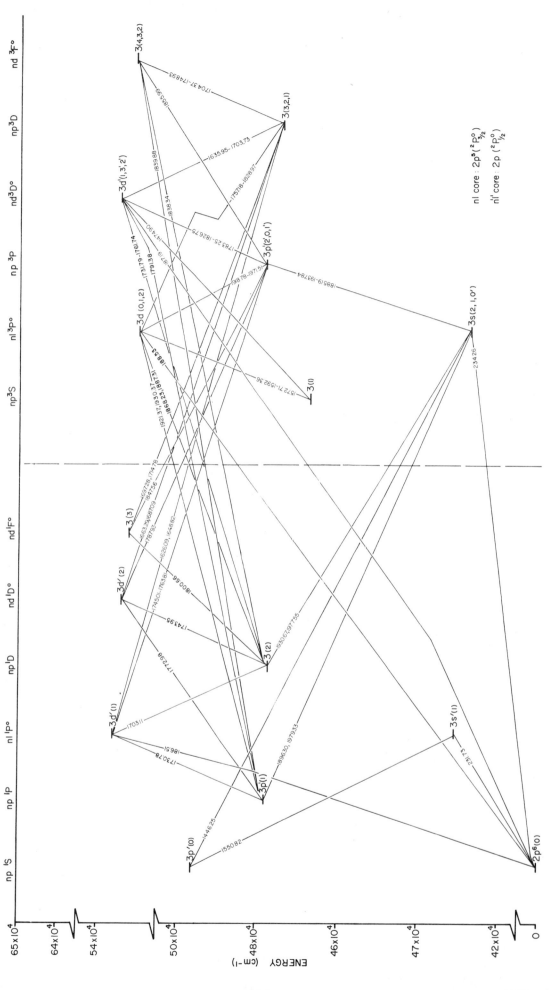

Mg III GROTRIAN DIAGRAM (10 electrons, Z=12)
(Ne I sequence, Configuration: $1s^2 2s^2 2p^5 nl$, Singlet & Triplet Terms)

Mg III ENERGY LEVELS (10 electrons, Z = 12)

(Ne I sequence, Configuration: $1s^2\,2s^2\,2p^5\,nl$, Singlet & Triplet Terms)

Mg III
SINGLET
TRIPLET

KEY

$[^2P^o_{3/2}]$ core
$[^2P^o_{1/2}]$ core

$3s(2,1,0)$
J (lowest to highest)
n
Energy Level (cm^{-1})

427852.1
426868.1
425640.3

Ionization Level
646410 cm^{-1}
(80.143 eV)

$[$ Mg III $1s^2\,2s^2\,2p^6\,{}^1S_0 \rightarrow$ Mg IV $1s^2\,2s^2\,2p^5\,{}^2P^o_{3/2}$ $]$

nl core $2p^5\,(^2P^o_{3/2})$
$n'l'$ core $2p^5\,(^2P^o_{1/2})$

393

Mg III
INTERMEDIATE
COUPLING
GROTRIAN
DIAGRAM

Mg III GROTRIAN DIAGRAM (10 electrons, Z = 12)

(Ne I sequence, Configuration: 1s² 2s² 2p⁵ nl, Intermediate Coupling)

For transitions see

Andersson, Physica Scripta,

3 203-210 (1971)

KEY

Parity

4 [3/2]° (2,1) FINAL J

J l Coupling

n

n l core = 2p⁵ (²P₃/₂)

n' l' core = 2p⁵ (²P₁/₂)

ENERGY (cm-1)

394

Mg III ENERGY LEVELS (10 electrons, Z=12)

(Ne I sequence, Intermediate Coupling, Configuration $1s^2 2s^2 2p^5 (^2P^o_{1/2}) n'l' [J,l] (J)$)

ns

611299. —— $6[\frac{1}{2}]^o(1)$

5912207 —— $5[\frac{1}{2}]^o(0,1)$
5909842

5487207 —— $4[\frac{1}{2}]^o(0,1)$
5480344

np

5701128 $\quad[\frac{1}{2}](0)$
5647316 —4 $[\frac{1}{2}](1)$
5646646 $\quad[\frac{3}{2}](2)$
5642898 $\quad[\frac{3}{2}](1)$

nd

632988. ——8 $[\frac{3}{2}]^o(1)$
628105. ——7 $[\frac{3}{2}]^o(1)$

620598 ——6 $[\frac{3}{2}]^o(1)$

608332. ——5 $[\frac{3}{2}]^o(1)$

5854662 $\quad[\frac{3}{2}]^o(1)$
5847379 $\quad[\frac{3}{2}]^o(2)$
5847334 —4 $[\frac{3}{2}]^o(3)$
5846285 $\quad[\frac{5}{2}]^o(2)$

nf

6090505 $\quad[\frac{5}{2}](2)$
6090457 $\quad[\frac{7}{2}](4)$
6090454 —5 $[\frac{7}{2}](3)$
6090452 $\quad[\frac{5}{2}](3)$

5867916 $\quad[\frac{5}{2}](2)$
5867851 $\quad[\frac{7}{2}](3)$
5867795 —4 $[\frac{7}{2}](4)$
5867779 $\quad[\frac{5}{2}](3)$

ng

6091116 $\quad[\frac{7}{2}]^o(4,3)$
6091114 —5 $[\frac{9}{2}](5,4)$

Legend box:

546531.6 —— 4 $[\frac{3}{2}]^o(2,1)$
5458204

J — l Coupling

n

Energy Levels (cm⁻¹)

Ionization Level
646410 cm⁻¹
(80.143 eV)

[Mg III $1s^2 2s^2 2p^6$ $^1S_0$ $\rightarrow$ Mg IV $1s^2 2s^2 2p^5$ $^2P^o_{3/2}$]

ENERGY (cm⁻¹)

65×10^4

60×10^4

55×10^4

50×10^4

45×10^4

40×10^4

0

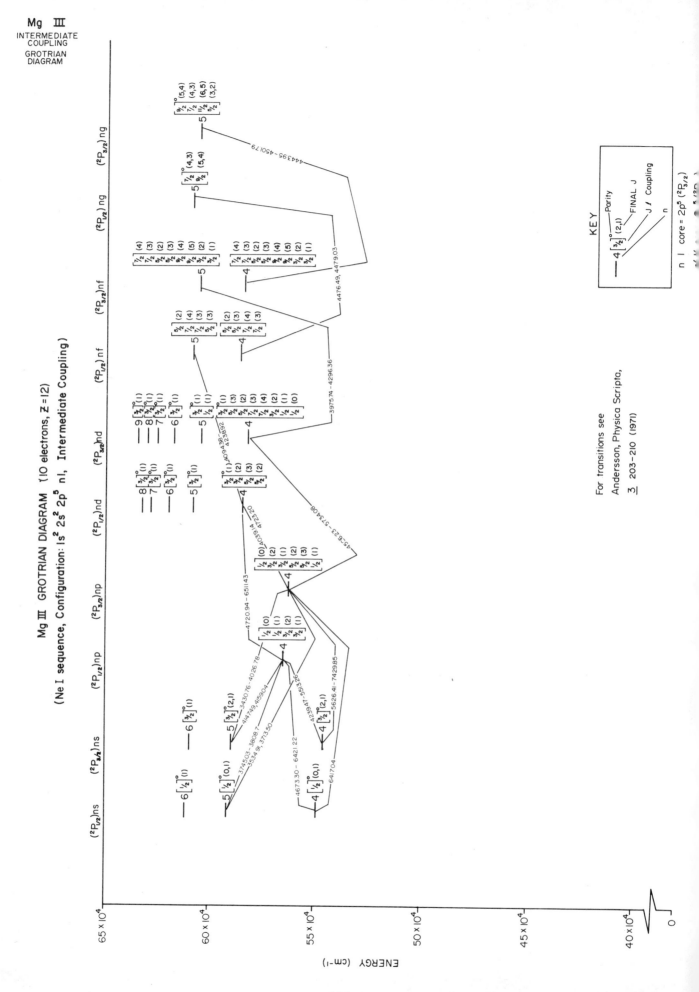

Mg III

INTERMEDIATE COUPLING

GROTRIAN DIAGRAM

Mg III GROTRIAN DIAGRAM (10 electrons, Z=12)

(Ne I sequence, Configuration: $1s^2\ 2s^2\ 2p^5\ nl$, Intermediate Coupling)

For transitions see

Andersson, Physica Scripta,

3 203–210 (1971)

KEY

Parity

FINAL J

J l Coupling

n

$4\left[\dfrac{3}{2}\right]^o (2,1)$

$n\ l\ \text{core} = 2p^5\ (^2P_{3/2})$

ENERGY (cm-1)

396

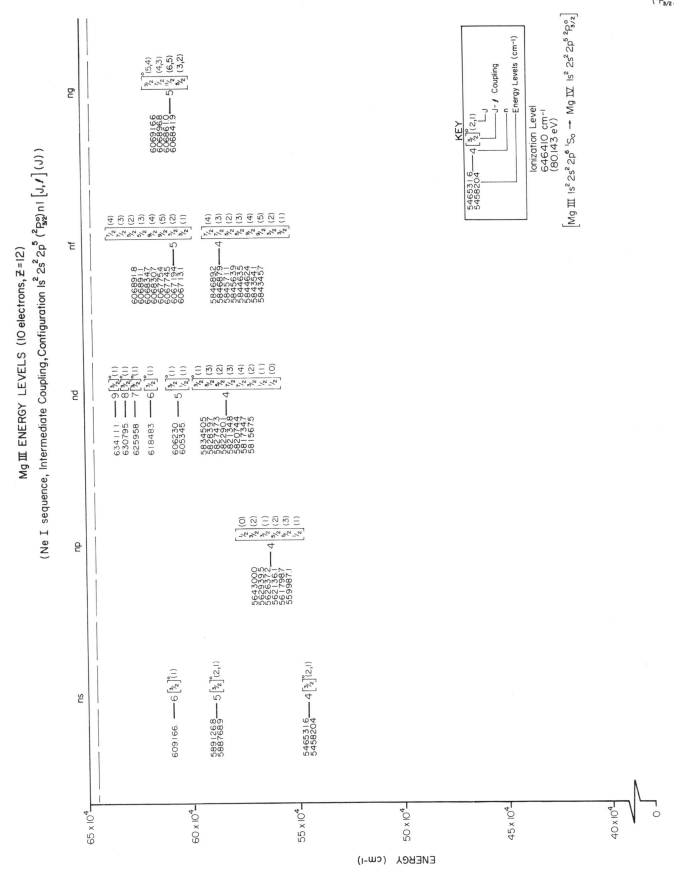

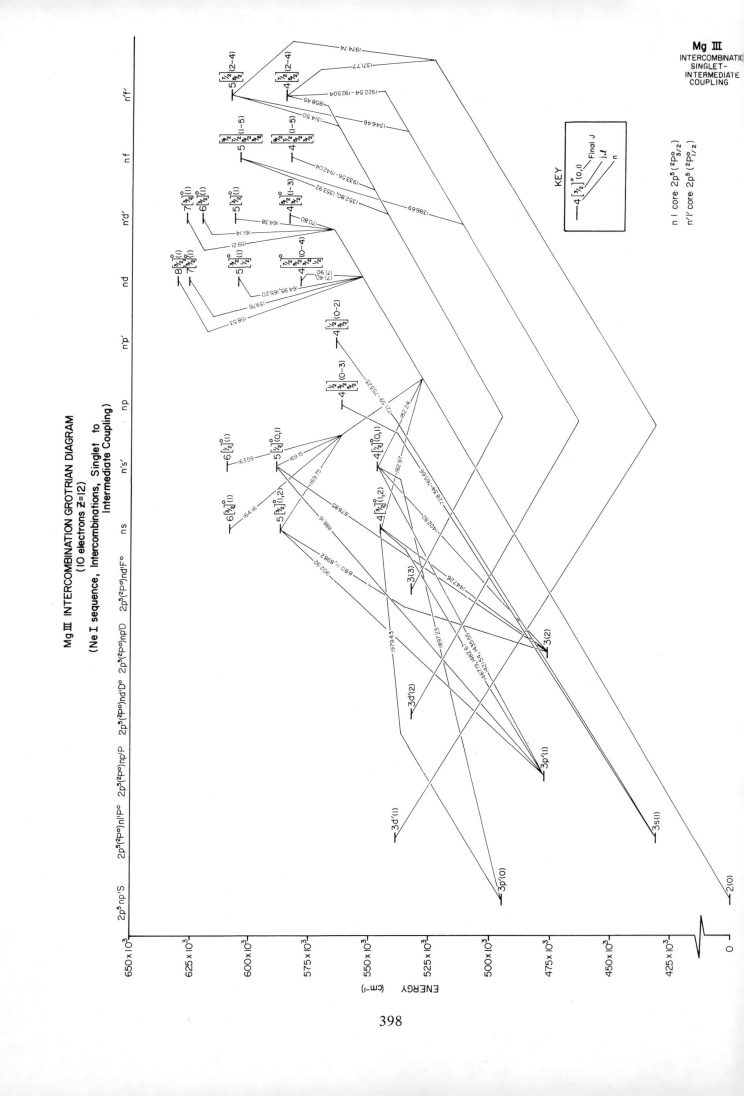

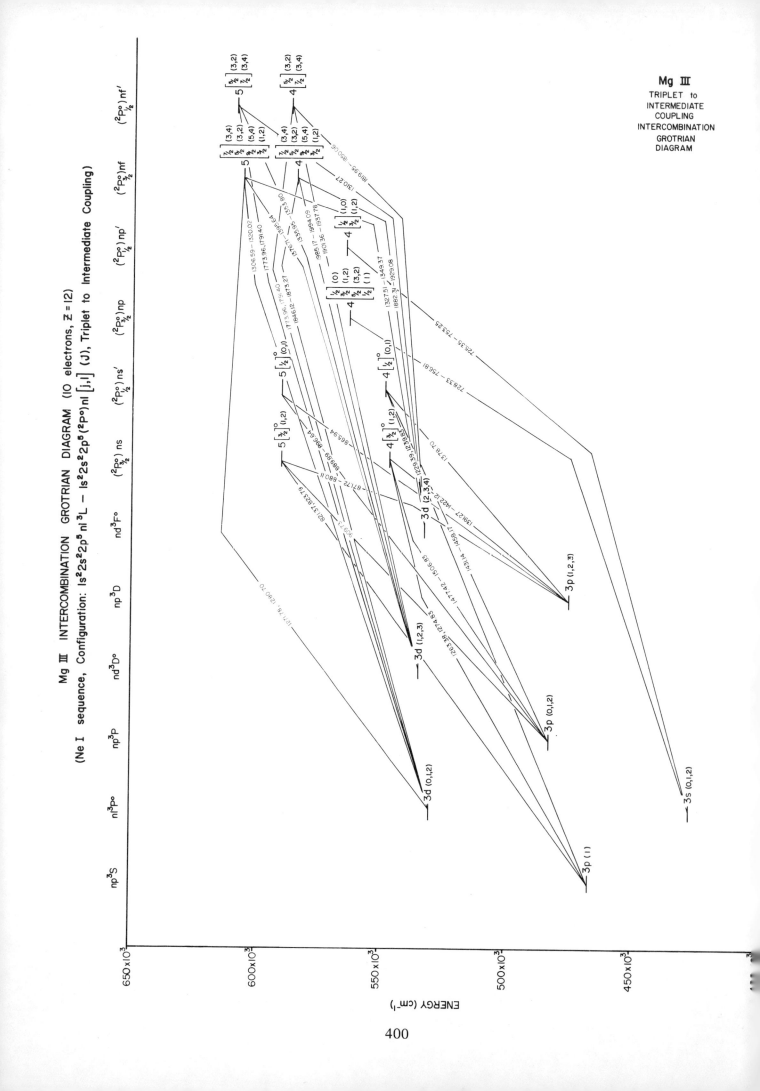

Mg III INTERCOMBINATION GROTRIAN DIAGRAM (10 electrons, Z = 12)

(Ne I sequence, Configuration: $1s^2 2s^2 2p^5 nl\ ^3L - 1s^2 2s^2 2p^5 (^2P^o)nl\ [j,]$ (J), Triplet to Intermediate Coupling)

Mg III
TRIPLET to
INTERMEDIATE
COUPLING
INTERCOMBINATION
GROTRIAN
DIAGRAM

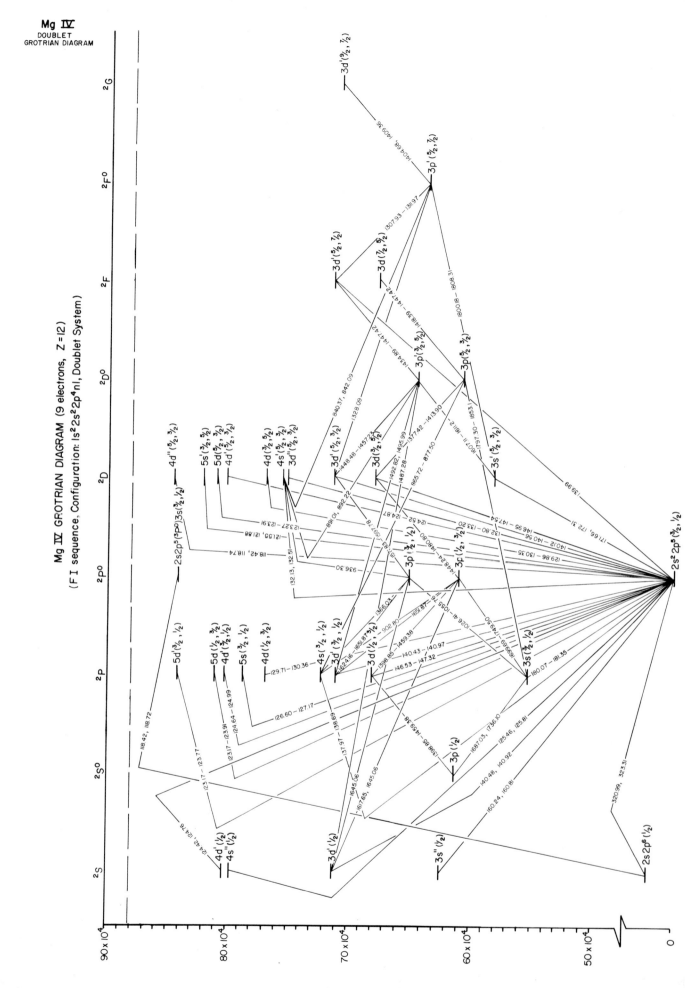

Mg IV GROTRIAN DIAGRAM (9 electrons, Z=12)

(F I sequence, Configuration: 1s²2s²2p⁴nl, Doublet System)

Mg IV
DOUBLET
GROTRIAN DIAGRAM

Mg IV ENERGY LEVEL (9 electrons, Z = 12)
(F I sequence, Configuration 1s²2s²2p⁴nl, Doublet System)

Mg IV
DOUBLET

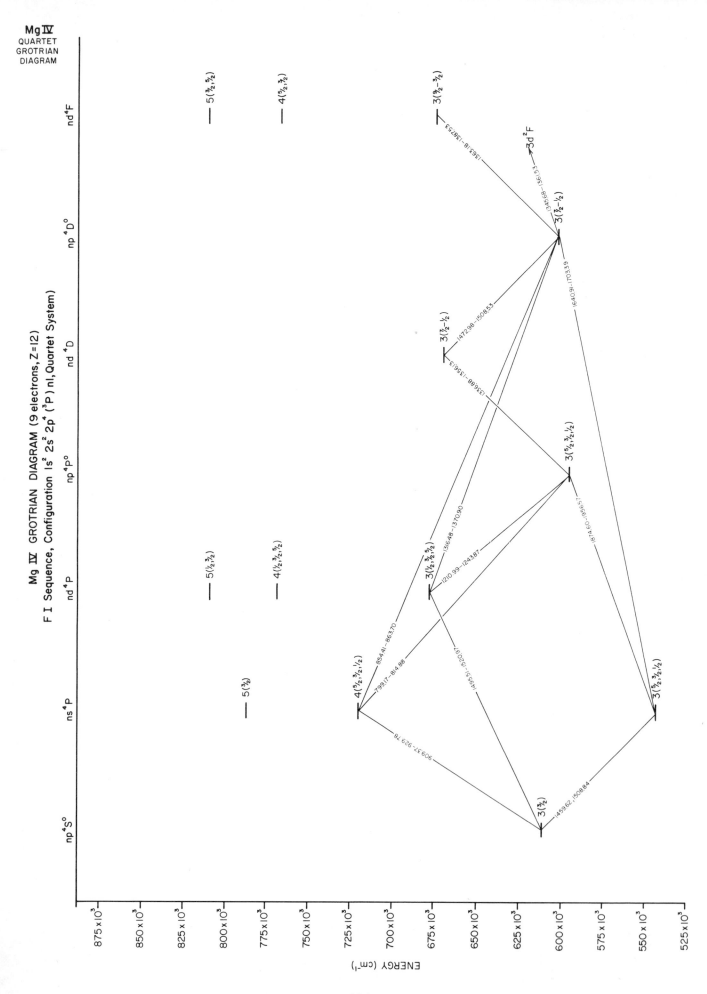

Mg IV QUARTET GROTRIAN DIAGRAM

Mg IV GROTRIAN DIAGRAM (9 electrons, Z = 12)
F I Sequence, Configuration 1s² 2s² 2p⁴ (³P) nl, Quartet System)

ENERGY (cm⁻¹)

404

Mg IV ENERGY LEVELS (9 electrons, Z=12)
(F I sequence, Configuration 1s² 2s² 2p⁴ (³P) nl, Quartet System)

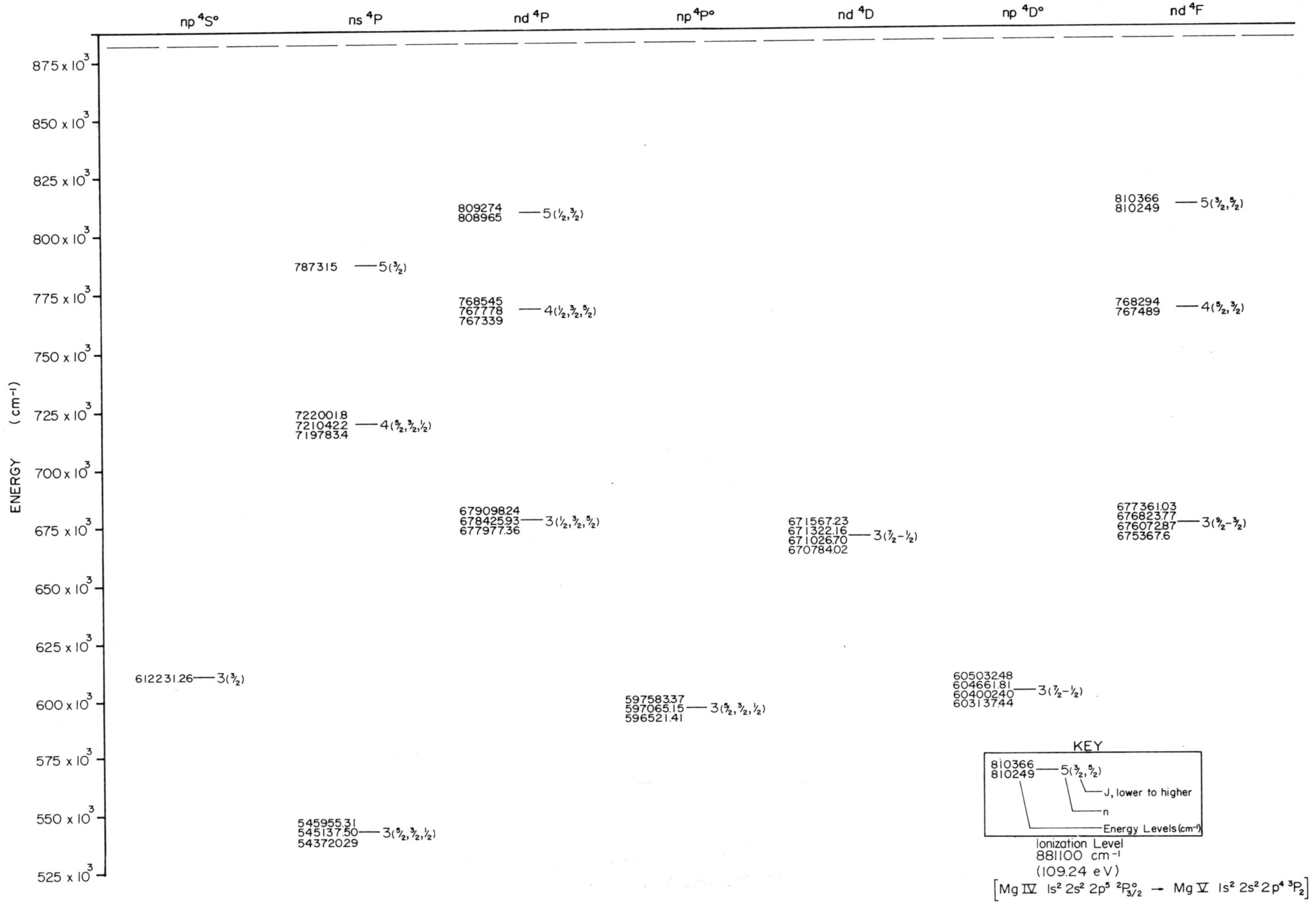

Mg IV
QUARTET

405

Mg IV INTERCOMBINATION GROTRIAN DIAGRAM (9 electrons, Z=12)
F I sequence, Configuration: 2p⁴(³P)nl, Doublet-Quartet System

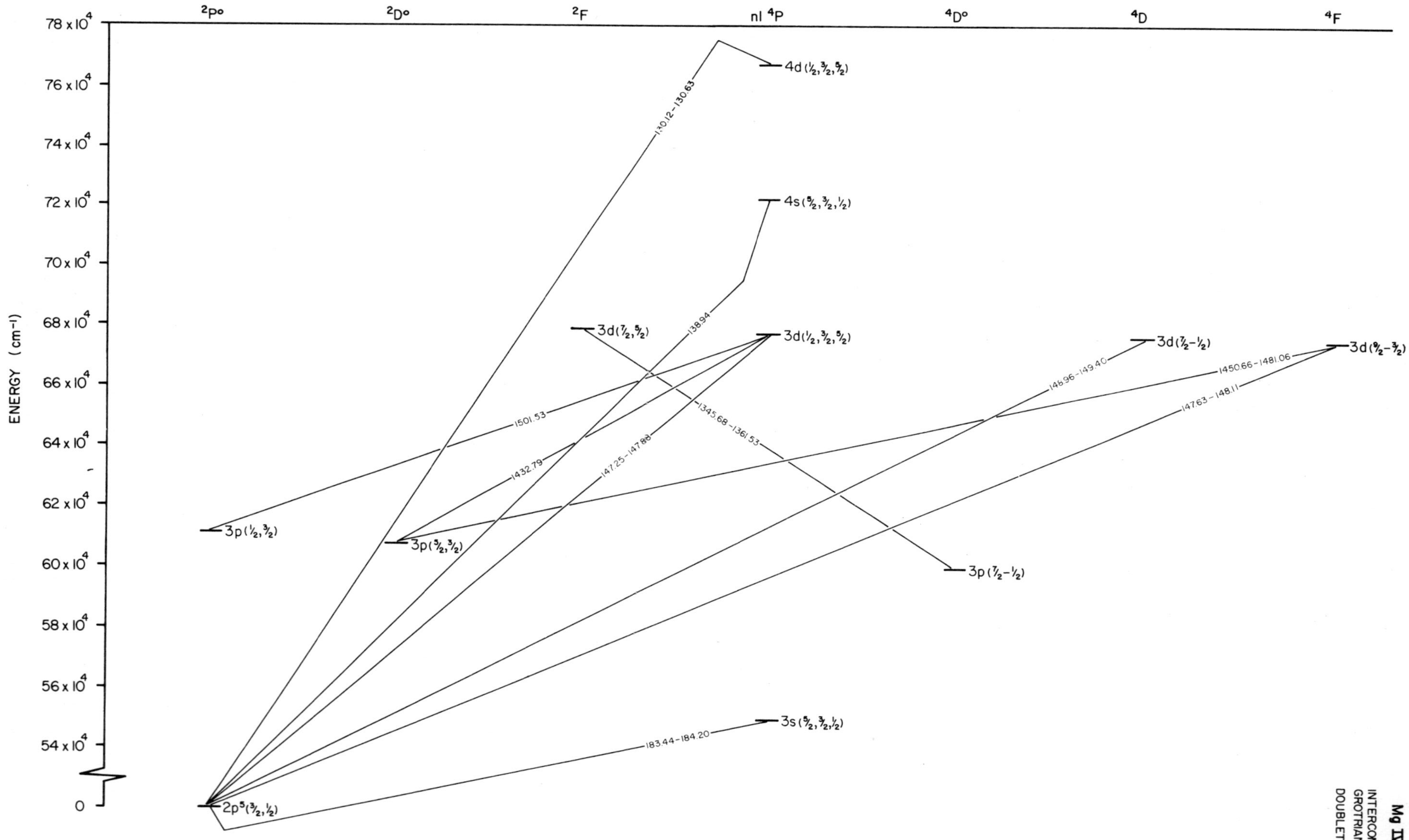

Mg IV
INTERCOMBINATION
GROTRIAN DIAGRAM
DOUBLET QUARTET

406

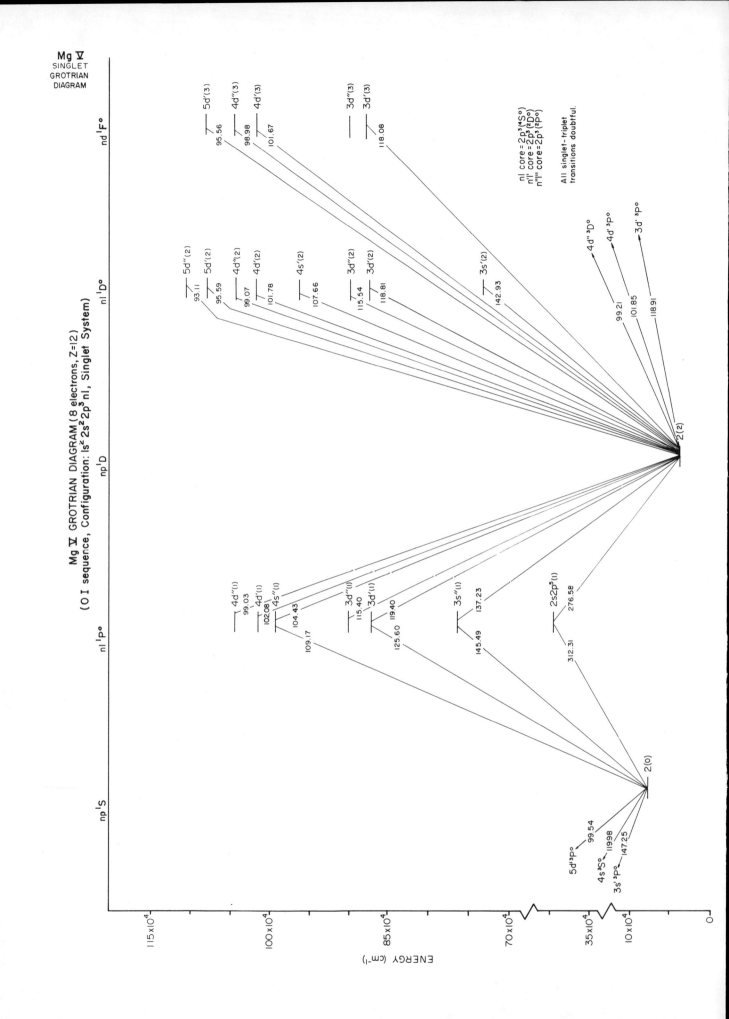

Mg Ⅴ
SINGLET
GROTRIAN
DIAGRAM

Mg Ⅴ GROTRIAN DIAGRAM (8 electrons, Z=12)
(O I sequence, Configuration: 1s² 2s² 2p³ nl, Singlet System)

nl core = 2p³(⁴S°)
n"l' core = 2p³(²D°)
n"'l" core = 2p³(²P°)

All singlet - triplet
transitions doubtful.

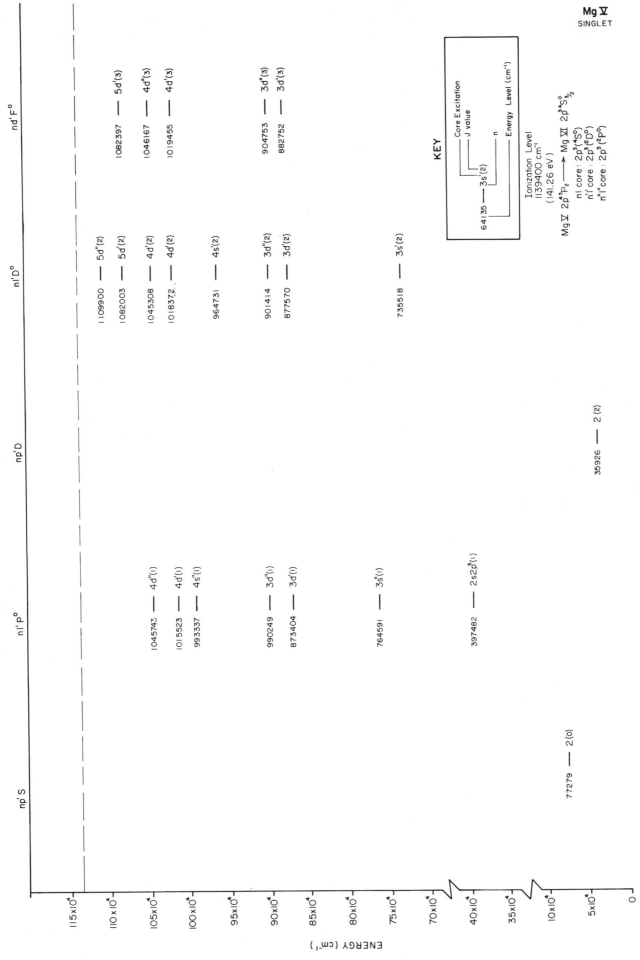

Mg V ENERGY LEVEL (8 electrons, Z = 12)

(OI Sequence, Configuration 1s²2s²2p³nl, Singlet Term System)

Mg **V**
SINGLET

409

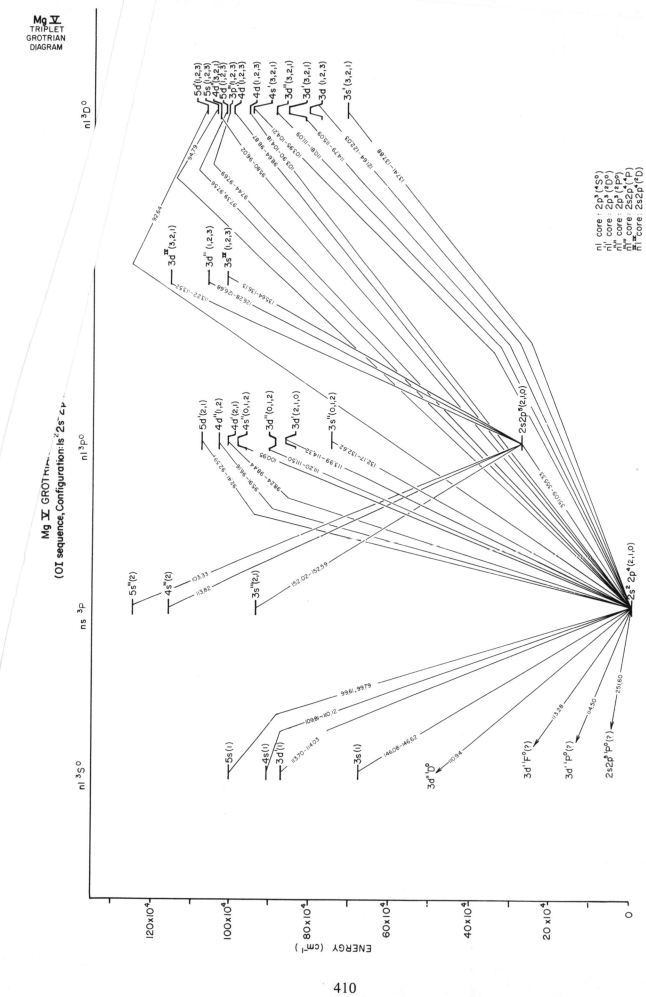

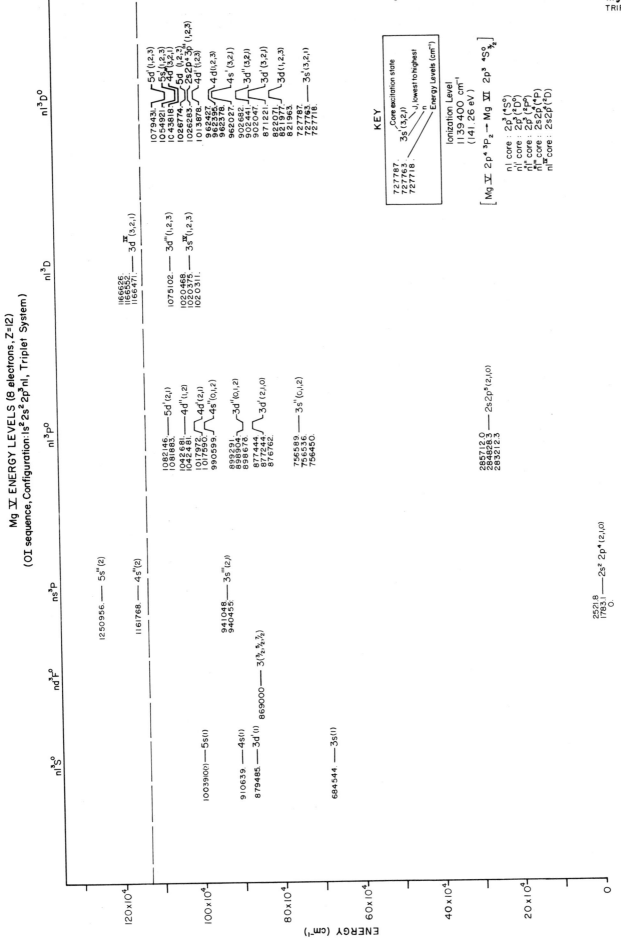

Mg V ENERGY LEVELS (8 electrons, Z=12)
(OI sequence, Configuration: 1s² 2s² 2p³ nl, Triplet System)

Mg VI GROTRIAN DIAGRAM (7 electrons, Z=12)
(N I sequence, Configuration: 1s²2s²2p²nl, Doublet System)

Mg VI
DOUBLET
GROTRIAN
DIAGRAM

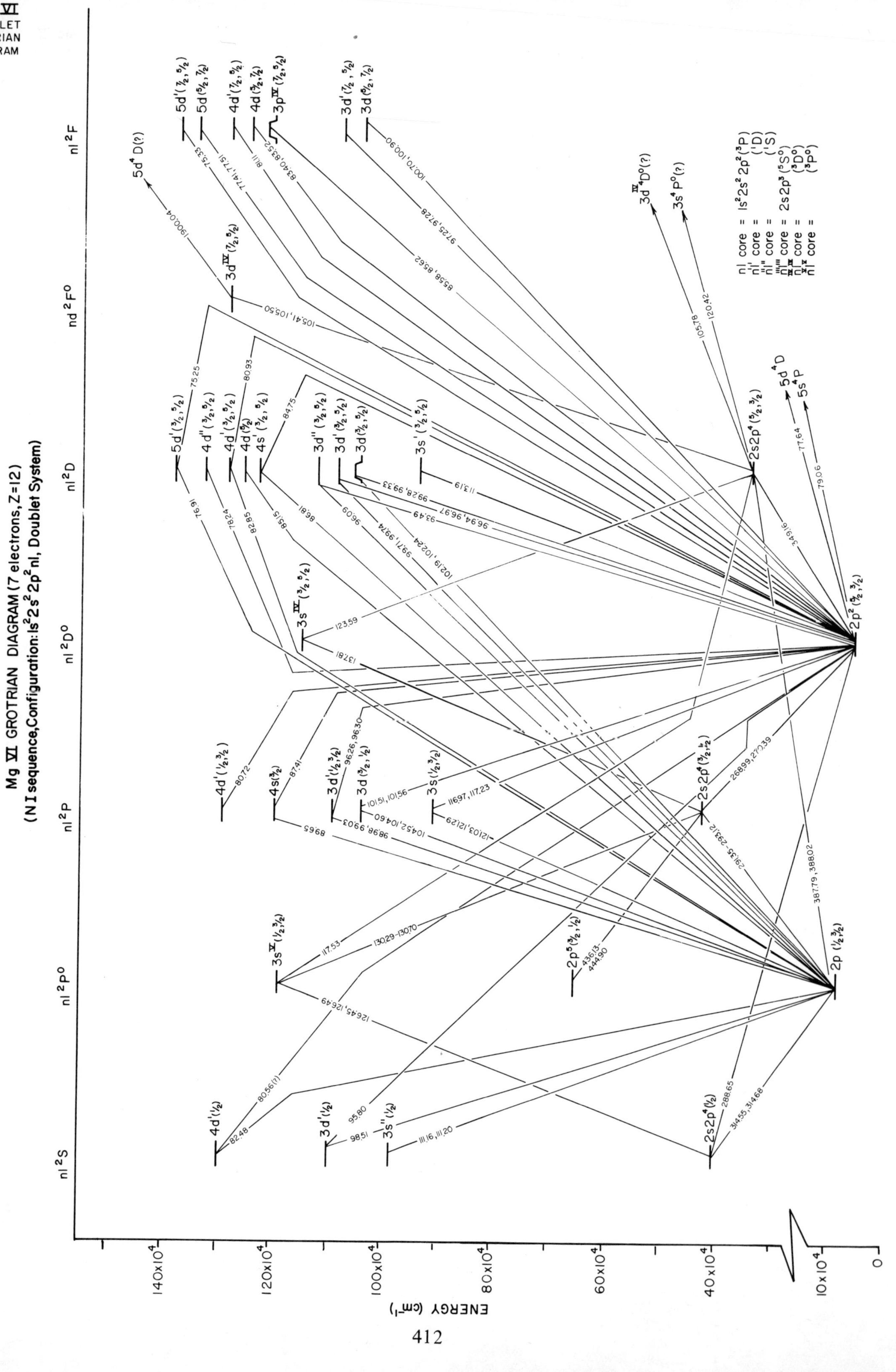

412

Mg VI ENERGY LEVELS (7 electrons, Z=12)
(N I sequence, Configuration: 1s² 2s² 2p² nl, Doublet System)

Mg VI
DOUBLET

nl ²S	nl ²P⁰	nl ²P	nl ²D⁰	nl ²D	nd ²F⁰	nl ²F

nl ²F column:
1382667. —— 5d'(⁷/₂,⁵/₂)
1347151. —— 5d'(⁵/₂,⁷/₂)
1345405.
1288199. —— 4d'(⁷/₂,⁵/₂)
1254243. —— 4d'(³/₂,¹/₂)
1252598.
1223804. ⎤ 3p IV (⁷/₂,⁵/₂)
1223169. ⎦
1083533. —— 3d'(⁷/₂,⁵/₂)
1083227.
1048274. —— 3d'(⁵/₂,⁷/₂)
1046507.

nd ²F⁰ column:
1290356. —— 3d IV (⁷/₂,⁵/₂)
1289495.

nl ²D column:
1384183. —— 5d'(³/₂,⁵/₂)
1333380. —— 4d''(³/₂,⁵/₂)
1290882. —— 4d'(³/₂,⁵/₂)
1258284. —— 4d(⁵/₂)
1235582. —— 4s'(³/₂,⁵/₂)
1124778. —— 3d''(³/₂,⁵/₂)
1086813. —— 3d'(³/₂,⁵/₂)
1086456.
1062506. ⎤ 3d(³/₂,⁵/₂)
1061943. ⎦
938723. —— 3s'(³/₂,⁵/₂)
341679. —— 2s2p⁴(⁵/₂,³/₂)
341646.

nl ²D⁰ column:
1150733. —— 3s IV (³/₂,⁵/₂)
55266. —— 2p(¹/₂,³/₂)
55245.

nl ²P column:
1294034. —— 4d'(¹/₂,³/₂)
1199360. —— 4s(³/₂)
1094141. —— 3d'(¹/₂,³/₂)
1093653.
1040567. —— 3d(³/₂,¹/₂)
1039950.
910191. —— 3s(¹/₂,³/₂)
908297.
427033. —— 2s2p⁴(³/₂,¹/₂)
425076.

nl ²P⁰ column:
1192527. —— 3s IV (¹/₂,³/₂)
1192221.
654382. —— 2p⁵(³/₂,¹/₂)
651762.
83927. —— 2p(¹/₂,³/₂)
83805.

nl ²S column:
1296416. —— 4d'(¹/₂)
1099073. —— 3d'(¹/₂)
983313. —— 3s''(¹/₂)
401714. —— 2s2p⁴(¹/₂)

KEY

Core excitation state

1223805. ⎤ 3p IV (⁷/₂,⁵/₂)
1223169. ⎦

J, Lower to higher
n
Energy Level (cm⁻¹)

Ionization Level
1504300 cm⁻¹
(186.50 electron volts)

nl' core =	1s² 2s² 2p²	(³P)
nl'' core =	"	(¹D)
nl''' core =	"	(¹S)
n'''',''''' core =	2s 2p³	(⁵S⁰)
IV core =	"	(³D⁰)
V,VI core =	"	(³P⁰)

[Mg VI 1s²2s²2p³ ⁴S³/₂⁰ ⟶ Mg VII 1s²2s²2p² ³P₀]

NOTE: All levels uncertain by ± y

ENERGY (cm⁻¹)

140×10⁴
120×10⁴
100×10⁴
80×10⁴
60×10⁴
40×10⁴
10×10⁴
0

Mg VI
QUARTET
GROTRIAN
DIAGRAM

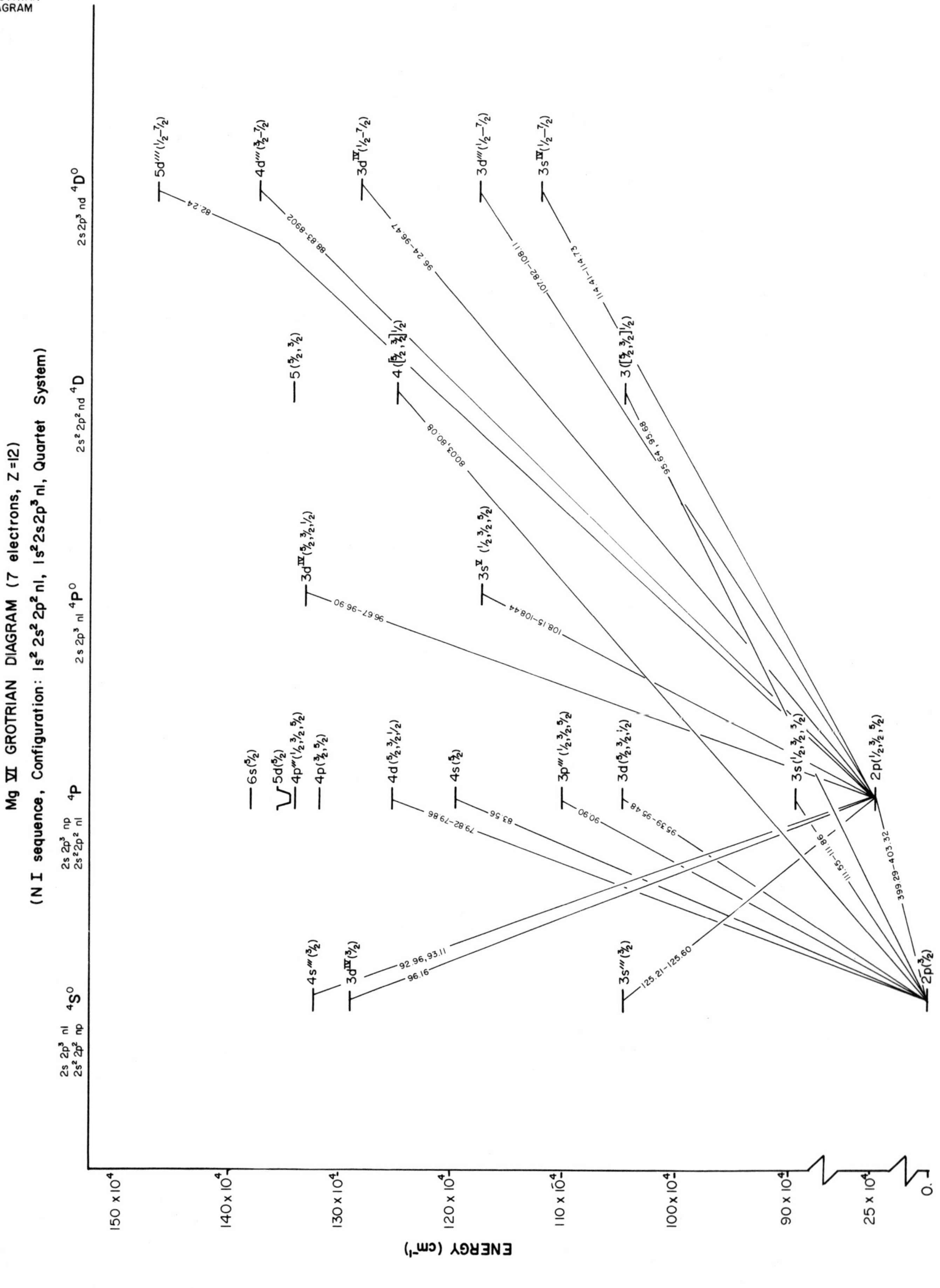

Mg VI GROTRIAN DIAGRAM (7 electrons, Z=12)
(N I sequence, Configuration: 1s² 2s² 2p² nl, 1s²2s2p³ nl, Quartet System)

Mg VI   ENERGY   LEVELS   (7 electrons, Z = 12)

(N I sequence, Configuration: $1s^2 2s^2 2p^2 nl$, $1s^2 2s 2p^3 nl$, Quartet System)

ENERGY (cm⁻¹)

$2s^2 2p^3$ $^4D°$

1463928 —— $5d'''(^1/_2 - ^7/_2)$

1373760 —— $4d''(^3/_2 - ^7/_2)$

1287044 —— $3d^{IV}(^1/_2 - ^7/_2)$

1175396 —— $3d'''(^1/_2 - ^7/_2)$

1122023 —— $3s^{IV}(^1/_2 - ^7/_2)$

$2s^2 2p^2$ nd ^4D

1342985 —— $5(^5/_2, ^7/_2)$

1249500 —— $4(^5/_2, ^3/_2, ^1/_2)$
1248829

1045620 —— $3([^5/_2, ^3/_2], ^1/_2)$
1045205

KEY

1120608 —— $3s^{II}(^1/_2, ^3/_2, ^5/_2)$ ← Core excitation
                   J, lowest to highest
                   n
                   Energy Level (cm⁻¹)

Ionization Level
1504300 cm⁻¹
(186.50 eV)

[Mg VI $1s^2 2s^2 2p^3$ $^4S°_{3/2}$ → Mg VII $1s^2 2s^2 2p^2$ $^3P_0$]

nl core = $2s^2 2p^2(^3P)$
n',l' core = $2s^2 2p^2(^1D)$
n'',l'' core = $2s^2 2p^2(^1S)$
n''',l''' core = $2s\ 2p^3(^5S°)$
n^{IV},l^{IV} core = $2s\ 2p^3(^3D°)$
n^{V},l^{V} core = $2s\ 2p^3(^3P°)$

$2s\ 2p^3$ nl $^4P°$

1282668 —— $3d^{IV}(^5/_2, ^3/_2, ^1/_2)$
1282398
1282028

1172608 —— $3s^{II}(^1/_2, ^3/_2, ^5/_2)$

$2s^2 2p^2$ nl ^4P
$2s^2 2p^2$ nl

1380643 —— $6s(^5/_2)$

1345550 $\lor$ $5d(^5/_2)$
1340950 —— $4p'''(^1/_2, ^3/_2, ^5/_2)$

1318670 —— $4p(^3/_2, ^5/_2)$
1317697

1252966 —— $4d(^5/_2, ^3/_2, ^1/_2)$
1252662
1252238

1196740 —— $4s(^5/_2)$

1100146 —— $3p'''(^1/_2, ^3/_2, ^5/_2)$

1048383 —— $3d(^5/_2, ^3/_2, ^1/_2)$
1047987
1047307

896443 —— $3s(^1/_2, ^3/_2, ^5/_2)$
894887
893943

250445 —— $2p(^1/_2, ^3/_2, ^5/_2)$
249578
247945

$2s\ 2p^3$ nl $^4S°$
$2s^2 2p^2$ np

1323609 —— $4s'''(^3/_2)$

1287889 —— $3d^{IV}(^3/_2)$

1046634 —— $3s'''(^3/_2)$

0 —— $2p(^3/_2)$

150×10^4
140×10^4
130×10^4
120×10^4
110×10^4
100×10^4
90×10^4
25×10^4
0

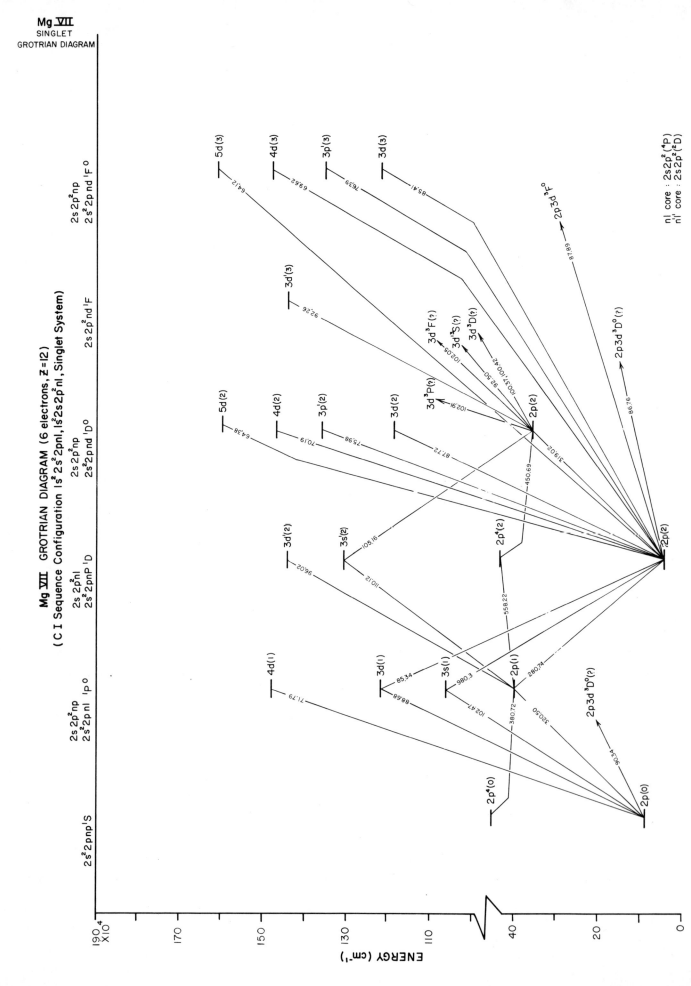

Mg VII GROTRIAN DIAGRAM (6 electrons, Z=12)
(C I Sequence Configuration 1s²2s²2pnl, 1s²2s2p²nl, Singlet System)

Mg VII
SINGLET
GROTRIAN DIAGRAM

ENERGY (cm⁻¹)

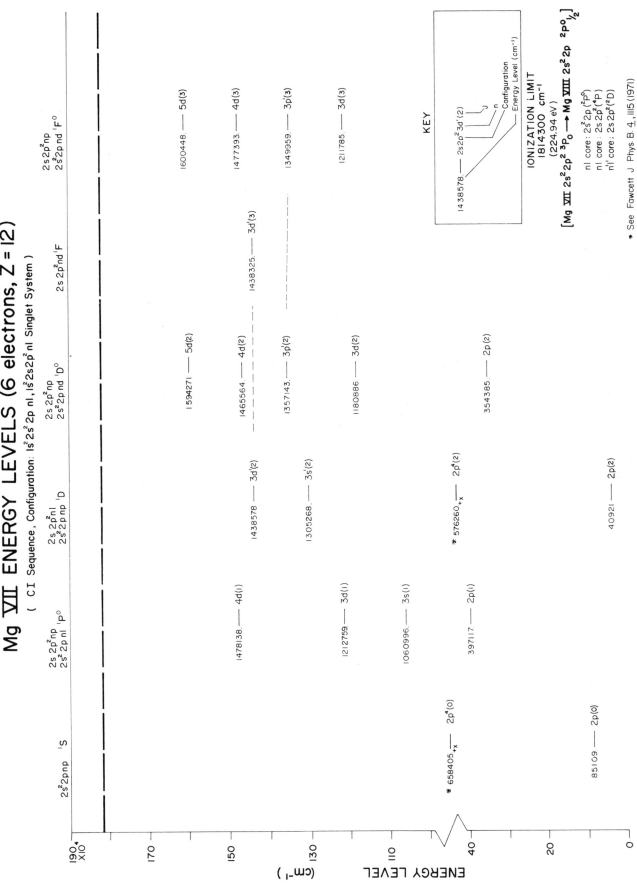

Mg VII ENERGY LEVELS (6 electrons, Z = 12)

(CI Sequence, Configuration: 1s² 2s² 2p nl, 1s² 2s2p² nl Singlet System)

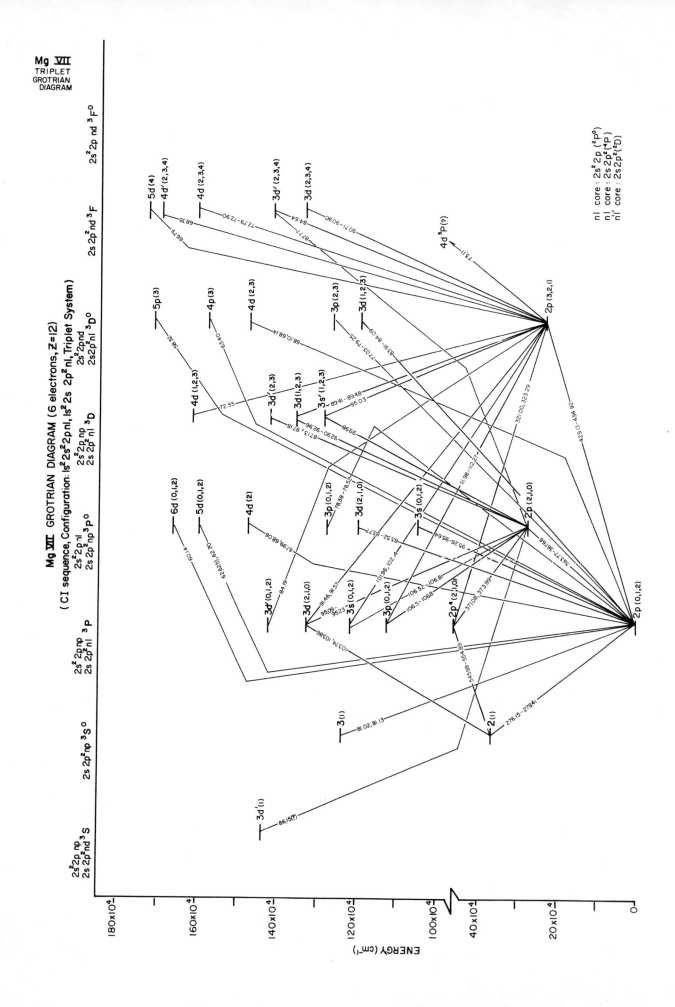

Mg VII GROTRIAN DIAGRAM (6 electrons, Z=12)

(CI sequence, Configuration: ls²2s²2pnl, ls²2s 2p²nl, Triplet System)

Mg VII
TRIPLET
GROTRIAN
DIAGRAM

nl core : 2s²2p (²P⁰)
n'l core: 2s 2p²(⁴P)
n'l' core : 2s2p²(²D)

418

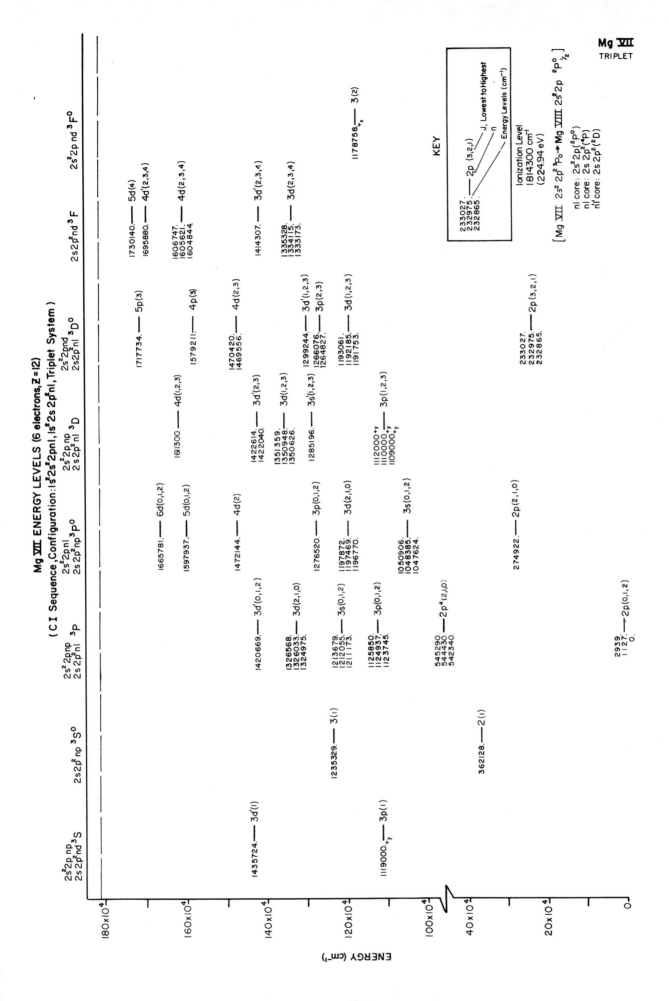

Mg VII ENERGY LEVELS (6 electrons, Z = 12)

(C I Sequence, Configuration: 1s²2s²2pnl, 1s²2s 2p³nl, Triplet System)

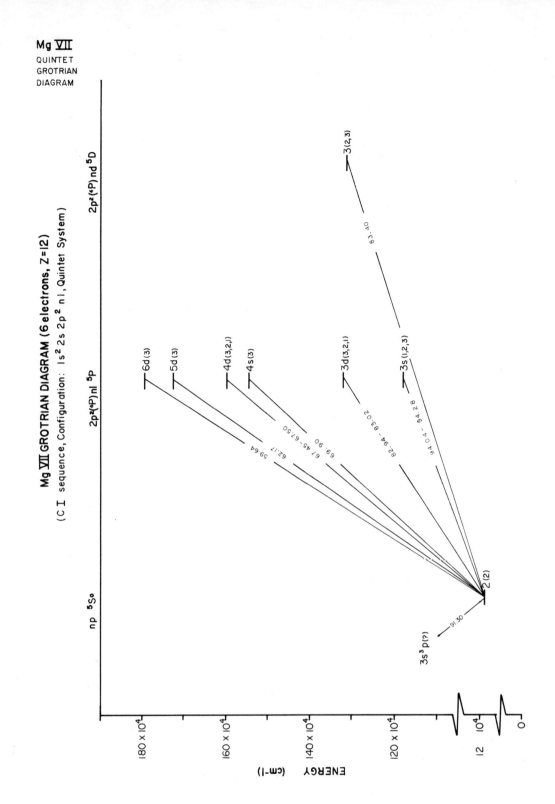

Mg VII
QUINTET
GROTRIAN
DIAGRAM

Mg VII GROTRIAN DIAGRAM (6 electrons, Z=12)
(C I sequence, Configuration: 1s² 2s 2p² nl, Quintet System)

Mg VII ENERGY LEVELS (6 electrons, Z=12)

(C I sequence, Configuration: 1s² 2s 2p² nl, Quintet System)

| | np ⁵S° | 2p²(⁴P) nl ⁵P | 2p²(⁴P) nd ⁵D |

KEY

1324311
1323889 ——— 3d(3,2,1) ——— J (lowest to highest)
1323222   n
                        Energy Levels (cm⁻¹)

Ionization Level
1814300 cm⁻¹
224.94 eV

[Mg VII 1s²2s²2p² ³P₀ → Mg VIII 1s² 2s² 2p ²P°₁/₂]

Note: all levels uncertain by ± Y

1745347 ——— 6d(3)

1727216 ——— 5d(3)

1601134
1600760 ——— 4d(3,2,1)
1600167

1549235 ——— 4s(3)

1324311
1323889 ——— 3d(3,2,1)
1323222

1317618 ——— 3(2,3)

1181963
1180484 ——— 3s(1,2,3)
1179696

118620 ——— 2(2)

ENERGY (cm⁻¹)

190 × 10⁴
180 × 10⁴
170 × 10⁴
160 × 10⁴
150 × 10⁴
140 × 10⁴
130 × 10⁴
120 × 10⁴
110 × 10⁴
12 × 10⁴
0

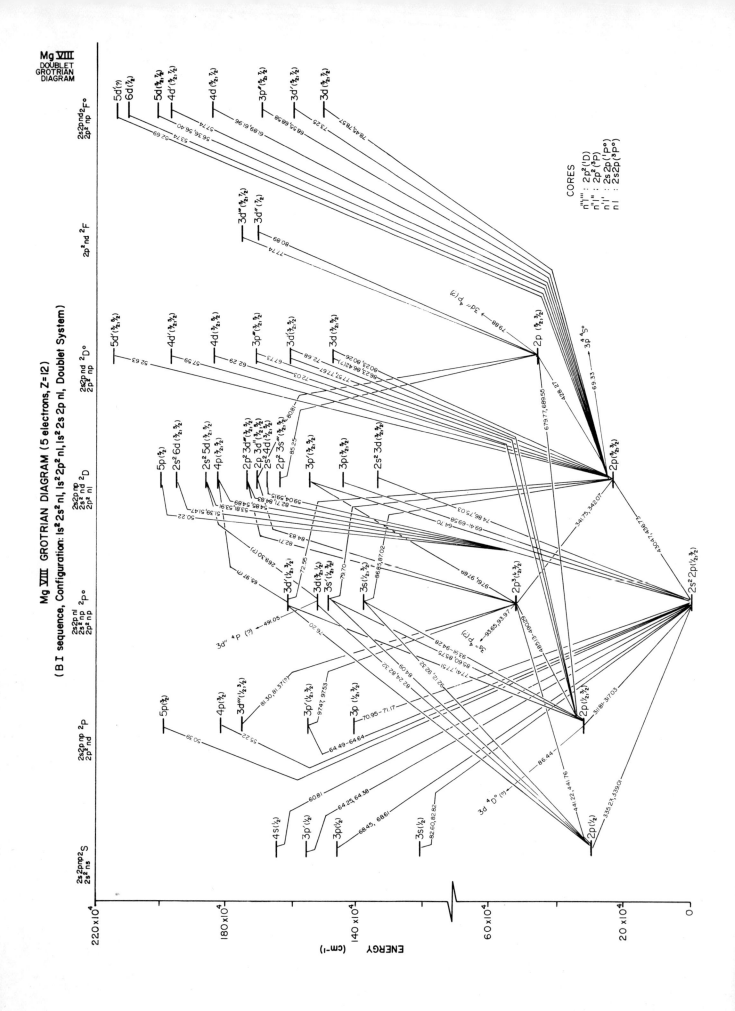

Mg VIII GROTRIAN DIAGRAM (5 electrons, Z=12)

(B I sequence, Configuration: 1s²2s²nl, 1s²2p²nl, 1s²2s 2p nl, Doublet System)

Mg VIII
DOUBLET
GROTRIAN
DIAGRAM

422

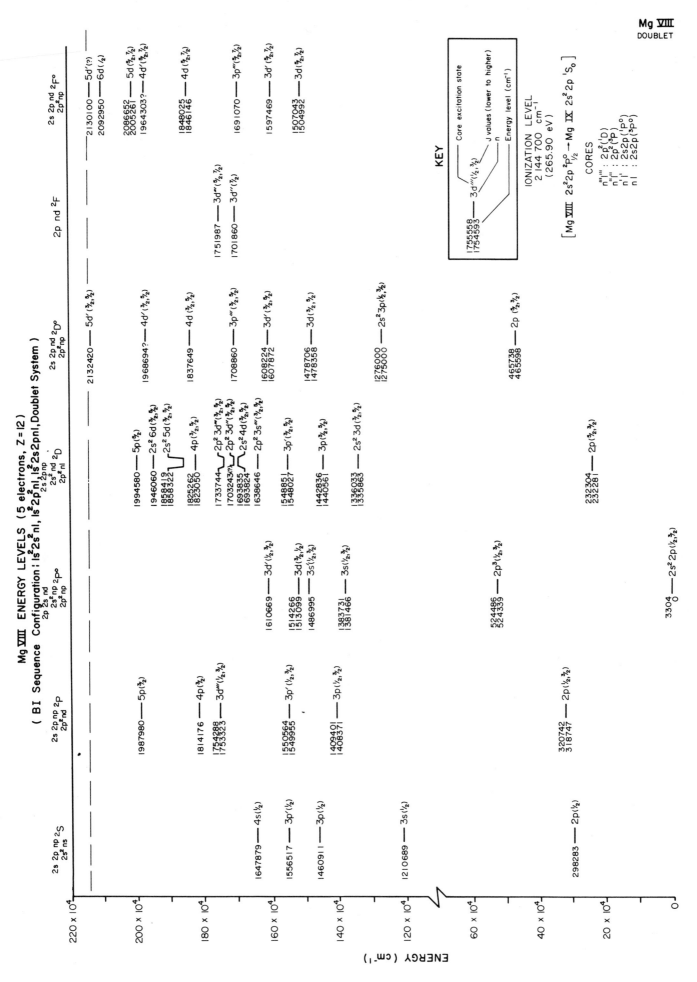

Mg VIII   ENERGY LEVELS   (5 electrons, Z = 12)

(BI Sequence Configuration ; 1s²2s²nl, 1s²2p²nl, 1s²2s2pnl, Doublet System)

KEY

Core excitation state

3d'''(¹/₂, ³/₂)

J values (lower to higher)
n
Energy level (cm⁻¹)

IONIZATION LEVEL
2 144 700 cm⁻¹
(265.90 eV)

[Mg VIII 2s²2p ²P°₁/₂ → Mg IX 2s² 2p ¹S₀]

CORES

n'''l''' : 2p²(¹D)
n''l'' : 2p²(³P)
n'l' : 2s2p(¹P°)
nl : 2s2p(³P°)

1755558
1754593

ENERGY (cm⁻¹)

423

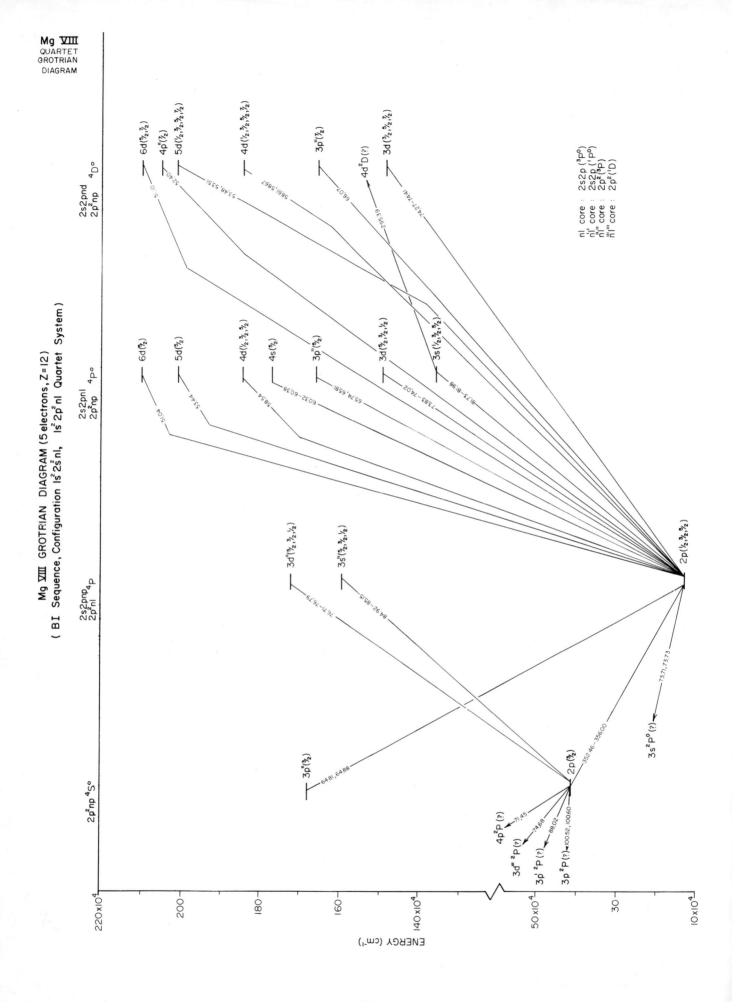

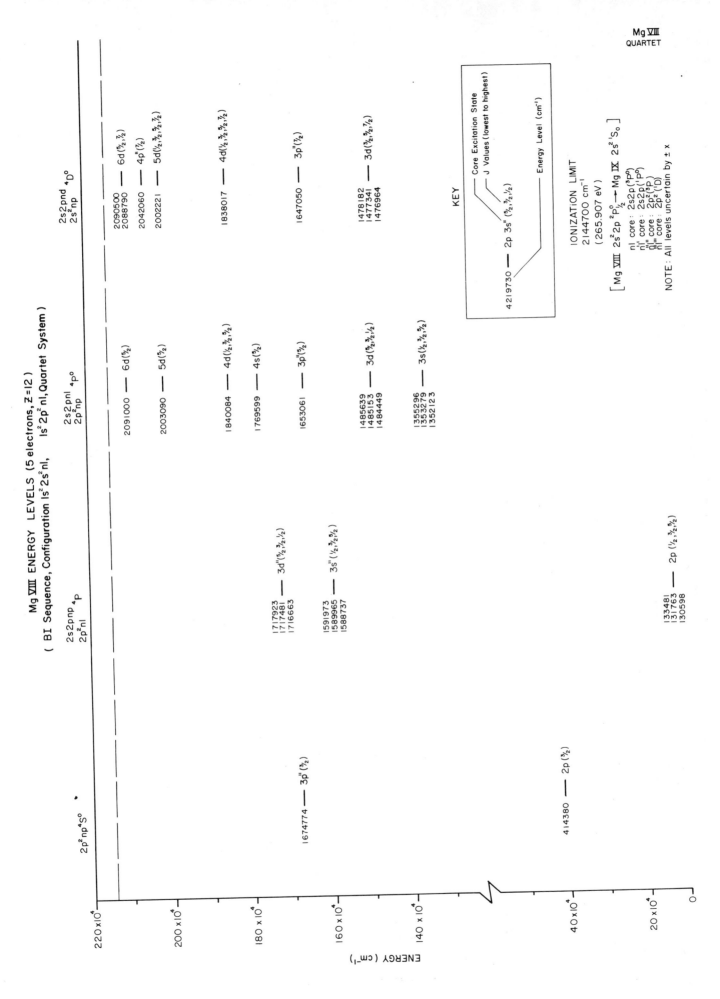

Mg VIII ENERGY LEVELS (5 electrons, Z = 12)
(BI Sequence, Configuration 1s² 2s² nl, 1s² 2p² nl, Quartet System)

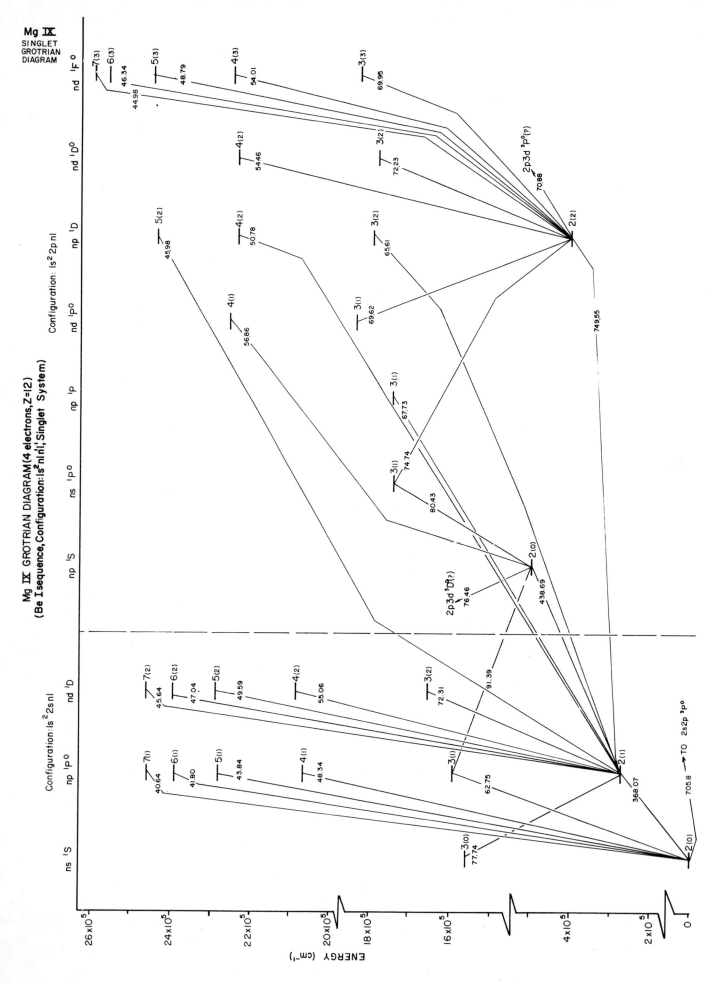

Mg IX GROTRIAN DIAGRAM (4 electrons, Z=12)
(Be I sequence, Configuration: 1s² nl n'l', Singlet System)

Mg IX
SINGLET
GROTRIAN
DIAGRAM

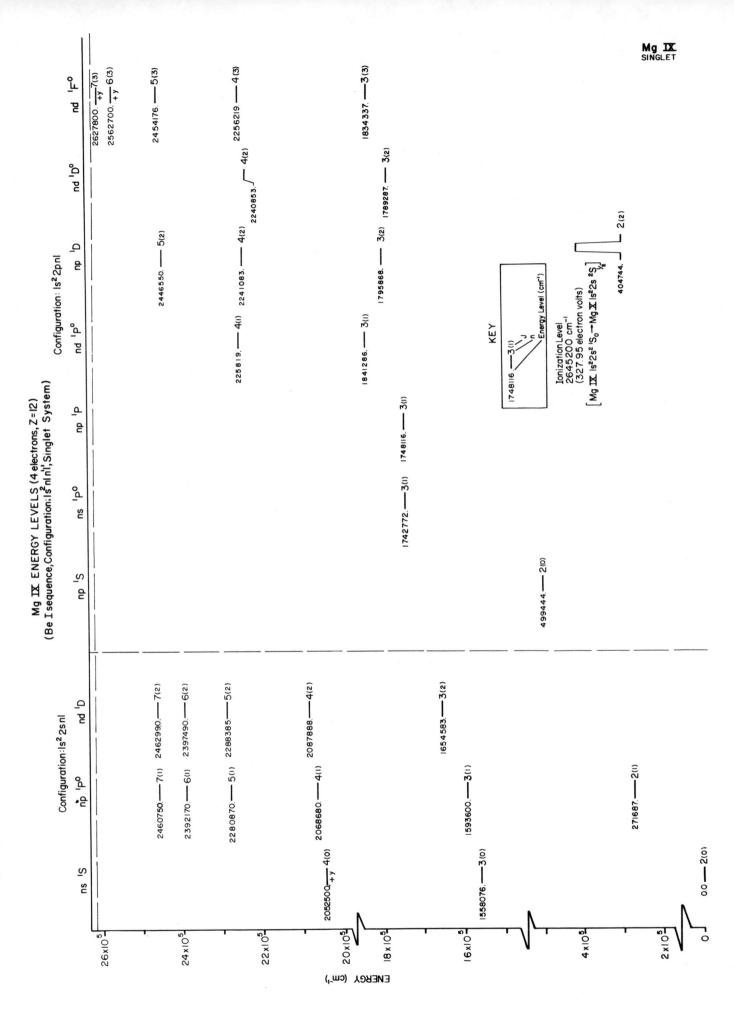

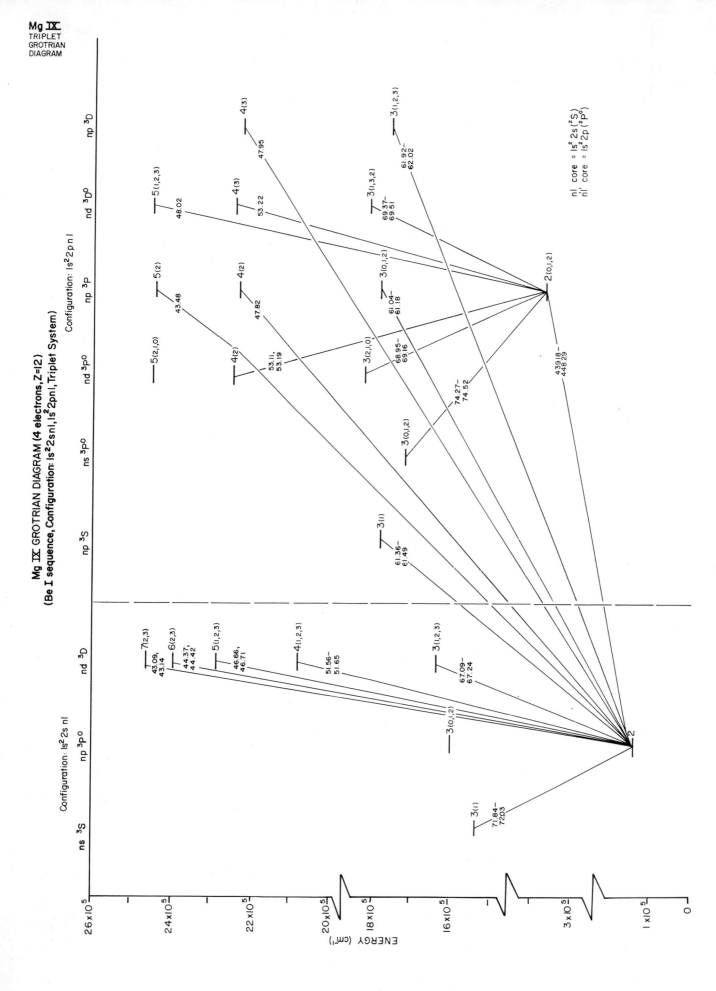

Mg IX
TRIPLET
GROTRIAN
DIAGRAM

Mg IX GROTRIAN DIAGRAM (4 electrons, Z=12)
(Be I sequence, Configuration: ls²2snl, ls²2pnl, Triplet System)

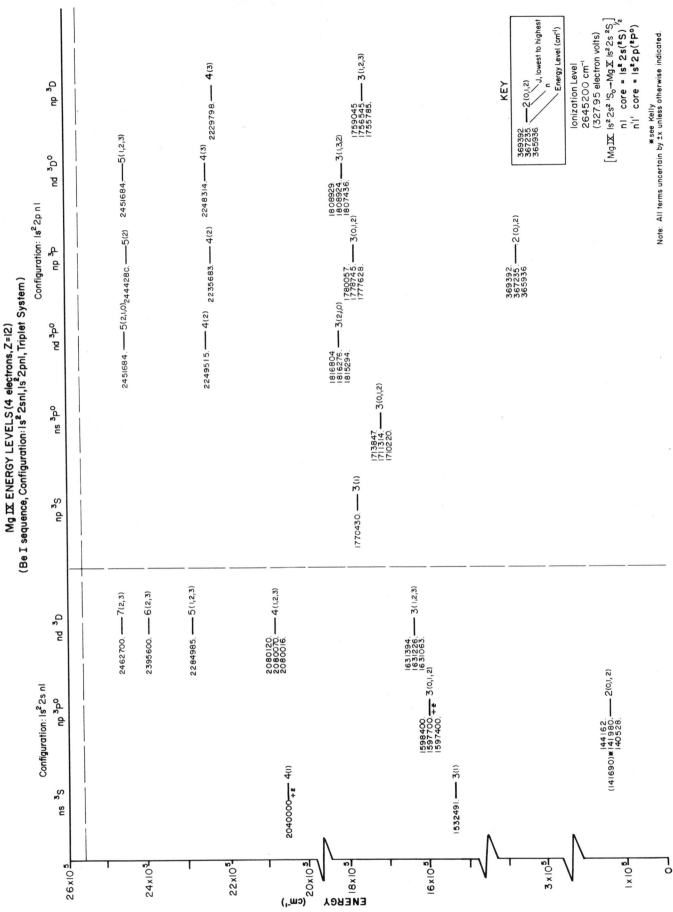

Mg IX ENERGY LEVELS (4 electrons, Z=12)
(Be I sequence, Configuration: 1s²2snl, 1s²2pnl, Triplet System)

Mg IX
TRIPLET

Mg **X** GROTRIAN DIAGRAM (3 electrons, Z=12)
(Li I sequence, Configuration: 1s²nl, Doublet)

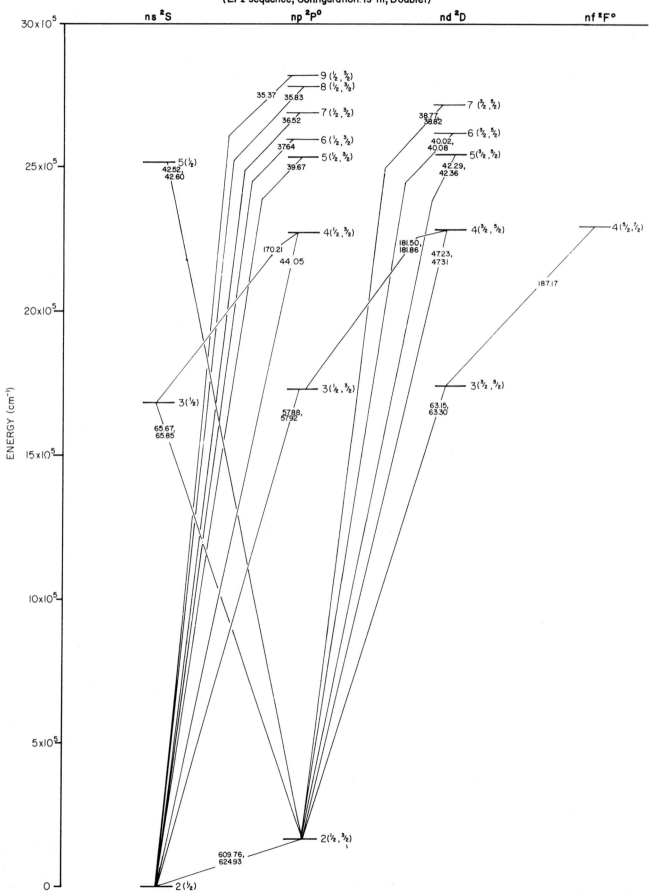

ns ²S

np ²P°

nd ²D

nf ²F°

30×10⁵

9 (½, ³⁄₂)
8 (½, ³⁄₂)
35.37   35.83

7 (³⁄₂, ⁵⁄₂)
38.77,
38.82

7 (½, ³⁄₂)
36.52

6 (½, ³⁄₂)
37.64

6 (³⁄₂, ⁵⁄₂)
40.02,
40.08

5 (½, ³⁄₂)
39.67

5 (³⁄₂, ⁵⁄₂)
42.29,
42.36

5 (½)
42.52,
42.60

25×10⁵

4 (½, ³⁄₂)
170.21
44.05

4 (³⁄₂, ⁵⁄₂)
181.50,
181.86
47.23,
47.31

4 (⁵⁄₂, ⁷⁄₂)
187.17

20×10⁵

3 (½)
65.67,
65.85

3 (½, ³⁄₂)
57.88,
57.92

3 (³⁄₂, ⁵⁄₂)
63.15,
63.30

15×10⁵

ENERGY (cm⁻¹)

10×10⁵

5×10⁵

2 (½, ³⁄₂)
609.76,
624.93

2 (½)

0

Mg X ENERGY LEVELS (3 electrons, Z=12)
(Li I sequence,   Configuration: 1s² nl, Doublet)

ns ²S   np ²P°   nd ²D   nf ²F°

30 × 10⁵

2827600. ———— 9(½,³/₂)
2791200. ———— 8(½,³/₂)

2739800. ———— 7(³/₂,⁵/₂)
2739300.
2738400. ———— 7(½,³/₂)

2659000. ———— 6(³/₂,⁵/₂)
2658500.
2656500. ———— 6(½,³/₂)

2524600. ———— 5(³/₂,⁵/₂)
2524300.
2511600. ———— 5(½)   2520900. ———— 5(½,³/₂)

25 × 10⁵

2277694. ———— 4(³/₂,⁵/₂)   *2278153+X ———— 4(⁵/₂,⁷/₂)
2277182.
2253000. ———— 4(½)   2270148. ———— 4(½,³/₂)

20 × 10⁵

1743880. ———— 3(³/₂,⁵/₂)
1743410.
1727832. ———— 3(½,³/₂)
1726519.
1682648. ———— 3(½)

15 × 10⁵

* see Fawcett, J. Phys. B. 4, 1115 (1971)

10 × 10⁵

KEY

1727832. ———— 3(½,³/₂)
1726519.
                                    J, Lower to Higher
                                  n
                        Energy Levels (cm⁻¹)

5 × 10⁵

Ionization Level
2964400. cm⁻¹
(36753 e V)
[Mg X 1s² 2s ²S₁/₂ → Mg XI 1s² ¹S₀]

163976. ———— 2(½,³/₂)
159929.

0 —   0.00 ———— 2(½)

ENERGY LEVEL (cm⁻¹)

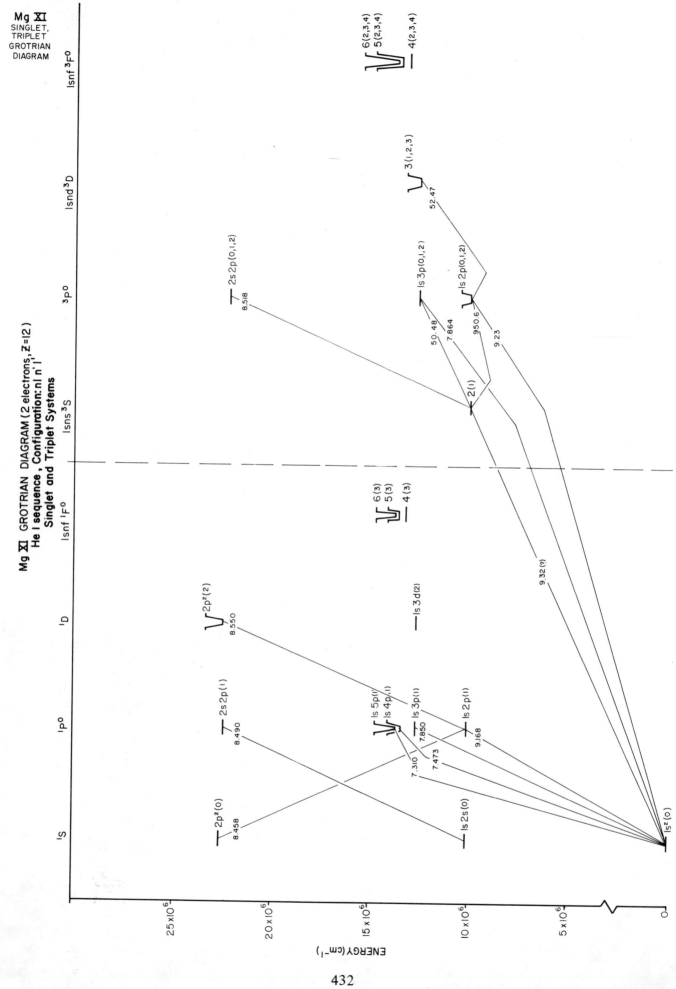

Mg XI
SINGLET,
TRIPLET
GROTRIAN
DIAGRAM

Mg XI GROTRIAN DIAGRAM (2 electrons, Z=12)
He I sequence, Configuration: nl n'l'
Singlet and Triplet Systems

432

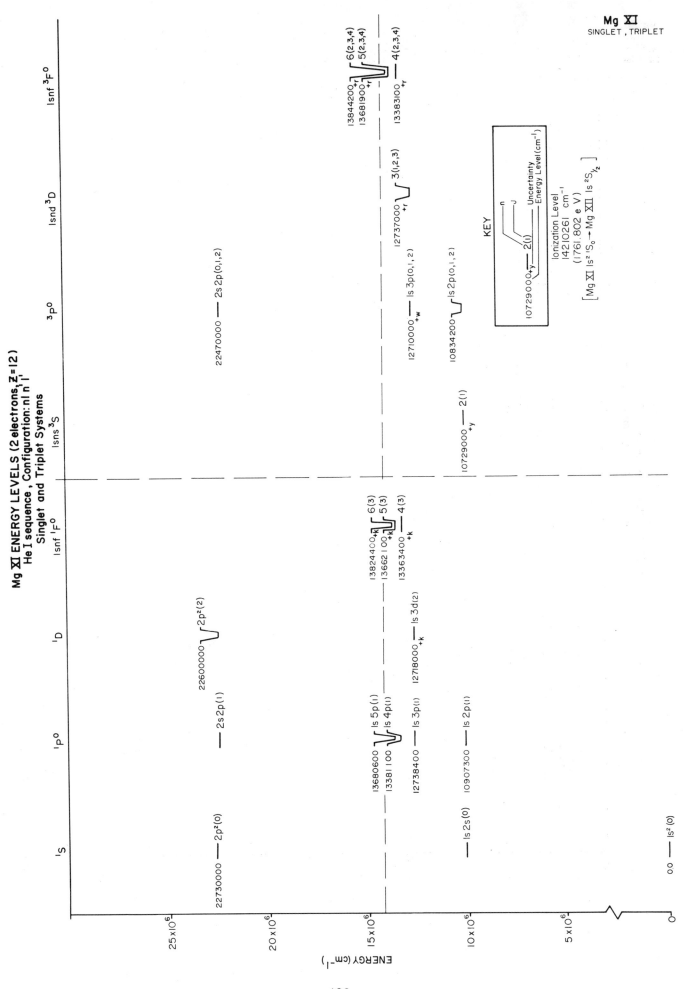

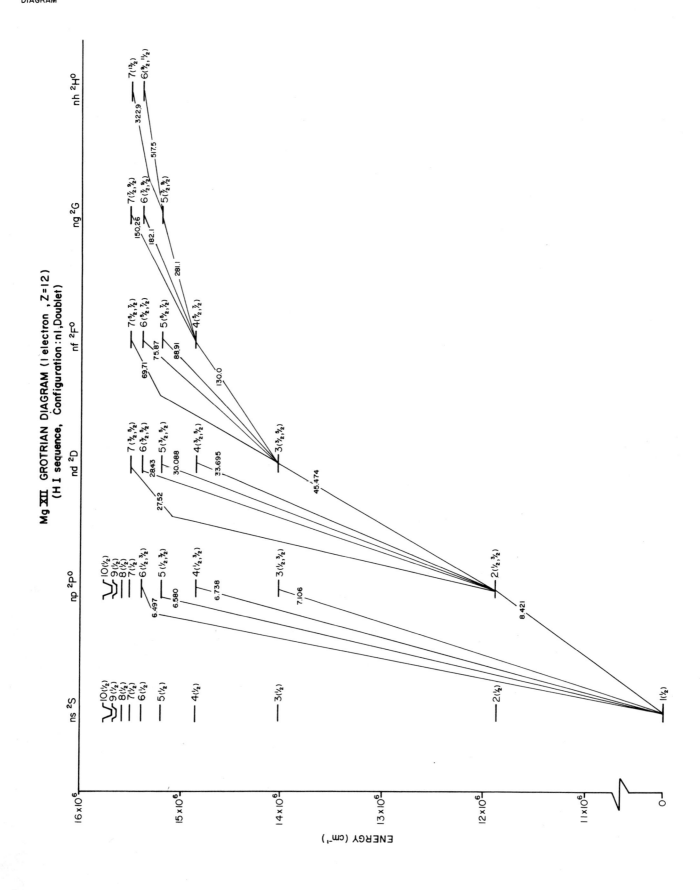

Mg XII
DOUBLET
GROTRIAN
DIAGRAM

Mg XII GROTRIAN DIAGRAM (1 electron , Z=12)
(H I sequence, Configuration : nl , Doublet)

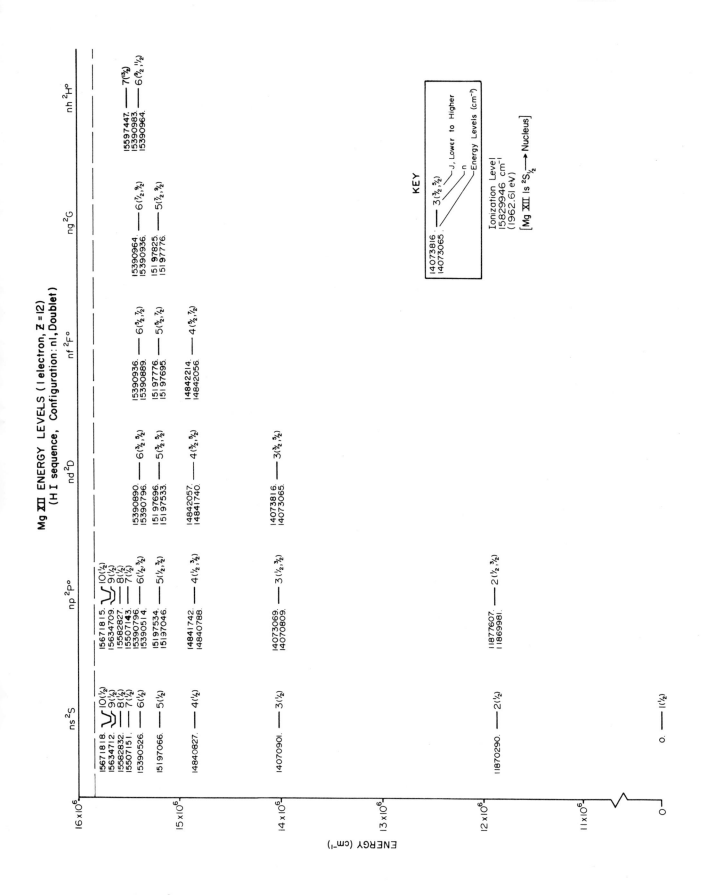

Mg XII ENERGY LEVELS (I electron, Z = 12)
(H I sequence, Configuration: nl, Doublet)

KEY

14073816 ——— 3(³⁄₂,⁵⁄₂)
14073065 ———

J, Lower to Higher
n
Energy Levels (cm⁻¹)

Ionization Level
15829946 cm⁻¹
(1962.61 eV)

[Mg XII Is ²S₁⁄₂ ⟶ Nucleus]

| Mg I | Z = 12 | 12 electrons |

I.S. Bowen, Ap. J. **132**, 1 (1960).

    Author gives transition wavelengths from nebular observations.

J.W. Swensson and G. Risberg, Ark. Fys. **31**, 237 (1966).

    Authors give line tables and identify those lines found in the solar spectrum.

| Mg II | Z = 12 | 11 electrons |

    Please see the general references.

| Mg III | Z = 12 | 10 electrons |

    Please see the general references.

    The intermediate-coupling designations are given by Andersson and Johannesson, Physica Scripta **3**, 203 (1971).

| Mg IV | Z = 12 | 9 electrons |

    Please see the general references.

| Mg V | Z = 12 | 8 electrons |

I.S. Bowen, Ap. J. **132**, 1 (1960).

    Author gives line tables of predicted and observed lines.

B. Edlén, Handbuch der Phys. **27**, 172 (1964).

    Author gives corrections to values in C.E. Moore's *Atomic Energy Levels.*

| Mg VI | Z = 12 | 7 electrons |

I.S. Bowen, Ap. J. **132**, 1 (1960).

    Author gives tables of observed lines.

B. Edlén, Handbuch der Phys. **27**, 172 (1964).

    Author gives corrections to values in C.E. Moore's *Atomic Energy Levels.*

Mg VII $Z = 12$ 6 electrons

I.S. Bowen, Ap. J. **132**, 1 (1960).

Author gives tables of lines calculated and observed.

B. Edlén, Handbuch der Phys. **27**, 172 (1964).

Author gives corrections to values in C.E. Moore's *Atomic Energy Levels*.

The following lines were omitted from the drawings:

$2s\,2p^3\,{}^1P^\circ - 2p^2\,({}^2D)\,3d'\,{}^3S$	92.50 (?)
$-2p^2\,({}^4P)\,3d\,{}^3D$	100.37 (?), 100.42 (?)
$-\quad\quad 3d\,{}^3F$	102.05 (?)
$-\quad\quad 3d\,{}^3P$	102.91 (?)
$2s\,2p^3\,{}^5S^\circ - 2p^2\,({}^4P)\,3s\,{}^3P$	91.30 (?)

Mg VIII $Z = 12$ 5 electrons

Please see the general references.

Mg IX $Z = 12$ 4 electrons

Please see the general references.

Mg X $Z = 12$ 3 electrons

Please see the general references.

Mg XI $Z = 12$ 2 electrons

Please see the general references.

Mg XII $Z = 12$ 1 electron

Please see the general references.

Aluminium (Al)

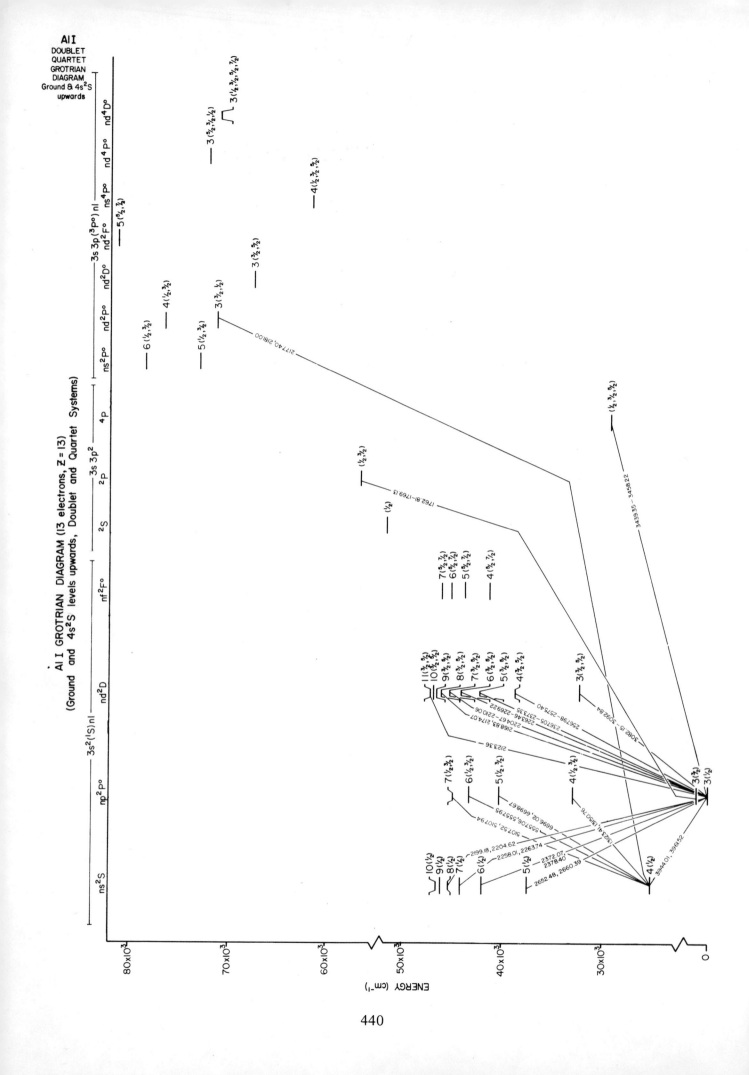

440

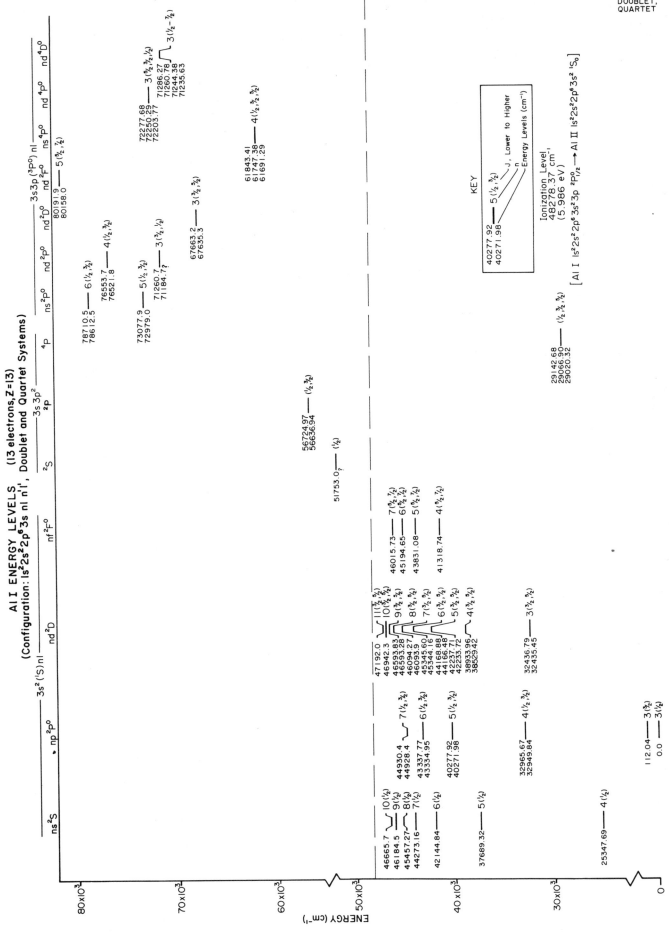

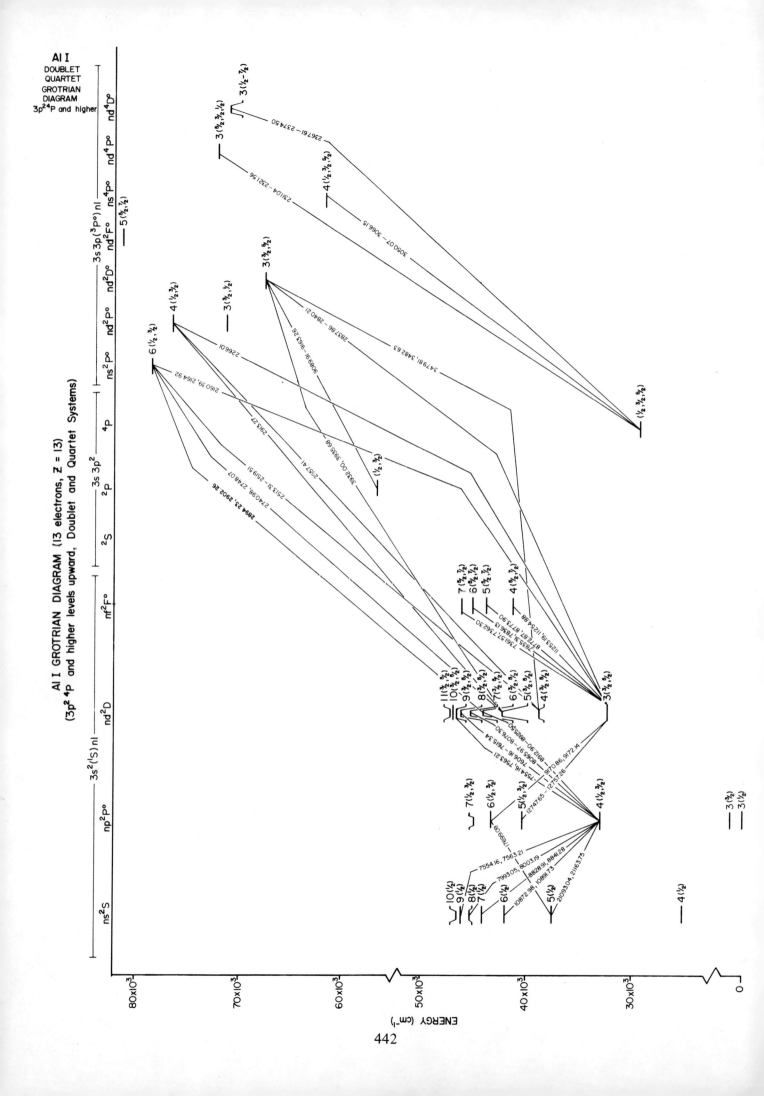

442

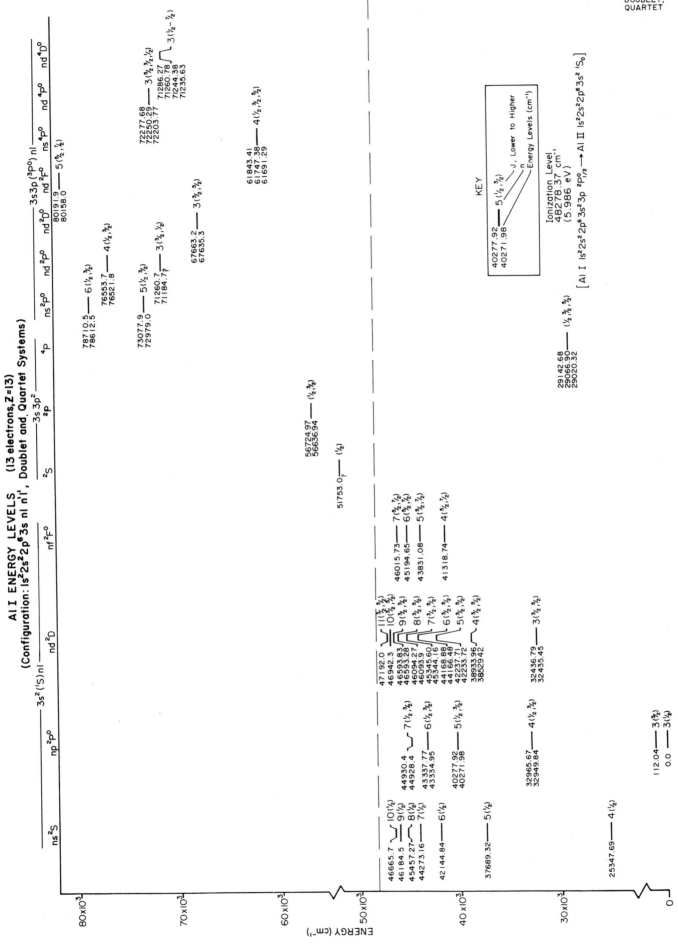

AL I ENERGY LEVELS   (13 electrons, Z=13)
(Configuration: 1s²2s²2p⁶3s nl n'l', Doublet and Quartet Systems)

AL I
DOUBLET,
QUARTET

ENERGY (cm⁻¹)

443

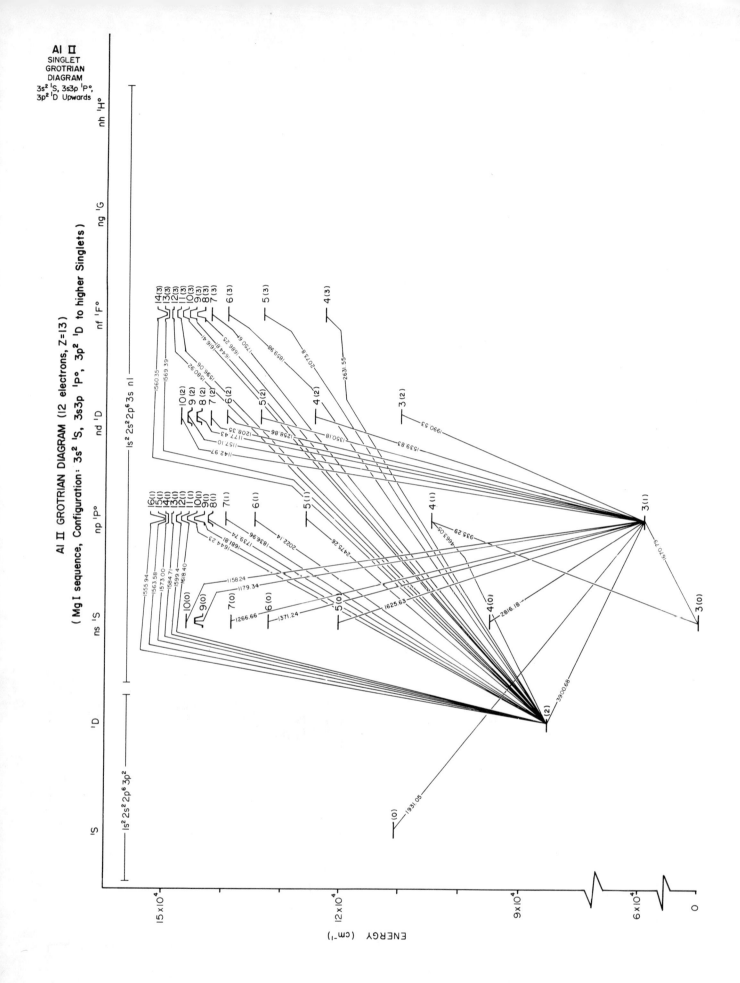

Al II
SINGLET
GROTRIAN
DIAGRAM
3s² ¹S, 3s3p ¹P°,
3p² ¹D Upwards

Al II GROTRIAN DIAGRAM (12 electrons, Z=13)
(Mg I sequence, Configuration: 3s² ¹S, 3s3p ¹P°, 3p² ¹D to higher Singlets)

ENERGY (cm⁻¹)

Al II ENERGY LEVELS (12 electrons, $Z=13$)
(Mg I sequence, Configuration: $1s^2 2s^2 2p^6 3s\, nl$, Singlet System)

Al II
SINGLET

Column headers: ^1S ^1D $ns\ ^1S$ $np\ ^1P^\circ$ $nd\ ^1D$ $nf\ ^1F^\circ$ $ng\ ^1G$ $nh\ ^1H^\circ$

Configuration: $1s^2 2s^2 2p^6 3p^2$ $1s^2 2s^2 2p^6 3s\, nl$

$nh\ ^1H^\circ$
- 147464.7 — 10(5)
- 146432.8 — 9(5)
- 144990.0 — 8(5)

$ng\ ^1G$
- 149252.9 — 13(4)
- 148217.6 — 11(4)
- 147451.0 — 10(4)
- 146414.5 — 9(4)
- 144964.7 — 8(4)
- 142849.2 — 7(4)
- 139588.7 — 6(4)
- 134181.2 — 5(4)

$nf\ ^1F^\circ$
- 150744.1 — 20(3)
- 148135.60 — 11(3)
- 147346.57 — 10(3)
- 146278.60 — 9(3)
- 144784.43 — 8(3)
- 142604.13 — 7(3)
- 139245.38 — 6(3)
- 133681.73 — 5(3)
- 123468.1 — 4(3)

$nd\ ^1D$
- 148132.7 — 11(2)
- 147343.2 — 10(2)
- 146274.4 — 9(2)
- 144780.2 — 8(2)
- 142607.0 — 7(2)
- 139286.8 — 6(2)
- 133914.05 — 5(2)
- 124794.27 — 4(2)
- 110089.90 — 3(2)

$np\ ^1P^\circ$
- 150007.6 — 17(1)
- 148004.35 — 12(1)
- 147270.26 — 11(1)
- 146299.82 — 10(1)
- 144941.03 — 9(1)
- 142961.06 — 8(1)
- 139918.91 — 7(1)
- 134919.33 — 6(1)
- 125868.92 — 5(1)
- 106918.2 — 4(1)
- 59852.04 — 3(1)

$ns\ ^1S$
- 149856.6 — 16(0)
- 148097.1 — 12(0)
- 147288.8 — 11(0)
- 146190.1 — 10(0)
- 144641.9 — 9(0)
- 142360.8 — 8(0)
- 138798.26 — 7(0)
- 132776.4 — 6(0)
- 121366.77 — 5(0)
- 9535050 — 4(0)
- 0.0 — 3(0)

^1D
- 85481.26 — (2)

^1S
- 111637.39(?) — (0)

KEY
110089.90 — 3(2)
n
J, l
Energy Level (cm^{-1})

Ionization Level
151860.4 cm^{-1}
(18.828 electron volts)
$[\text{Al II } 1s^2 2s^2 2p^6 3s^2\ ^1S_0 \rightarrow \text{Al III } 1s^2 2s^2 2p^6 3s\ ^2S_{1/2}]$

Energy axis: 15×10^4, 12×10^4, 9×10^4, 6×10^4, 0

ENERGY (cm^{-1})

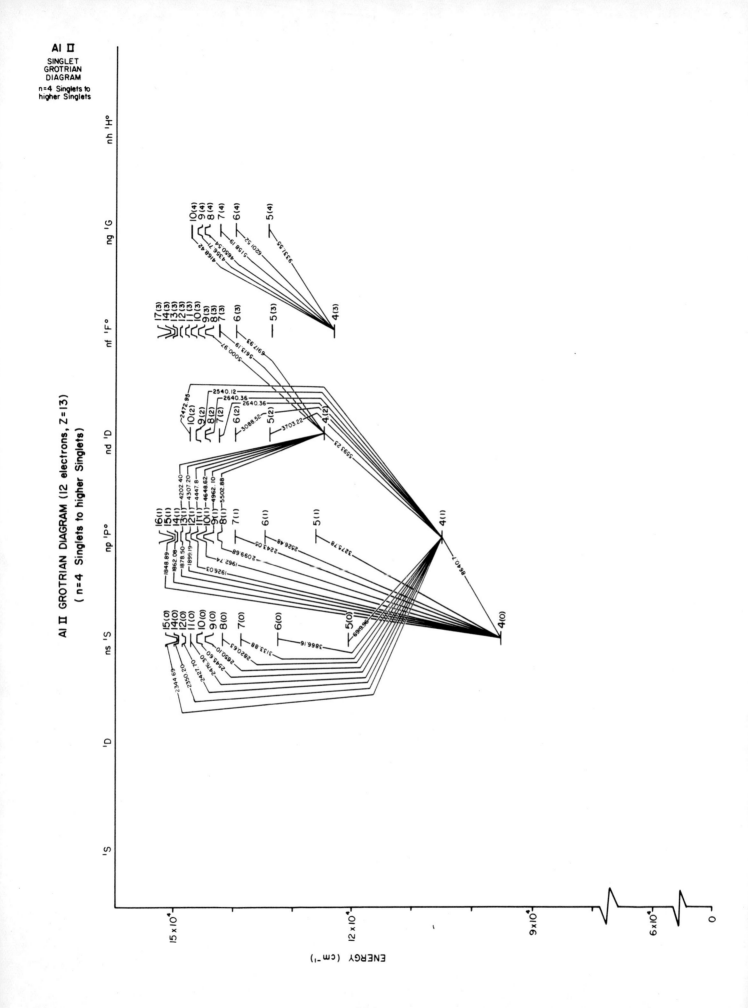

AI II GROTRIAN DIAGRAM (12 electrons, Z=13)
(n=4 Singlets to higher Singlets)

AI II
SINGLET
GROTRIAN
DIAGRAM
n=4 Singlets to
higher Singlets

ENERGY (cm⁻¹)

446

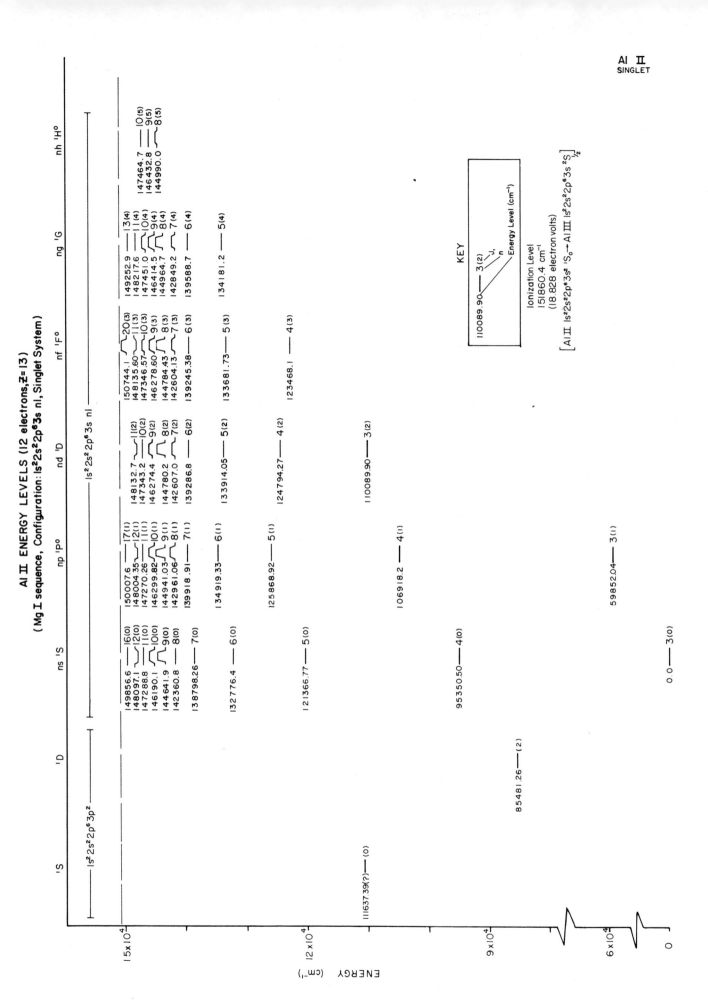

Al II ENERGY LEVELS (12 electrons, Z=13)
(Mg I sequence, Configuration: 1s²2s²2p⁶3s nl, Singlet System)

Al II
SINGLET

447

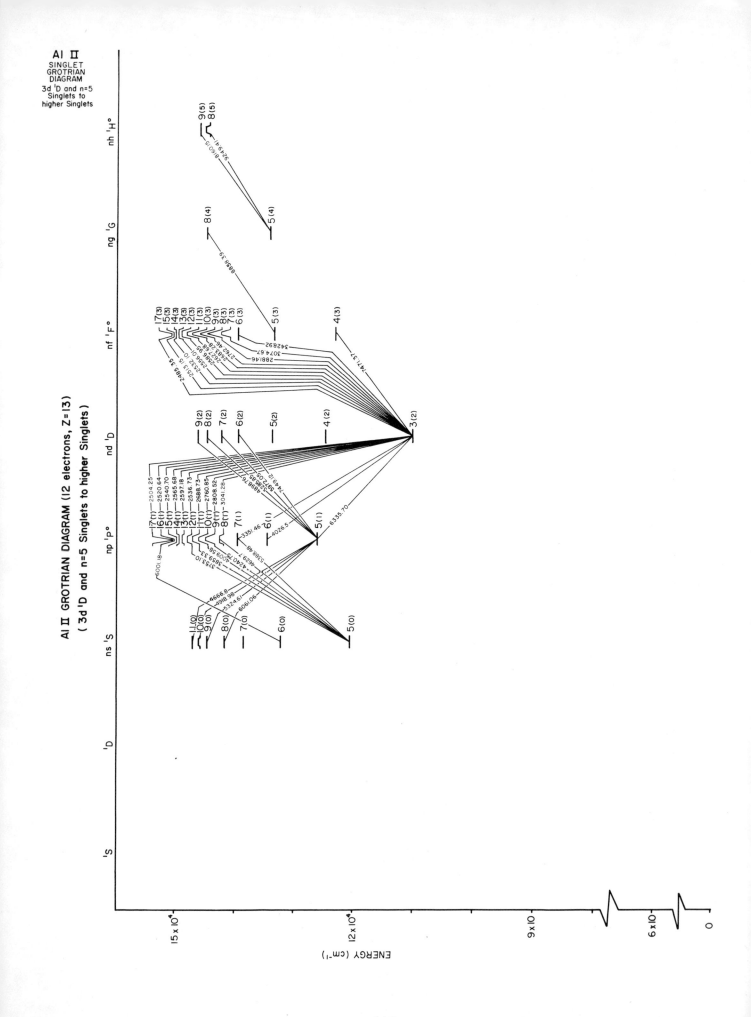

Al II ENERGY LEVELS (12 electrons, Z=13)
(Mg I sequence, Configuration: 1s²2s²2p⁶3s nl, Singlet System)

¹S	¹D	ns ¹S	np ¹P°	nd ¹D	nf ¹F°	ng ¹G	nh ¹H°

1s²2s²2p⁶3p² ——

1s²2s²2p⁶3s nl

nh ¹H°:
147464.7 —— 10(5)
146432.8 —— 9(5)
144990.0 —— 8(5)

ng ¹G:
149252.9 —— 13(4)
148217.6 —— 11(4)
147451.0 —— 10(4)
146414.5 —— 9(4)
144964.7 —— 8(4)
142849.2 —— 7(4)
139588.7 —— 6(4)
134181.2 —— 5(4)

nf ¹F°:
150744.1 —— 20(3)
148135.60 —— 11(3)
147346.57 —— 10(3)
146278.60 —— 9(3)
144784.43 —— 8(3)
142604.13 —— 7(3)
139245.38 —— 6(3)
133681.73 —— 5(3)
123468.1 —— 4(3)

nd ¹D:
148132.7 —— 11(2)
147343.2 —— 10(2)
146274.4 —— 9(2)
144780.2 —— 8(2)
142607.0 —— 7(2)
139286.8 —— 6(2)
133914.05 —— 5(2)
124794.27 —— 4(2)
110089.90 —— 3(2)

np ¹P°:
150007.6 —— 17(1)
148004.35 —— 12(1)
147270.26 —— 11(1)
146299.82 —— 10(1)
144941.03 —— 9(1)
142961.06 —— 8(1)
139918.91 —— 7(1)
134919.33 —— 6(1)
125868.92 —— 5(1)
106918.2 —— 4(1)
59852.04 —— 3(1)

ns ¹S:
149856.6 —— 16(0)
148097.1 —— 12(0)
147288.8 —— 11(0)
146190.1 —— 10(0)
144641.9 —— 9(0)
142360.8 —— 8(0)
138798.26 —— 7(0)
132776.4 —— 6(0)
121366.77 —— 5(0)
95350.50 —— 4(0)
0.0 —— 3(0)

¹D:
85481.26 —— (2)

¹S:
111637.39(?) —— (0)

KEY

110089.90 — 3(2)
                J
                n
Energy Level (cm⁻¹)

Ionization Level
151860.4 cm⁻¹
(18.828 electron volts)
[Al II 1s²2s²2p⁶3s² ¹S₀ → Al III 1s²2s²2p⁶3s ²S₁/₂]

ENERGY (cm⁻¹)

15×10⁴
12×10⁴
9×10⁴
6×10⁴
0

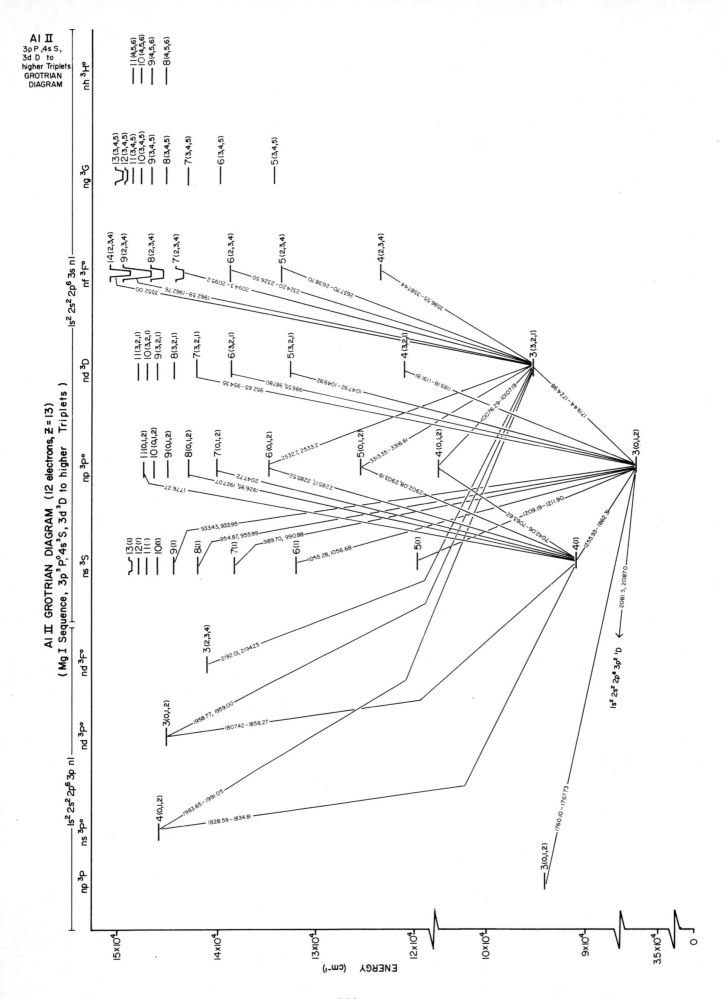

AI II
3p P, 4s S,
3d D to
higher Triplets
GROTRIAN
DIAGRAM

AI II GROTRIAN DIAGRAM (12 electrons, Z = 13)
(Mg I Sequence, 3p ³P°, 4s ³S, 3d ³D to higher Triplets)

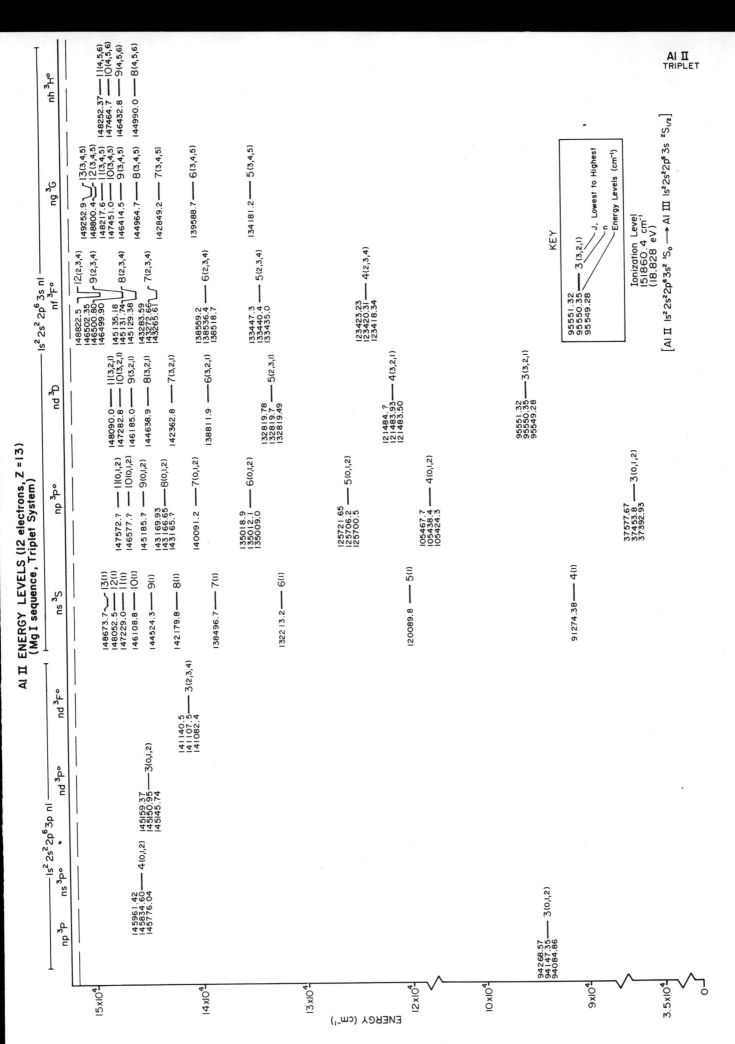

AI II ENERGY LEVELS (12 electrons, Z=13)
(Mg I sequence, Triplet System)

AI II
TRIPLET

451

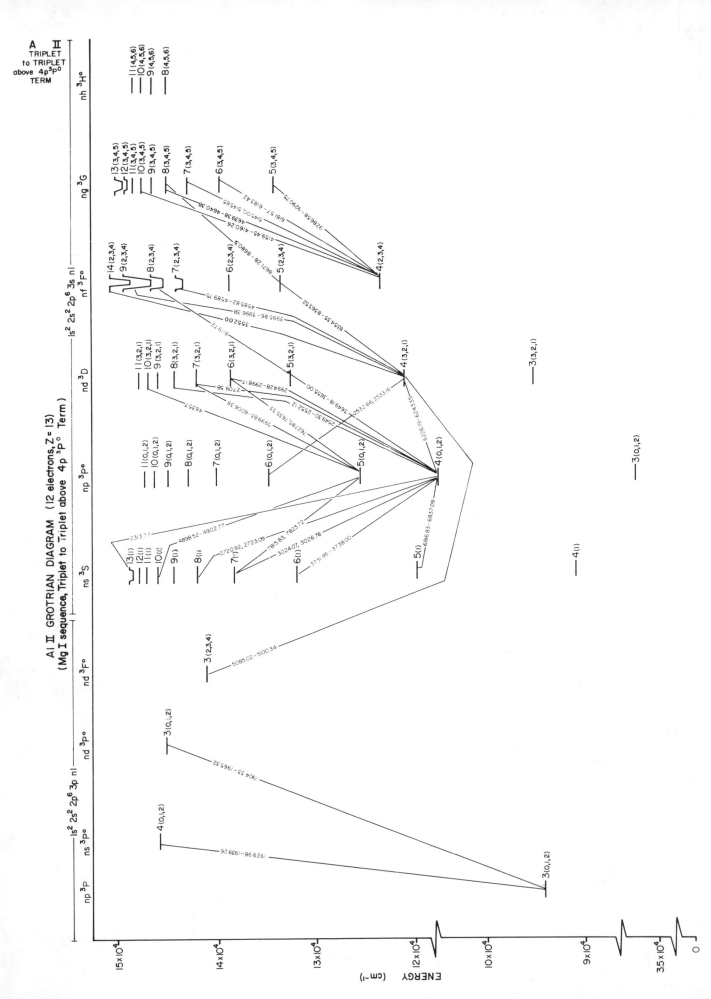

Al II GROTRIAN DIAGRAM (12 electrons, Z = 13)
(Mg I sequence, Triplet to Triplet above 4p³P° Term)

452

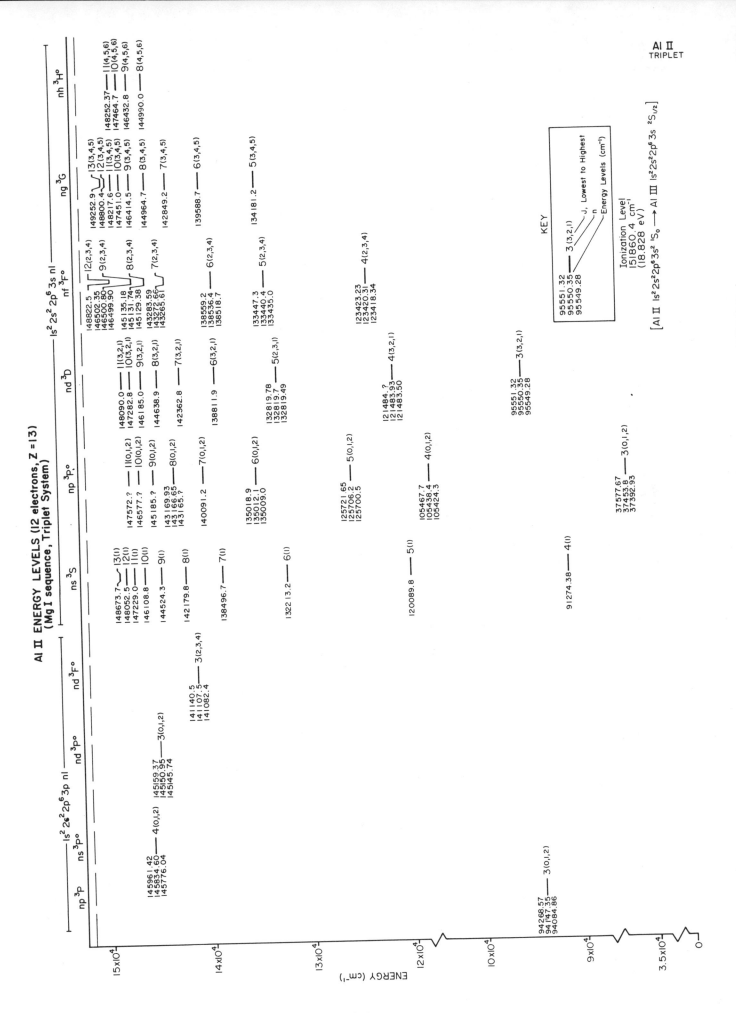

Al II ENERGY LEVELS (12 electrons, Z =13)
(Mg I sequence, Triplet System)

AI III
DOUBLET
GROTRIAN
DIAGRAM

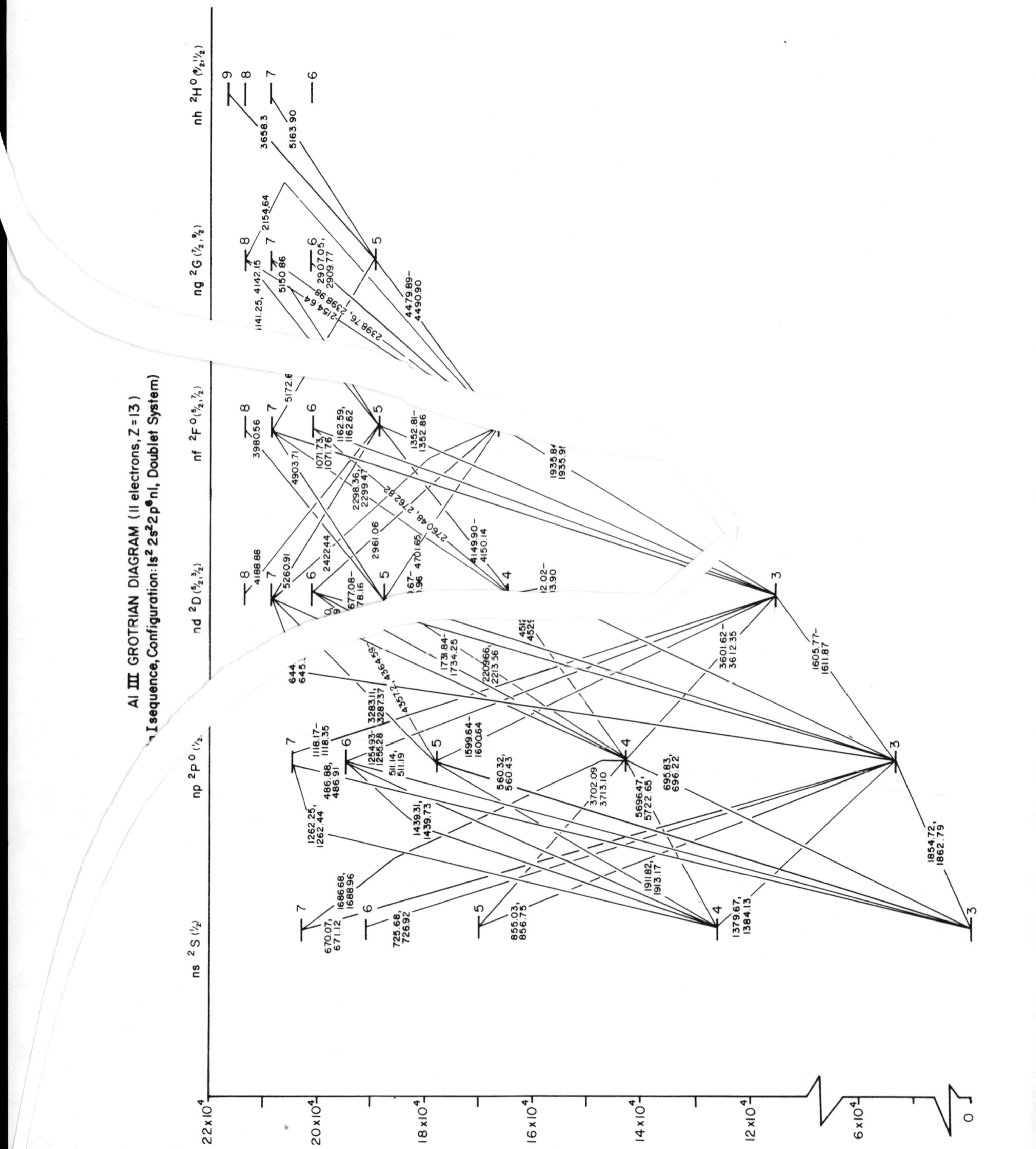

AI III GROTRIAN DIAGRAM (11 electrons, Z=13)
(Al I sequence, Configuration: 1s² 2s²2p⁶nl, Doublet System)

454

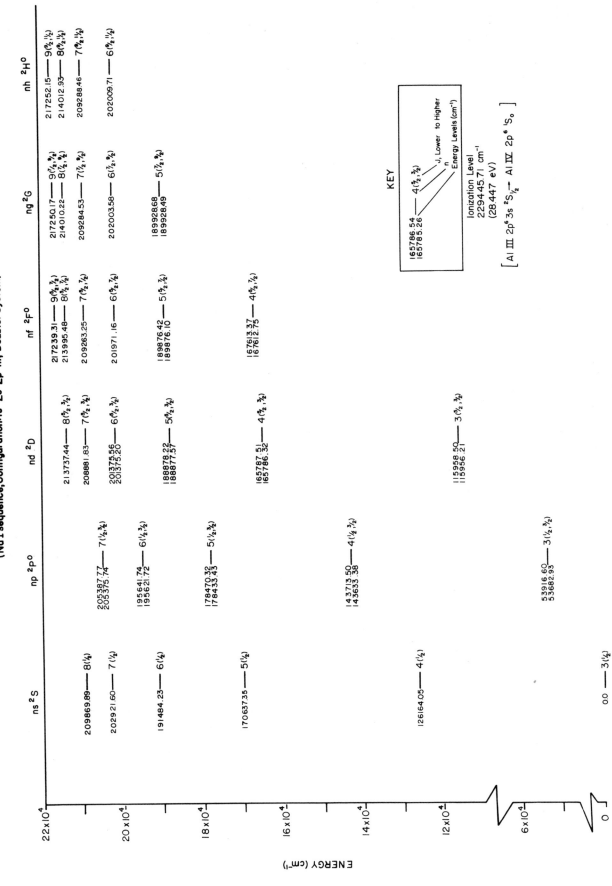

Al III ENERGY LEVELS (11 electrons, Z=13)
(Na I sequence, Configuration: 1s²2s²2p⁶nl, Doublet System)

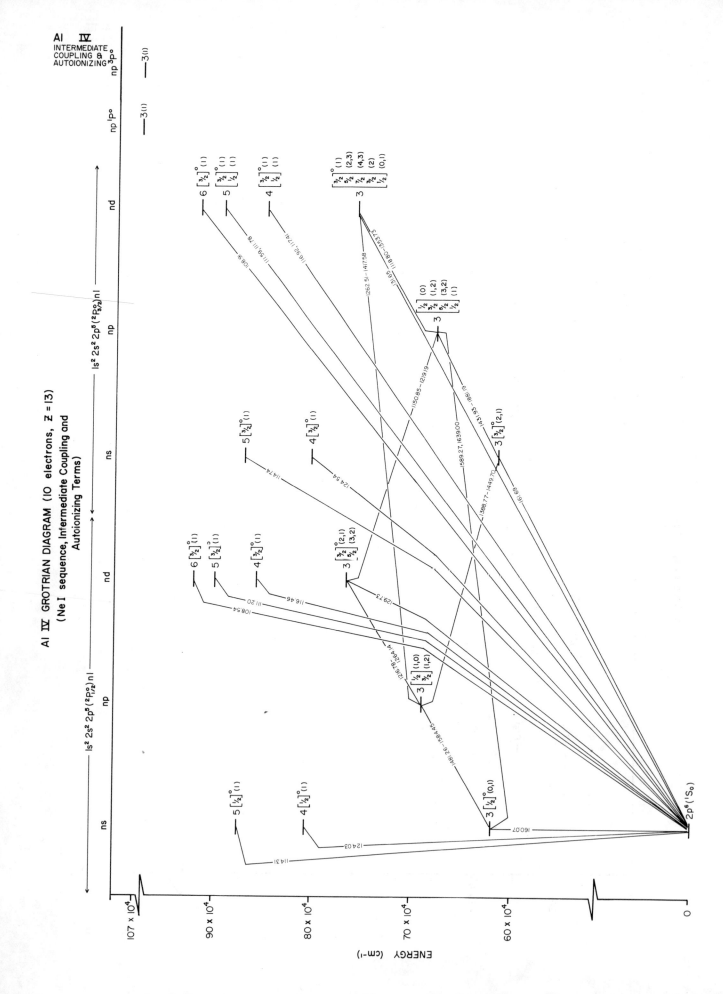

AI IV GROTRIAN DIAGRAM (10 electrons, Z = 13)
(Ne I sequence, Intermediate Coupling and Autoionizing Terms)

ENERGY (cm-1)

AI IV ENERGY LEVELS (10 electrons, Z = 13)
(Ne I sequence, Intermediate Coupling and Autoionizing Terms)

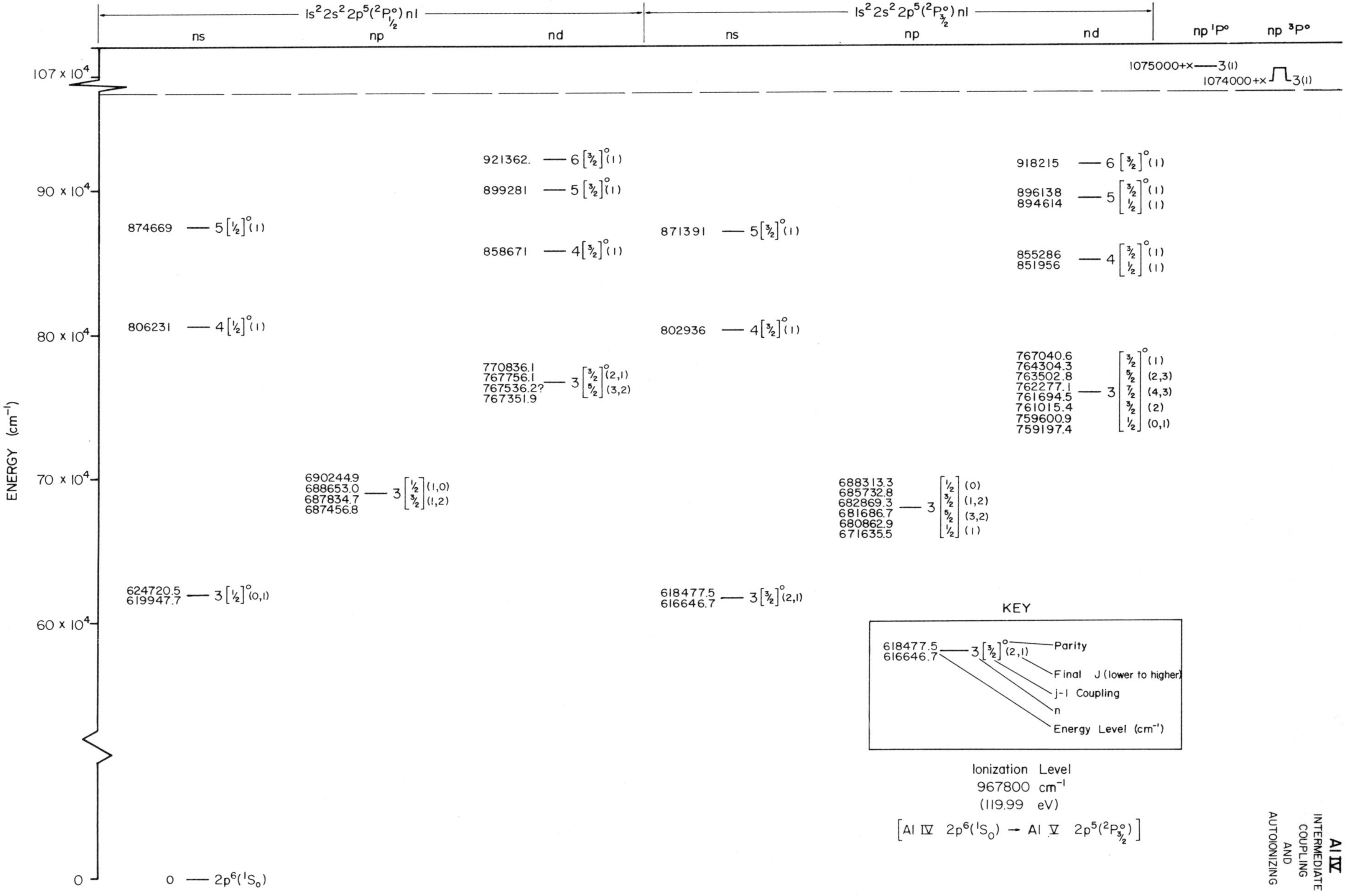

AI IV
INTERMEDIATE
COUPLING
AND
AUTOIONIZING

ENERGY (cm⁻¹)

457

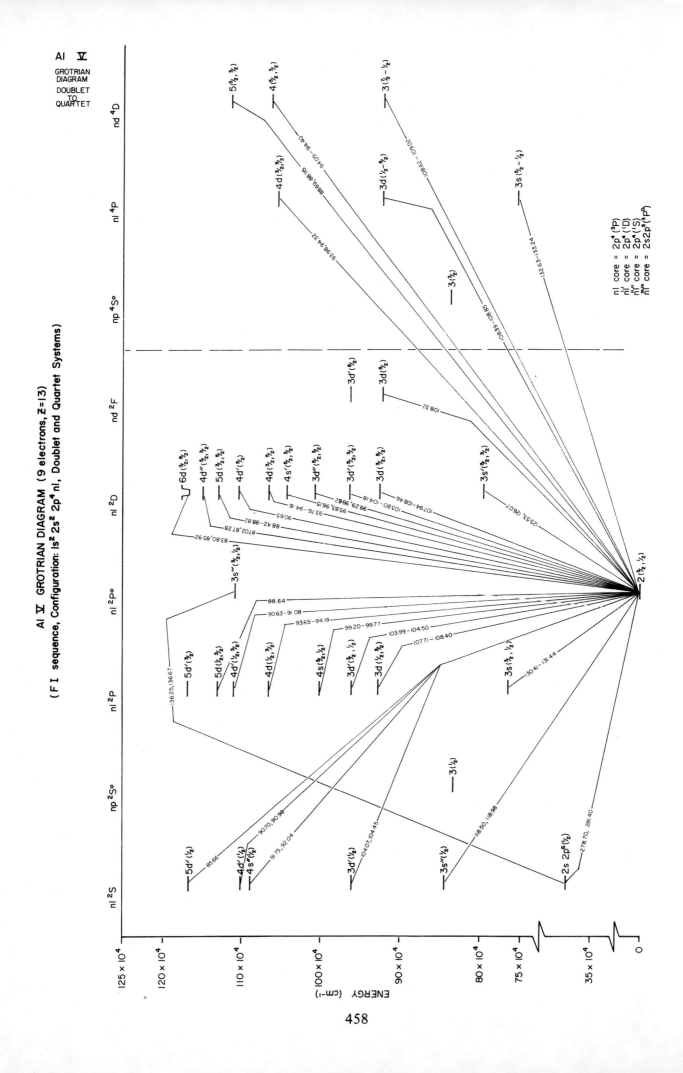

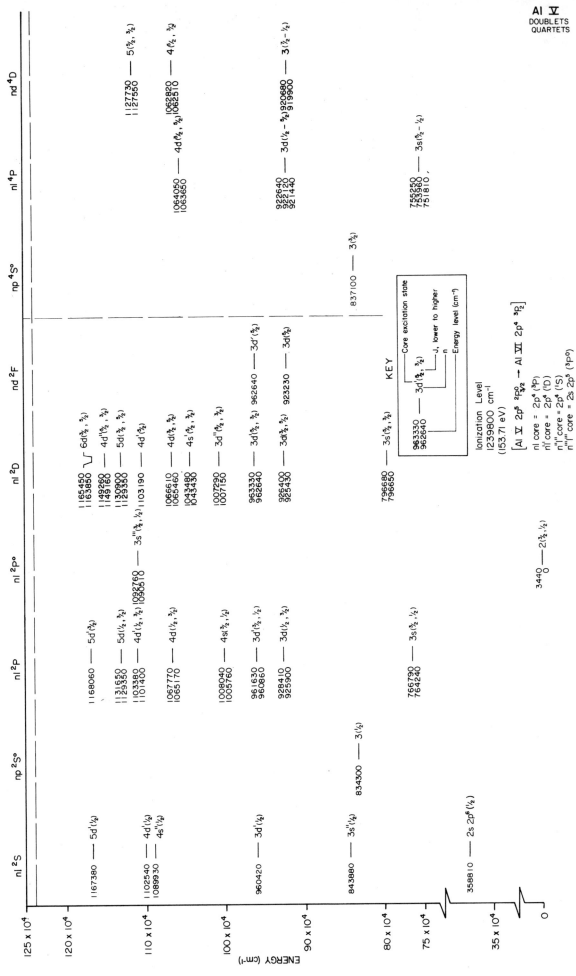

AI V ENERGY LEVELS (9 electrons, Z=13)

(F I sequence, Configuration: ls² 2s² 2p⁴ nl, Doublet and Quartet Systems)

KEY

Core excitation state

963330 — 3d'(⁵⁄₂, ³⁄₂)
962640

J, lower to higher
n
Energy level (cm⁻¹)

Ionization Level
1239800 cm⁻¹
(153.71 eV)

[AI V 2p⁵ 2p⁰ ³/₂ → AI VI 2p⁴ ³P₂]

nl core = 2p⁴ (³P)
n'l' core = 2p⁴ (¹D)
n''l'' core = 2p⁴ (¹S)
n'''l''' core = 2s 2p⁵ (³P°)

ENERGY (cm⁻¹)

459

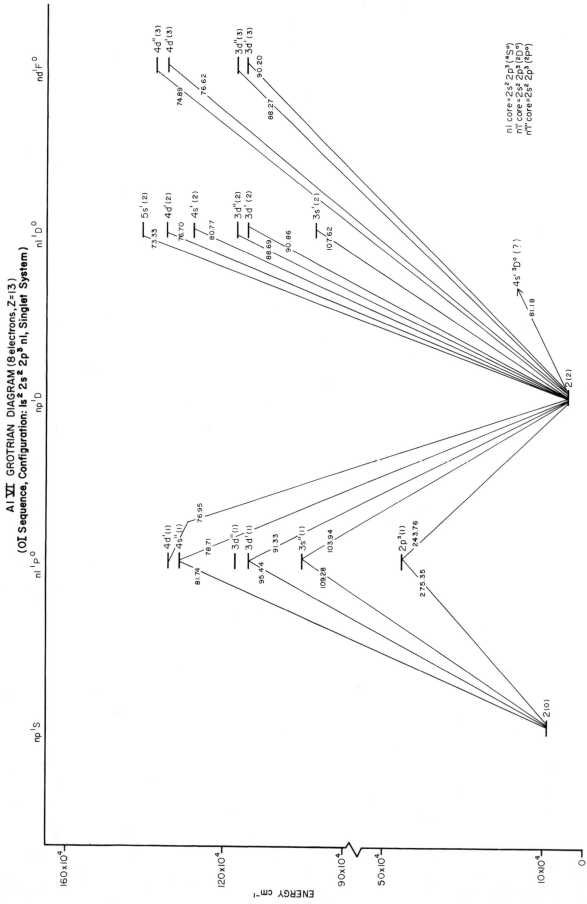

Al VI
SINGLET
GROTRIAN
DIAGRAM

Al VI GROTRIAN DIAGRAM (8 electrons, Z=13)
(OI Sequence, Configuration: 1s² 2s² 2p³ nl, Singlet System)

nl core=2s² 2p³(⁴S°)
n'l' core=2s² 2p³(²D°)
n'l'' core=2s² 2p³(²P°)

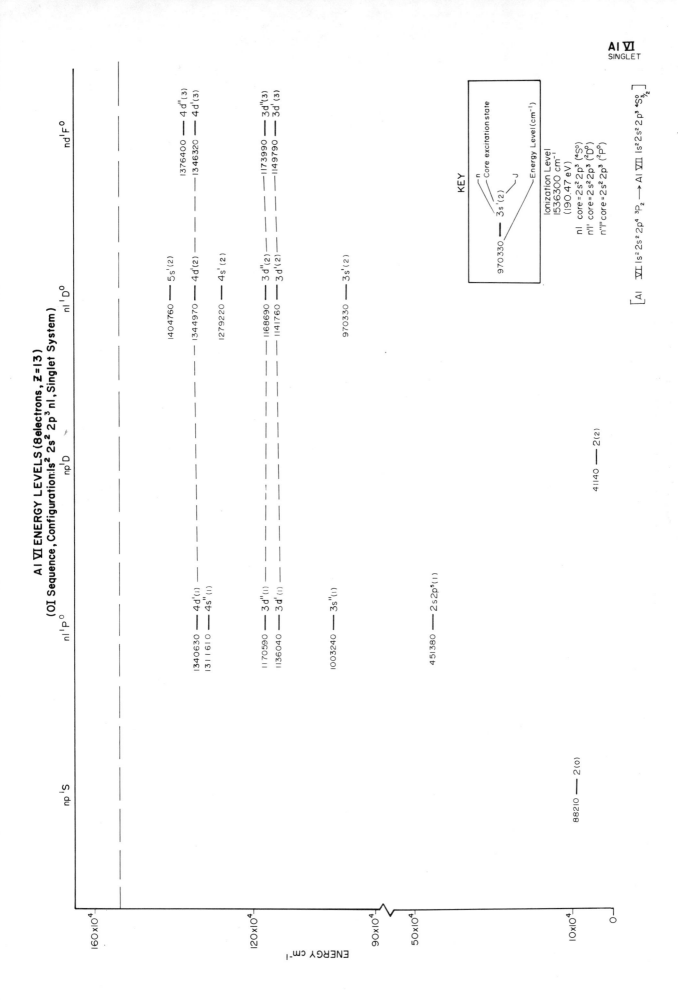

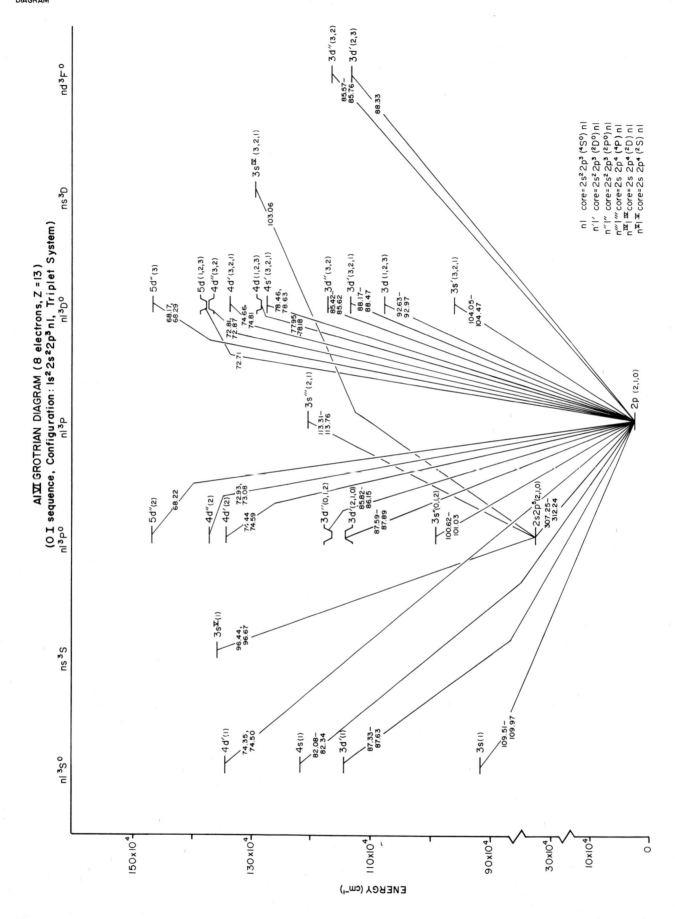

AI VI GROTRIAN DIAGRAM (8 electrons, Z = 13)
(O I sequence, Configuration: 1s²2s²2p³nl, Triplet System)

AI VI
TRIPLET
GROTRIAN
DIAGRAM

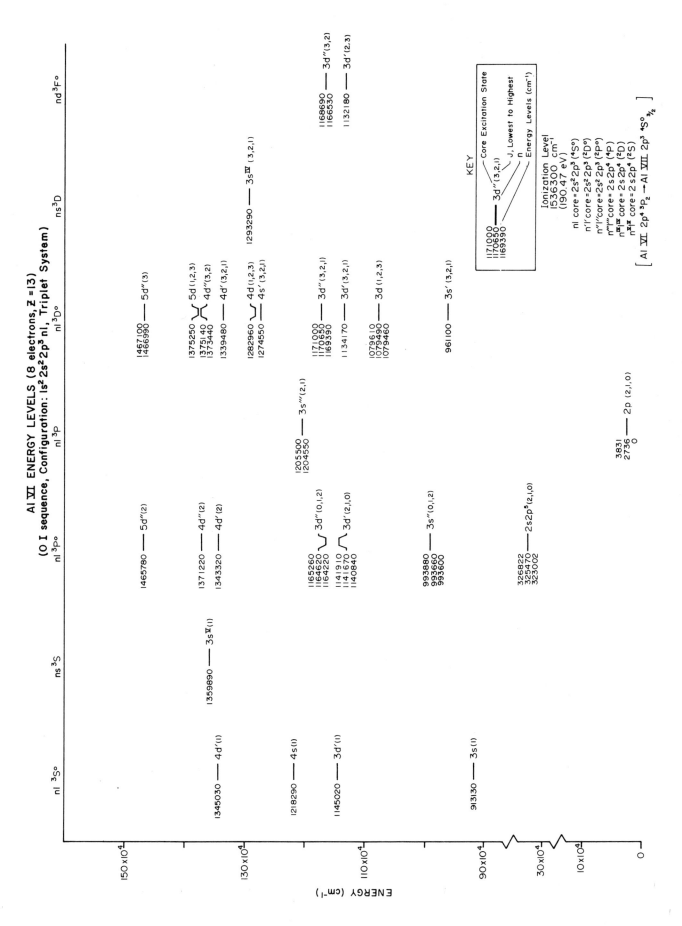

AI VI ENERGY LEVELS (8 electrons, Z = 13)
(O I sequence, Configuration: 1s²2s²2p³nl, Triplet System)

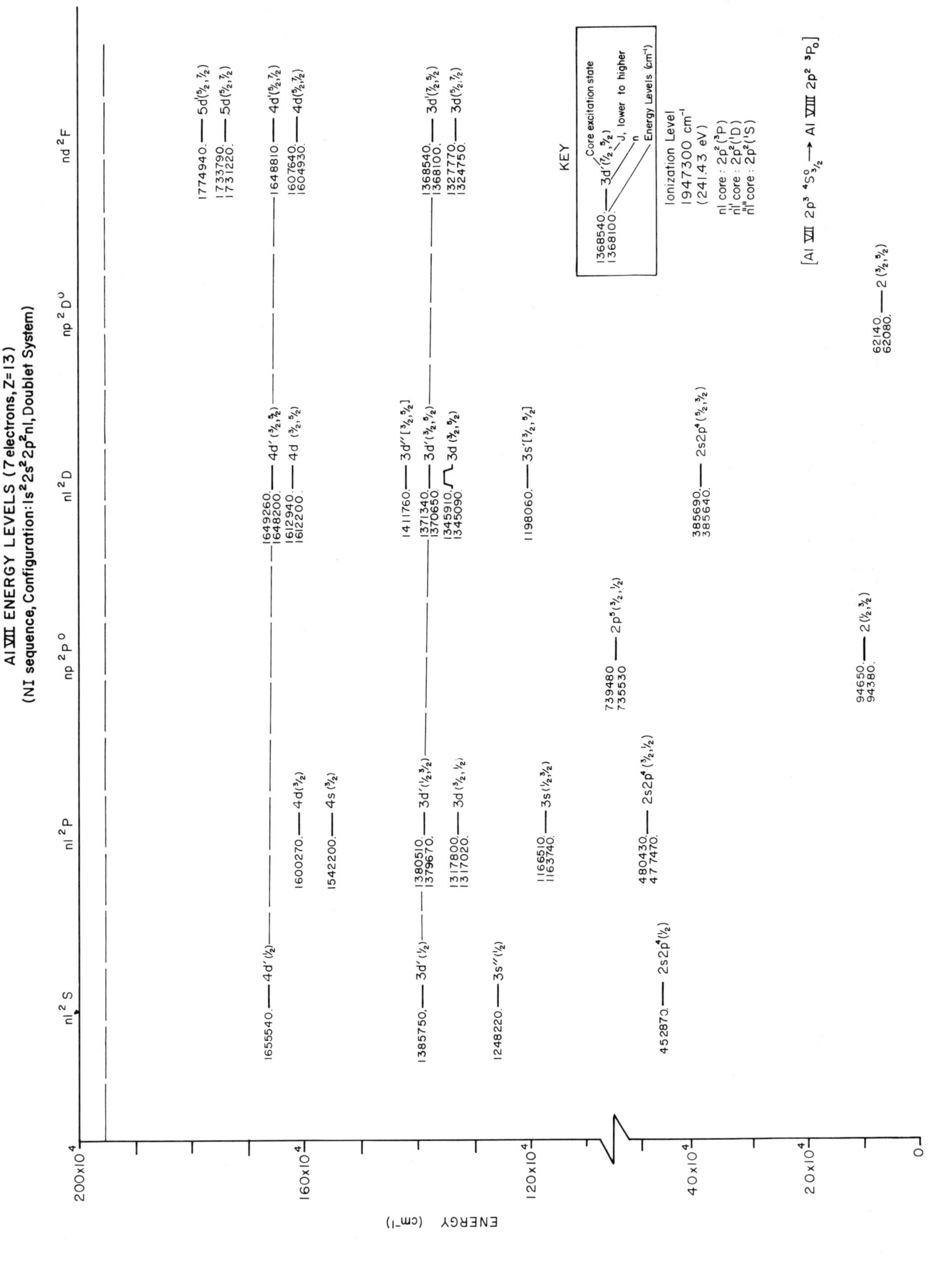
Al VII ENERGY LEVELS (7 electrons, Z=13)
(NI sequence, Configuration: 1s²2s²2p²nl, Doublet System)
nl ²S
nl ²P
np ²P⁰
nl ²D
np ²D⁰
nd ²F
165540. —— 4d'(½)
1385750. —— 3d'(½)
1248220. —— 3s''(½)
452870. —— 2s2p⁴(½)
1600270. —— 4d(³⁄₂)
1542200. —— 4s(³⁄₂)
1380510. 1379670. —— 3d'(½,³⁄₂)
1317800. 1317020. —— 3d(³⁄₂,½)
1165510. 1163740. —— 3s(½,³⁄₂)
480430. 477470. —— 2s2p⁴(³⁄₂,½)
739480 735530 —— 2p⁵(³⁄₂,½)
94650. 94380. —— 2(½,³⁄₂)
1649260. 1648200. —— 4d'(³⁄₂,⁵⁄₂)
1612940. 1612200. —— 4d(³⁄₂,⁵⁄₂)
1411760. —— 3d''[³⁄₂,⁵⁄₂]
1371340. 1370650. —— 3d'(³⁄₂,⁵⁄₂)
1345910. 1345090. —— 3d(³⁄₂,⁵⁄₂)
1198060. —— 3s'(³⁄₂,⁵⁄₂)
385690. 385640. —— 2s2p⁴(⁵⁄₂,³⁄₂)
62140. 62080. —— 2(³⁄₂,⁵⁄₂)
1774940. —— 5d'(⁵⁄₂,⁷⁄₂)
1733790. 1731220. —— 5d(⁵⁄₂,⁷⁄₂)
1648810 —— 4d'(⁵⁄₂,⁷⁄₂)
1607640. 1604930. —— 4d(⁵⁄₂,⁷⁄₂)
1368540. 1368100. —— 3d'(⁷⁄₂,⁵⁄₂)
1327770. 1324750. —— 3d(⁵⁄₂,⁷⁄₂)
KEY
1368540. 1368100. —— 3d'(⁷⁄₂,⁵⁄₂)
Core excitation state
J, lower to higher
n
Energy Levels (cm⁻¹)
Ionization Level
1947300 cm⁻¹
(241.43 eV)
nl core : 2p²(³P)
nl' core : 2p²(¹D)
nl'' core : 2p²(¹S)
[Al VII 2p³ ⁴S°₃/₂ → Al VIII 2p² ³P₀]
200x10⁴
160x10⁴
120x10⁴
40x10⁴
20x10⁴
0
ENERGY (cm⁻¹)

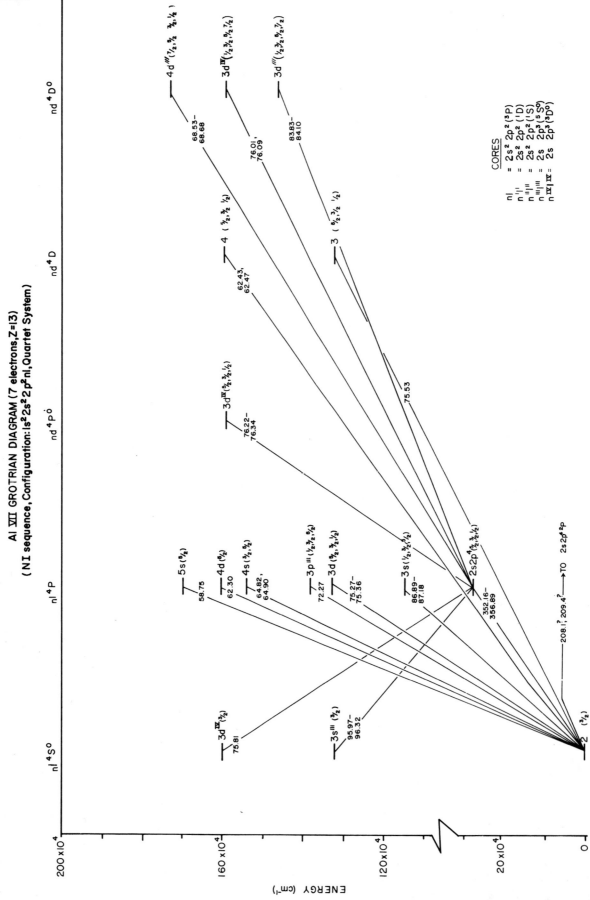

Al VII GROTRIAN DIAGRAM (7 electrons, Z=13)
(N I sequence, Configuration: 1s²2s²2p²nl, Quartet System)

Al VII
QUARTET
GROTRIAN
DIAGRAM

CORES

nl   = 2s² 2p² (³P)
nⁱⁱ   = 2s² 2p² (¹D)
nⁱⁱⁱ   = 2s² 2p² (¹S)
nⁱᵛ   = 2s 2p³ (⁵Sᵒ)
nⁱᵛ,ⁱᵛ = 2s 2p³ (³Dᵒ)

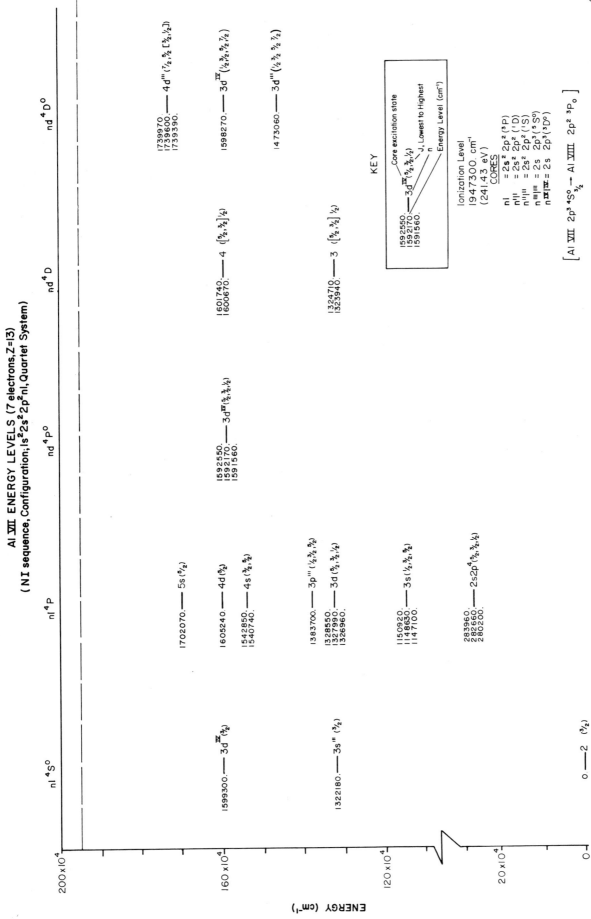

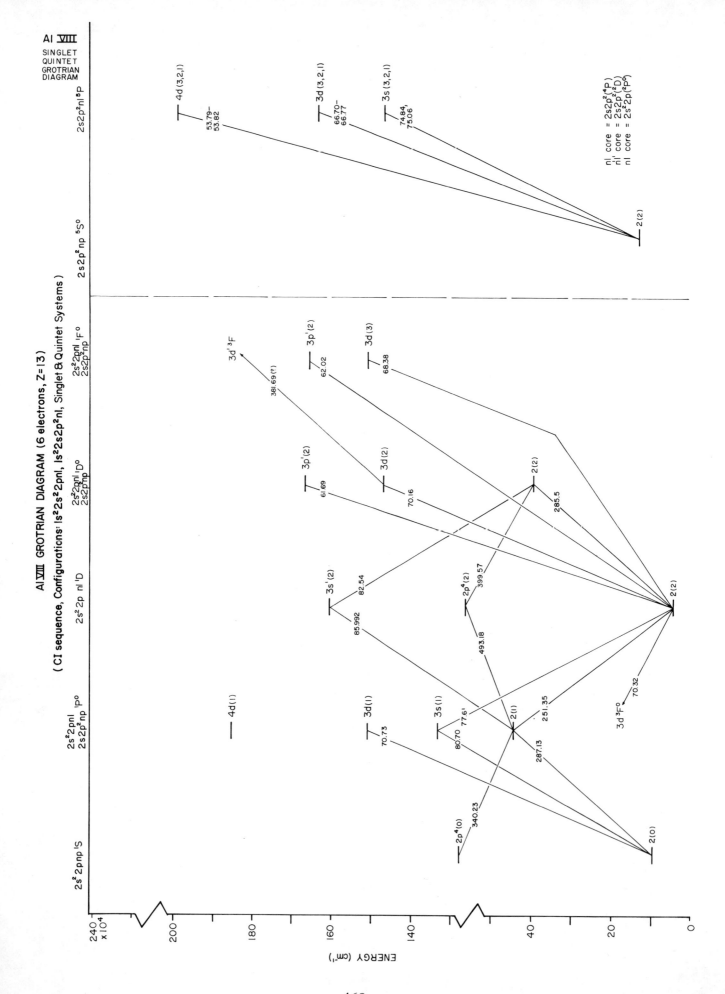

Al VIII GROTRIAN DIAGRAM (6 electrons, Z=13)
(CI sequence, Configurations: 1s²2s²2pnl, 1s²2s2p²nl, Singlet & Quintet Systems)

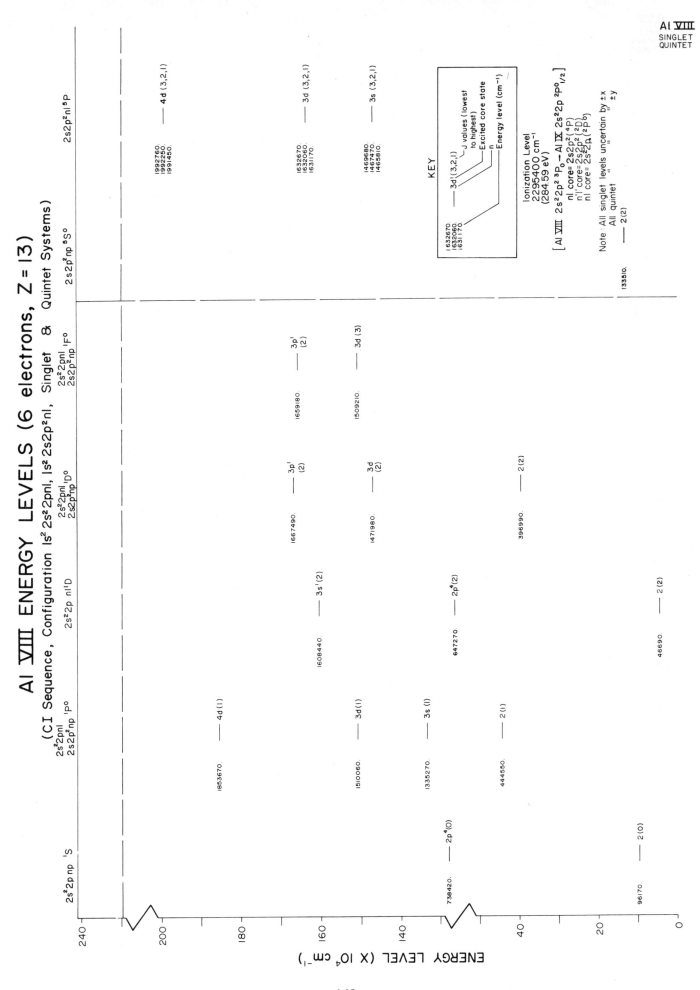

Al VIII ENERGY LEVELS (6 electrons, Z = 13)

(CI Sequence, Configuration 1s² 2s² 2pnl, 1s² 2s2p²nl, Singlet & Quintet Systems)

ENERGY LEVEL (X 10⁴ cm⁻¹)

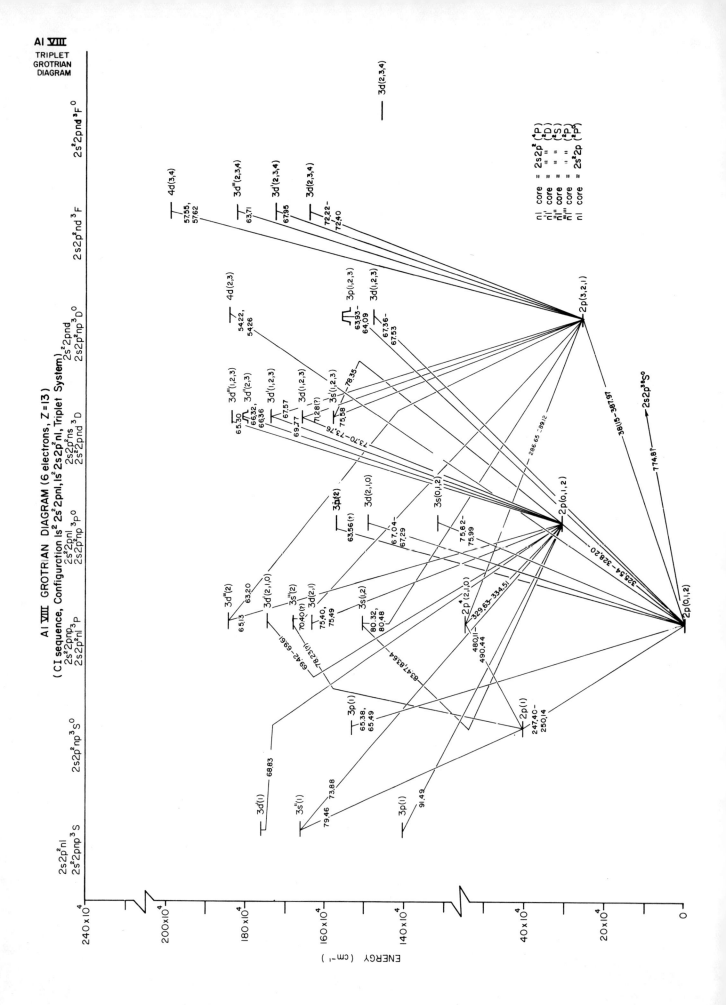

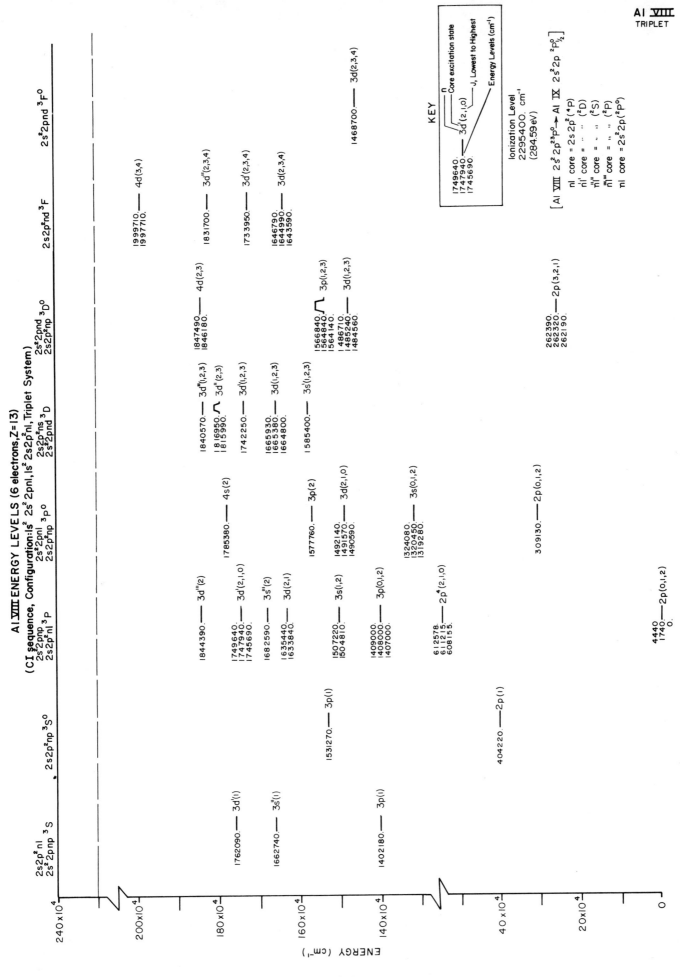

AI VIII ENERGY LEVELS (6 electrons, Z=13)

(CI sequence, Configuration: 1s² 2s² 2pnl, 1s² 2s2p²nl, Triplet System)

AI VIII
TRIPLET

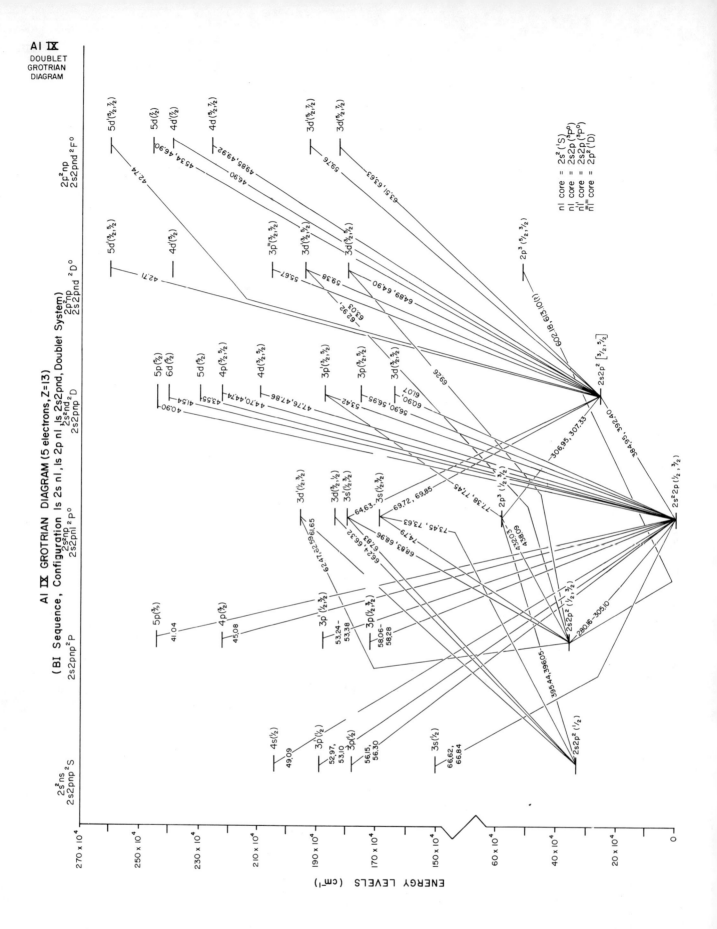

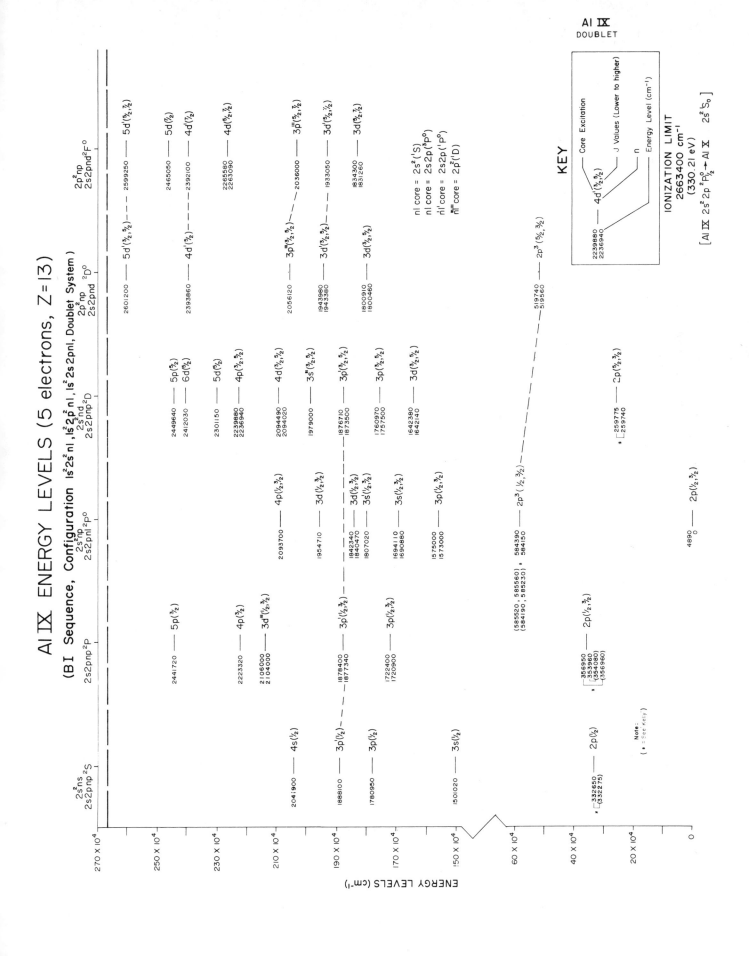

Al IX ENERGY LEVELS (5 electrons, Z=13)

(BI Sequence, Configuration 1s²2s²nl, 1s²2p²nl, 1s²2s2pnl, Doublet System)

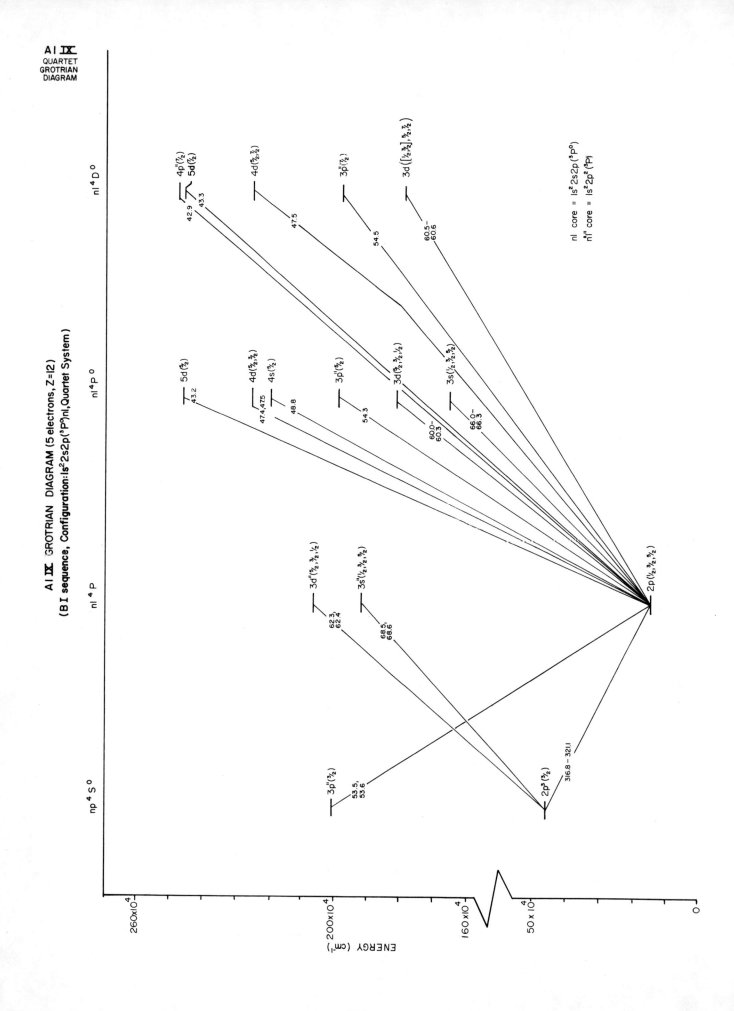

AI IX ENERGY LEVELS (5 electrons, Z=12)

(BI Sequence, Configuration Is²2s2p(³P°)nl Quartet System)

AI IX
QUARTET

ENERGY (cm⁻¹)

np⁴S°	nl⁴P	nl⁴P°	nl¹D°

270 × 10⁴

2479970. —— 4p"(7/2)
2462040. —— 5d(7/2)

250 × 10⁴

2463200. —— 5d(5/2)

230 × 10⁴

2256950. —— 4d(5/2, 3/2)
2256240.
2201140. —— 4s(5/2)

2254250. —— 4d(5/2, 7/2)
2253750.

2067100. —— 3d"(5/2, 3/2, 1/2)
2066350.
2065270.

210 × 10⁴

2017670. —— 3p³(3/2)

1991700. —— 3p"(5/2)

1986800. —— 3p"(7/2)

1921100. —— 3s"(1/2, 3/2, 5/2)
1918850.
1917920.

190 × 10⁴

1814090. —— 3d(5/2, 3/2, 1/2)
1808530.
1807490.

1800980. —— 3d([1/2,3/2],5/2,7/2)
1799490.
1799090.

1662340. —— 3s(1/2, 3/2, 5/2)
1659350.
1657690.

170 × 10⁴

50 × 10⁴

461985. —— 2p³(3/2)

150505. —— 2p(1/2, 3/2, 5/2)
148015.
146310.

30 × 10⁴

10 × 10⁴

0

KEY

n —— Core Excitation State

3d"(5/2, 3/2, 1/2) —— J (lowest to highest)

2067100. —— Energy Level (cm⁻¹)
2066350.
2065270.

nl core = Is²2s2p(³P°)
n"l" core = Is²2p²(³P)

IONIZATION LIMIT
2663400 cm⁻¹
(330.21 eV)

[AI IX 2s² 2p ²P° — AI X 2s² ¹S₀]

Note: All terms uncertain by ±x

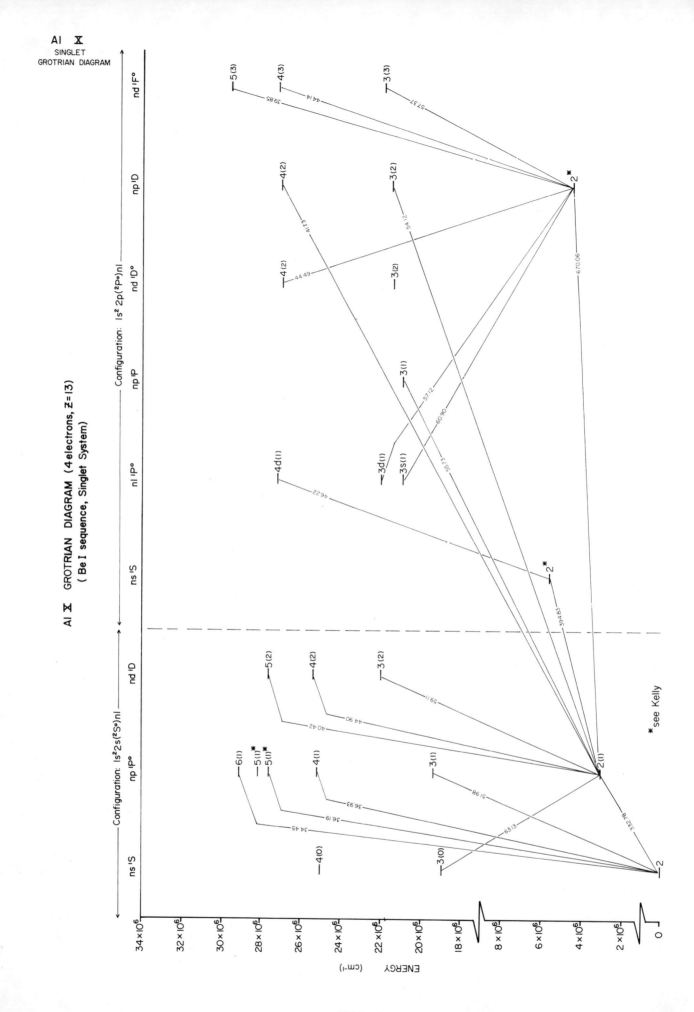

AI X

SINGLET
GROTRIAN DIAGRAM

AI X GROTRIAN DIAGRAM (4 electrons, Z=13)
(Be I sequence, Singlet System)

ENERGY (cm⁻¹)

*see Kelly

476

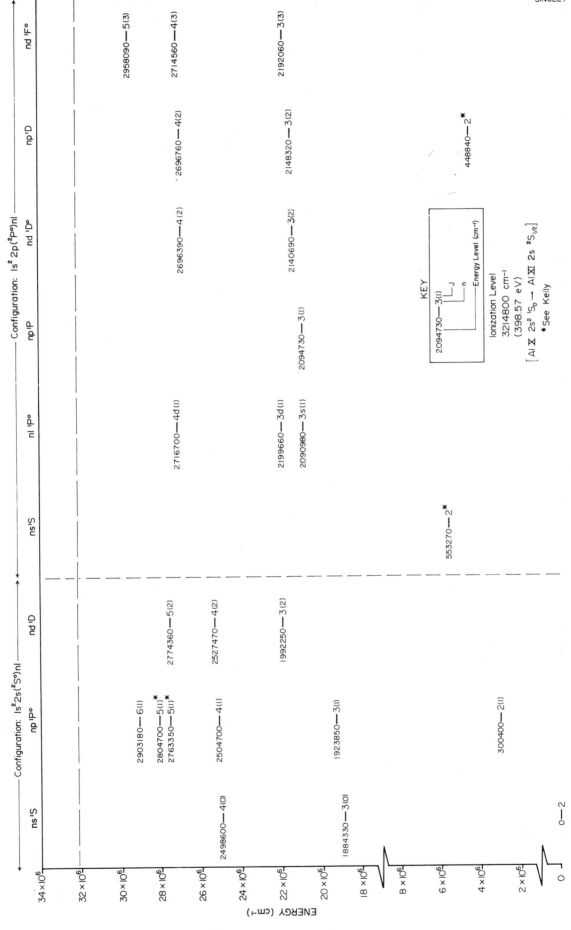

Al X ENERGY LEVELS (4 electrons, Z = 13)
(Be I sequence, Singlet System)

Al X
SINGLET

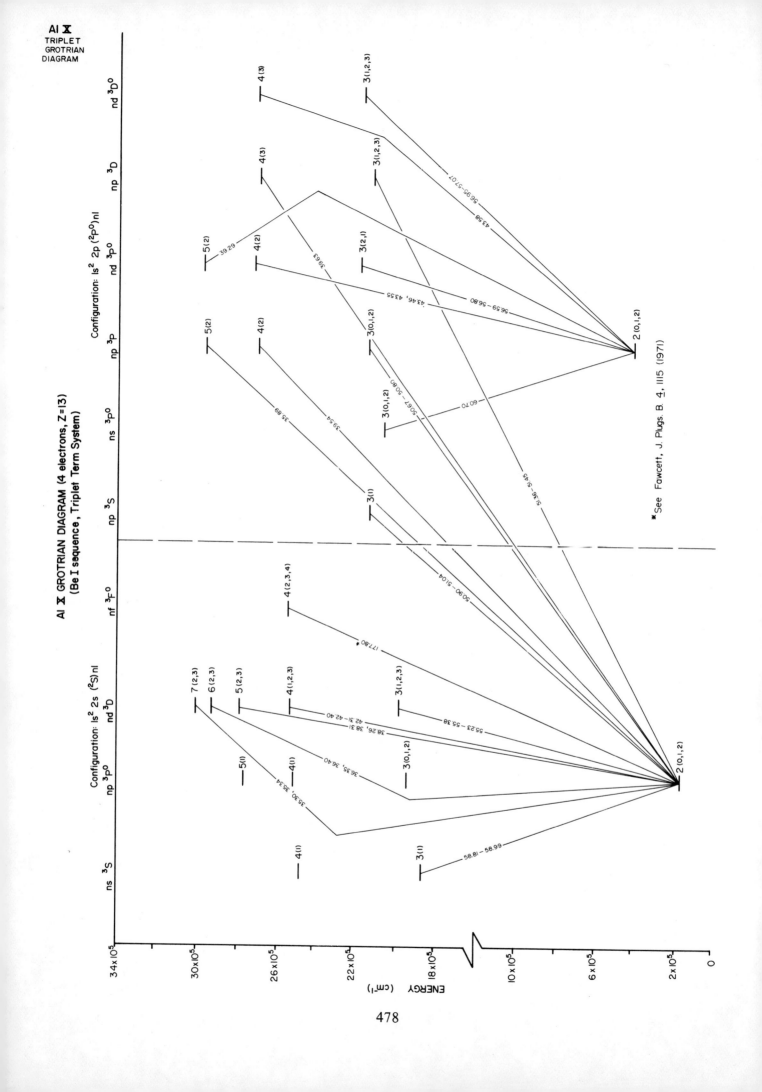

Al X
TRIPLET
GROTRIAN
DIAGRAM

Al X GROTRIAN DIAGRAM (4 electrons, Z=13)
(Be I sequence, Triplet Term System)

Configuration: 1s² 2s (²S)nl

Configuration: 1s² 2p (²P⁰)nl

*See Fawcett, J. Phys. B. 4, 1115 (1971)

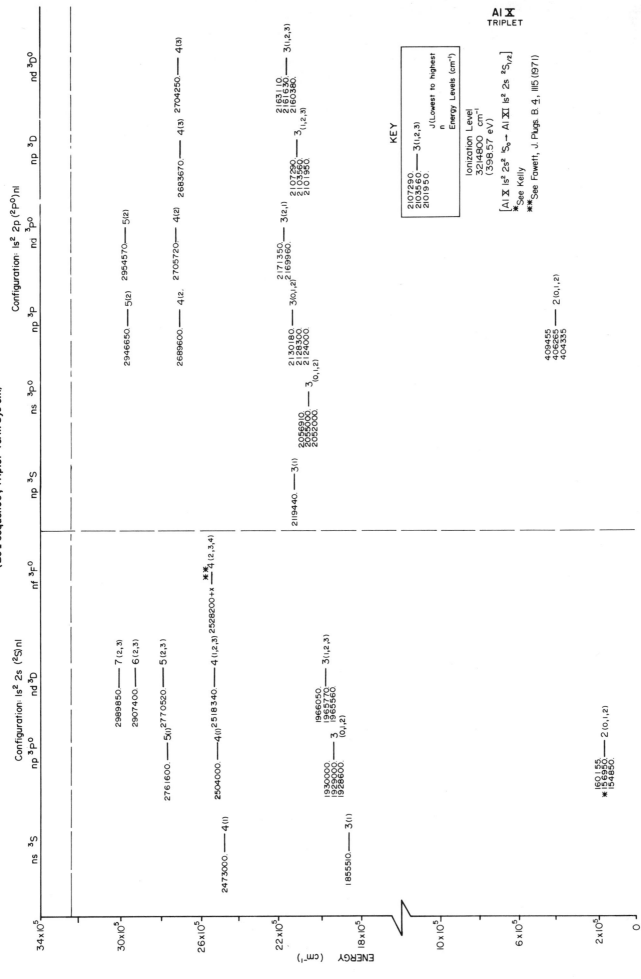

Al X ENERGY LEVELS (4 electrons, Z =13)
(Be I sequence, Triplet Term System)

479

Al XI GROTRIAN DIAGRAM (3 electrons, Z = 13)
(Li I sequence, Configuration: Is²nl, Doublet System)

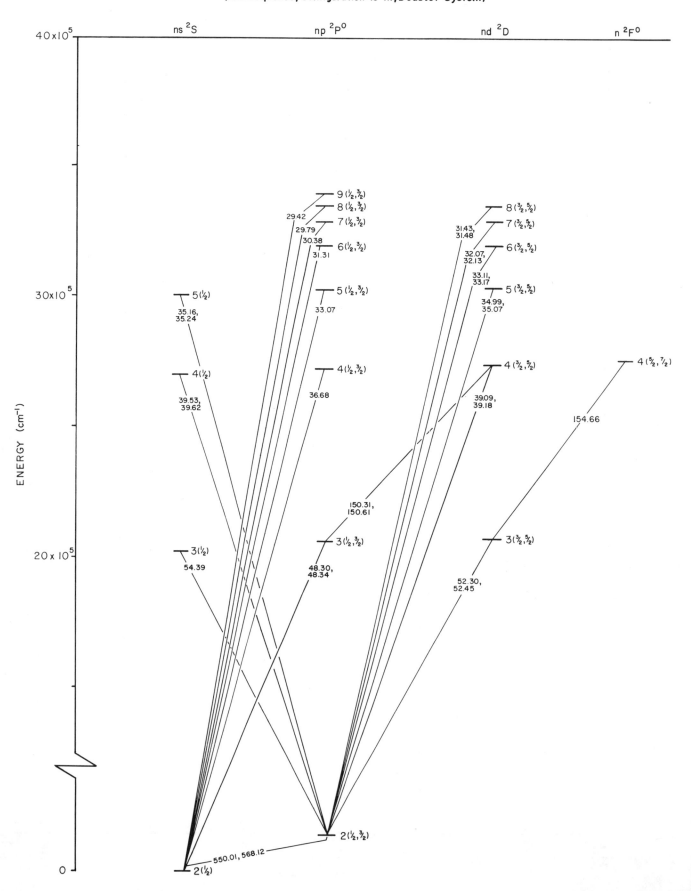

480

Al XI ENERGY LEVELS (3 electrons, Z=13)
(Li I sequence, Configuration: 1s² nl)
(Doublet System)

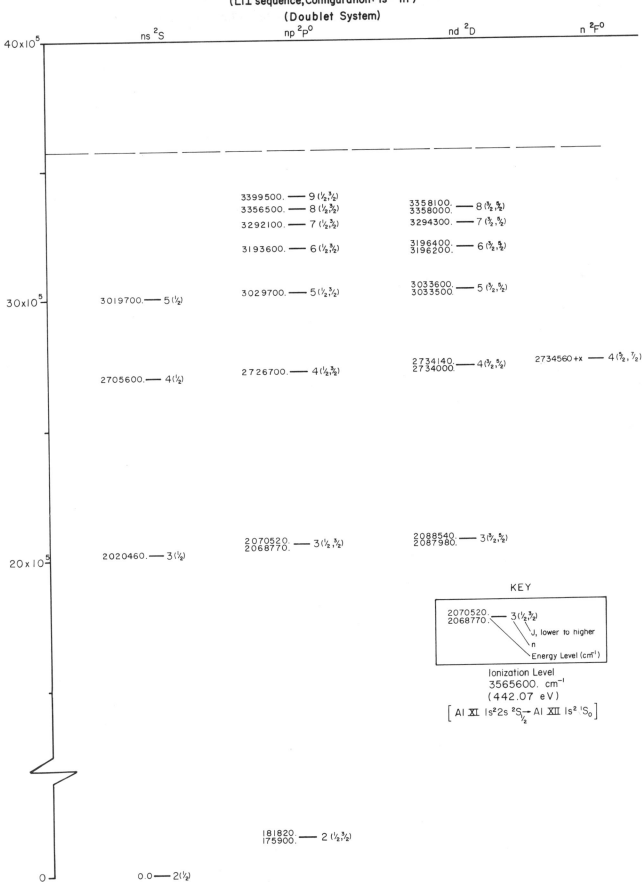

ENERGY (cm⁻¹)

ns ²S   np ²P⁰   nd ²D   n ²F⁰

40x10⁵

3399500. —— 9 (½,³⁄₂)
3356500. —— 8 (½,³⁄₂)
3292100. —— 7 (½,³⁄₂)

3358100.
3358000. —— 8 (³⁄₂,⁵⁄₂)
3294300. —— 7 (³⁄₂,⁵⁄₂)

3193600. —— 6 (½,³⁄₂)

3196400.
3196200. —— 6 (³⁄₂,⁵⁄₂)

30x10⁵

3019700. —— 5(½)

3029700. —— 5(½,³⁄₂)

3033600.
3033500. —— 5 (³⁄₂,⁵⁄₂)

2705600. —— 4(½)

2726700. —— 4(½,³⁄₂)

2734140.
2734000. —— 4(³⁄₂,⁵⁄₂)

2734560+x —— 4(⁵⁄₂,⁷⁄₂)

2070520.
2068770. —— 3(½,³⁄₂)

2088540.
2087980. —— 3(³⁄₂,⁵⁄₂)

2020460. —— 3(½)

20x10⁵

KEY

2070520.
2068770. —— 3(½,³⁄₂)
J, lower to higher
n
Energy Level (cm⁻¹)

Ionization Level
3565600. cm⁻¹
(442.07 eV)
[Al XI 1s²2s ²S₁/₂ → Al XII 1s² ¹S₀]

181820.
175900. —— 2 (½,³⁄₂)

0   0.0 —— 2(½)

481

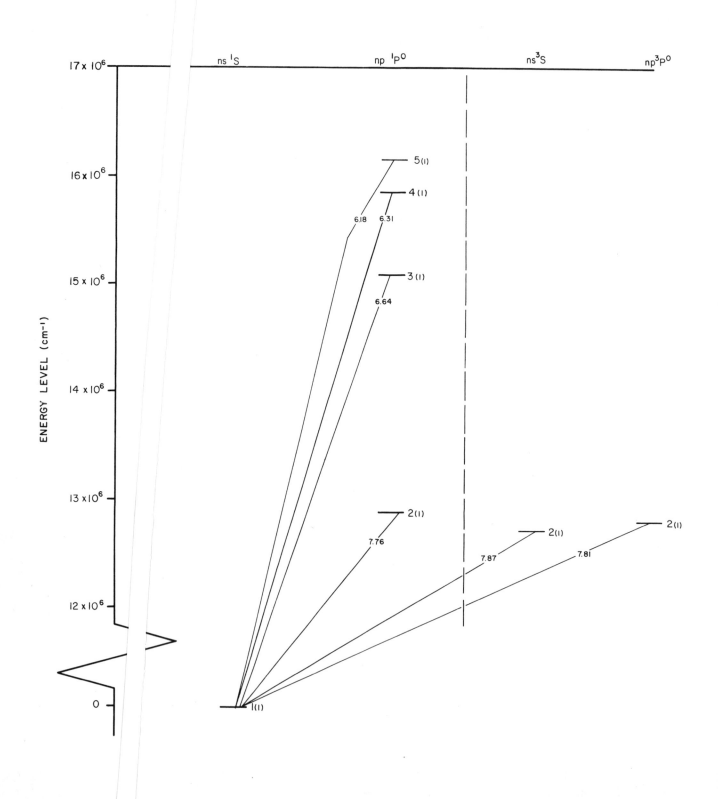

Al XII GROTRIAN DIAGRAM (2 electrons, Z = 13)
He I Sequence, Configuration : 1s nl, Singlets & Triplet System)

Al XII ENERGY LEVELS (2 electrons, Z = 13)
(He I sequence, Configuration: 1s nl)
(Singlet and Triplet)

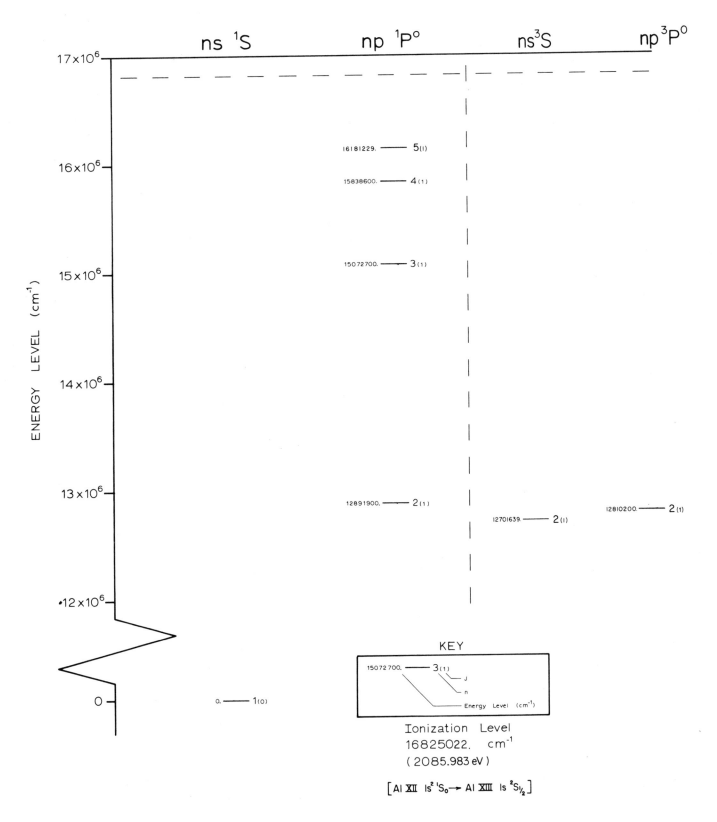

ns ^{1}S   np ^{1}P^o ns ^{3}S   np ^{3}P^o

ENERGY LEVEL (cm^{-1})

17 × 10^6

16181229. ——— 5 (1)

16 × 10^6 15838600. ——— 4 (1)

15072700. ——— 3 (1)

15 × 10^6

14 × 10^6

13 × 10^6 12891900. ——— 2 (1)   12810200. ——— 2 (1)

12701639. ——— 2 (1)

•12 × 10^6

KEY

0. ——— 1 (0)

15072700. ——— 3 (1)
                          J
                          n
                    Energy Level (cm^{-1})

Ionization Level
16825022. cm^{-1}
(2085.983 eV)

[Al XII 1s^2 ^{1}S$_0$ → Al XIII 1s ^{2}S$_{1/2}$]

AI XIII GROTRIAN DIAGRAM (I electron, Z = 13)
(H I sequence, Configuration: nl, Doublet System)

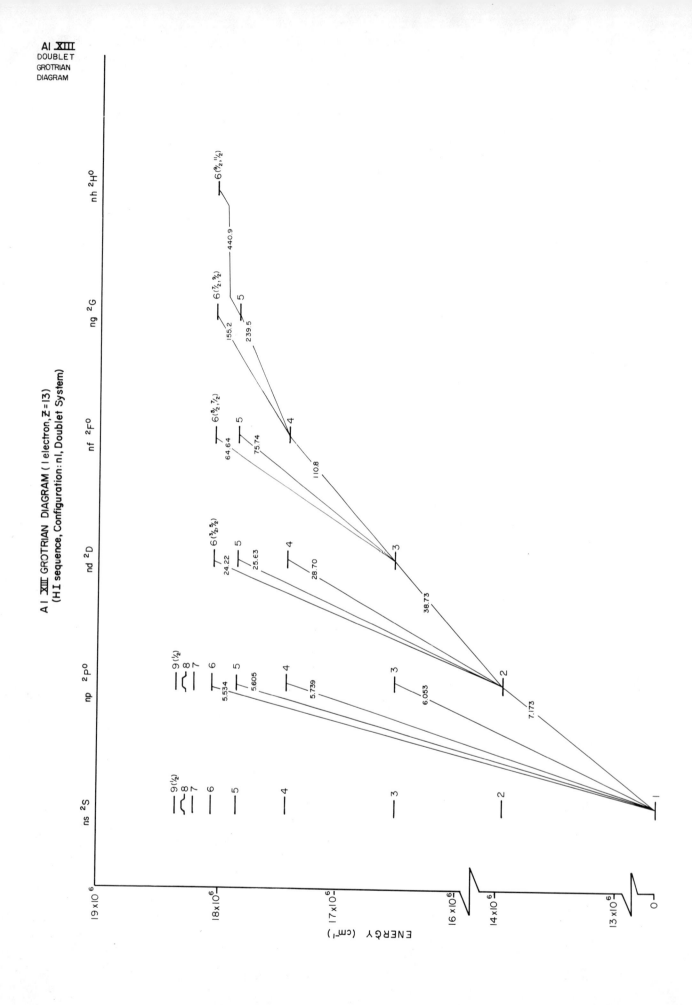

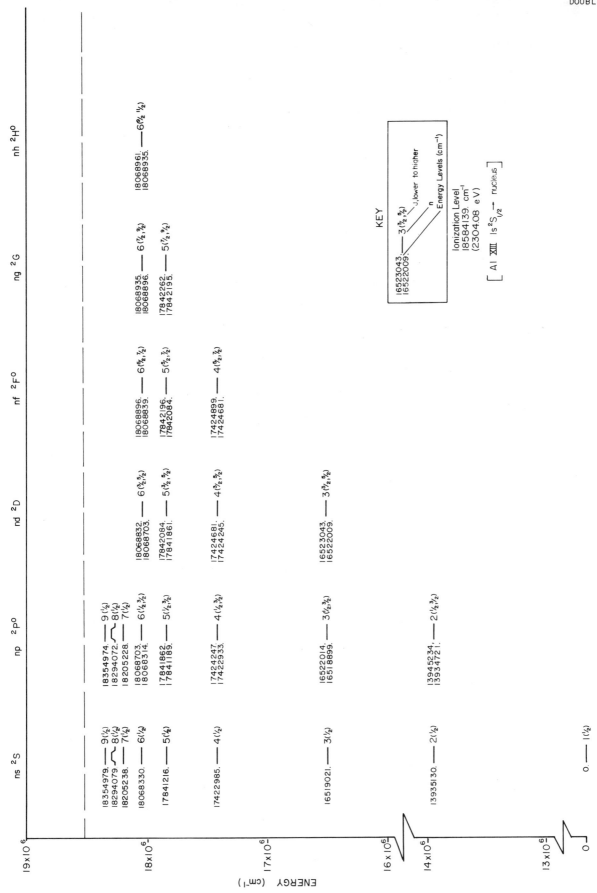

AI XIII ENERGY LEVELS (I electron, Z=13)
(H I sequence, Configuration; nl, Doublet System)

Al I $Z = 13$ 13 electrons

K.B.S. Eriksson and H.B.S. Isberg, Ark. Fys. **33**, 593 (1967).

Authors give line and energy level tables in comparison of their results with Penkin and Shabanova [Opt. and Spectros. **18**, 425 (1925)]

N.P. Penkin and L.N. Shabanova, Opt. and Spectros. **18**, 425 (1925).

Authors give line and energy level tables for lines observed in the region 2000–2300 Å.

E.W.H. Selwyn, Proc. Phys. Soc. **41**, 392 (1929).

Author gives a table of lines observed in the region 1600–2100 Å.

M. Shimauchi, Sci. of Light **7**, 101 (1958).

Author gives tables of lines observed in air, nitrogen, oxygen, and argon atmospheres.

M. Shimauchi, Sci. of Light **12**, 31 (1962).

Author gives a line table and spectrograms of observed lines in a helium atmosphere.

S. Weniger, Ann. d'Astrophysics Suppl. **28**, 117 (1965).

Author gives tables of lines observed in the region 4000–2100 Å studied under variations of temperature.

T. Yamashita, Sci. of Light **14**, 28 (1965).

Author gives line and term tables from lines observed in the region 2120–2080 Å.

Al II $Z = 13$ 12 electrons

S. Bashkin, W.S. Bickel, H.D. Dieselman, and J.B. Schroeder, J. Opt. Soc. Amer. **57**, 1395 (1967).

Authors give wavelengths of lines observed in the region 4140–4670 Å.

E.W.H. Selwyn, Proc. Phys. Soc. **41**, 392 (1929).

Author gives a table of lines observed in the region 1600–2100 Å.

S. Weniger, Ann. d'Astrophysics Suppl. **28**, 117 (1965).

Author gives tables of lines observed in the region 4000–2100 Å studied under variations of temperature.

Note: The third of the Grotrian diagrams for singlets is designated 3d ^{1}D and $n = 5$ singlets to higher singlets. The correct designation is 3d ^{1}D and singlets with $n \geqslant 5$ to higher singlets.

Al III $Z = 13$ 11 electrons

S. Bashkin, W.S. Bickel, H.D. Dieselman, and J.B. Schroeder, J. Opt. Soc. Amer. **57**, 1395 (1967).

Authors give wavelengths of lines observed in the region 4140–4670 Å.

E.W.H. Selwyn, Proc. Phys. Soc. **41**, 392 (1929).

Author gives a table of lines observed in the region 1600–2100 Å.

Al IV $Z = 13$ 10 electrons

Please see the general references. The intermediate-coupling designations are taken from Moore, *Atomic Energy Levels*, Vol. I, N.B.S. Circular 467 (1949).

Al V $Z = 13$ 9 electrons

Please see the general references.

Al VI $Z = 13$ 8 electrons

I.S. Bowen, Ap. J. **132**, 1 (1960).

Author gives tables of lines observed and calculated for forbidden transitions.

B. Edlén, Handbuch der Phys. **27**, 172 (1964).

Author gives corrections to values in C.E. Moore's *Atomic Energy Levels.*

Al VII $Z = 13$ 7 electrons

I.S. Bowen, Ap. J. **132**, 1 (1960).

Author gives tables of observed lines of forbidden transitions.

B. Edlén, Handbuch der Phys. **27**, 172 (1964).

Author gives corrections to values in C.E. Moore's *Atomic Energy Levels.*

Al VIII $Z = 13$ 6 electrons

Please see the general references.

Al IX $Z = 13$ 5 electrons

L.J. Shamey, J. Opt. Soc. Amer. **61**, 942 (1971).

Author gives line and energy level tables of calculated values.

Al X $Z = 13$ 4 elecrons

Please see the general references.

Al XI $Z = 13$ 3 electrons

Please see the general references.

Al XII $Z = 13$ 2 electrons

Please see the general references.

Please see the general references.

Silicium (Si)

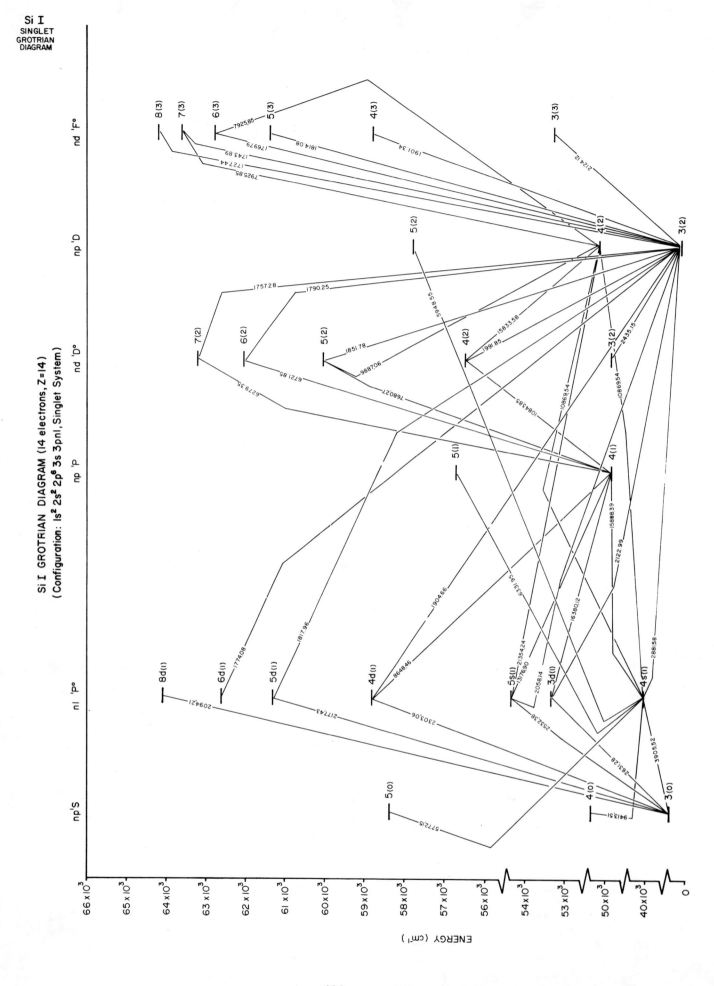

Si I
SINGLET
GROTRIAN
DIAGRAM

Si I GROTRIAN DIAGRAM (14 electrons, Z=14)
(Configuration: 1s² 2s² 2p⁶ 3s 3pnl, Singlet System)

490

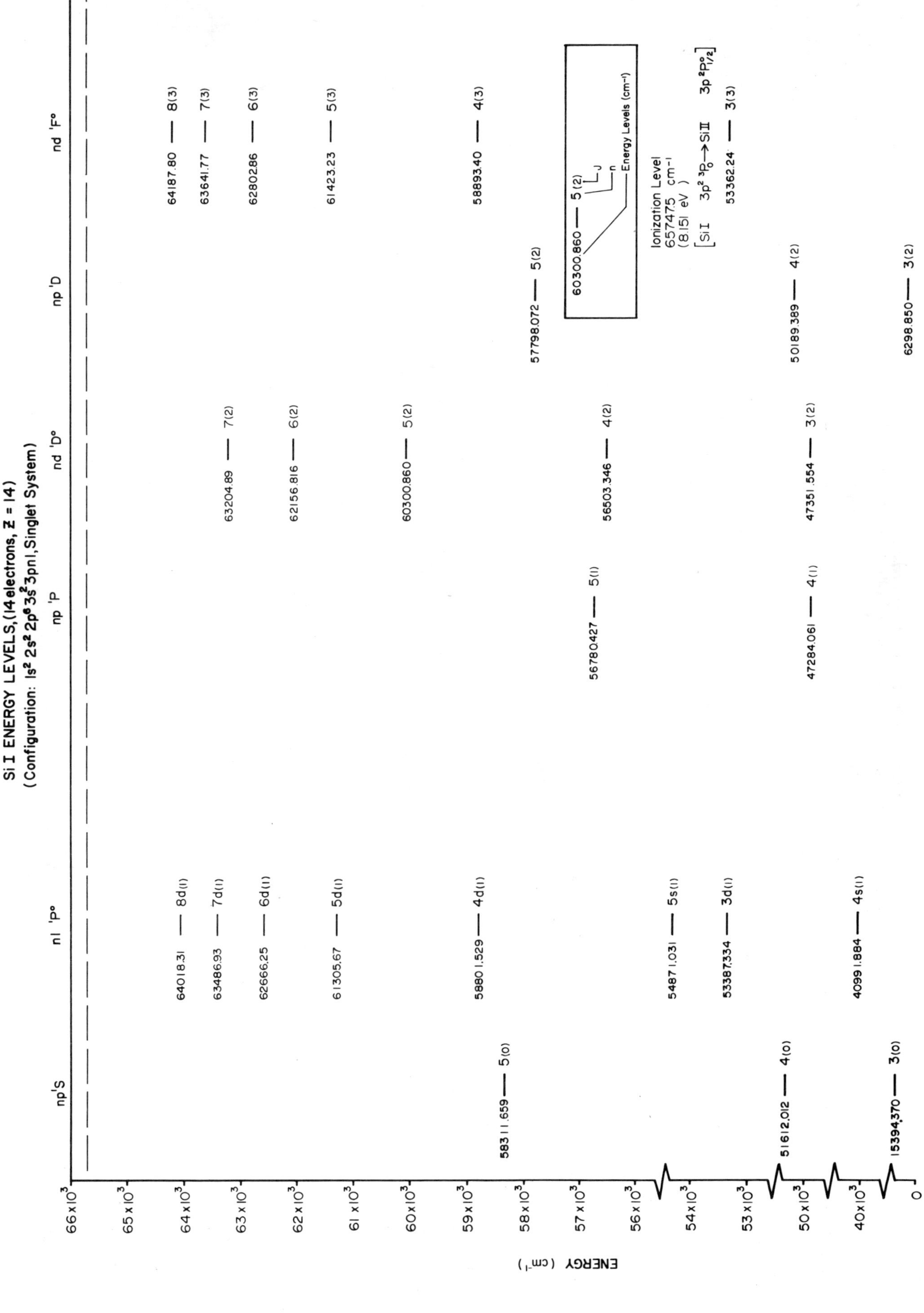

Si I ENERGY LEVELS, (14 electrons, Z = 14)
(Configuration: $1s^2 2s^2 2p^6 3s^2 3pnl$, Singlet System)
np 'S
nl 'P°
np 'P
nd 'D°
np 'D
nd 'F°
ENERGY (cm⁻¹)
66 x 10³
65 x 10³
64 x 10³
63 x 10³
62 x 10³
61 x 10³
60 x 10³
59 x 10³
58 x 10³
57 x 10³
56 x 10³
54 x 10³
53 x 10³
50 x 10³
40 x 10³
0
58311.659 — 5(0)
51612.012 — 4(0)
15394.370 — 3(0)
64018.31 — 8d(1)
63486.93 — 7d(1)
62666.25 — 6d(1)
61305.67 — 5d(1)
58801.529 — 4d(1)
54871.031 — 5s(1)
53387.334 — 3d(1)
40991.884 — 4s(1)
56780427 — 5(1)
47284.061 — 4(1)
63204.89 — 7(2)
62156.816 — 6(2)
60300.860 — 5(2)
56503.346 — 4(2)
47351.554 — 3(2)
57798.072 — 5(2)
50189.389 — 4(2)
6298.850 — 3(2)
64187.80 — 8(3)
63641.77 — 7(3)
62802.86 — 6(3)
61423.23 — 5(3)
58893.40 — 4(3)
53362.24 — 3(3)
60300.860 — 5(2)
J
n
Energy Levels (cm⁻¹)
Ionization Level
65747.5 cm⁻¹
(8.151 eV)
[Si I 3p² ³P₀ → Si II 3p ²P°₁/₂]

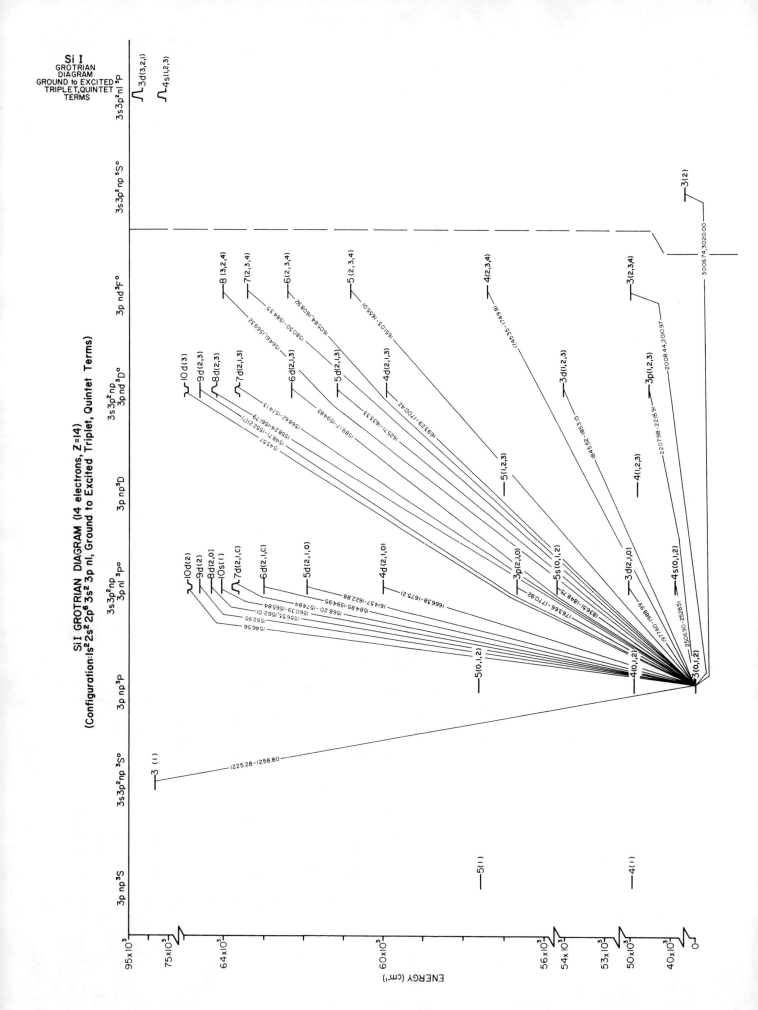

492

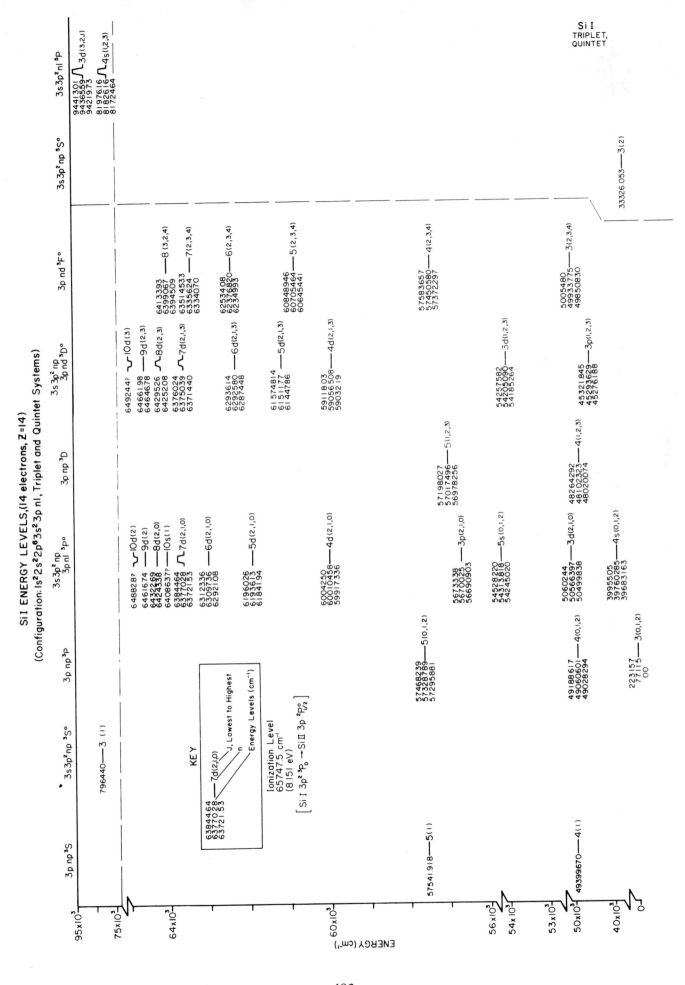

SiI ENERGY LEVELS,(14 electrons, Z=14)
(Configuration: 1s²2s²2p⁶3s²3p nl, Triplet and Quintet Systems)

SiI
TRIPLET,
QUINTET

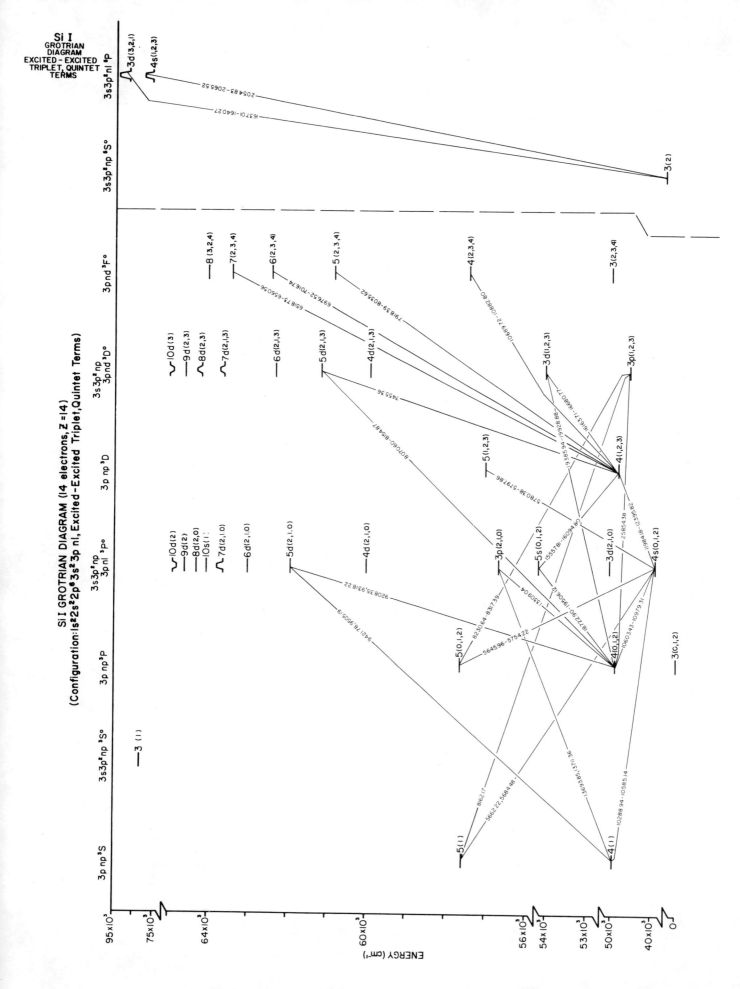

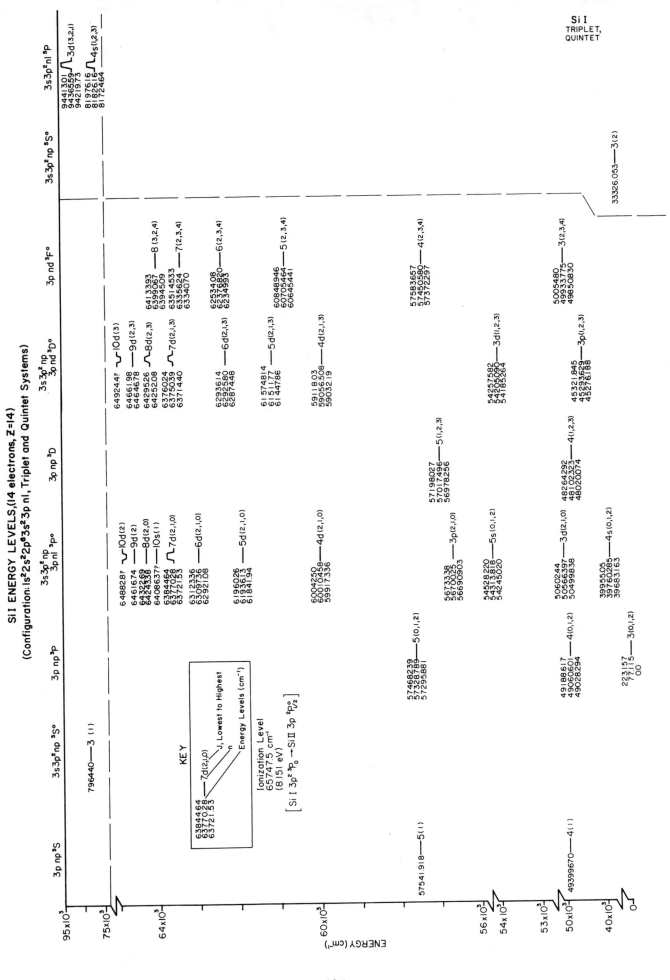

Si I ENERGY LEVELS,(14 electrons, Z=14)
(Configuration: 1s²2s²2p⁶3s²3p nl, Triplet and Quintet Systems)

495

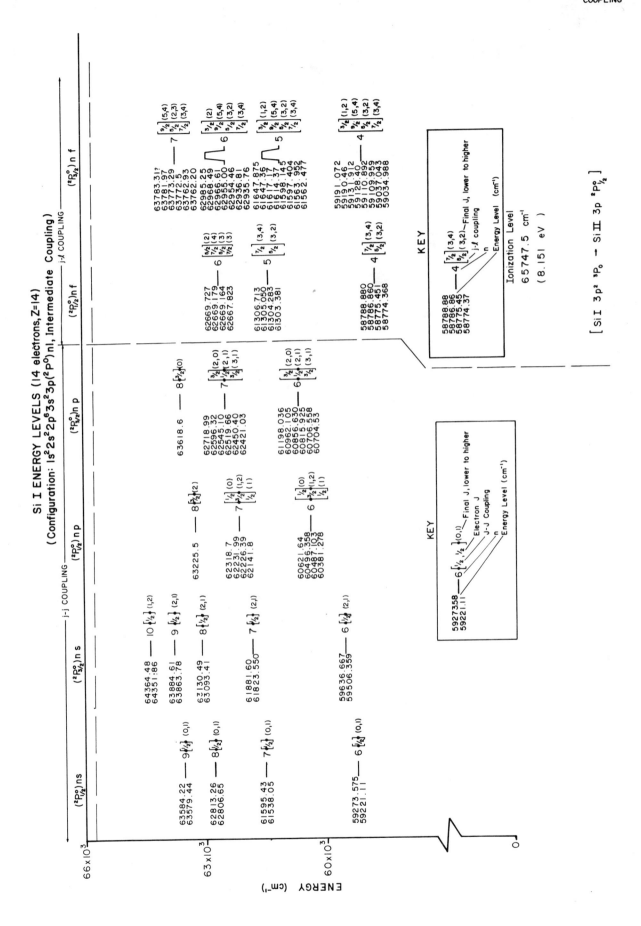

Si I ENERGY LEVELS (14 electrons, Z=14)
(Configuration: $1s^2 2s^2 2p^6 3s^2 3p(^2P^o)nl$, Intermediate Coupling)

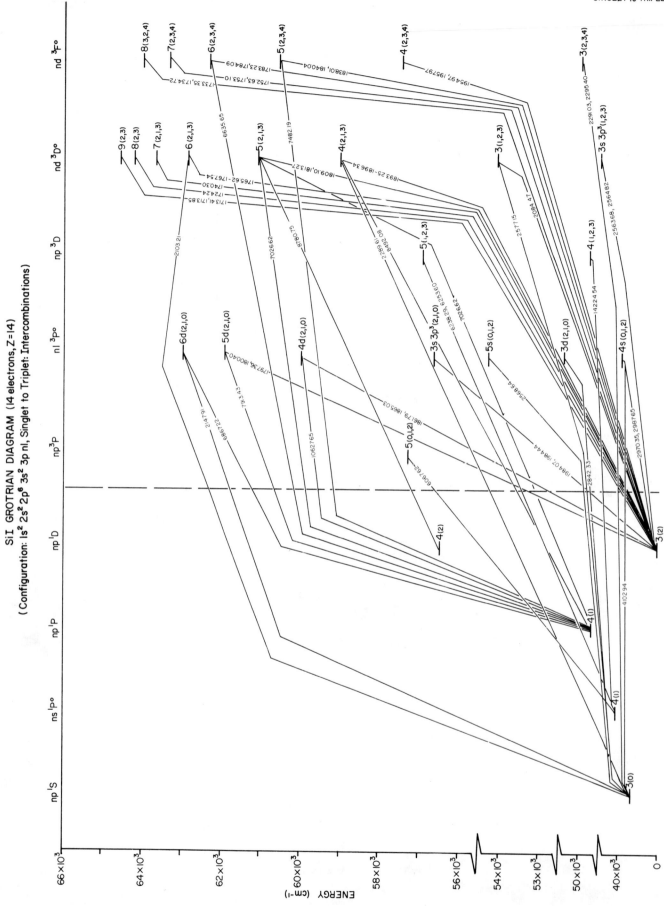

SiI GROTRIAN DIAGRAM (14 electrons, Z=14)

(Configuration: 1s² 2s² 2p⁶ 3s² 3p nl, Singlet to Triplet: Intercombinations)

Si I
INTERCOMBINATION
GROTRIAN
DIAGRAM
SINGLET to TRIPLET

498

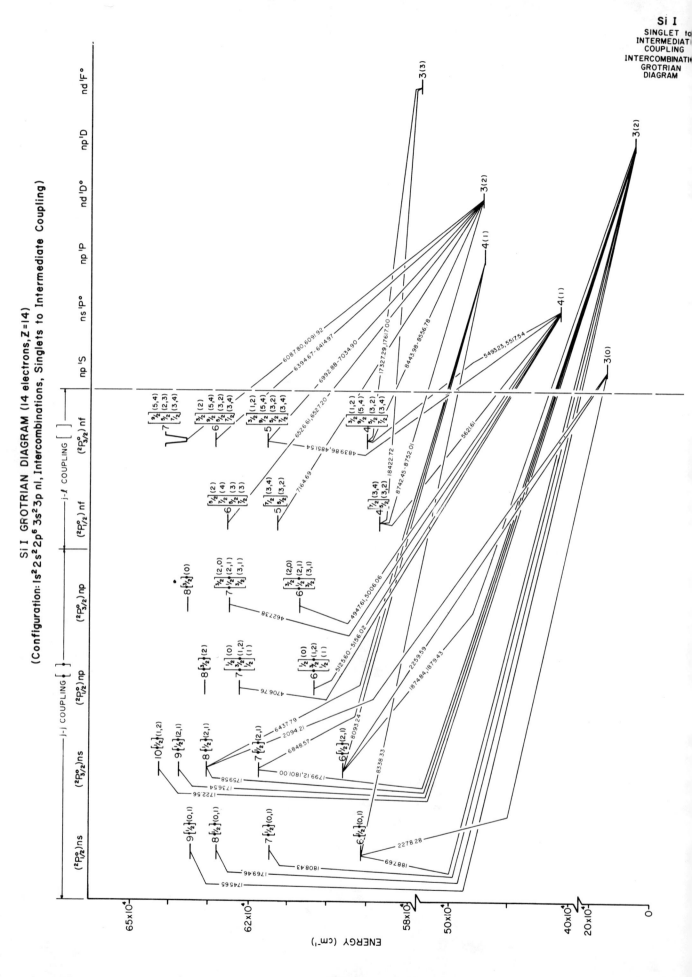

Si I

SINGLET to
INTERMEDIAT[E]
COUPLING
INTERCOMBINATI[ON]
GROTRIAN
DIAGRAM

Si I GROTRIAN DIAGRAM (14 electrons, Z=14)

(Configuration: 1s²2s²2p⁶3s²3p nl, Intercombinations, Singlets to Intermediate Coupling)

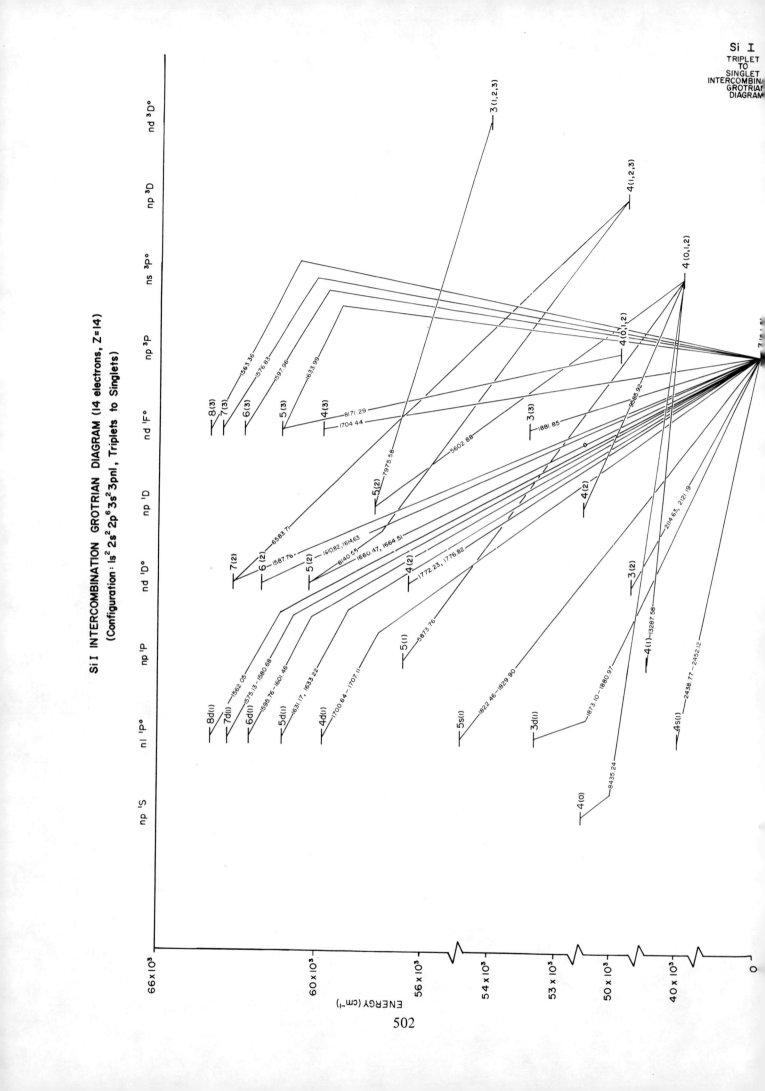

Si I INTERCOMBINATION GROTRIAN DIAGRAM (14 electrons, Z=14)
(Configuration: 1s² 2s² 2p⁶ 3s² 3pnl, Triplets to Singlets)

Si I
TRIPLET
TO
SINGLET
INTERCOMBINATION
GROTRIAN
DIAGRAM

502

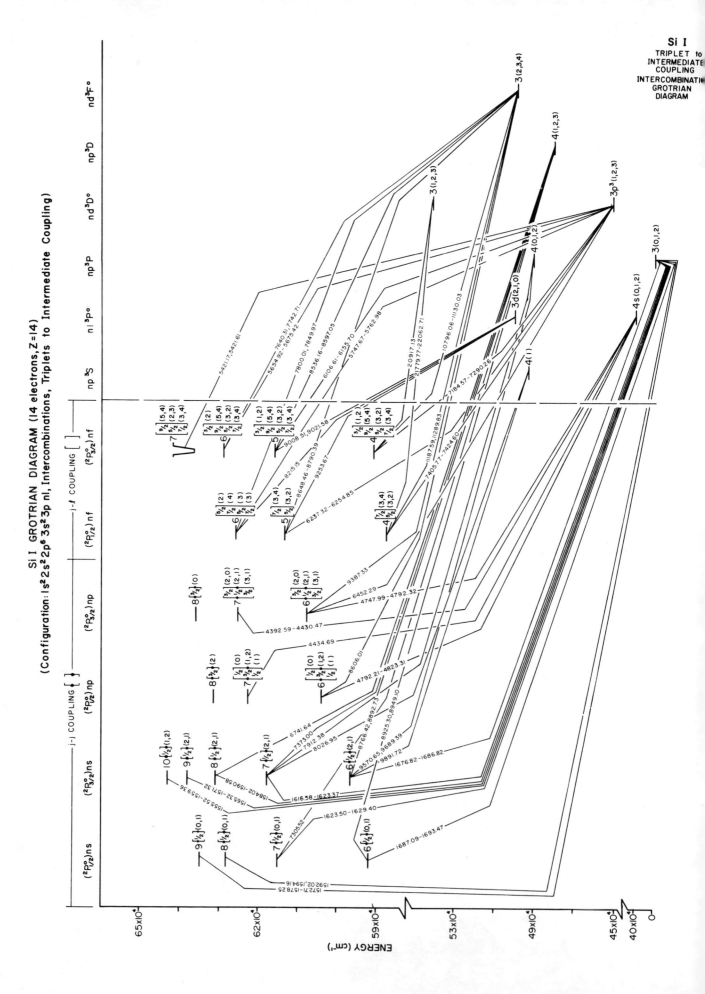

Si I GROTRIAN DIAGRAM (14 electrons, Z=14)

(Configuration: 1s²2s²2p⁶3s²3p nl, Intercombinations, Triplets to Intermediate Coupling)

Si I
TRIPLET to
INTERMEDIATE
COUPLING
INTERCOMBINATION
GROTRIAN
DIAGRAM

ENERGY (cm⁻¹)

504

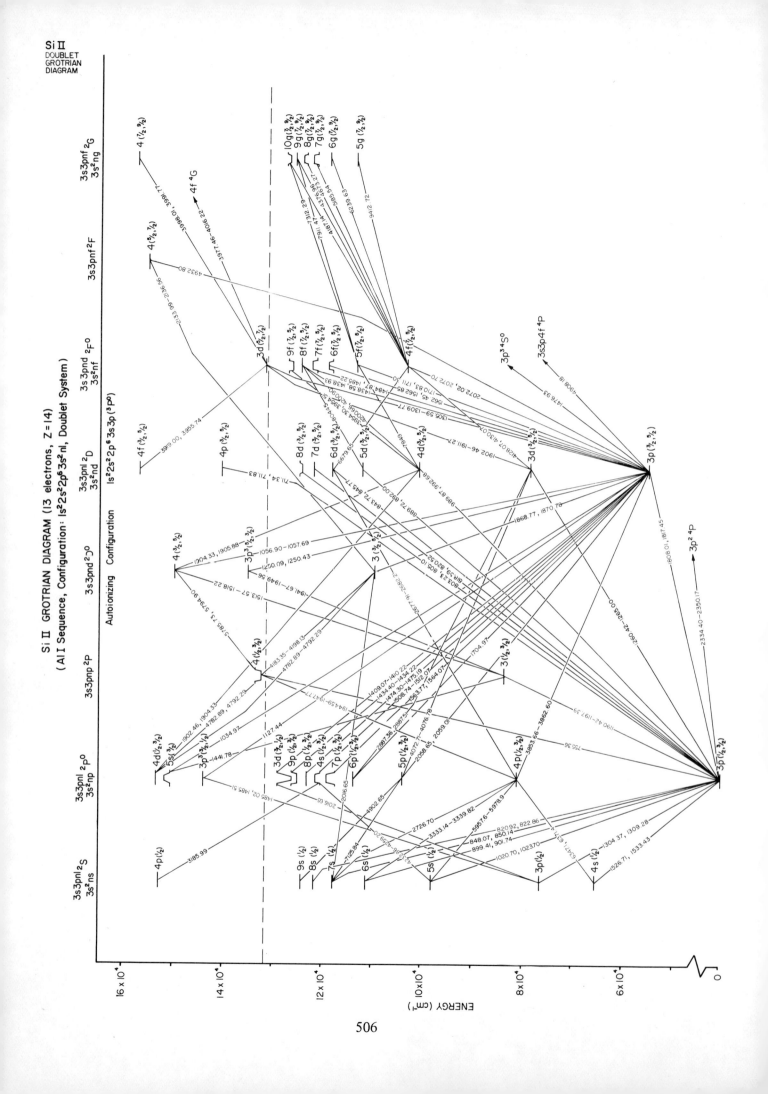

Si II
DOUBLET

Si II ENERGY LEVELS (13 electrons, Z=14)
(Al I Sequence, Configuration: 1s²2s²2p⁶3s²nl, Doublet System)

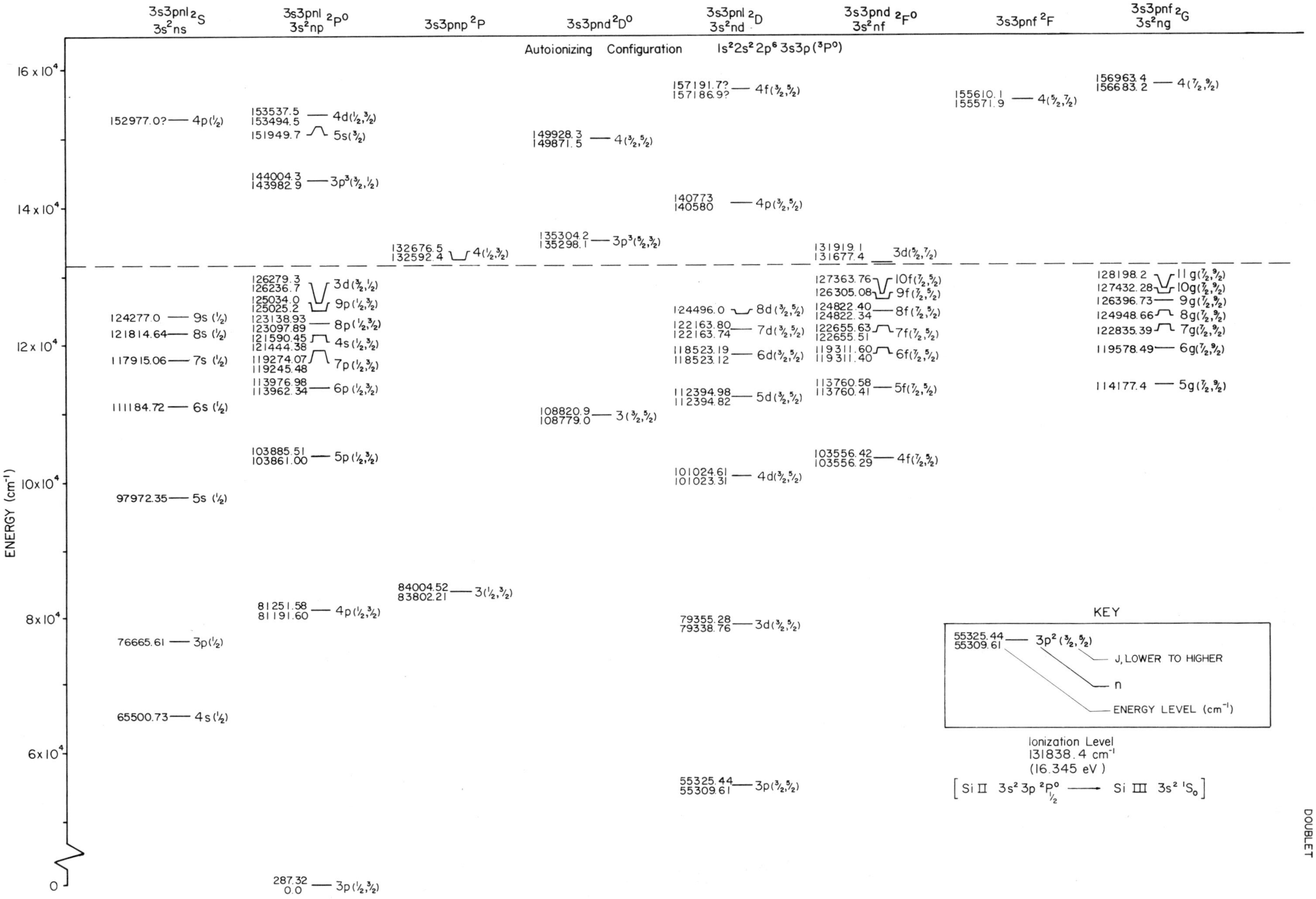

507

Si II
QUARTET
GROTRIAN DIAGRAM

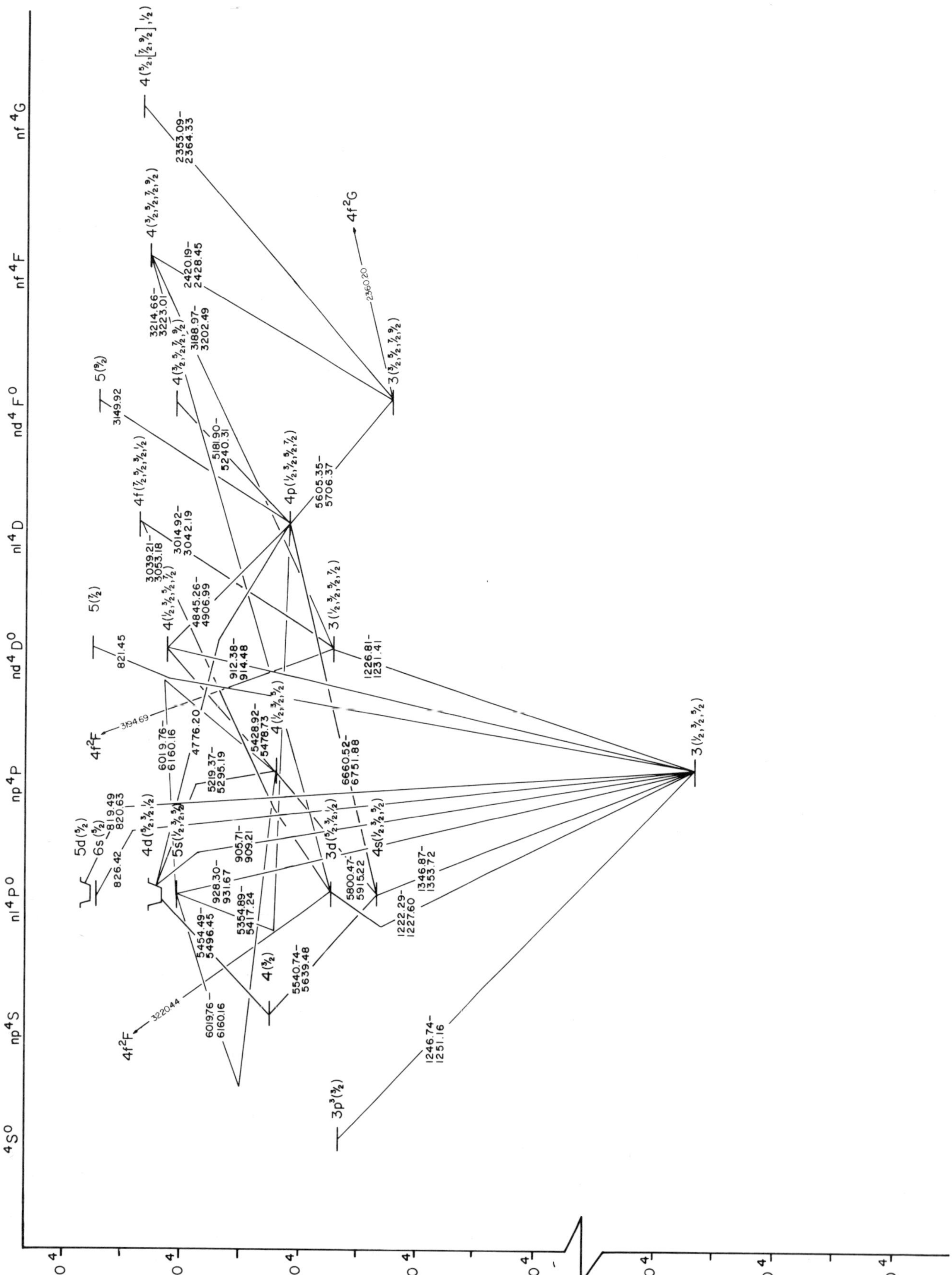

Si II GROTRIAN DIAGRAM (13 electrons, Z=14)
(Al I sequence, Configuration: 1s²2s²2p⁶3s 3pnl, Quartet System)

508

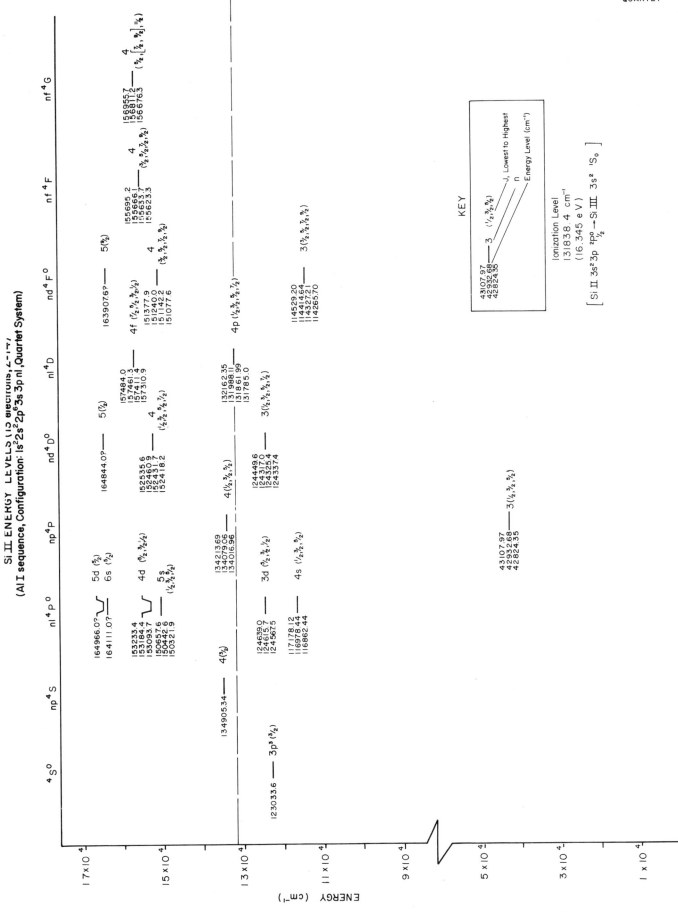

Si II
QUARTET

Si II ENERGY LEVELS (13 electrons, Z=14)
(Al I sequence, Configuration: 1s²2s²2p⁶3s 3p nl, Quartet System)

509

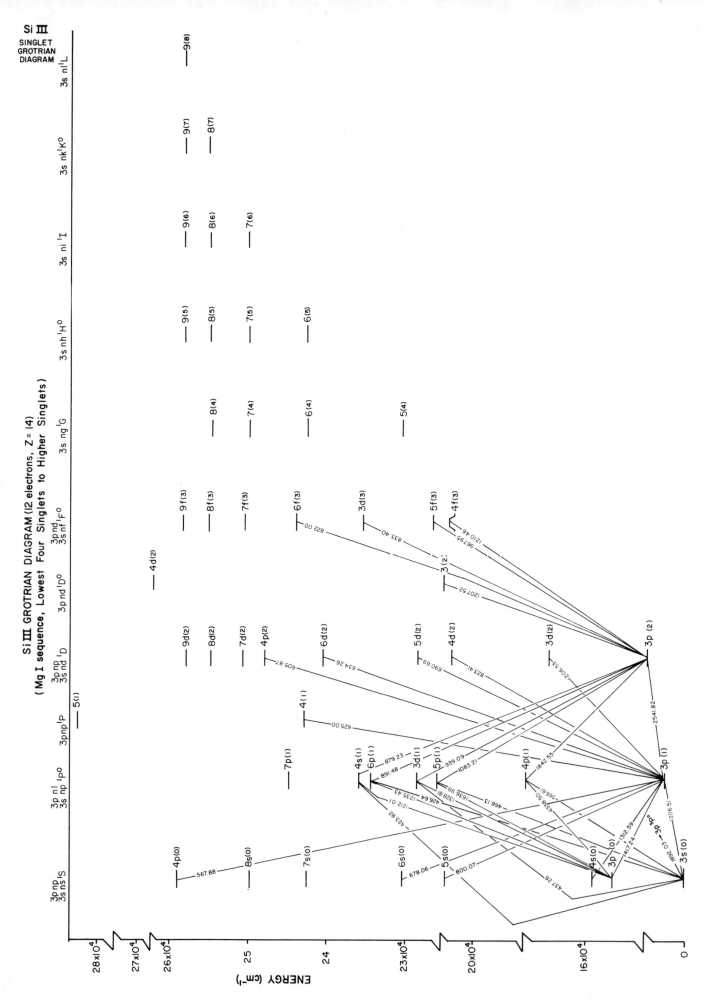

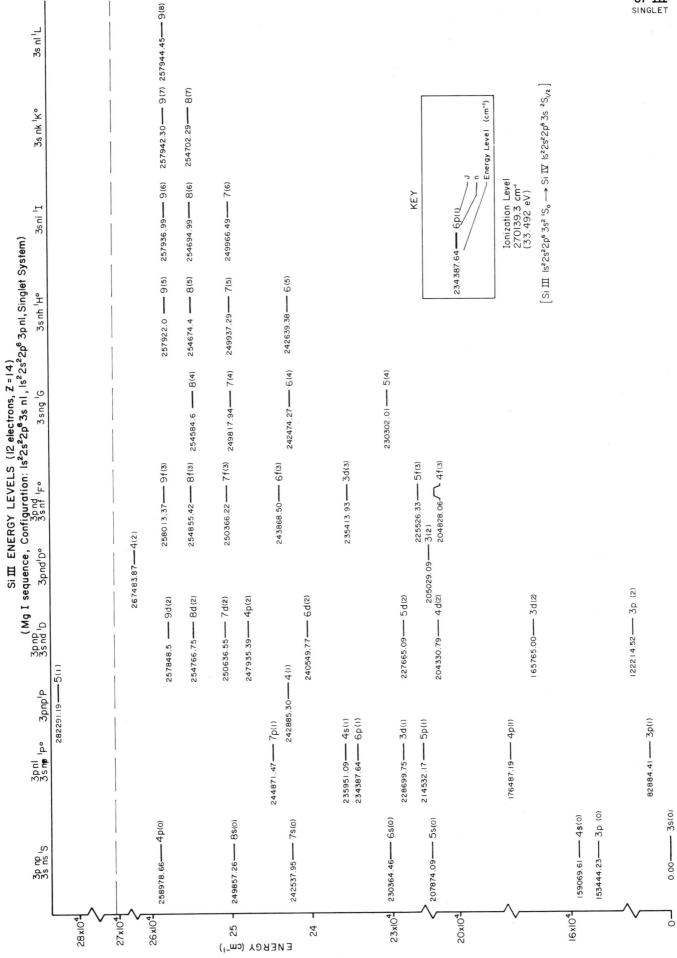

Si III ENERGY LEVELS (12 electrons, Z = 14)

(Mg I sequence, Configuration: 1s²2s²2p⁶3s nl, 1s²2s²2p⁶ 3pnl, Singlet System)

Si III
SINGLET

KEY

234387.64 —— 6p(1) — J
— n
— Energy Level (cm⁻¹)

Ionization Level
270139.3 cm⁻¹
(33.492 eV)

[Si III 1s²2s²2p⁶ 3s² ¹S₀ ⟶ Si IV 1s²2s²2p⁶ 3s ²S₁/₂]

ENERGY (cm⁻¹)

511

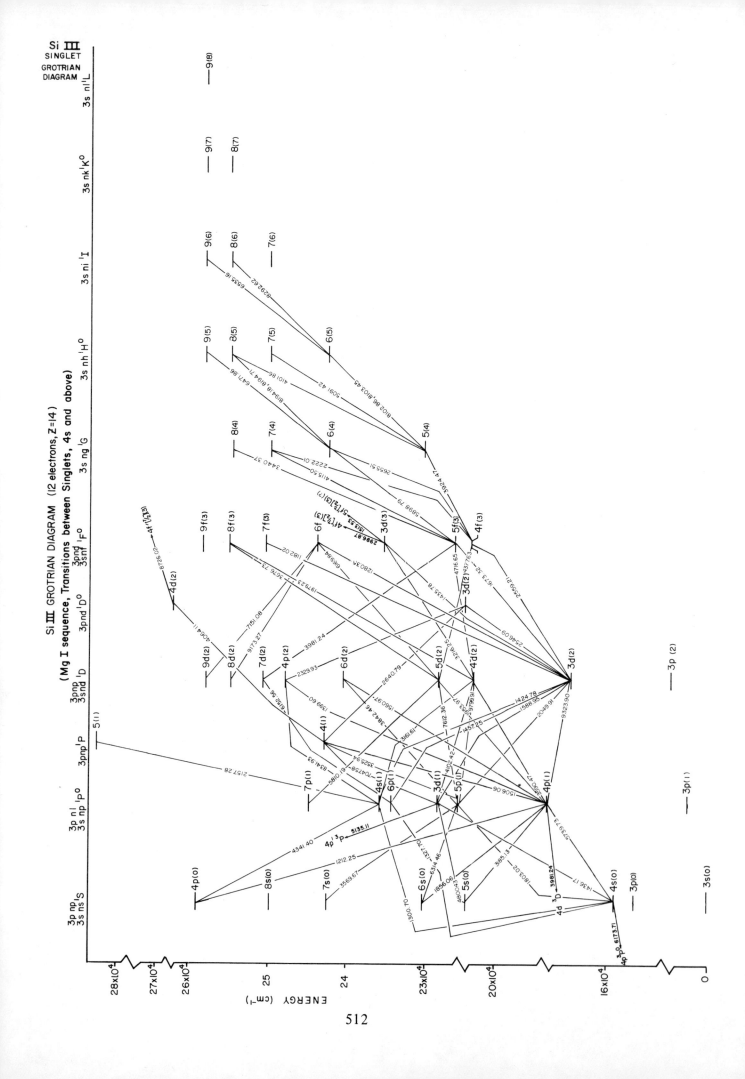

Si III
SINGLET
GROTRIAN
DIAGRAM

Si III GROTRIAN DIAGRAM (12 electrons, Z=14)
(Mg I sequence, Transitions between Singlets, 4s and above)

512

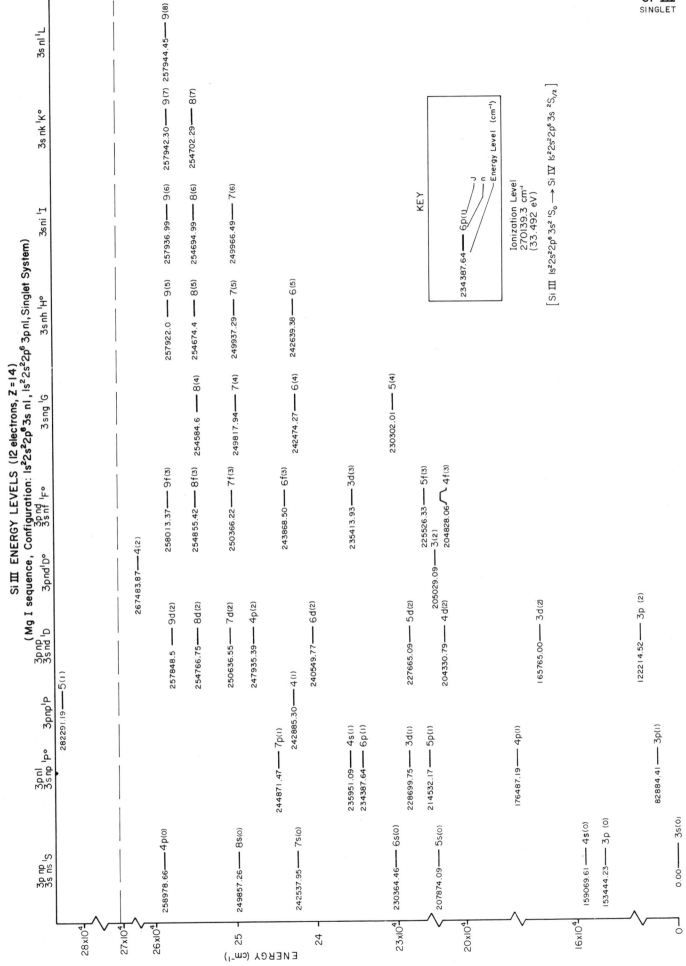

Si III ENERGY LEVELS (12 electrons, Z=14)

(Mg I sequence, Configuration: 1s²2s²2p⁶3s nl, 1s²2s²2p⁶ 3pnl, Singlet System)

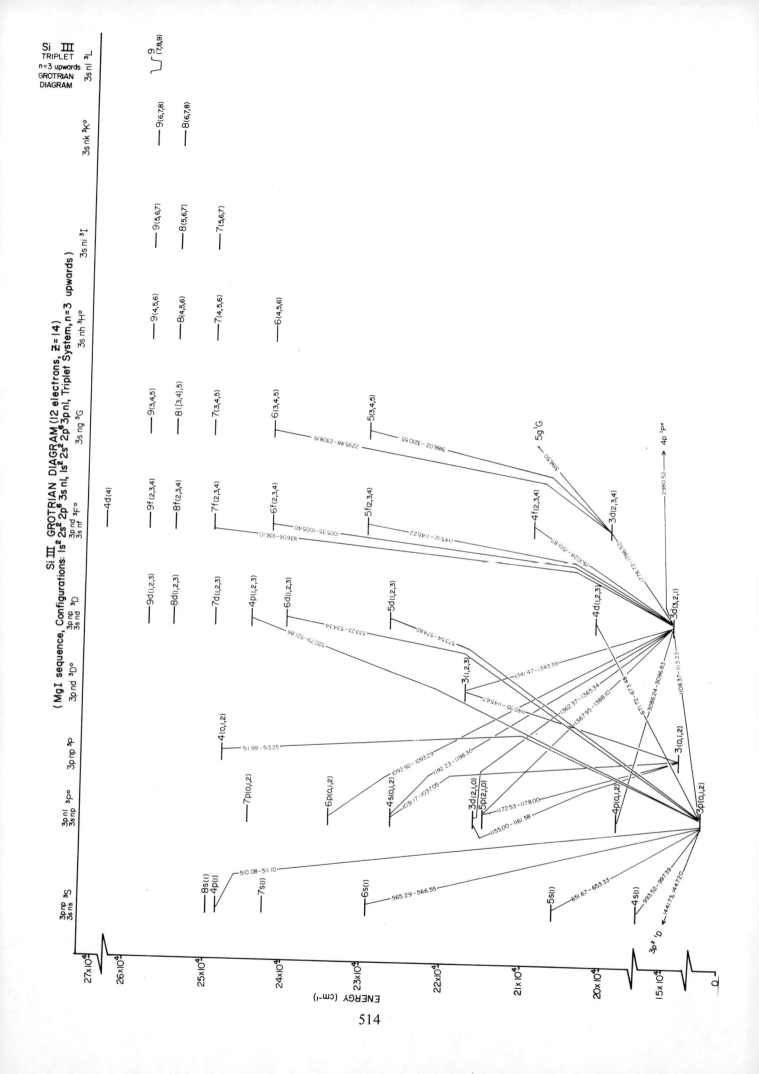

Si III GROTRIAN DIAGRAM (12 electrons, Z=14)

(MgI sequence, Configurations: 1s² 2s² 2p⁶ 3s nl, 1s² 2s² 2p⁶ 3pnl, Triplet System, n=3 upwards)

Si III
TRIPLET
n=3 upwards
GROTRIAN
DIAGRAM

ENERGY (cm-1)

514

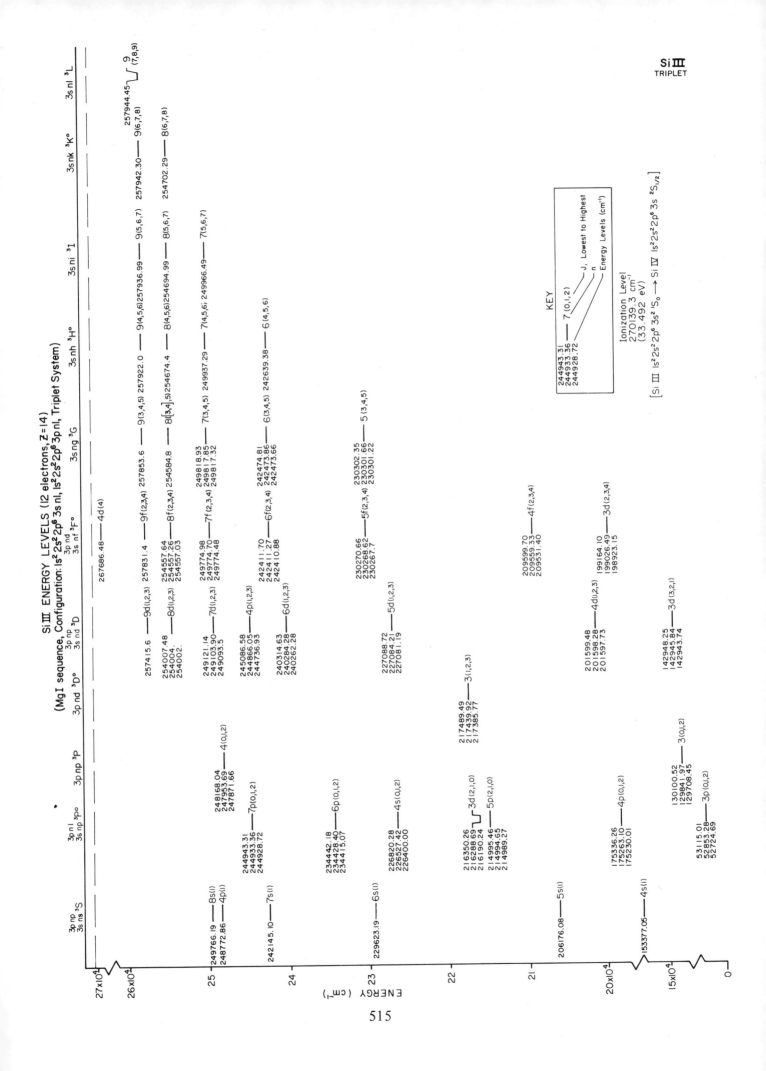

Si III ENERGY LEVELS (12 electrons, Z=14)

(Mg I sequence, Configuration: 1s²2s²2p⁶ 3s nl, 1s²2s²2p⁶3pnl, Triplet System)

ENERGY (cm⁻¹)

515

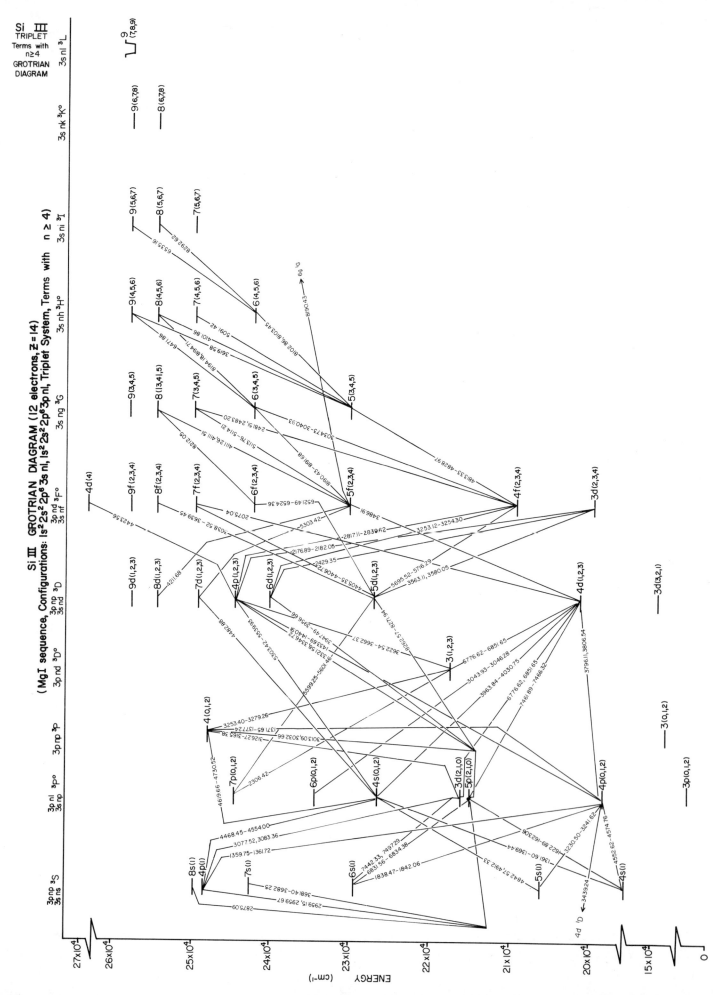

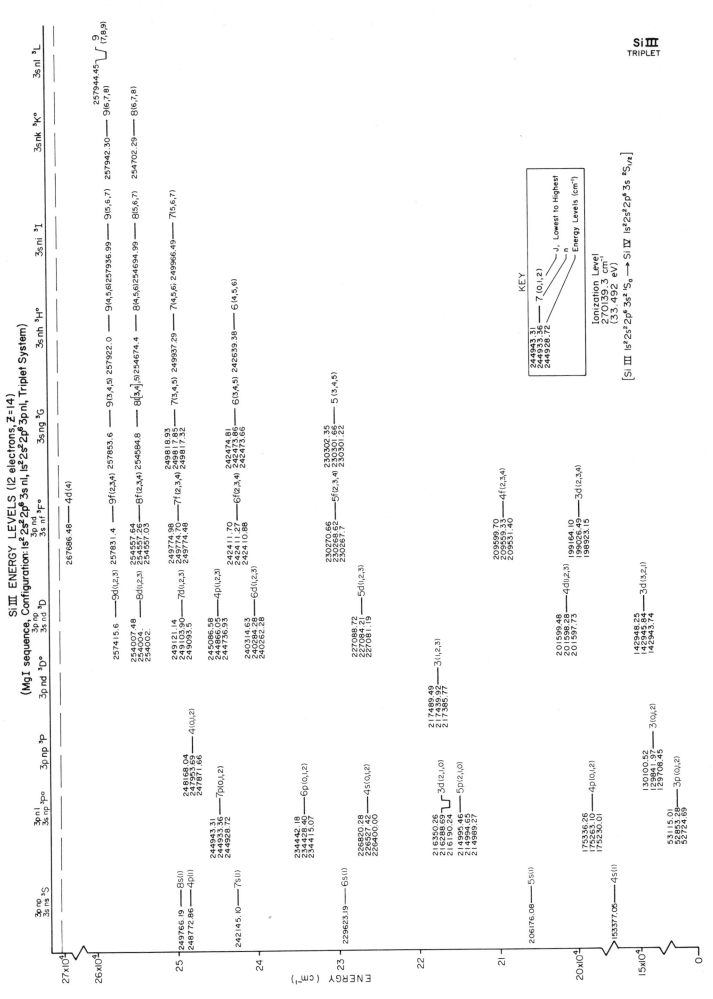

SiIII ENERGY LEVELS (12 electrons, Z=14)
(MgI sequence, Configuration: 1s² 2s² 2p⁶ 3s nl, 1s²2s²2p⁶ 3pnl, Triplet System)

SiIII
TRIPLET

KEY

244943.31
244933.36 ——— 7 (0,1,2) ——— J, Lowest to Highest
244928.72 ——— n
                         Energy Levels (cm⁻¹)

Ionization Level
270139.3 cm⁻¹
(33.492 eV)

[Si III 1s²2s²2p⁶ 3s² ¹S₀ → Si IV 1s²2s²2p⁶ 3s ²S₁/₂]

ENERGY (cm⁻¹)

517

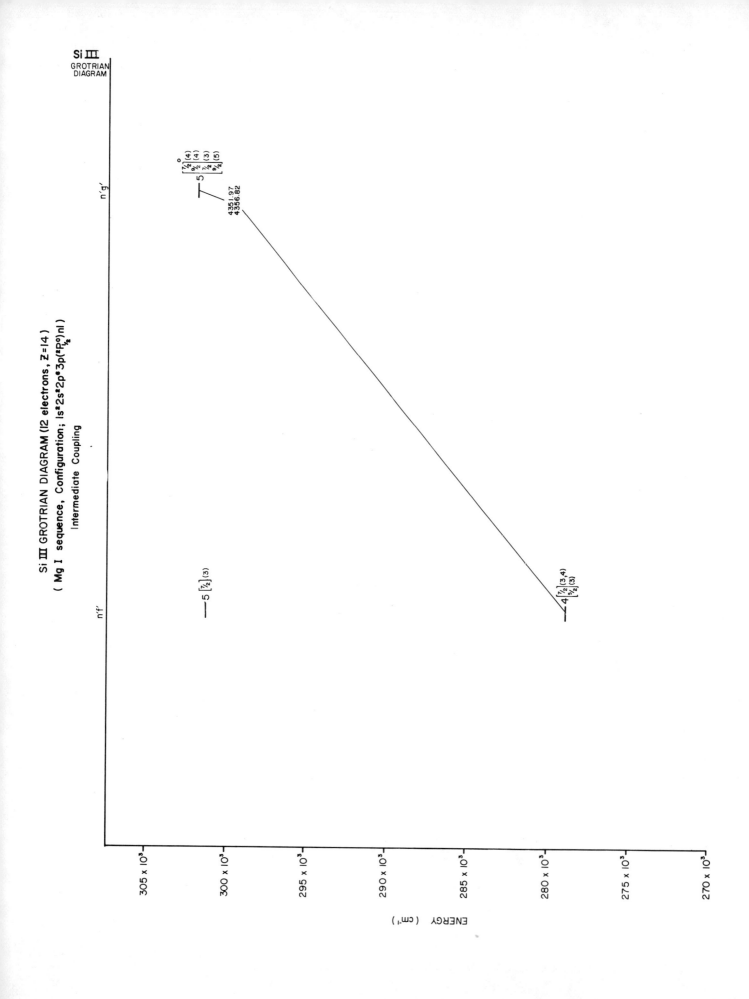

518

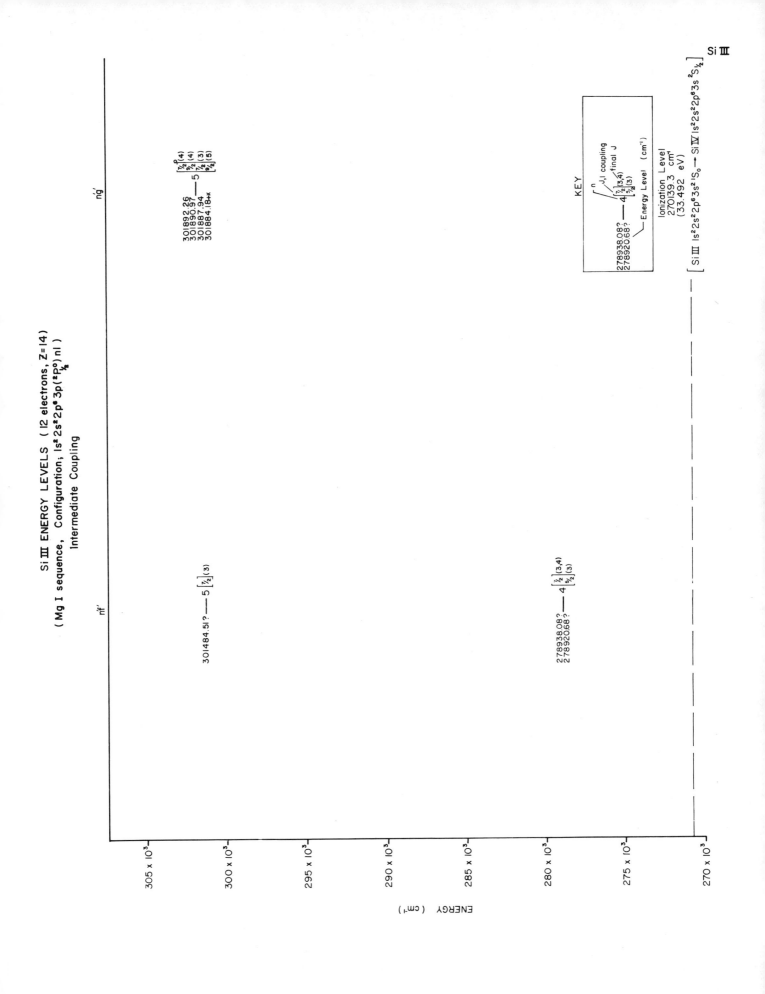

Si III ENERGY LEVELS (12 electrons, Z=14)

(Mg I sequence, Configuration; $1s^2 2s^2 2p^6 3p(^2P^o_{3/2})nl$)

Intermediate Coupling

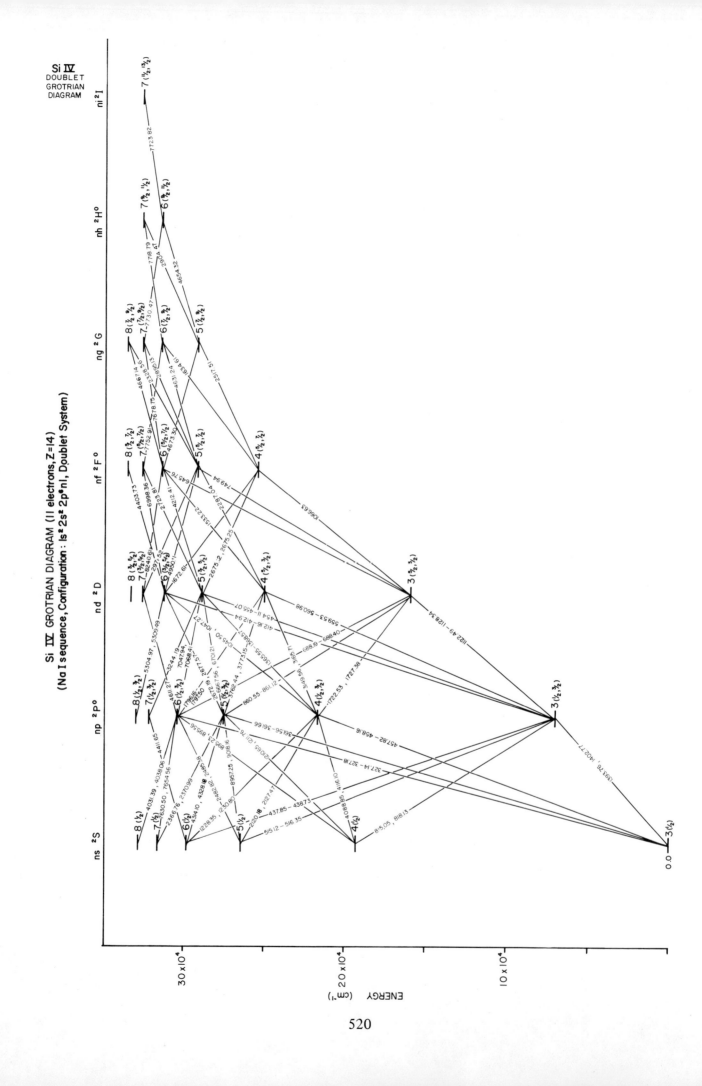

Si IV GROTRIAN DIAGRAM (11 electrons, Z=14)
(Na I sequence, Configuration : 1s² 2s² 2p⁶nl, Doublet System)

Si IV
DOUBLET
GROTRIAN
DIAGRAM

520

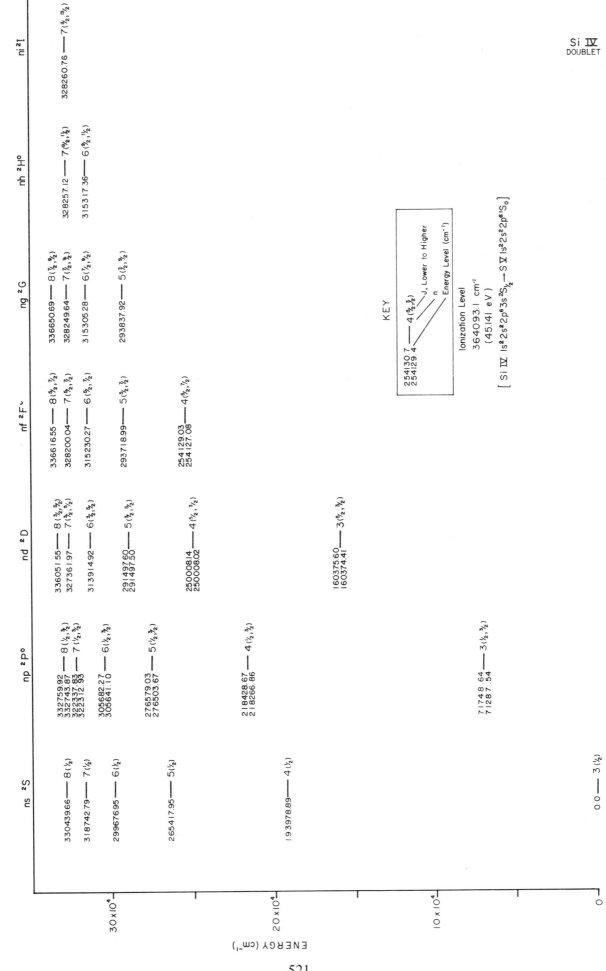

Si IV ENERGY LEVELS (11 electrons, Z=14)
(Na I sequence, Configuration: 1s² 2s² 2p⁶ nl, Doublet System)

Si IV
DOUBLET

ENERGY (cm⁻¹)

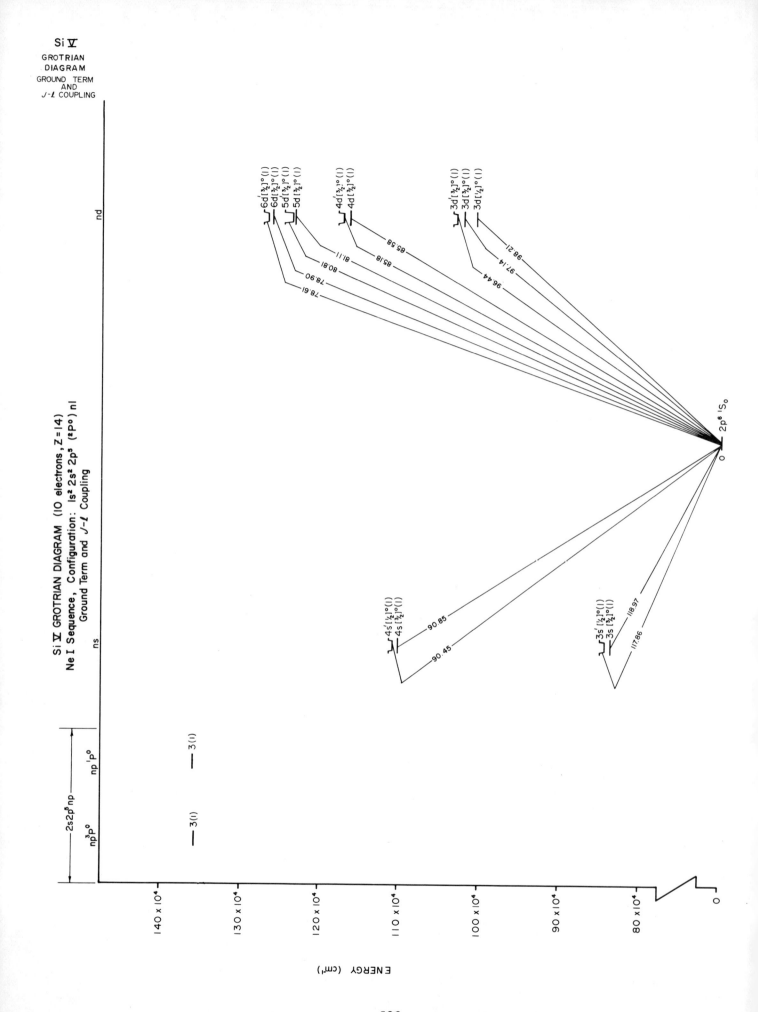

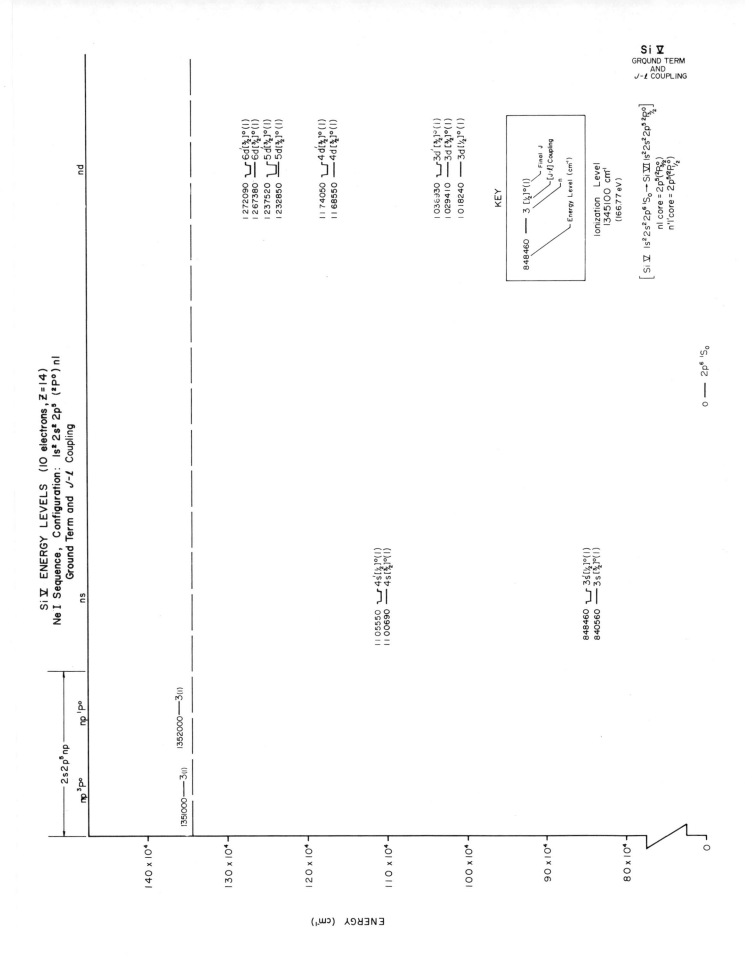

Si Ⅴ ENERGY LEVELS (10 electrons, Z=14)
Ne I Sequence, Configuration: 1s² 2s² 2p⁵ (²P°) nl
Ground Term and J-ℓ Coupling

Si Ⅴ
GROUND TERM
AND
J-ℓ COUPLING

ns

nd

np ³P°   np ¹P°

2s²2p⁵np

1351000 ——— 3(1)   1352000 ——— 3(1)

1272090 ⌐ 6d'[³⁄₂]°(1)
1267380 ∟ 6d[³⁄₂]°(1)
1237520 ⌐ 5d'[³⁄₂]°(1)
1232850 ∟ 5d[³⁄₂]°(1)

1174050 ⌐ 4d'[³⁄₂]°(1)
1168550 ∟ 4d[³⁄₂]°(1)

1105550 ⌐ 4s'[½]°(1)
1100690 ∟ 4s[³⁄₂]°(1)

1036930 ⌐ 3d'[³⁄₂]°(1)
1029410 ∟ 3d[³⁄₂]°(1)
1018240 ——— 3d[½]°(1)

KEY

848460 ——— 3 [½]°(1)
                    └ Final J
              └ [J-ℓ] Coupling
          └ n
      └ Energy Level (cm⁻¹)

Ionization Level
1345100 cm⁻¹
(166.77 eV)

[Si Ⅴ 1s²2s²2p⁶ ¹S₀ → Si Ⅵ 1s²2s²2p⁵ ²P°₃⁄₂]
nl core = 2p⁵(²P°₃⁄₂)
n'l'core = 2p⁵(²P°₁⁄₂)

848460 ⌐ 3s'[½]°(1)
840560 ∟ 3s[³⁄₂]°(1)

0 ——— 2p⁶ ¹S₀

ENERGY (cm⁻¹)

140 × 10⁴

130 × 10⁴

120 × 10⁴

110 × 10⁴

100 × 10⁴

90 × 10⁴

80 × 10⁴

0

Si VI
DOUBLET
GROTRIAN
DIAGRAM

Si VI GROTRIAN DIAGRAM (9 electrons, Z=14)
(F I sequence, Configuration: 1s² 2s² 2p⁴ (³P) nl, Doublet System)

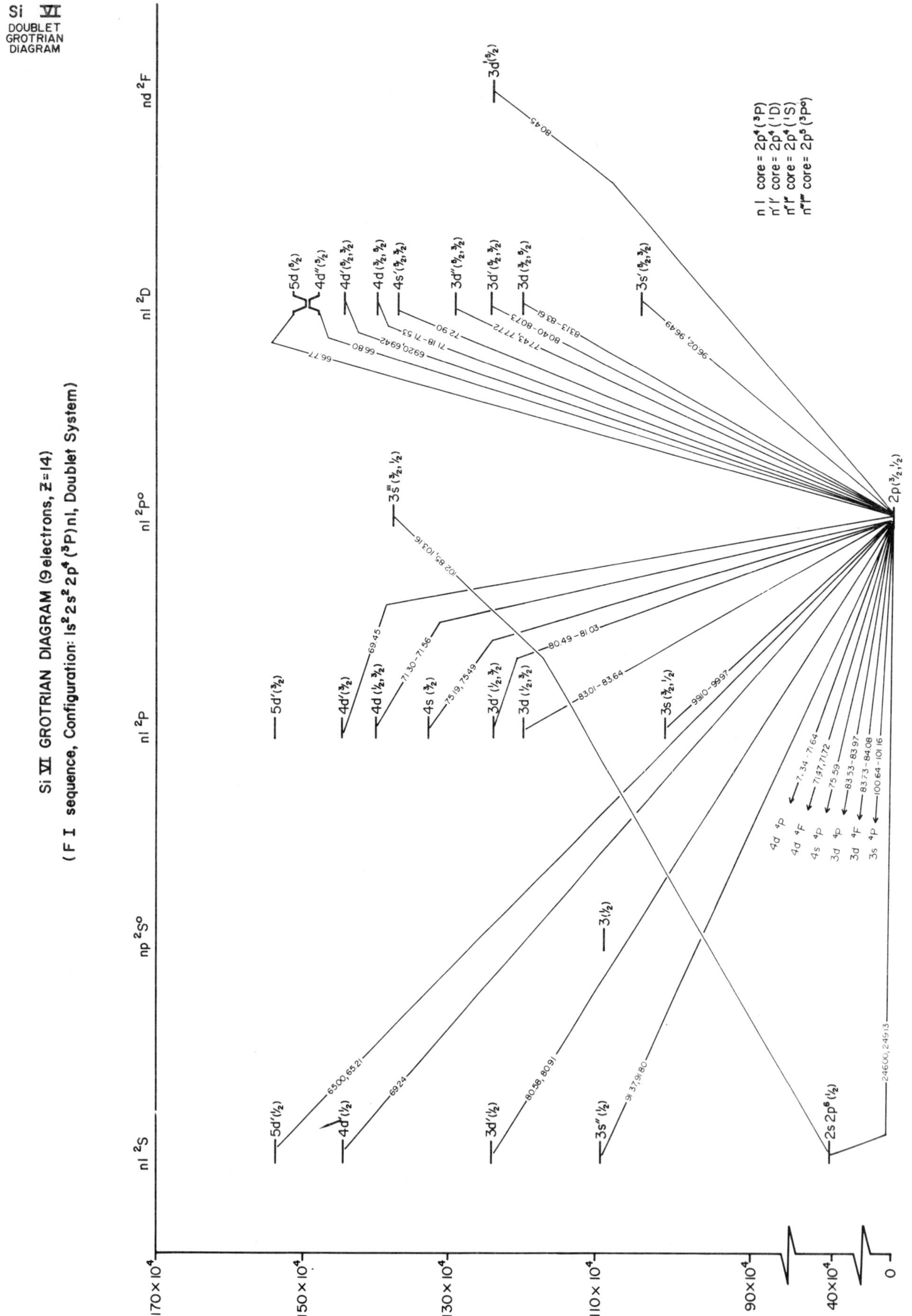

524

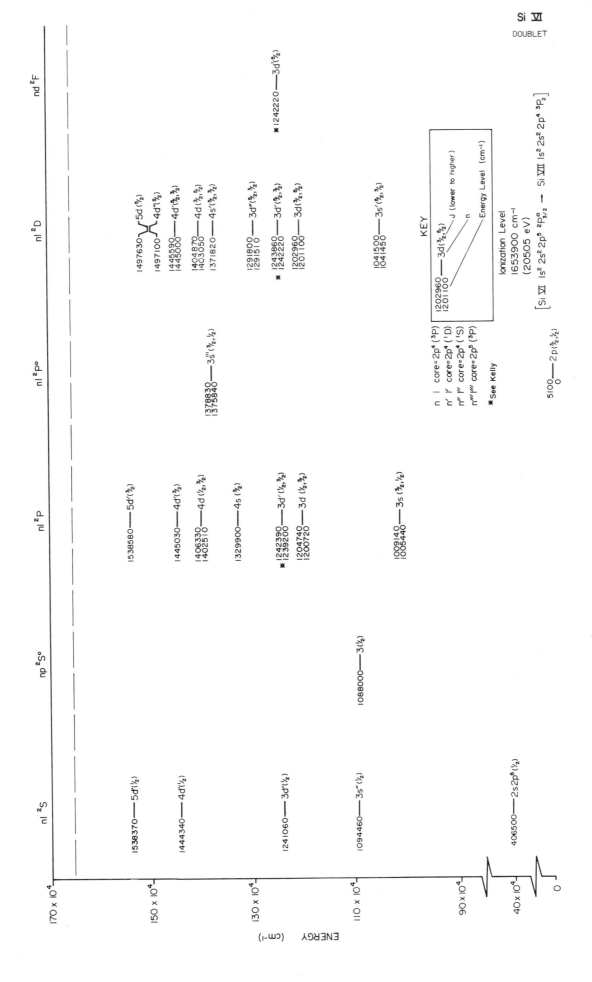

Si Ⅵ ENERGY LEVELS (9 electrons, Z=14)

(F Ⅰ sequence, Configuration: 1s² 2s² 2p⁴(³P)nl, Doublet System)

Si Ⅵ

DOUBLET

525

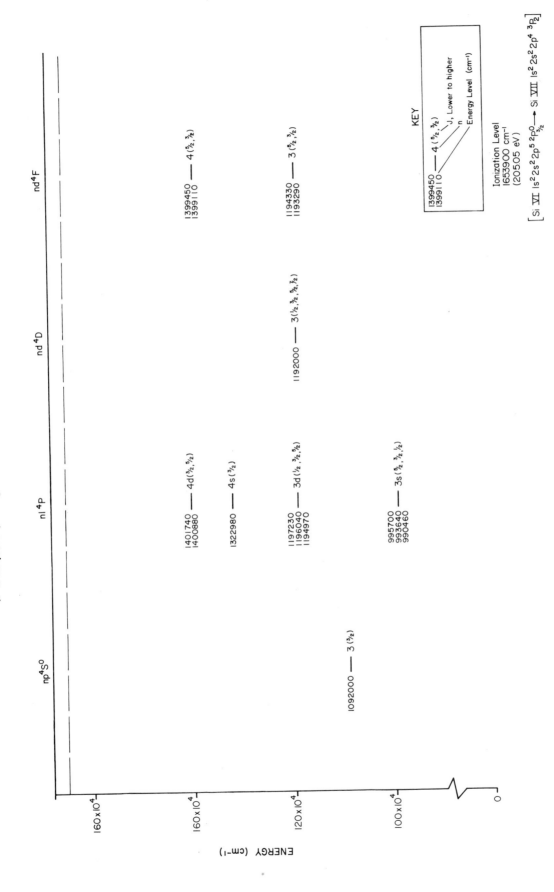

Si VI ENERGY LEVELS (9 electrons, Z=14)
(F I Sequence, Configuration: 1s²2s²2p⁴(³P)nl, Quartet System)

Si VI
QUARTET

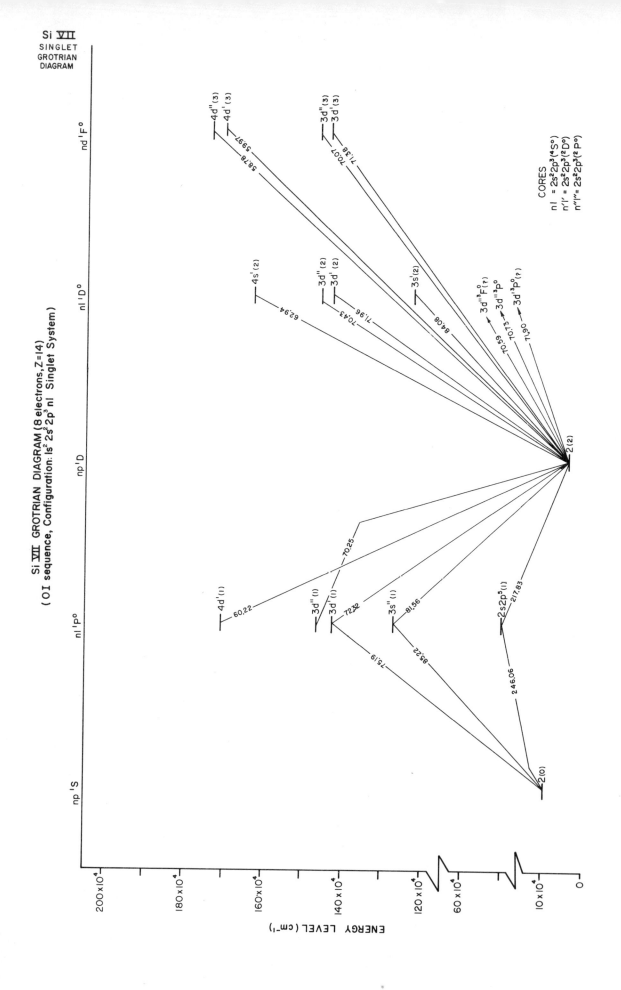

Si VII
SINGLET
GROTRIAN
DIAGRAM

Si VII GROTRIAN DIAGRAM (8 electrons, Z=14)
(OI sequence, Configuration: ls² 2s² 2p³ nl Singlet System)

CORES
nl = 2s²2p³(⁴S°)
n'l' = 2s²2p³(²D°)
n"l" = 2s²2p³(²P°)

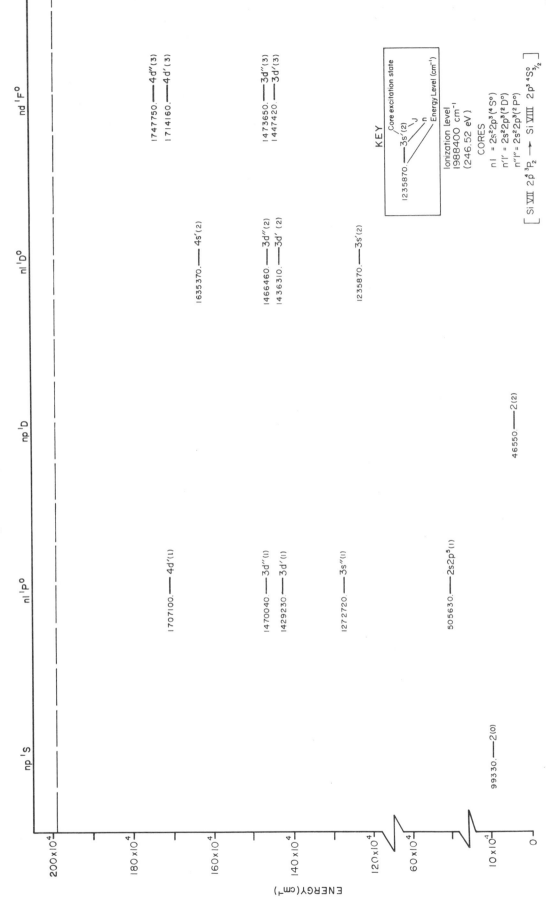

Si VII ENERGY LEVELS (8 electrons, Z = 14)
(O I sequence, Configuration: 1s²2s²2p³nl, Singlet System)

Si VII
SINGLET

KEY

Core excitation state
1235870. ——— 3s'(2)
J
n
Energy Level (cm⁻¹)

Ionization level
1988400 cm⁻¹
(246.52 eV)

CORES
nl = 2s²2p³(⁴S°)
n'l' = 2s²2p³(²D°)
n''l'' = 2s²2p³(²P°)

[Si VII 2p⁴ ³P₂ ⟶ Si VIII 2p³ ⁴S°₃/₂]

NOTE: All levels uncertain by ± x.

np ¹S

99330. ——— 2(0)

nl ¹P°

1707100. ——— 4d'(1)

1470040 ——— 3d''(1)
1429230 ——— 3d'(1)

1272720 ——— 3s''(1)

505630. ——— 2s2p⁵(1)

np ¹D

46550 ——— 2(2)

nl ¹D°

1635370. ——— 4s'(2)

1466460 ——— 3d''(2)
1436310. ——— 3d'(2)

1235870. ——— 3s'(2)

nd ¹F°

1747750. ——— 4d''(3)
1714160. ——— 4d'(3)

1473650. ——— 3d''(3)
1447420. ——— 3d'(3)

1235870. ——— 3s'(2)

ENERGY (cm⁻¹)

200×10⁴
180×10⁴
160×10⁴
140×10⁴
120×10⁴
60×10⁴
10×10⁴
0

Si VII TRIPLET GROTRIAN DIAGRAM

Si VII GROTRIAN DIAGRAM (8 electrons, Z=14)
(OI sequence, Configuration: 1s²2s²2pnl, Triplet System)

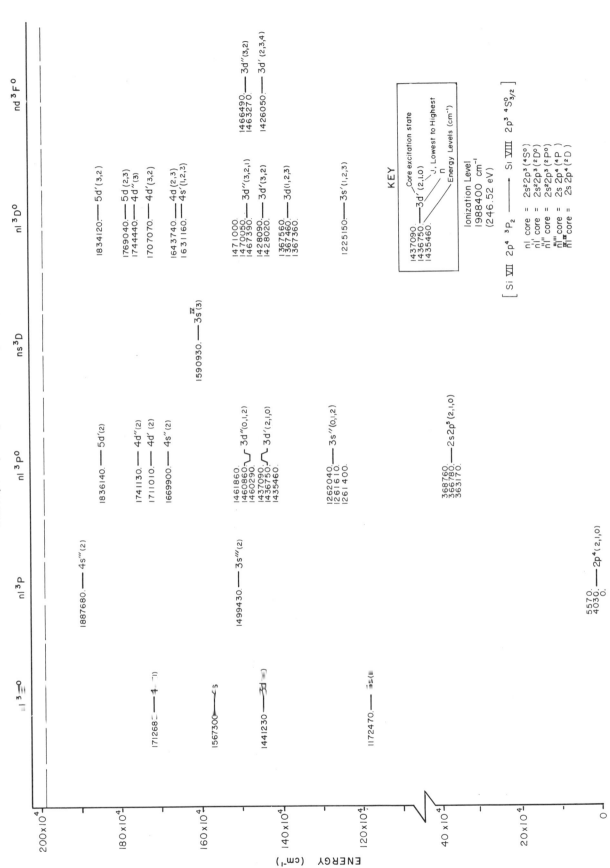

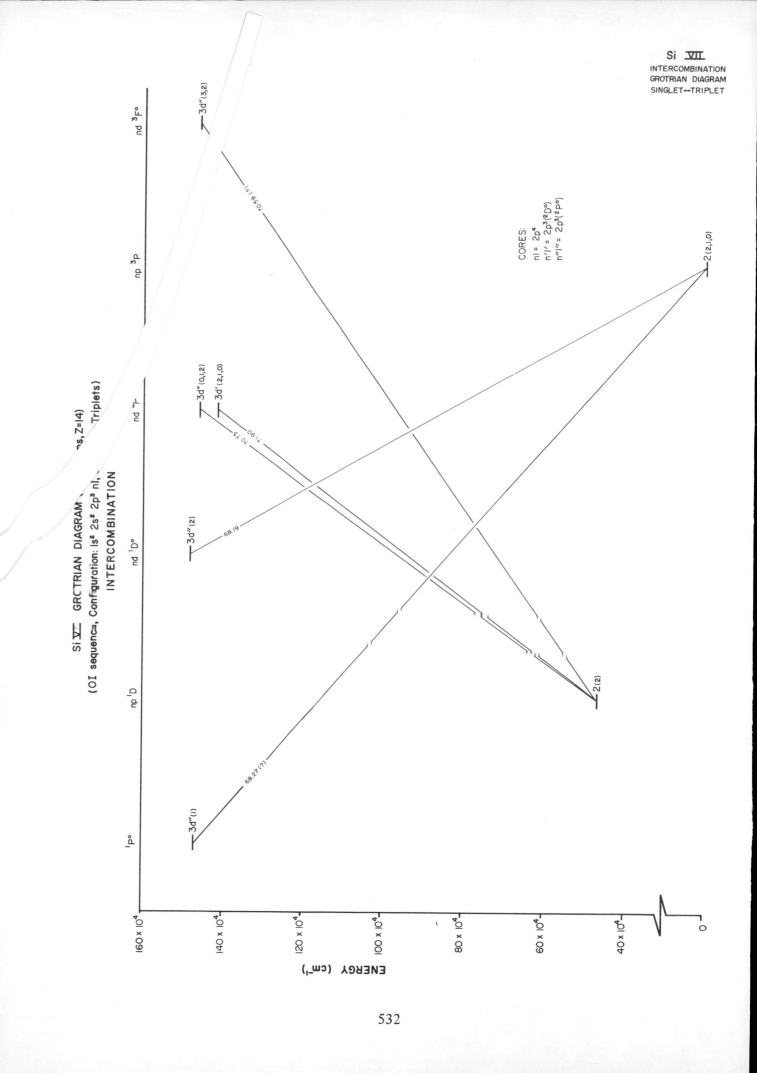

Si VII   GROTRIAN DIAGRAM

(OI sequence, Configuration: 1s² 2s² 2p³ nl,   Triplets)

INTERCOMBINATION

CORES:
nl = 2p⁴
n′l′ = 2p³(²D°)
n″l″ = 2p³(²P°)

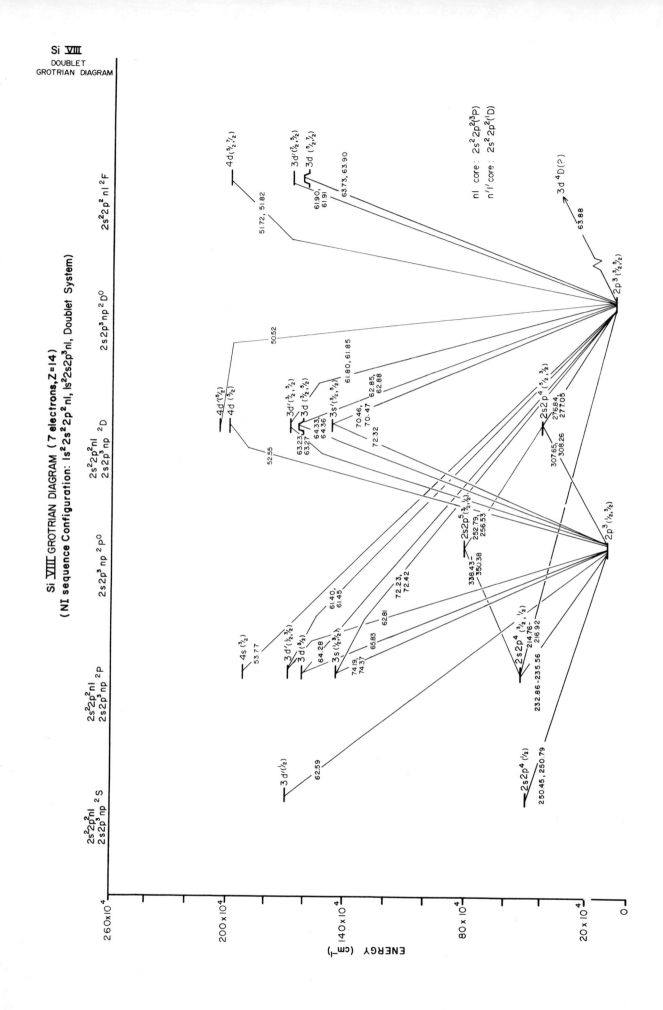

Si VIII GROTRIAN DIAGRAM (7 electrons, Z=14)
(NI sequence Configuration: 1s²2s²2p²nl, 1s²2s2p³nl, Doublet System)

534

Si VIII ENERGY LEVELS (7 electrons, Z=14)
(NI sequence, Configuration: 1s²2s²2p²nl, 1s²2s2p³nl, Doublet System)

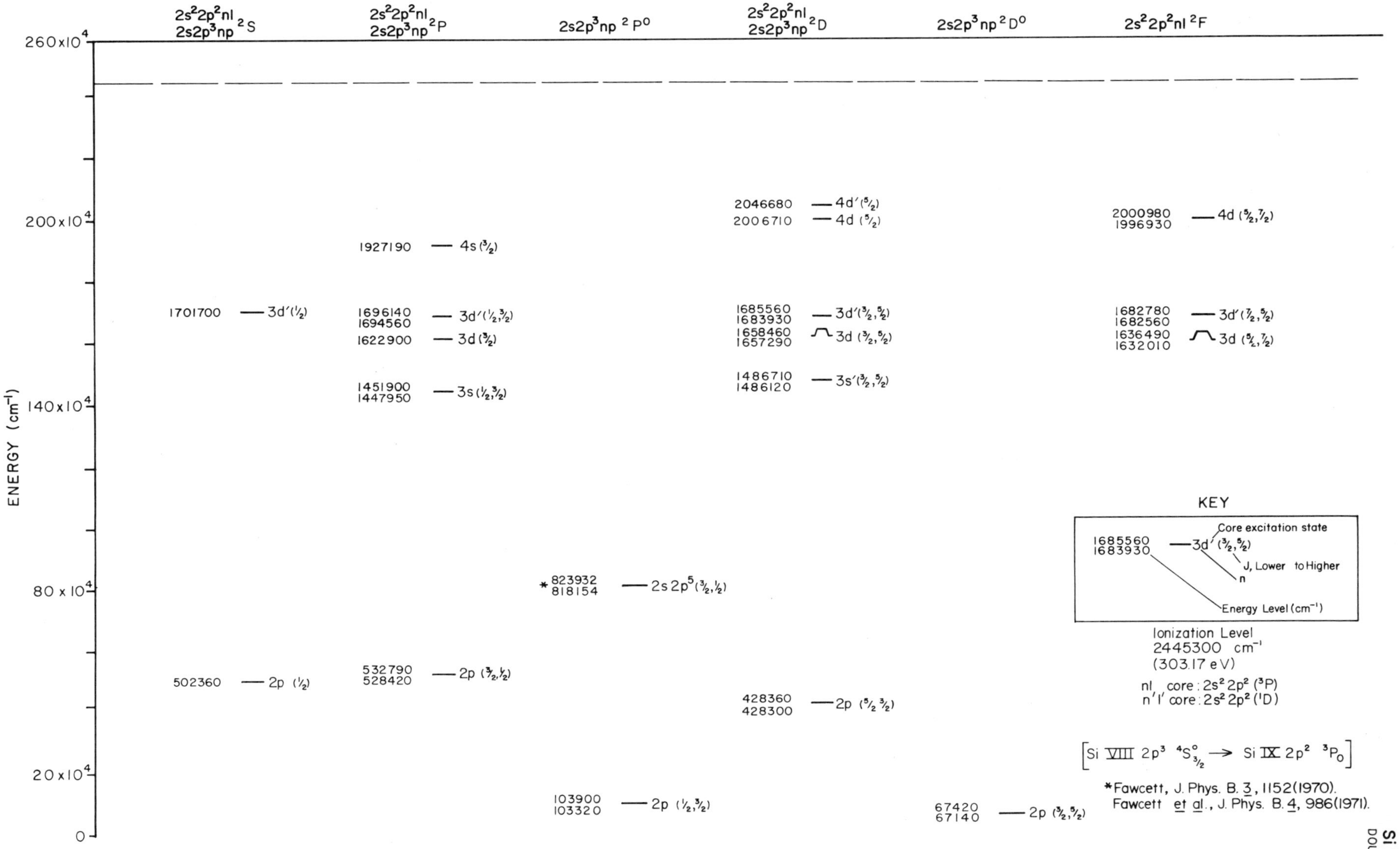

Si VIII
DOUBLET

535

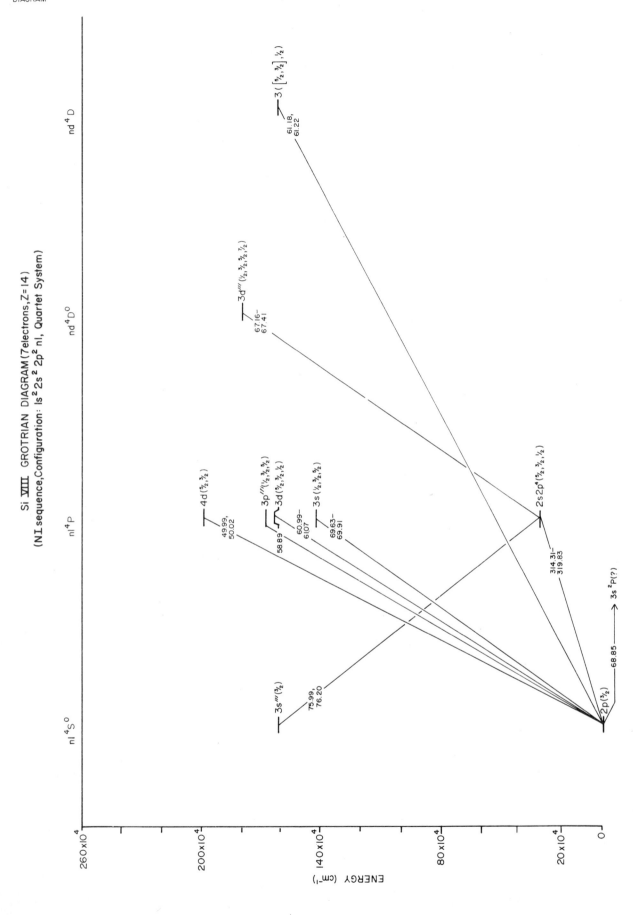

Si VIII
QUARTET
GROTRIAN
DIAGRAM

Si VIII GROTRIAN DIAGRAM (7 electrons, Z=14)
(N I sequence, Configuration: 1s² 2s² 2p² nl, Quartet System)

Si VIII ENERGY LEVELS (7 electrons, Z=14)
(NI sequence, Configuration: 1s² 2s² 2p² nl, Quartet System)

nl ⁴S°   nl ⁴P   nd ⁴D°   nd ⁴D

1628660.——3s‴(³/₂)

2000520.——4d(⁵/₂,³/₂)
1999240.

1698230.——3p‴(¹/₂,³/₂,⁵/₂)
1639640.——3d(⁵/₂,³/₂,¹/₂)
1638830.
1637470.

1436120.——3s(¹/₂,³/₂,⁵/₂)
1432870.
1430510.

318160.——2s2p⁴(⁵/₂,³/₂,¹/₂)
316260.
312670.

1801710.——3d‴(¹/₂,³/₂,⁵/₂,⁷/₂)

1634660.——3([⁵/₂,³/₂],¹/₂)
1633370.

KEY
                    Core excitation state
1639640.——3d‴(⁵/₂,³/₂,¹/₂) J, Lowest to Highest
1638830.   n
1637470.   Energy Levels(cm⁻¹)

Ionization Level
2445300 cm⁻¹
(303.17 eV)

nl core = 2s² 2p² (³P)
nl'core = 2s 2p³ (⁵S°)
[Si VIII 1s²2s²2p³ ⁴S°_{3/2}→Si IX 1s²2s²2p² ³P₀]

0 ——2p (³/₂)

260×10⁴
200×10⁴
140×10⁴
80×10⁴
20×10⁴
0

ENERGY cm⁻¹

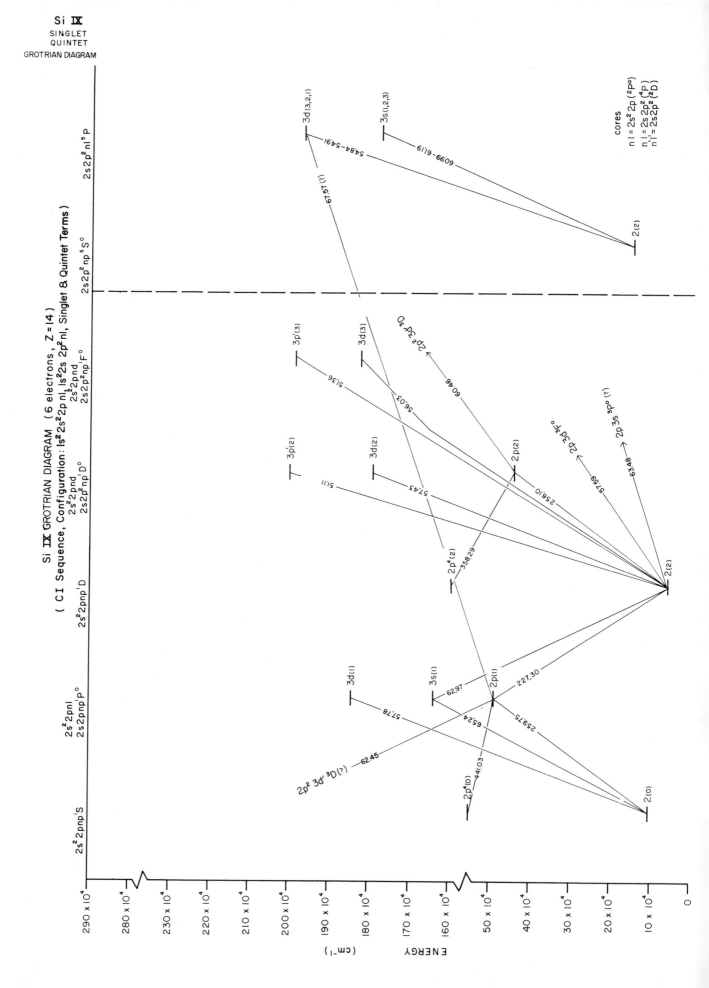

Si IX
SINGLET
QUINTET
GROTRIAN DIAGRAM

Si IX GROTRIAN DIAGRAM (6 electrons, Z=14)
(CI Sequence, Configuration: 1s²2s²2p nl, 1s²2s 2p²nl, 2p²nl, Singlet & Quintet Terms)

538

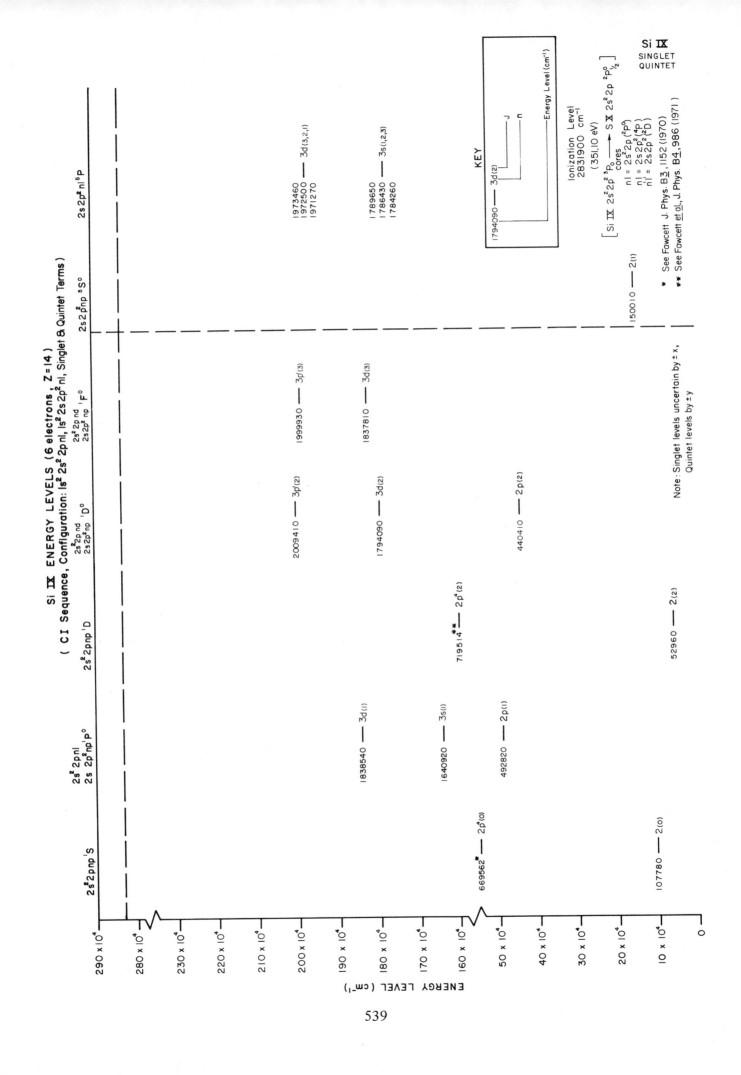

Si IX ENERGY LEVELS (6 electrons, Z=14)

(CI Sequence, Configuration: 1s² 2s² 2pnl, 1s²2s2p²nl, Singlet & Quintet Terms)

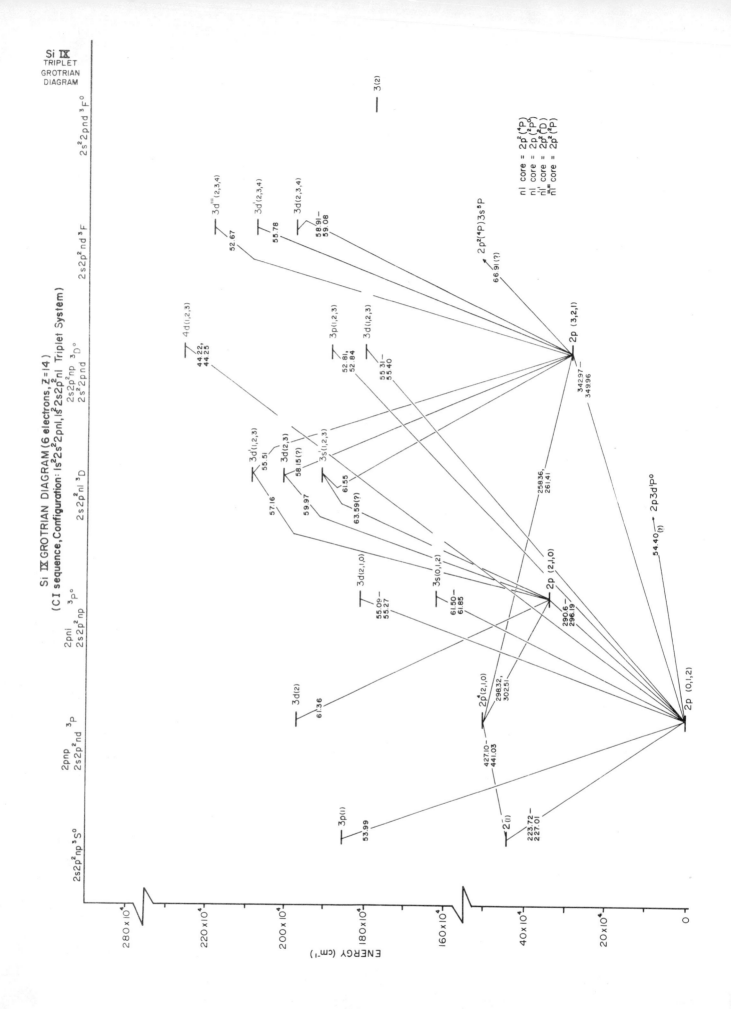

Si IX
TRIPLET
GROTRIAN
DIAGRAM

Si IX GROTRIAN DIAGRAM (6 electrons, Z=14)
(C I sequence, Configuration: 1s²2s²2pnl, 1s²2s2p²nl Triplet System)

nl core = 2p² (⁴P)
nl' core = 2p² (²Pᵒ)
nl'' core = 2p² (²D)
nl''' core = 2p² (²P)

540

ENERGY (cm⁻¹)

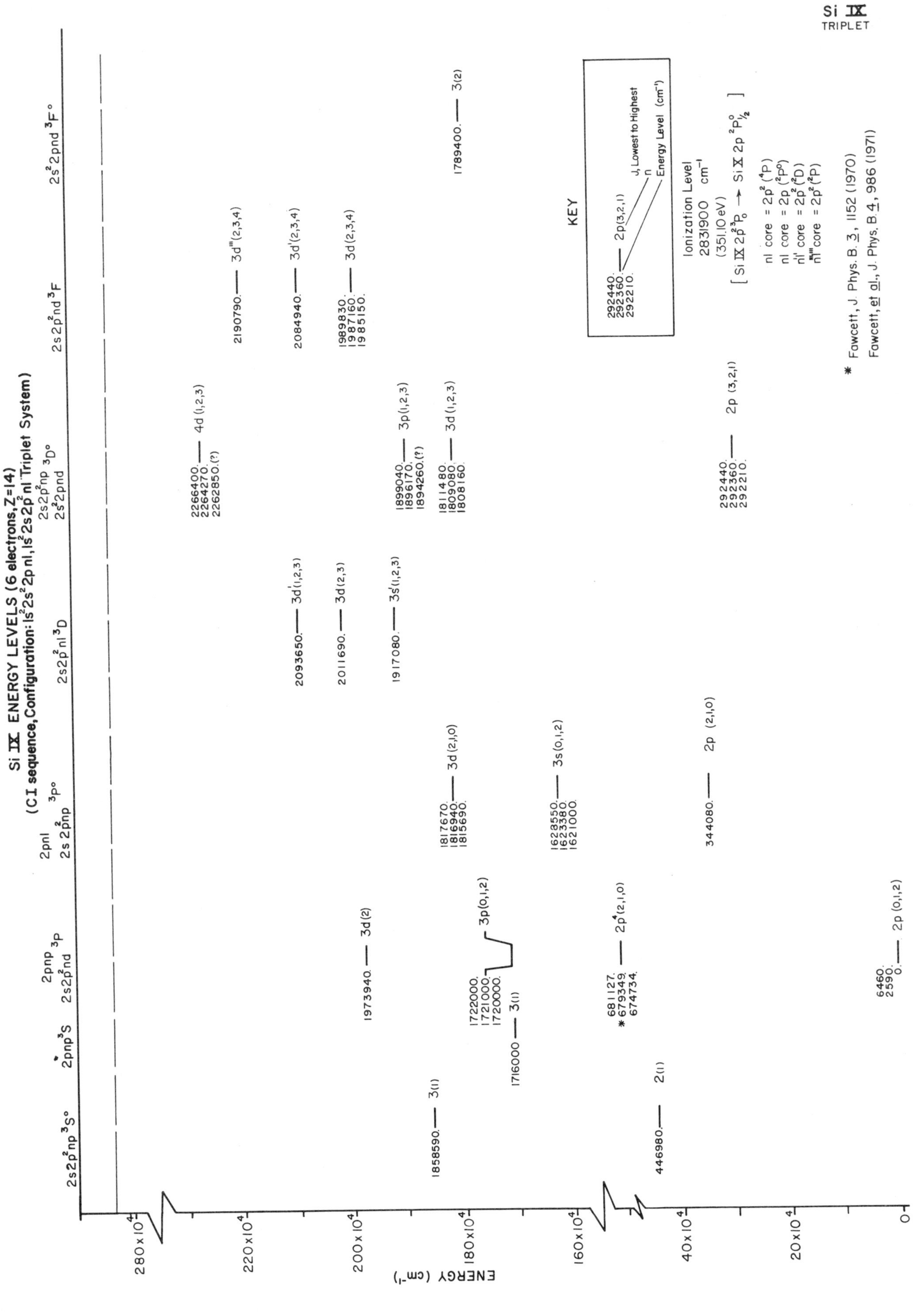

Si IX TRIPLET
Si IX ENERGY LEVELS (6 electrons, Z=14)
(C I sequence, Configuration: $1s^22s^22pnl, 1s^22s2p^2nl$ Triplet System)
$2s2p^2np\ ^3S^o$
$^*2pnp\ ^3S$
2pnp $2s2p^2nd\ ^3P$
2pnl $2s2p^2np\ ^3P^o$
$2s2p^2nl\ ^3D$
$2s2p^2np\ ^3D^o$
$2s^22pnd$
$2s^22pnd\ ^3F$
$2s^22pnd\ ^3F^o$
ENERGY (cm^{-1})
280×10^4
220×10^4
200×10^4
180×10^4
160×10^4
40×10^4
20×10^4
0
1858590. 3(1)
446980. 2(1)
1716000 3(1)
1722000.
1721000.
1720000.
3p(0,1,2)
681127.
*679349.
674734.
$2p^4$(2,1,0)
6460.
2590.
0.
2p (0,1,2)
1973940. 3d(2)
1817670.
1816940.
1815690.
3d (2,1,0)
1628550.
1623380.
1621000.
3s(0,1,2)
344080. 2p (2,1,0)
2093650. 3d'(1,2,3)
2011690. 3d(2,3)
1917080. 3s'(1,2,3)
2266400.
2264270.
2262850.(?)
4d (1,2,3)
1899040.
1896170.
1894260.(?)
3p(1,2,3)
1811480.
1809080.
1808160.
3d (1,2,3)
292440.
292360.
292210.
2p (3,2,1)
2190790. 3d'''(2,3,4)
2084940. 3d'(2,3,4)
1989830.
1987160.
1985150.
3d (2,3,4)
1789400. 3(2)
KEY
292440.
292360.
292210.
2p(3,2,1)
J, Lowest to Highest
n
Energy Level (cm^{-1})
Ionization Level
2831900 cm^{-1}
(351.10 eV)
[Si IX $2p^2\ ^3P_0$ → Si X $2p\ ^2P^o_{1/2}$]
nl core = $2p^2\ (^4P)$
nl core = $2p\ (^2P^o)$
nl' core = $2p^2\ (^2D)$
nl''' core = $2p^2\ (^2P)$
* Fawcett, J. Phys. B. 3, 1152 (1970)
Fawcett, et al., J. Phys. B. 4, 986 (1971)

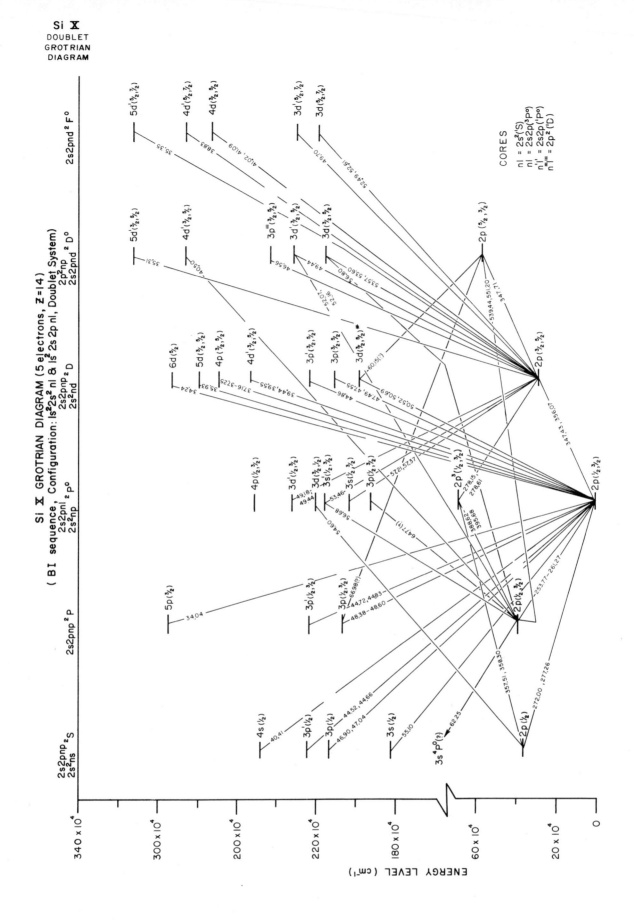

Si X ENERGY LEVELS (5 electrons, Z=14)

(B I Sequence, Configuration: 1s²2s²nl & 1s²2s2pnl, Doublet System)

Si X
DOUBLET

Column: 2s2pnp ²S / 2s²ns ²S

2481800 — 4s(½)
2246300 — 3p'(½)
2132500 — 3p(½)
1822000 — 3s(½)
367650 — 2p(½)

Column: 2s2pnp ²P

2944700 — 5p(³/₂)
2237600 / 2236200 — 3p'(½,³/₂)
2066600 / 2064600 — 3p(½,³/₂)
394000 / 389740 — 2p(½,³/₂)

Column: 2s2pnl / 2s²np ²P°

2534500 / 2533800 — 4p(½,³/₂)
2321100 — 3d'(½,³/₂)
2201770 / 2199190 — 3d(³/₂,½)
2158290 — 3s'(½,³/₂)
2035810 / 2032000 — 3s(½,³/₂)
1940000 / 1938400 — 3p(½,³/₂)
644940 / 644560 — 2p³(½,³/₂)
6990 / 0 — 2p(½,³/₂)

Column: 2s2pnp / 2s²nd ²D

2927700 — 6d(⁵/₂)
2790000 — 5d(³/₂,⁵/₂)
2695300 — 4p(³/₂,⁵/₂)
2535300 — 4d(³/₂,⁵/₂)
2236400 / 2233000 — 3p'(³/₂,⁵/₂)
2110260 / 2105800 — 3p(³/₂,⁵/₂)
1979730 / 1979260 — 3d(³/₂,⁵/₂) *
287830 — 2p(³/₂,⁵/₂)

Column: 2p²np / 2s2pnd ²D°

3116400 — 5d'(³/₂,⁵/₂)
2665200 — 4d'(³/₂,⁵/₂)
2435400 / 2435000+z — 3p'''(³/₂,⁵/₂)
2311360 / 2310230 — 3d'(³/₂,⁵/₂)
2154440 / 2153680 — 3d(³/₂,⁵/₂)
574600 / 574360 — 2p(⁵/₂,³/₂)

Column: 2s2pnd ²F°

3116400 — 5d'(⁵/₂,⁷/₂)
2863200 — 4d'(⁵/₂,⁷/₂)
2725500 / 2721800 — 4d(³/₂,⁵/₂)
2299860 — 3d'(⁵/₂,⁷/₂)
2193140 / 2188570 — 3d(⁵/₂,⁷/₂)

KEY

n — core Excitation State
nl(³/₂,⁵/₂) — J values (lower to higher energy)
3d'(³/₂,⁵/₂)
2311360
2310230 — Energy level (cm⁻¹)

IONIZATION LIMIT
3237800 cm⁻¹
(401.43 eV)
cores: nl = 2s² (¹S)
nl' = 2s 2p (³P°)
nl'' = 2s 2p (¹P°)
nl'''' = 2p² (¹D)

[Si X 2s²2p ²P° → Si XI 2s² ¹S₀]

* see Kelly

ENERGY (cm⁻¹)

340 × 10⁴
300 × 10⁴
260 × 10⁴
220 × 10⁴
180 × 10⁴
60 × 10⁴
20 × 10⁴
0

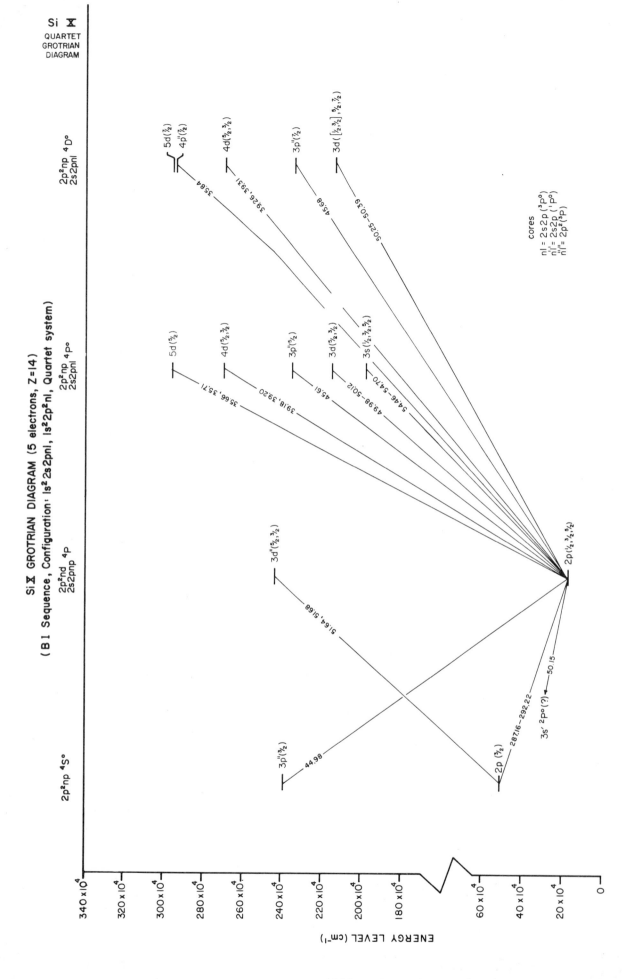

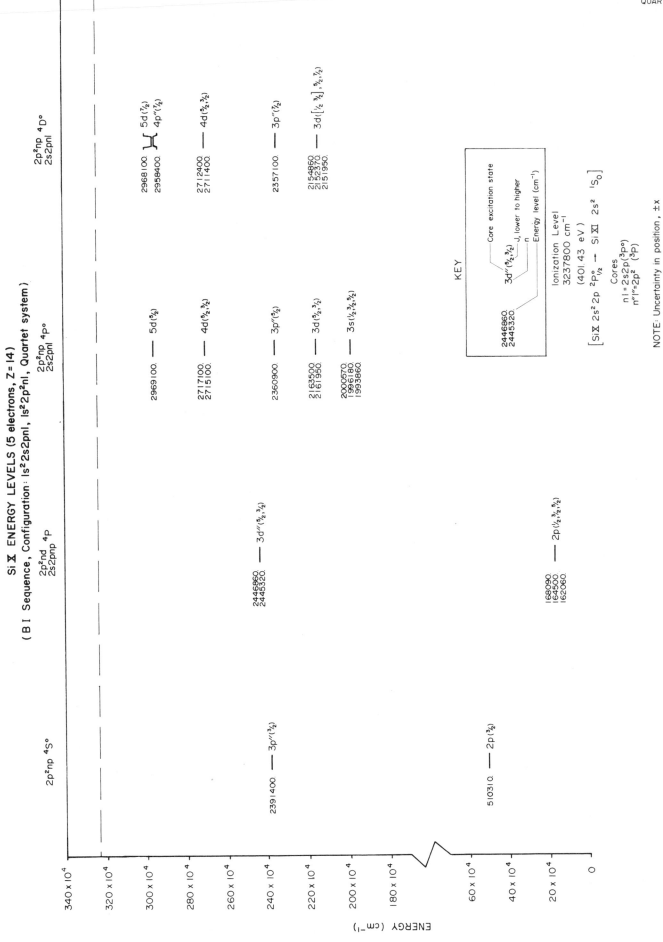

Si X ENERGY LEVELS (5 electrons, Z=14)

(B I Sequence, Configuration: 1s²2s2pnl, 1s²2p²nl, Quartet system)

Si X
QUARTET

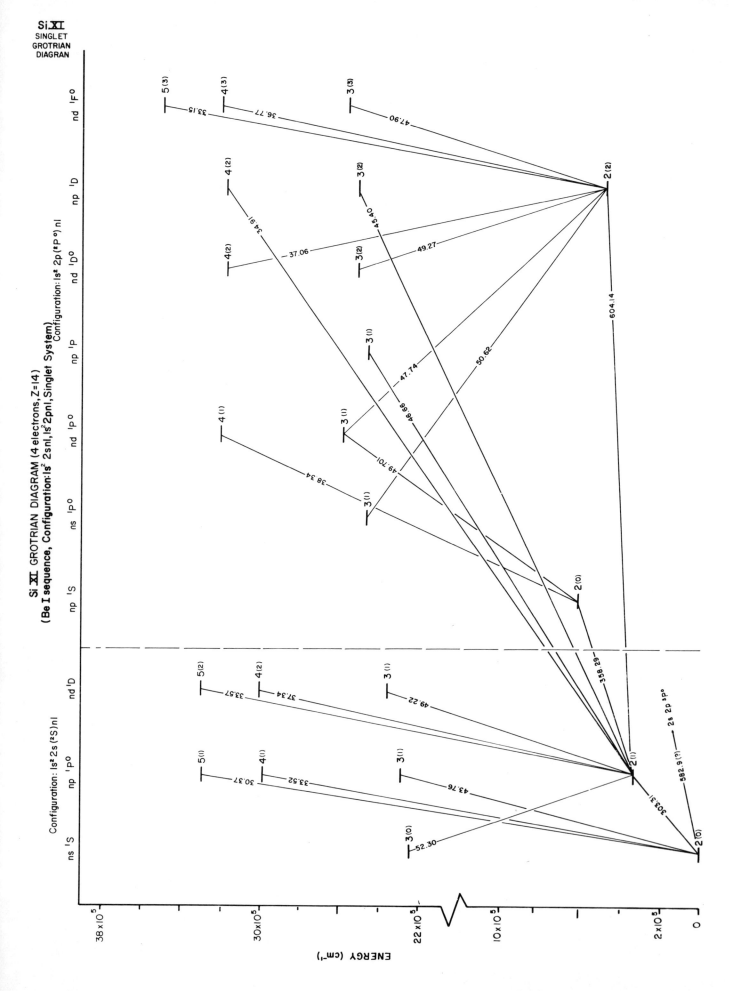

Si XI GROTRIAN DIAGRAM (4 electrons, Z=14)
(Be I sequence, Configuration:1s² 2snl,1s²2pnl,Singlet System)

Si XI
SINGLET
GROTRIAN
DIAGRAN

546

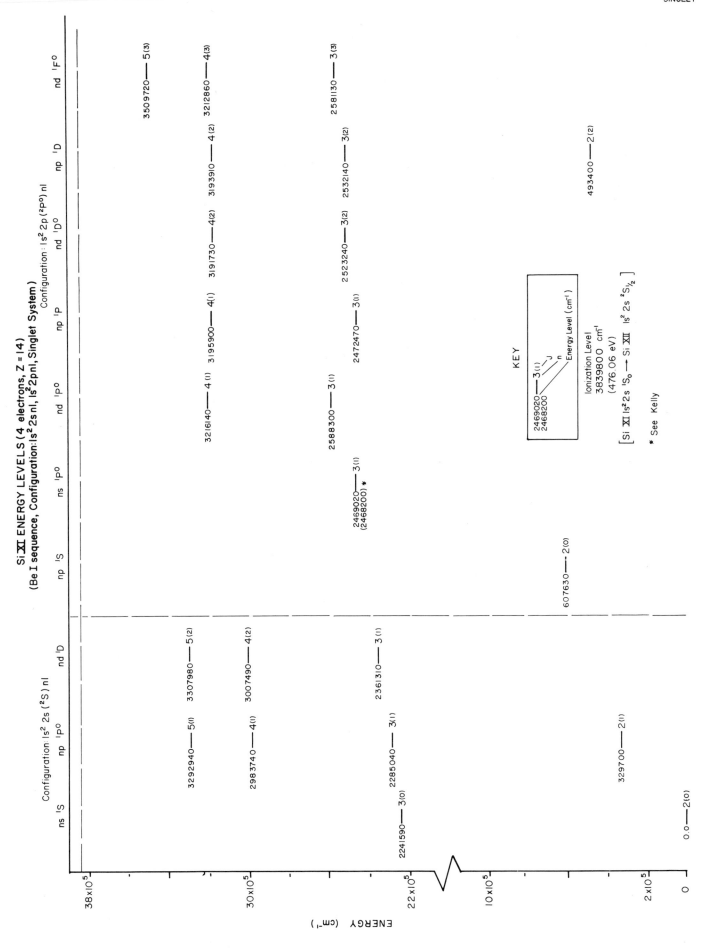

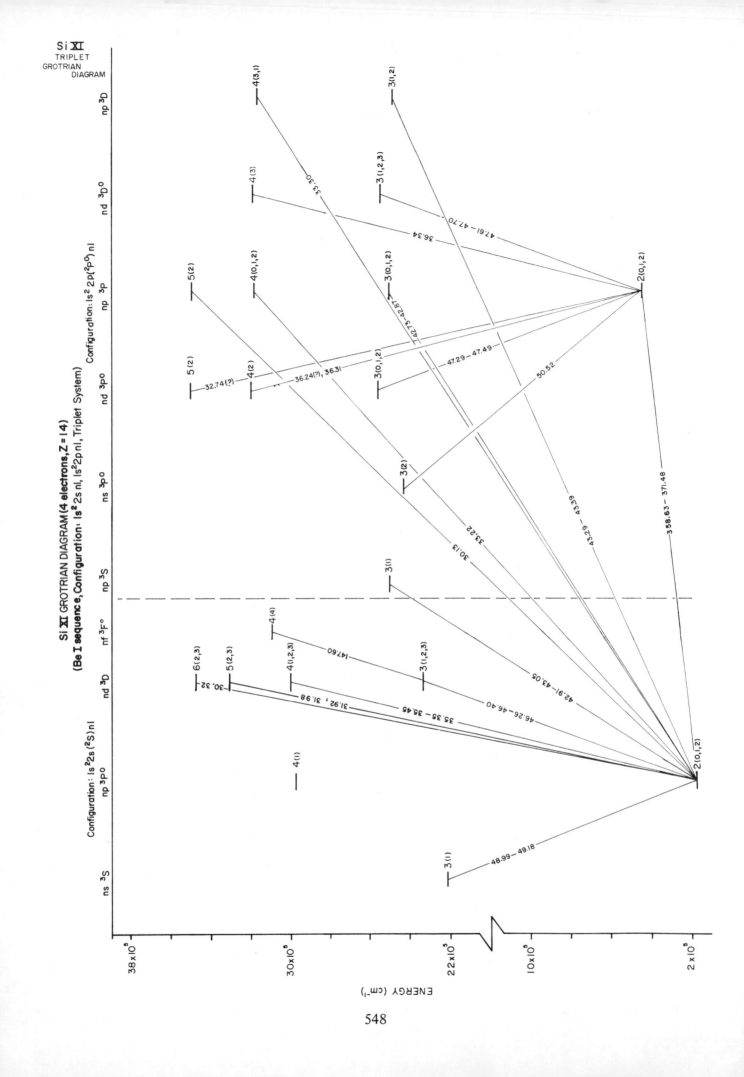

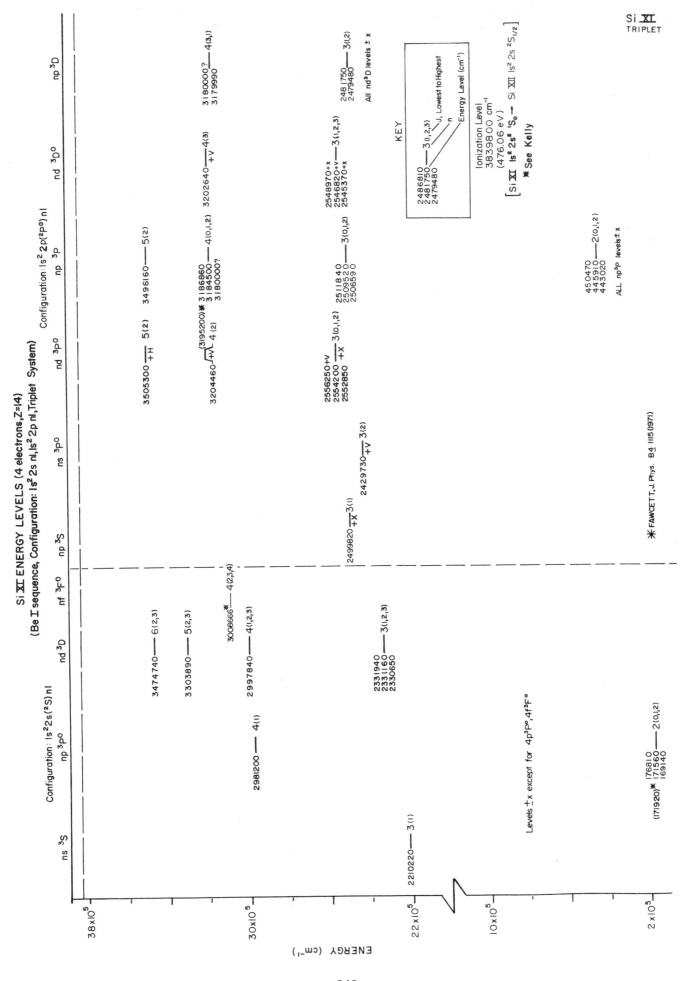

Si XI ENERGY LEVELS (4 electrons, Z=14)

(Be I sequence, Configuration: 1s² 2s nl, 1s² 2p nl, Triplet System)

Si XI
TRIPLET

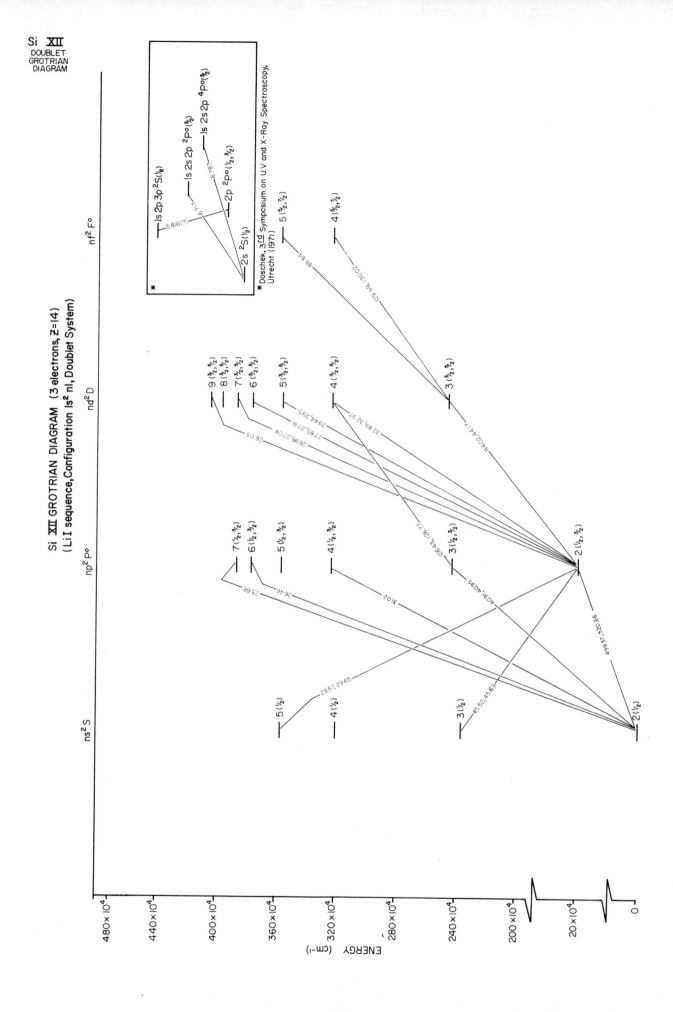

Si **XII**
DOUBLET
GROTRIAN
DIAGRAM

Si **XII** GROTRIAN DIAGRAM (3 electrons, Z=14)
(Li I sequence, Configuration 1s² nl, Doublet System)

* Doschek, 3rd Symposium on U.V. and X-Ray Spectroscopy, Utrecht (1971)

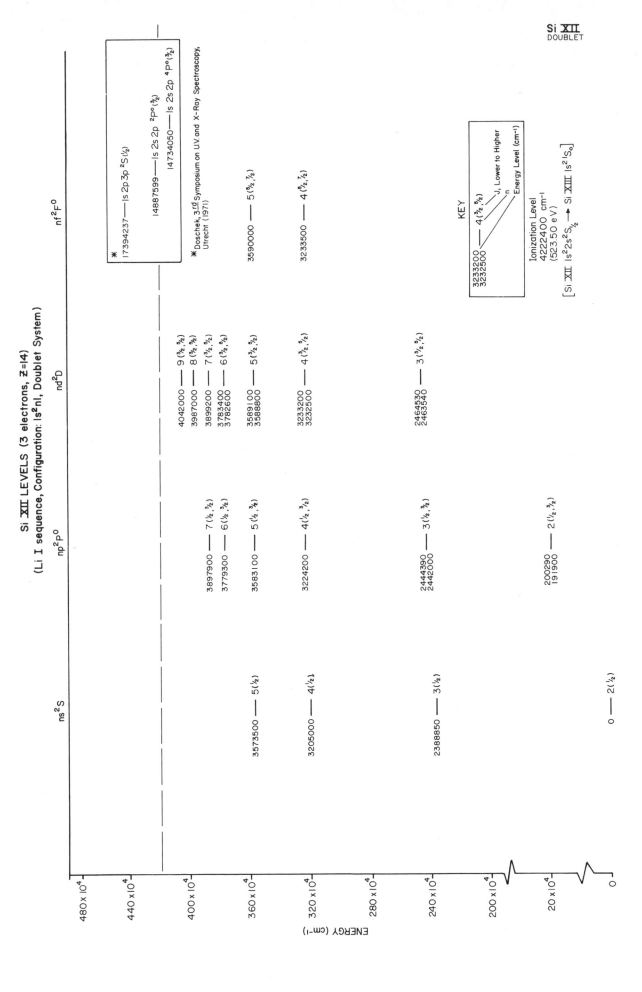

Si XII LEVELS (3 electrons, Z=14)
(Li I sequence, Configuration: 1s²nl, Doublet System)

Si XII
DOUBLET

551

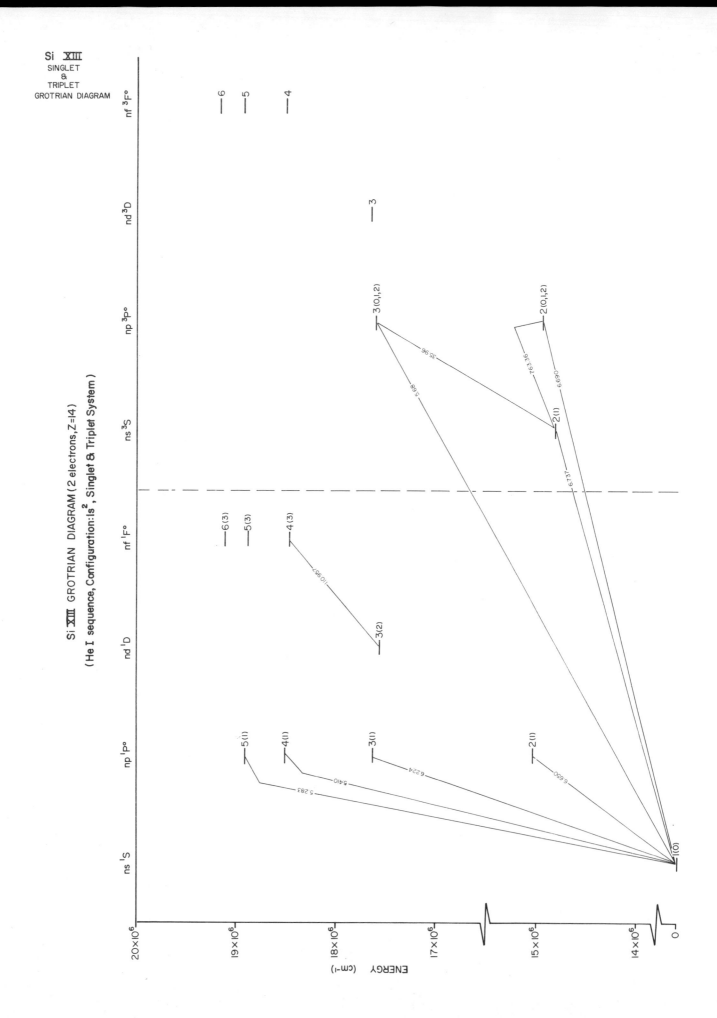

Si XIII
SINGLET
&
TRIPLET
GROTRIAN DIAGRAM

Si XIII GROTRIAN DIAGRAM (2 electrons, Z=14)

(He I sequence, Configuration: ls², Singlet & Triplet System)

552

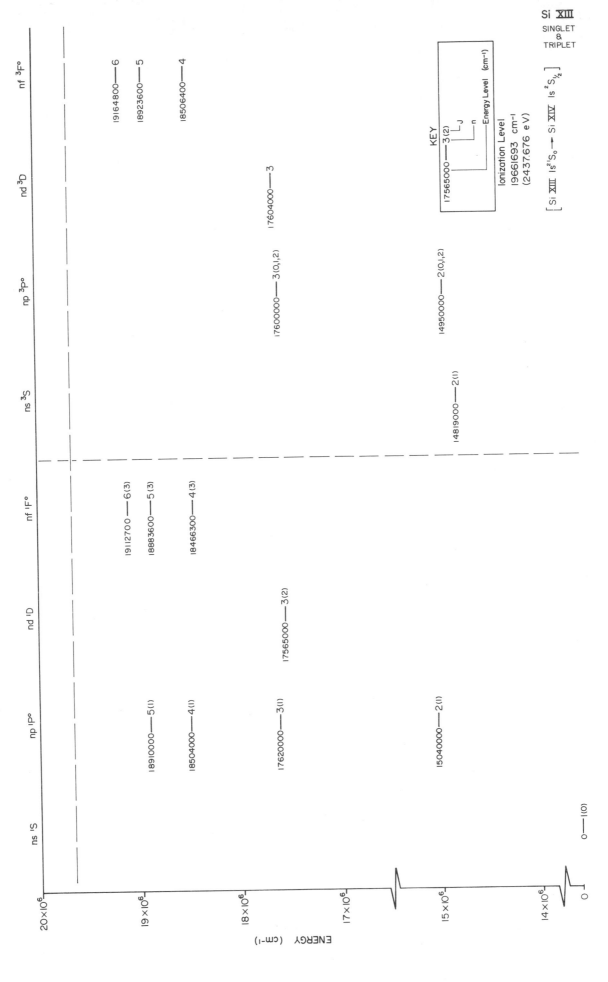

Si XIII ENERGY LEVELS (2 electrons, Z=14)

(He I sequence, Configuration: 1s², Singlet & Triplet System)

553

Si XIV GROTRIAN DIAGRAM (1 electron, Z=14)

(H I sequence, Configuration: nl, Doublet System)

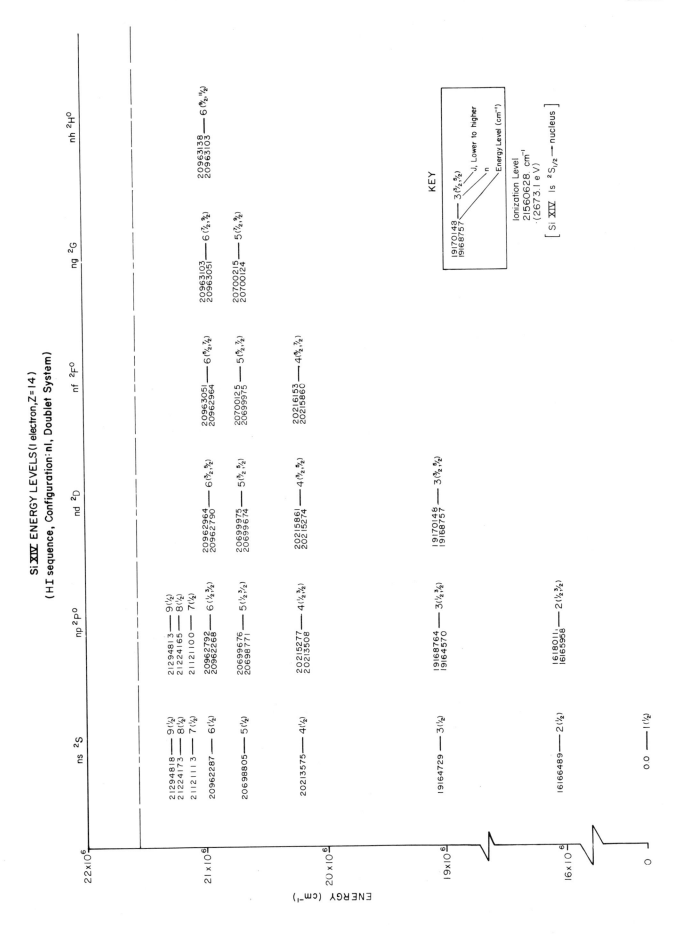

Si XIV ENERGY LEVELS (1 electron, Z=14)
(H I sequence, Configuration: nl, Doublet System)

Si I $Z = 14$ 14 electrons

I.S. Bowen, Ap. J. **132**, 1 (1960).

Author gives line tables for observed lines of forbidden transitions.

D.L. Lambert and B. Warner, Mon. Not. R. Astr. Soc. **138**, 213 (1968).

Authors give line tables for lines observed in the solar spectrum.

D.L. Lambert and B. Warner, Mon. Not. R. Astr. Soc. **139**, 35 (1968).

Authors give line tables of absorption lines predicted and of those affirmed in the solar spectrum in the range 25 578–2935 Å. Energy level tables are included.

U. Litzén, Ark. Fys. **31**, 453 (1966).

Author gives line and energy level tables for lines observed by Litzén, and Radziemski and Andrew.

Wavelength corrections: On the Excited–Excited Triplet–Quintet diagram, for 3p 4p ^{3}S–3s 3p^3 ^{3}P° 13693.85–13711.36, please read 13613.94–13711.36.
The following lines were omitted from the drawings:

3p^2 ^{1}D–3s 3p^3 ^{3}D°	2563.68, 2564.82
3p^2 ^{3}P–3s 3p^3 ^{3}S°	1255.28–1258.80
–3s 3p^3 ^{3}P°	1763.66–1770.92
–3s 3p^3 ^{3}D°	2207.98–2218.91
–3s 3p^3 ^{5}S°	3006.74–3020.00

Correction to key: In brackets {a,b}, a denotes j of the core and b denotes electron j.

Si II $Z = 14$ 13 electrons

D.L. Lambert and B. Warner, Mon. Not. R. Astr. Soc. **138**, 213 (1968).

Authors give line tables for lines observed in the solar spectrum.

The following lines were omitted from the drawings:

3p^3 ^{2}D° – 4f ^{4}F	4908.18
3d ^{2}F° – 4f ^{4}G	3977.46, 4016.22

Si III $Z = 14$ 12 electrons

Please see the general references.

The intermediate-coupling designations are taken from Moore, *Atomic Energy Levels*, N.B.S. Circular 467, Vol. I (1949).

Si IV $Z = 14$ 11 electrons

Please see the general references.

Si V $Z = 14$ 10 electrons

Please see the general references.
The intermediate-coupling designations are taken from Moore, *Atomic Energy Levels*, N.B.S. Circular 467, Vol. I (1949).

Si VI $Z = 14$ 9 electrons

Please see the general references.

Si VII $Z = 14$ 8 electrons

B. Edlén, Handbuch der Phys. **27**, 172 (1964).
Author gives corrections to values in C.E. Moore's *Atomic Energy Levels*.

Si VIII $Z = 14$ 7 electrons

Please see the general references.

Si IX $Z = 14$ 6 electrons

Please see the general references.

Si X $Z = 14$ 5 electrons

L.J. Shamey, J. Opt. Soc. Amer. **61**, 942 (1971).
Author gives a line table and a table of calculated energy levels.

Si XI $Z = 14$ 4 electrons

Please see the general references.

Si XII $Z = 14$ 3 electrons

Please see the general references.

Si XIII $Z = 14$ 2 electrons

Please see the general references.

Si XIV $Z = 14$ 1 electron

Please see the general references.

Phosphorus (P)

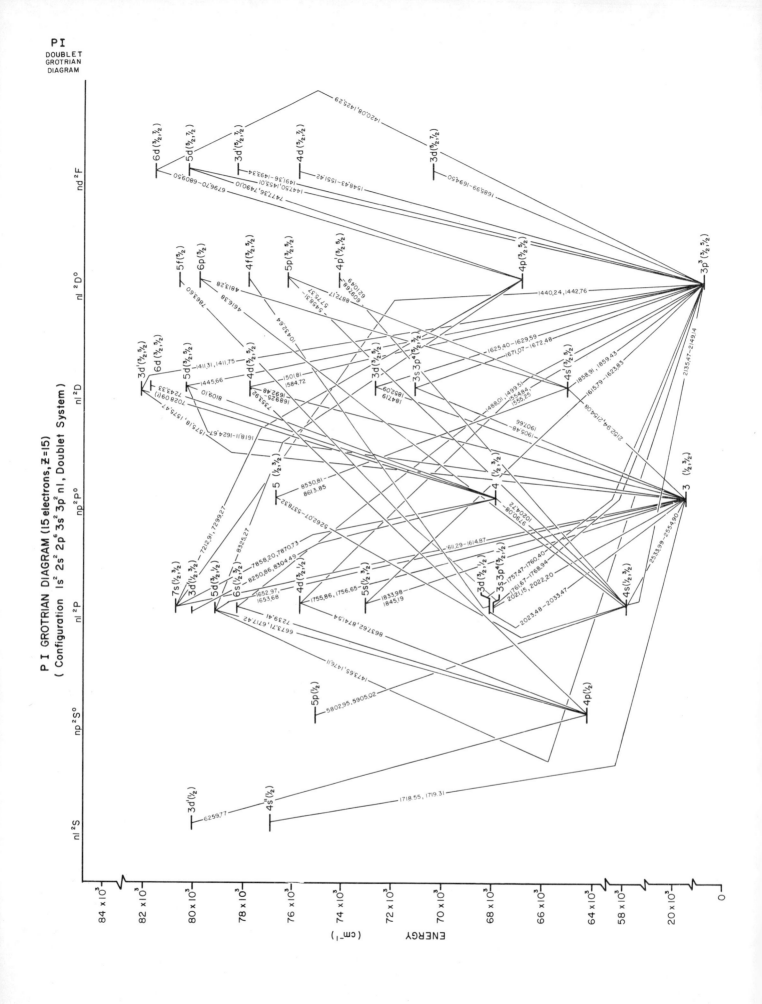

P I GROTRIAN DIAGRAM (15 electrons, Z=15)
(Configuration 1s² 2s² 2p⁶ 3s² 3p² nl, Doublet System)

PI
DOUBLET
GROTRIAN
DIAGRAM

560

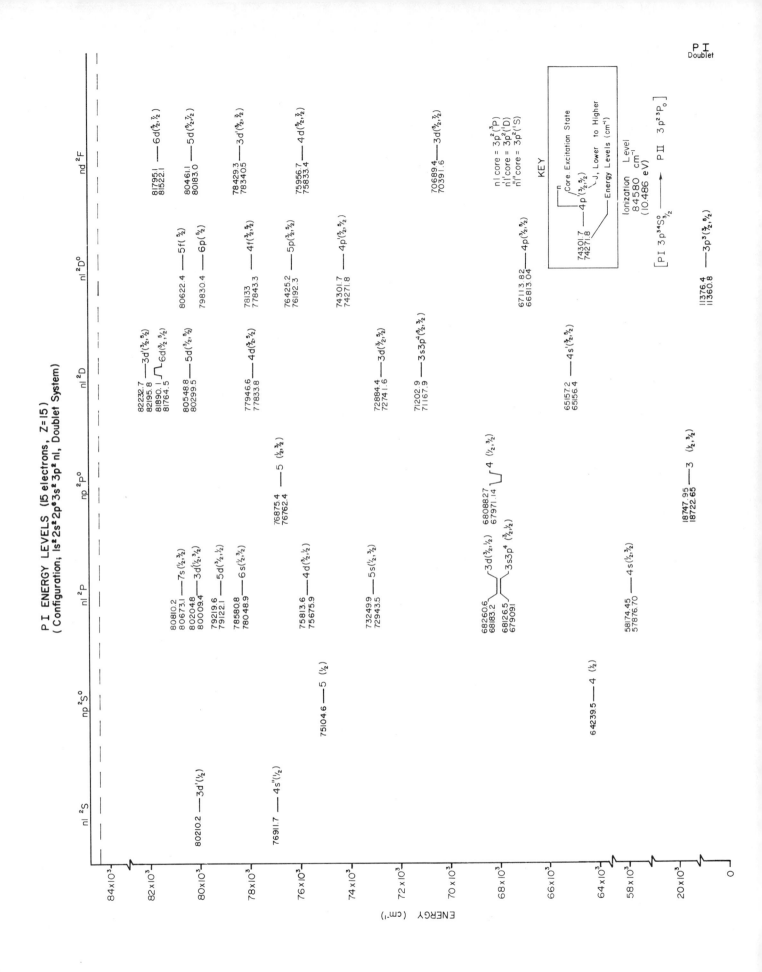

P I ENERGY LEVELS (15 electrons, Z=15)
(Configuration; 1s² 2s² 2p⁶ 3s² 3p² nl, Doublet System)

561

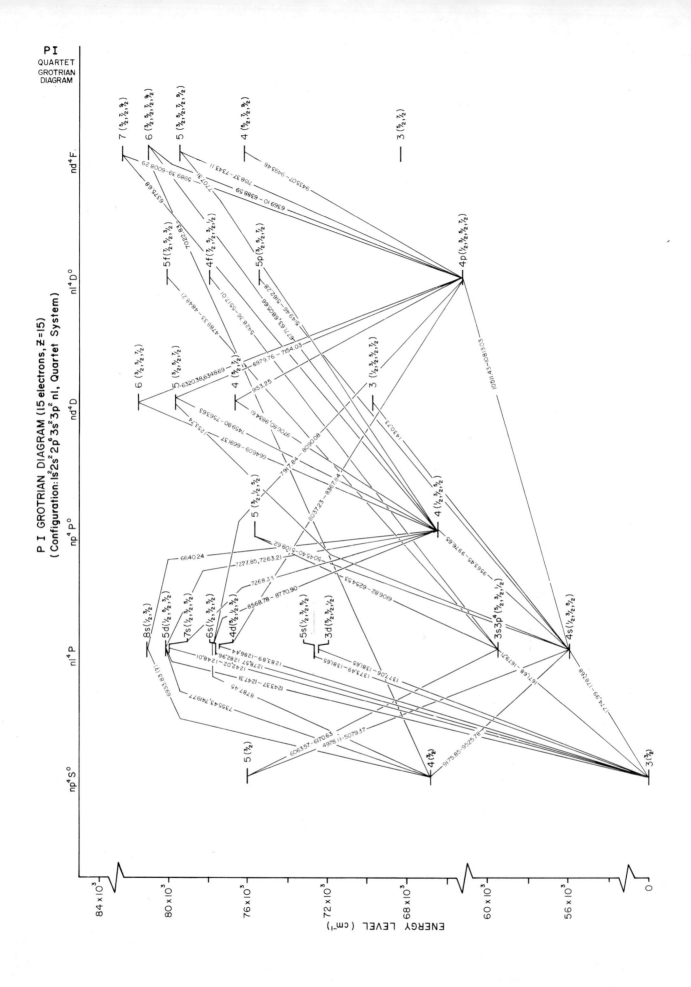

P I GROTRIAN DIAGRAM (15 electrons, Z=15)
(Configuration: 1s²2s²2p⁶3s²3p²nl, Quartet System)

PI
QUARTET
GROTRIAN
DIAGRAM

ENERGY LEVEL (cm⁻¹)

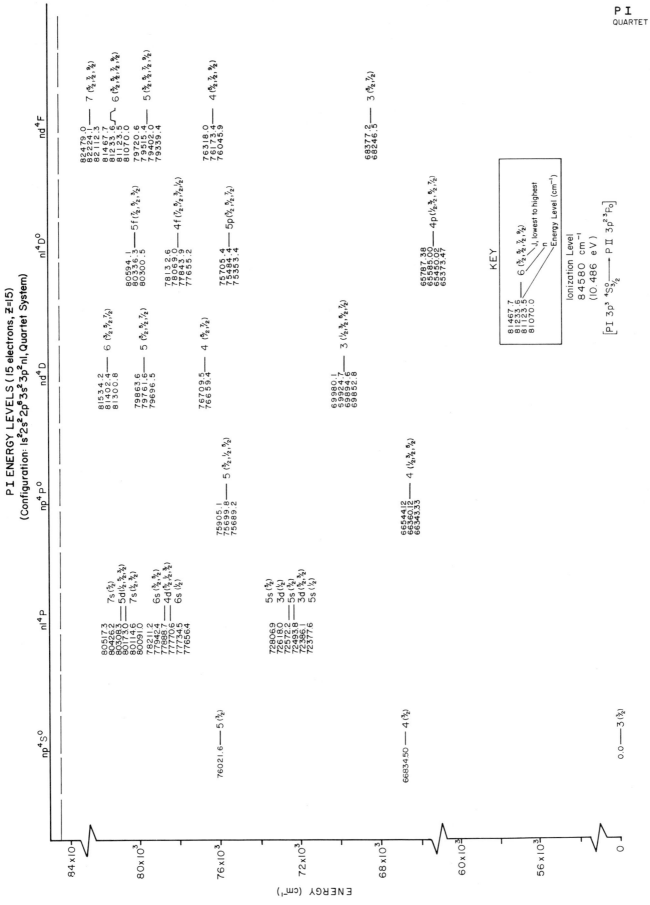

PI ENERGY LEVELS (15 electrons, Z=15)
(Configuration: 1s²2s²2p⁶3s²3p²nl, Quartet System)

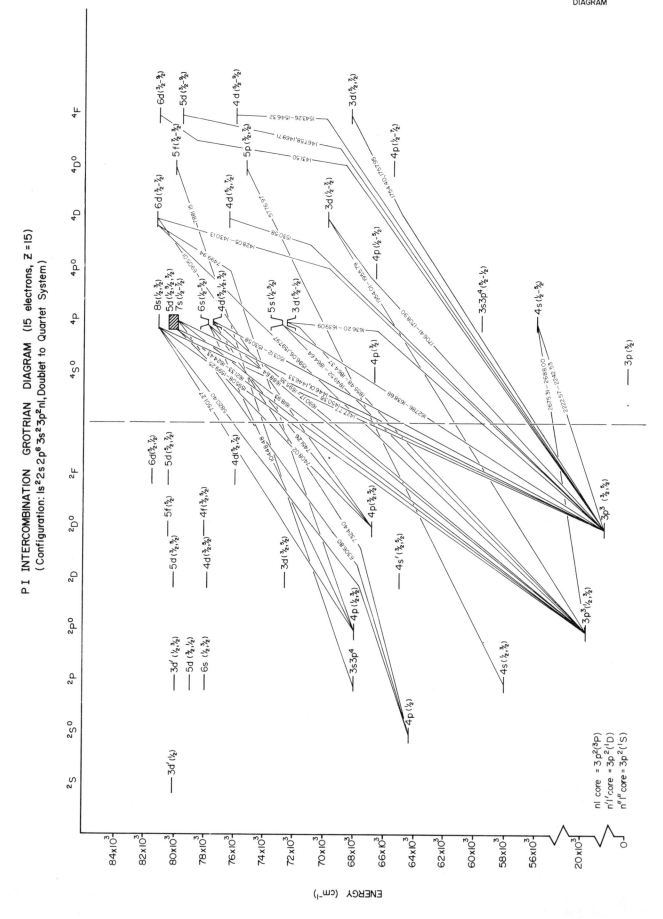

P I INTERCOMBINATION GROTRIAN DIAGRAM (15 electrons, Z = 15)
(Configuration: 1s²2s 2p⁶ 3s² 3p²nl, Doublet to Quartet System)

ENERGY (cm⁻¹)

nl core = 3p²(³P)
n'l'core = 3p²(¹D)
n''l''core = 3p²(¹S)

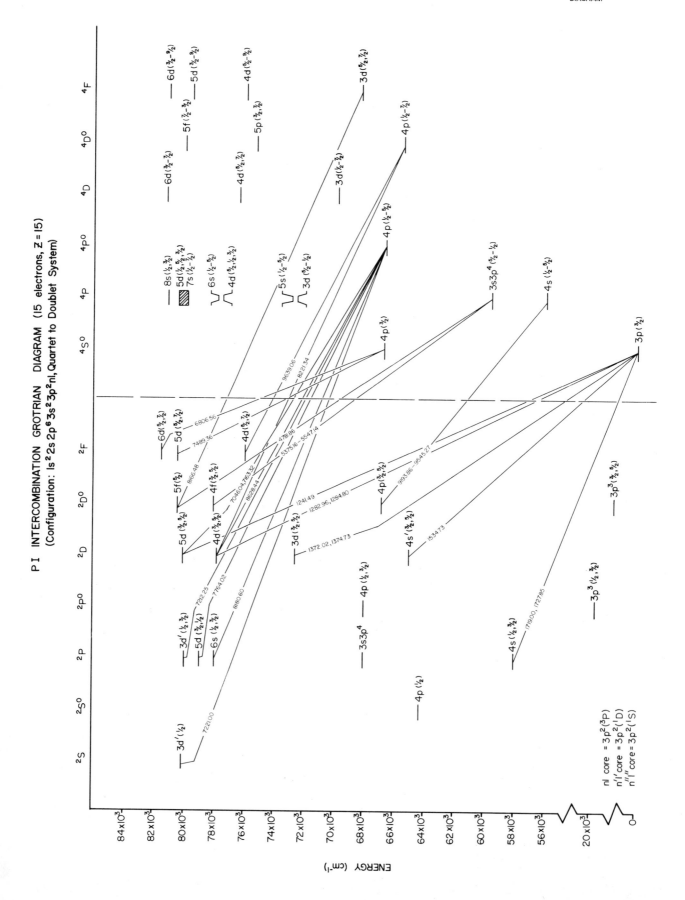

P I INTERCOMBINATION GROTRIAN DIAGRAM (15 electrons, Z = 15)
(Configuration: $1s^2 2s 2p^6 3s^2 3p^2 nl$, Quartet to Doublet System)

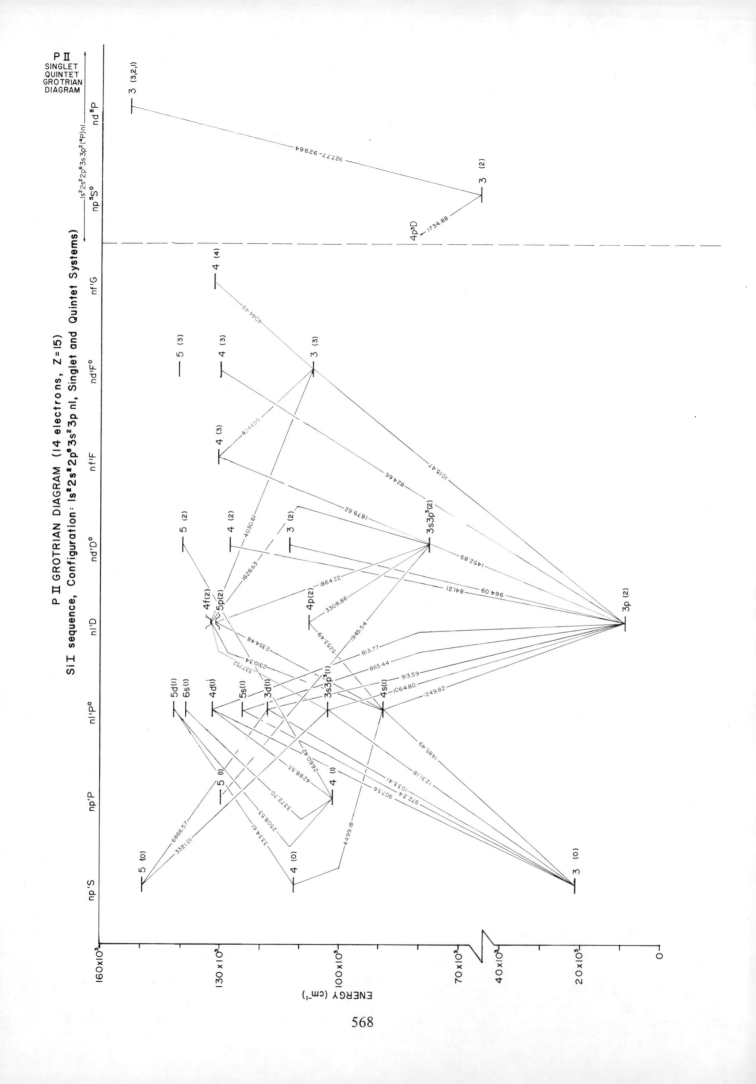

P II GROTRIAN DIAGRAM (14 electrons, Z=15)
Si I sequence, Configuration: 1s²2s²2p⁶3s²3p nl, Singlet and Quintet Systems)

P II
SINGLET
QUINTET
GROTRIAN
DIAGRAM

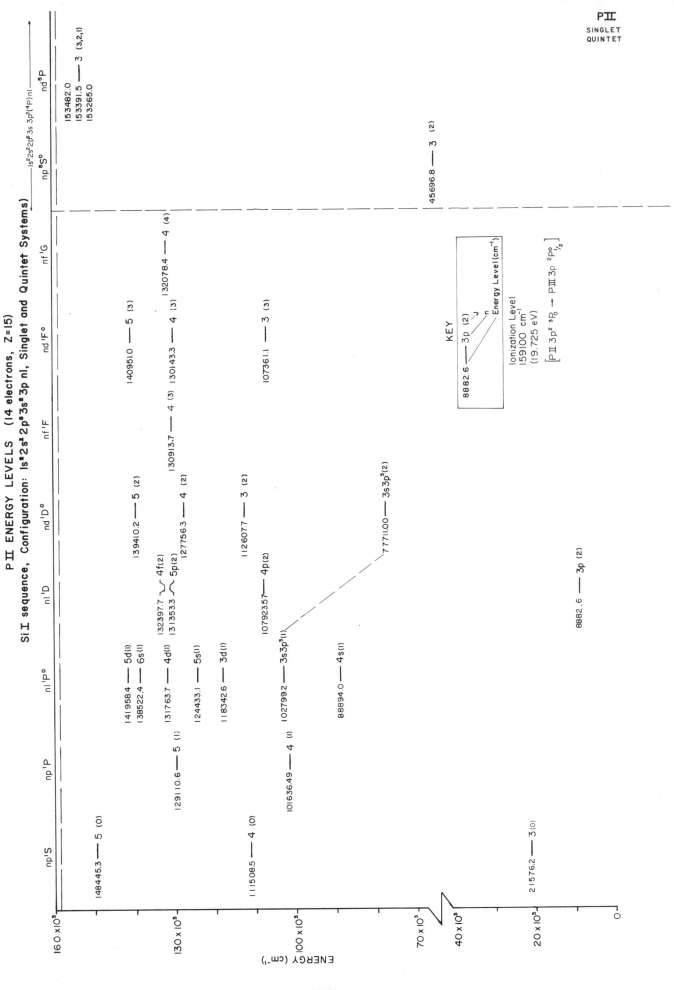

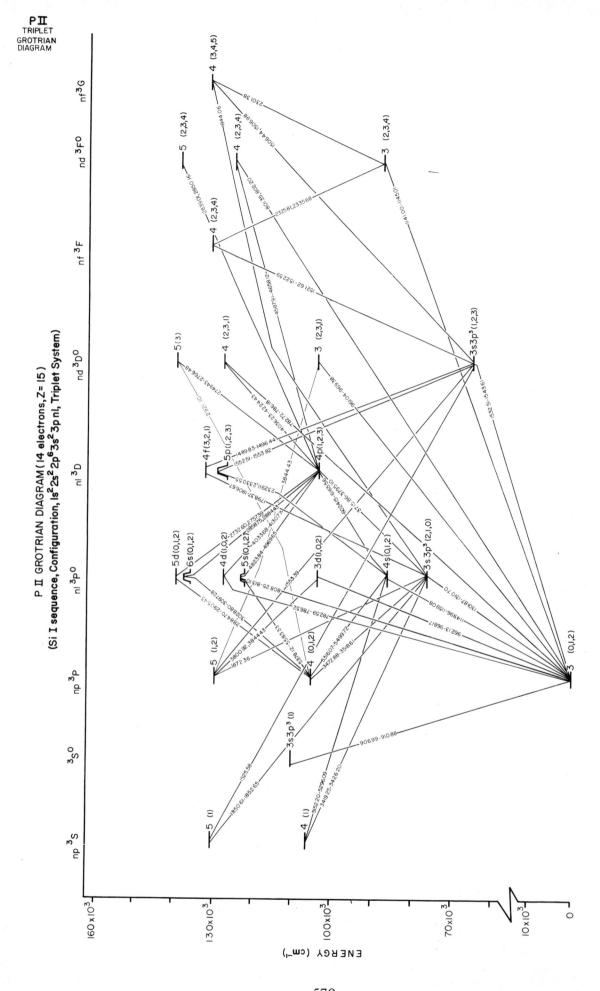

P II
TRIPLET
GROTRIAN
DIAGRAM

P II GROTRIAN DIAGRAM (14 electrons, Z=15)
(Si I sequence, Configuration, $ls^2 2s^2 2p^6 3s^2 3p\,nl$, Triplet System)

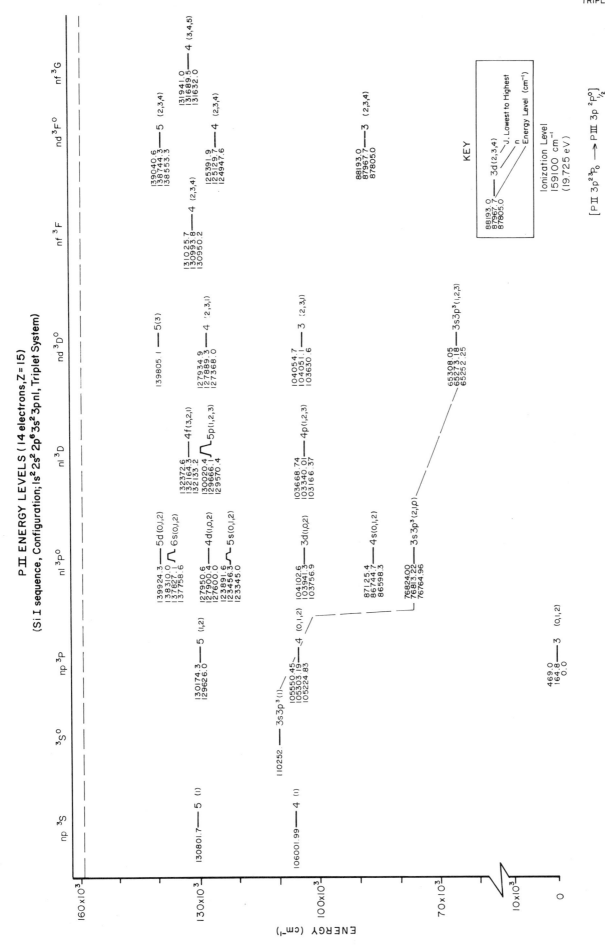

P II ENERGY LEVELS (14 electrons, Z=15)
(Si I sequence, Configuration; 1s² 2s² 2p⁶ 3s² 3pnl, Triplet System)

571

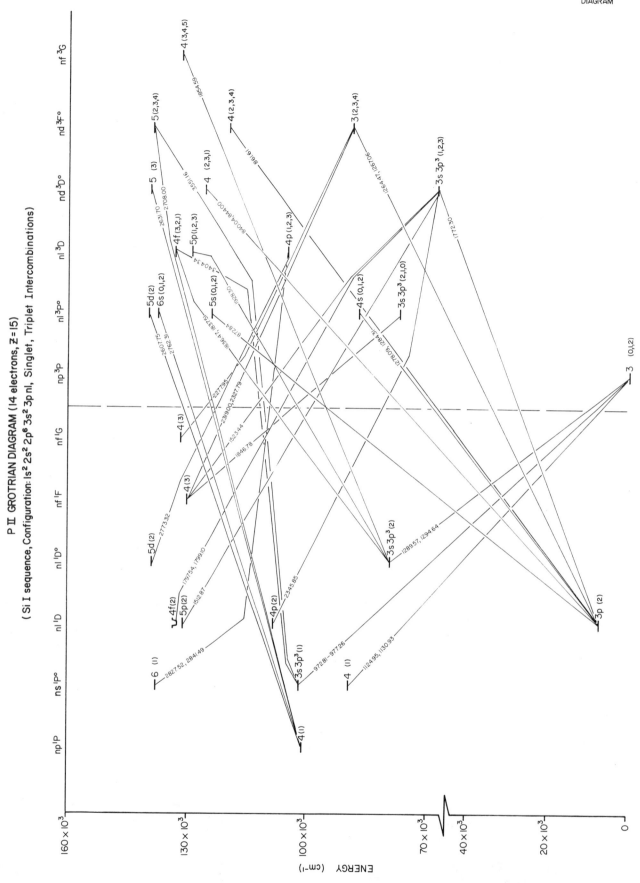

P II GROTRIAN DIAGRAM (14 electrons, Z = 15)

(Si I sequence, Configuration: 1s² 2s² 2p⁶ 3s² 3pnl, Singlet, Triplet Intercombinations)

P II
SINGLET
TRIPLET
INTERCOMBINATION
GROTRIAN
DIAGRAM

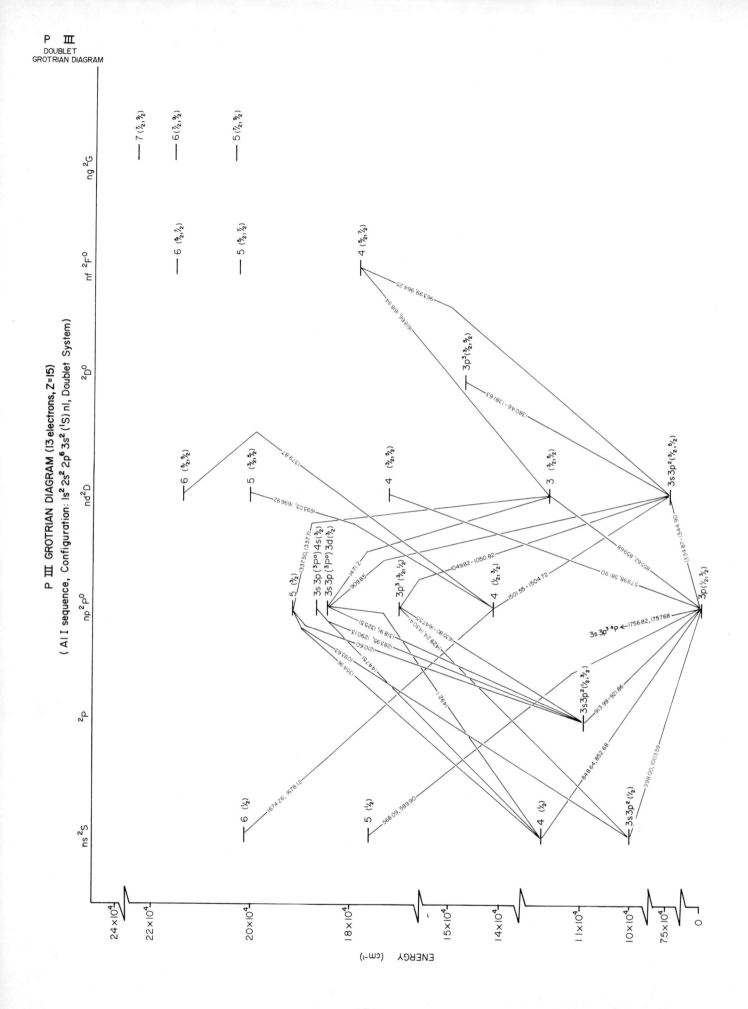

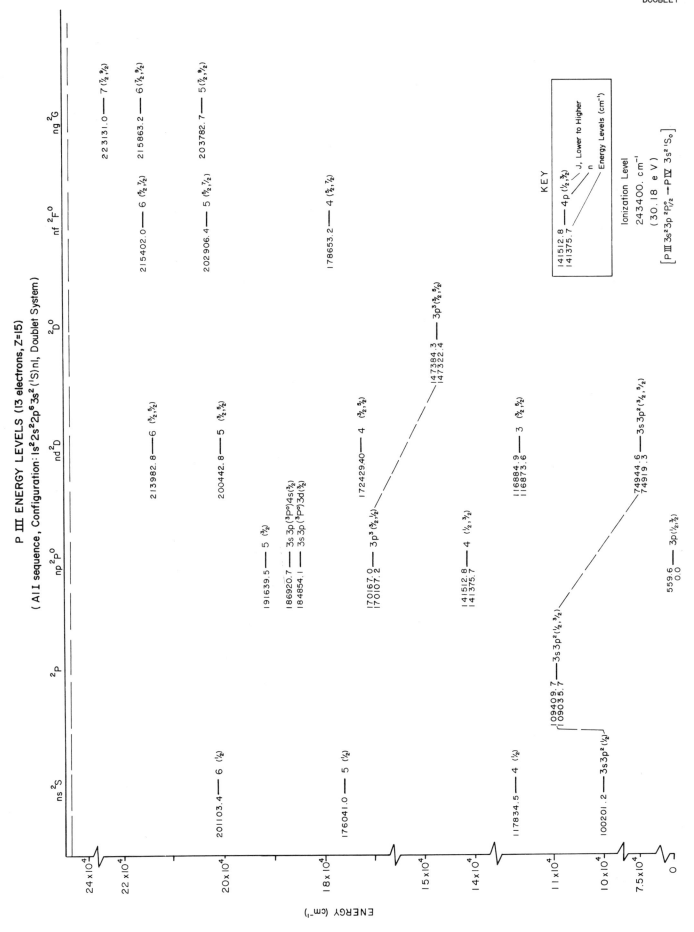

P III ENERGY LEVELS (13 electrons, Z=15)

(Al I sequence, Configuration: 1s² 2s² 2p⁶ 3s² (¹S) nl, Doublet System)

P Ⅲ GROTRIAN DIAGRAM (13 electrons, Z = 15)

(Al I sequence, Configuration: $1s^2\,2s^2\,2p^6\,3s\,3p\,(^3P^o)nl$, Quartet System and Autoionizing States)

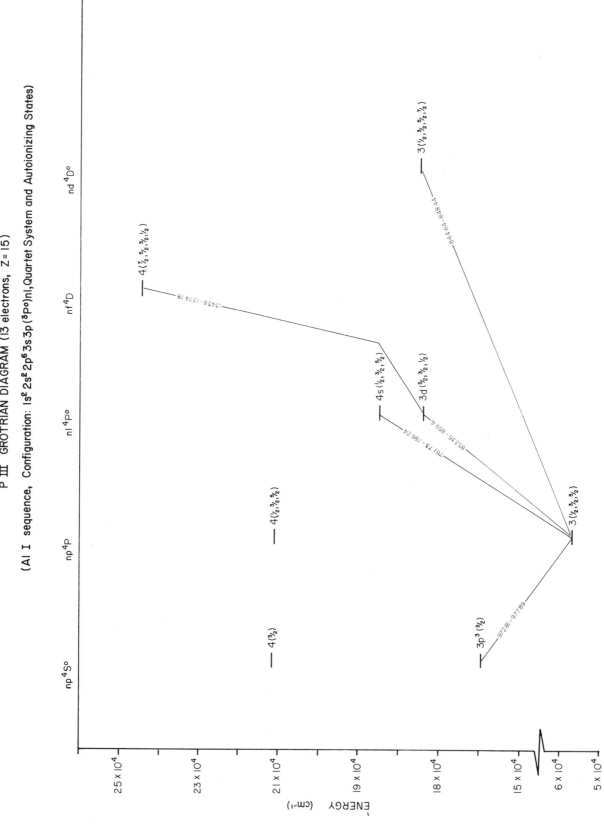

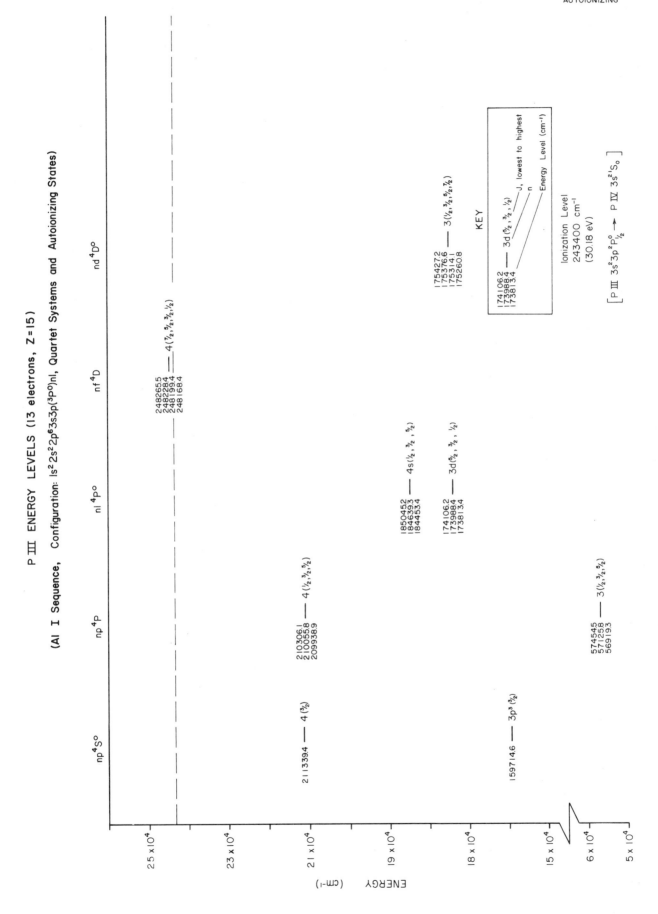

P III ENERGY LEVELS (13 electrons, Z=15)

(Al I Sequence, Configuration: 1s²2s²2p⁶3s3p(³P°)nl, Quartet Systems and Autoionizing States)

np ⁴S°

np ⁴P

nl ⁴P°

nf ⁴D

nd ⁴D°

211339.4 —— 4(³/₂)

159714.6 —— 3p³(³/₂)

210306.1
210055.8 —— 4(½,³/₂,⁵/₂)
209938.9

57454.5
57125.8 —— 3(½,³/₂,⁵/₂)
56919.3

185045.2
184669.3 —— 4s(½,³/₂,⁵/₂)
184453.4

174106.2
173988.4 —— 3d(⁵/₂,³/₂,½)
173813.4

248265.5
248228.4 —— 4(⁷/₂,⁵/₂,³/₂,½)
248199.4
248168.4

175427.2
175376.6 —— 3(½,³/₂,⁵/₂,⁷/₂)
175314.1
175260.8

KEY

174106.2
173988.4 —— 3d(⁵/₂,³/₂,½)
173813.4

3(½,³/₂,⁵/₂,½) J, lowest to highest
n
Energy Level (cm⁻¹)

Ionization Level
243400 cm⁻¹
(30.18 eV)

[P III 3s²3p²P°₁/₂ ⟶ P IV 3s²¹S₀]

ENERGY (cm⁻¹)

25 x10⁴
23 x10⁴
21 x10⁴
19 x10⁴
18 x10⁴
15 x10⁴
6 x10⁴
5 x10⁴

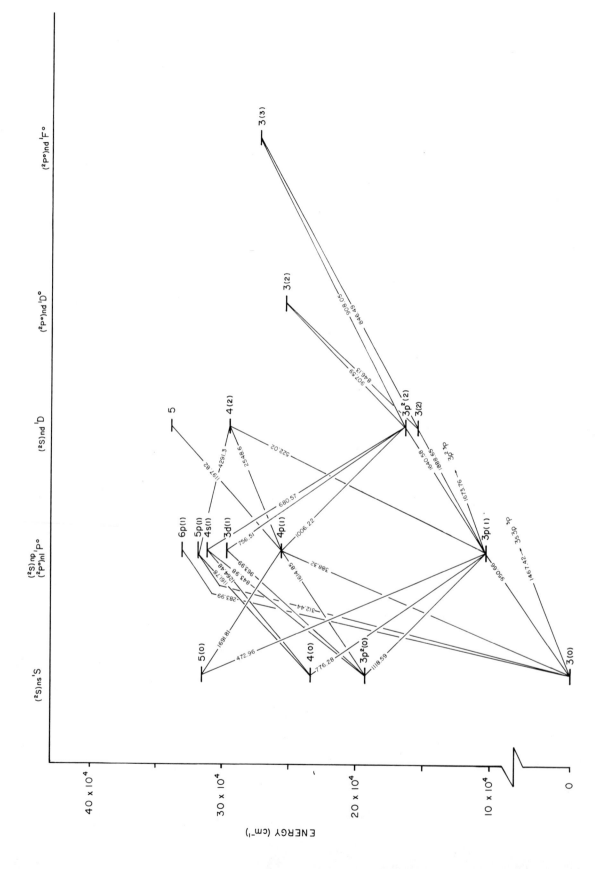

P IV
SINGLET
GROTRIAN
DIAGRAM

P IV GROTRIAN DIAGRAM (12 electrons, Z = 15)

(Mg I sequence, Configuration: Is² 2s² 2p⁶ 3l (²L)nl, Singlet System)

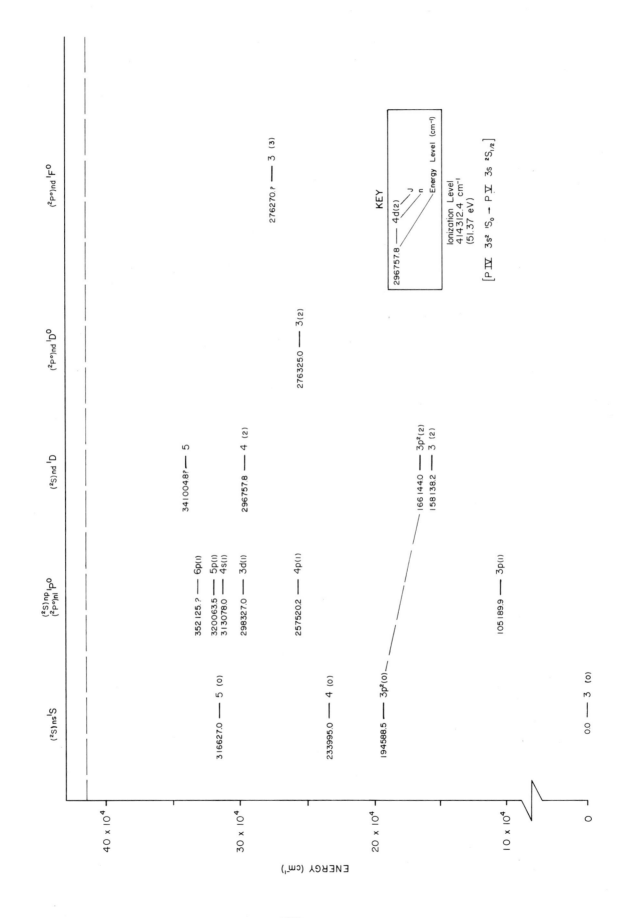

P IV ENERGY LEVELS (12 electrons Z = 15)
(Mg I Sequence, Configuration: $1s^2 2s^2 2p^6 3l$ (^{2}L)nl, Singlet System)

P IV GROTRIAN DIAGRAM (12 electrons, Z = 15)
((Mg I sequence, Configuration: $1s^2 2s^2 2p^6 3s(^2S)nl$, $1s^2 2s^2 2p^6 3p(P^o)nl$, Triplet System)

P IV
TRIPLET
GROTRIAN
DIAGRAM

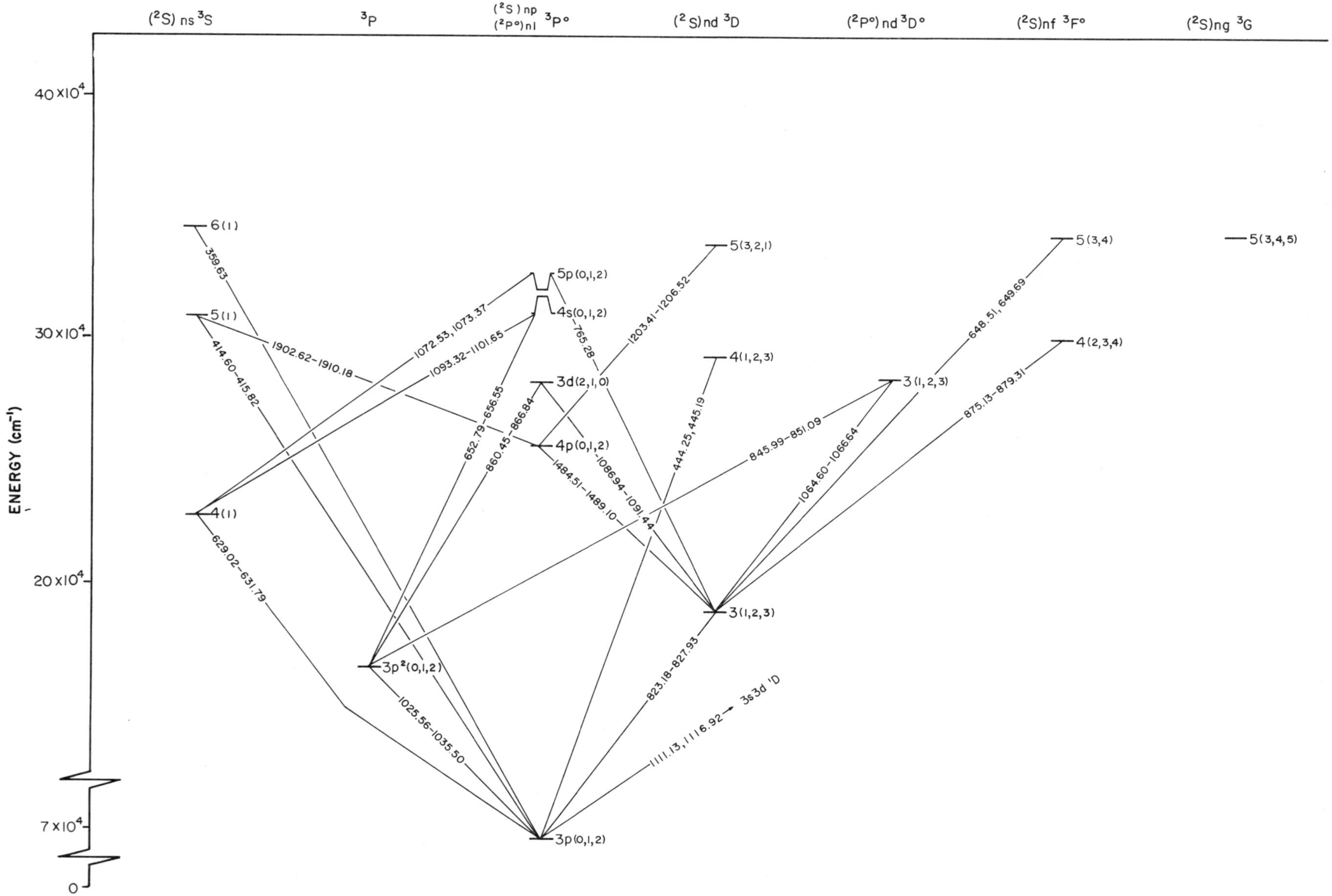

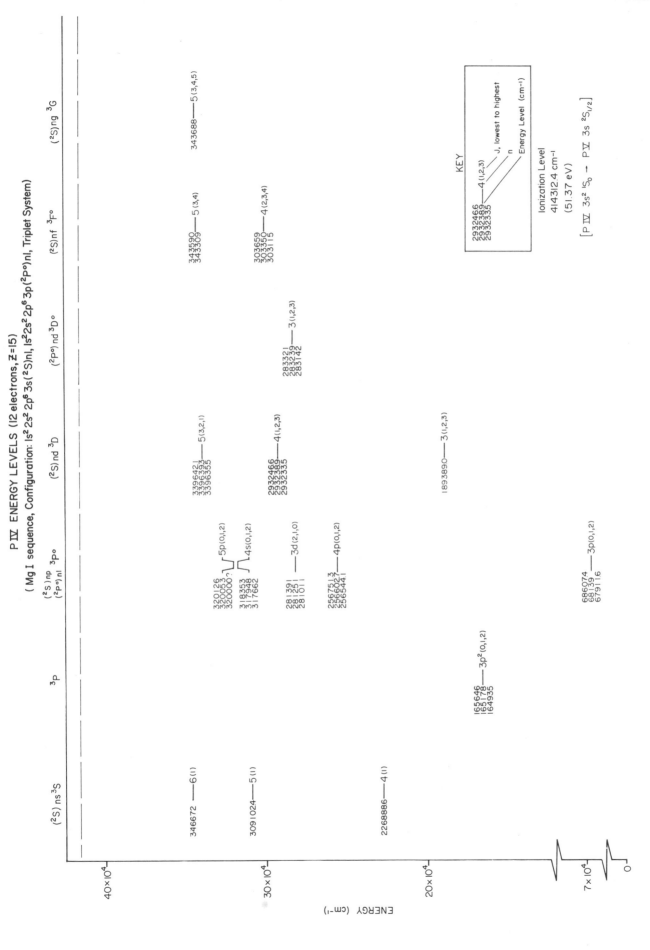

P IV ENERGY LEVELS (12 electrons, Z=15)

(Mg I sequence, Configuration: 1s² 2s² 2p⁶ 3s(²S)nl, 1s²2s² 2p⁶ 3p(²P°)nl, Triplet System)

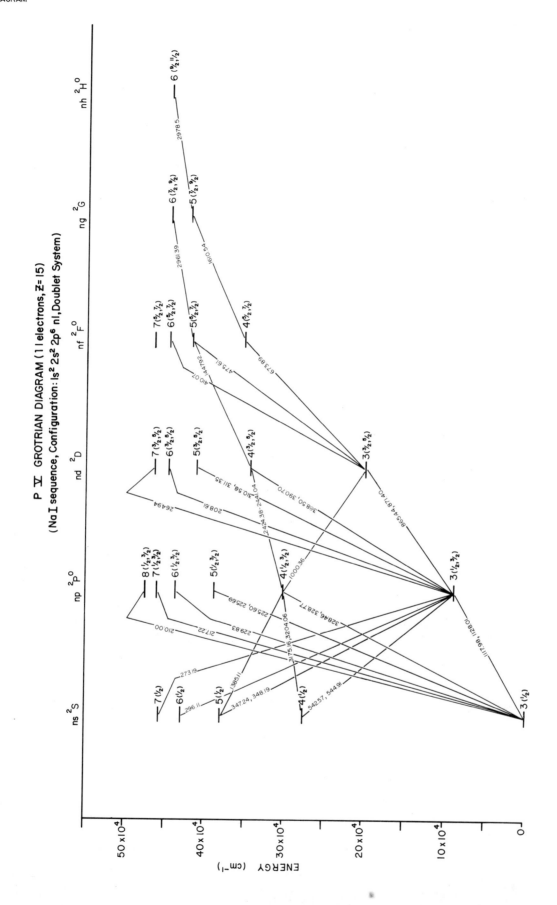

P V GROTRIAN DIAGRAM (11 electrons, Z=15)
(Na I sequence, Configuration: 1s² 2s² 2p⁶ nl, Doublet System)

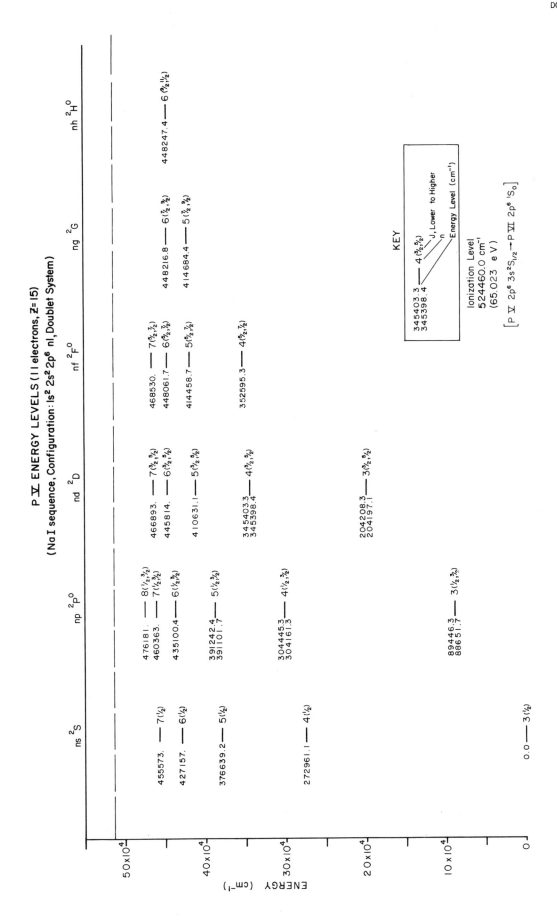

P V ENERGY LEVELS (11 electrons, Z=15)
(Na I sequence, Configuration: 1s² 2s² 2p⁶ nl, Doublet System)

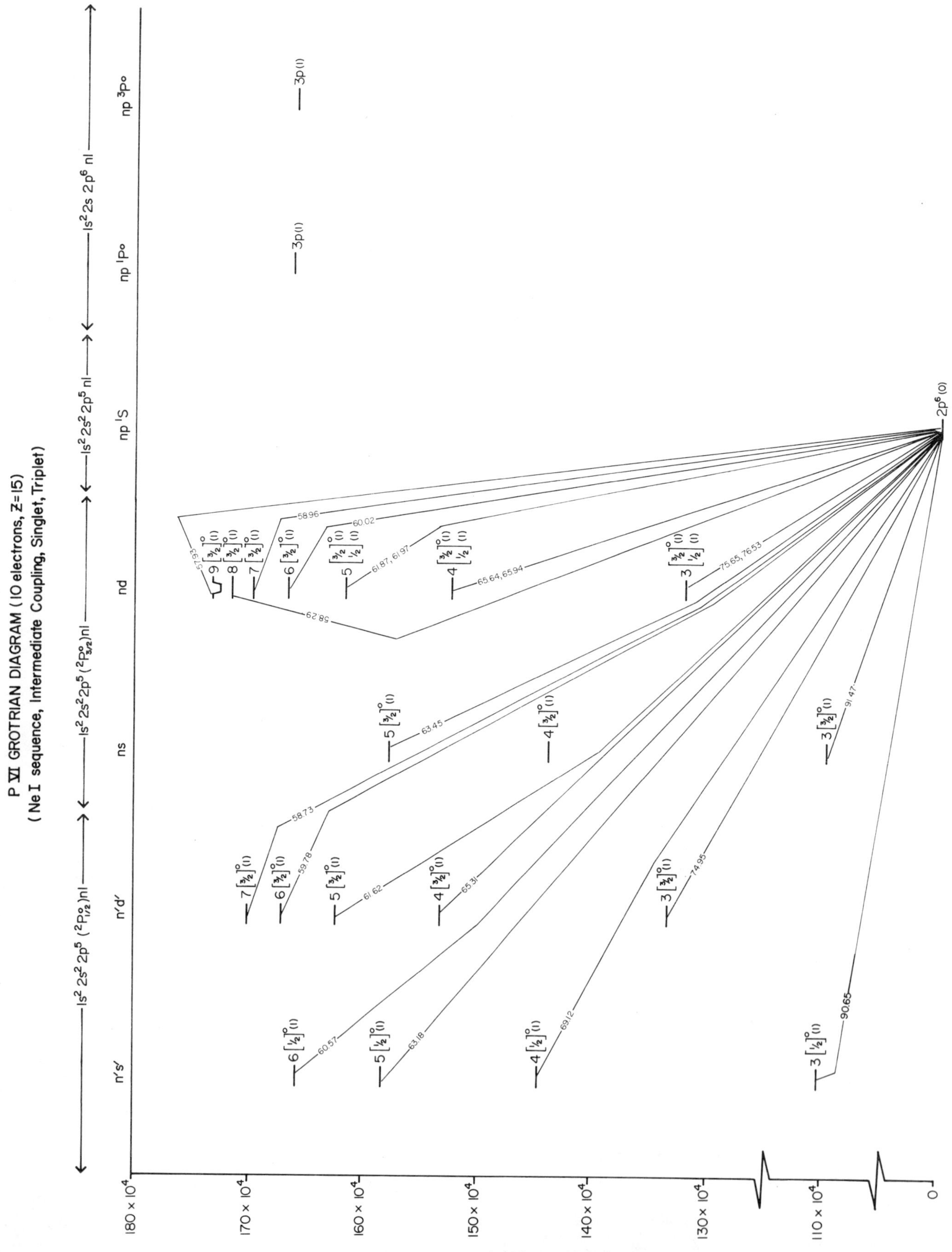

P VI GROTRIAN DIAGRAM (10 electrons, Z=15)
(Ne I sequence, Intermediate Coupling, Singlet, Triplet)

P VI
SINGLET
TRIPLET
INTERMEDIATE COUPLING
GROTRIAN DIAGRAM

584

P VI ENERGY LEVELS (10 electrons, Z=15)
(Ne I sequence, Intermediate Coupling, Singlet, Triplet)

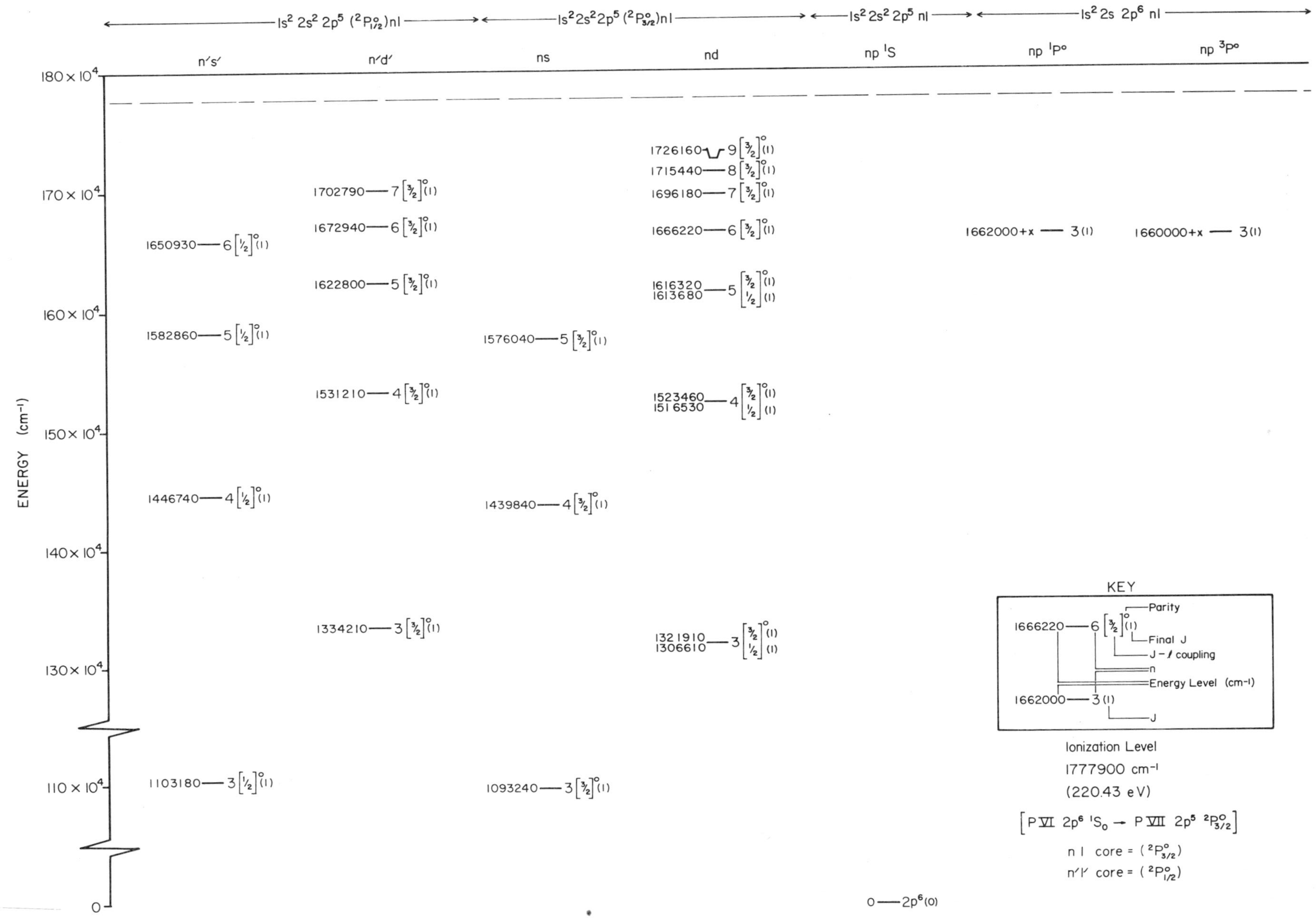

ENERGY (cm⁻¹)

Column headings:

$1s^2 2s^2 2p^5 (^2P^o_{1/2}) nl$ — $1s^2 2s 2p^5 (^2P^o_{3/2}) nl$ — $1s^2 2s^2 2p^5 nl$ — $1s^2 2s 2p^6 nl$

n's' — n'd' — ns — nd — np ¹S — np ¹P° — np ³P°

KEY

Parity
Final J
J–ℓ coupling
n
Energy Level (cm⁻¹)
J

$166220 \quad 6\left[\tfrac{3}{2}\right]^o_{(1)}$

$162000 \quad 3_{(1)}$

Ionization Level
1777900 cm⁻¹
(220.43 eV)

$\left[P\,VII\ 2p^6\ ^1S_0 \rightarrow P\,VII\ 2p^5\ ^2P^o_{3/2} \right]$

nℓ core = ($^2P^o_{3/2}$)
n'ℓ' core = ($^2P^o_{1/2}$)

Energy levels

n's':
1650930 — 6 $\left[\tfrac{1}{2}\right]^o_{(1)}$
1582860 — 5 $\left[\tfrac{1}{2}\right]^o_{(1)}$
1446740 — 4 $\left[\tfrac{1}{2}\right]^o_{(1)}$
1103180 — 3 $\left[\tfrac{1}{2}\right]_{(1)}$

n'd':
1702790 — 7 $\left[\tfrac{3}{2}\right]^o_{(1)}$
1672940 — 6 $\left[\tfrac{3}{2}\right]_{(1)}$
1622800 — 5 $\left[\tfrac{3}{2}\right]^o_{(1)}$
1531210 — 4 $\left[\tfrac{3}{2}\right]^o_{(1)}$
1334210 — 3 $\left[\tfrac{3}{2}\right]_{(1)}$

ns:
1576040 — 5 $\left[\tfrac{3}{2}\right]^o_{(1)}$
1439840 — 4 $\left[\tfrac{3}{2}\right]^o_{(1)}$
1093240 — 3 $\left[\tfrac{3}{2}\right]^o_{(1)}$

nd:
1726160 — 9 $\left[\tfrac{3}{2}\right]^o_{(1)}$
1715440 — 8 $\left[\tfrac{3}{2}\right]^o_{(1)}$
1696180 — 7 $\left[\tfrac{3}{2}\right]_{(1)}$
166220 — 6 $\left[\tfrac{3}{2}\right]^o_{(1)}$
1616320 — 5 $\left[\tfrac{3}{2}\right]_{(1)}$
1613680 — $\left[\tfrac{1}{2}\right]_{(1)}$
1523460 — 4 $\left[\tfrac{3}{2}\right]^o_{(1)}$
1516530 — $\left[\tfrac{1}{2}\right]_{(1)}$
1321910 — 3 $\left[\tfrac{3}{2}\right]^o_{(1)}$
1306610 — $\left[\tfrac{1}{2}\right]_{(1)}$

np ¹P°:
1662000+x — 3 (1)

np ³P°:
1660000+x — 3 (1)

np ¹S (baseline):
0 — 2p⁶ (0)

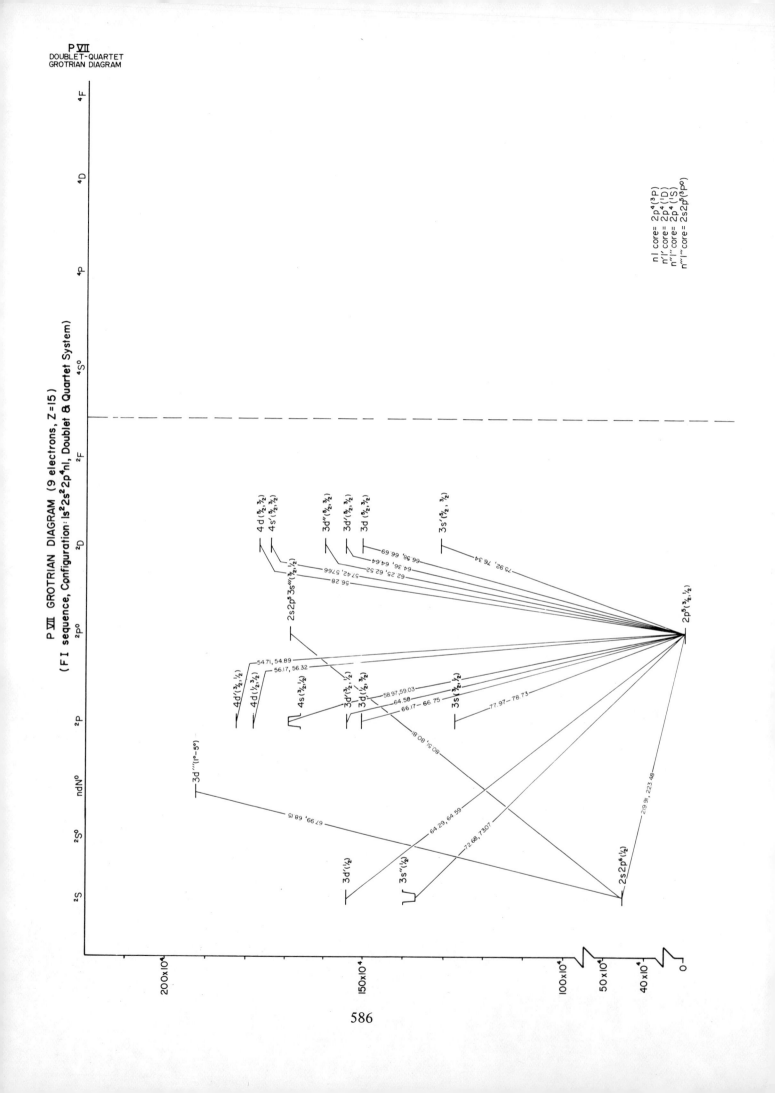

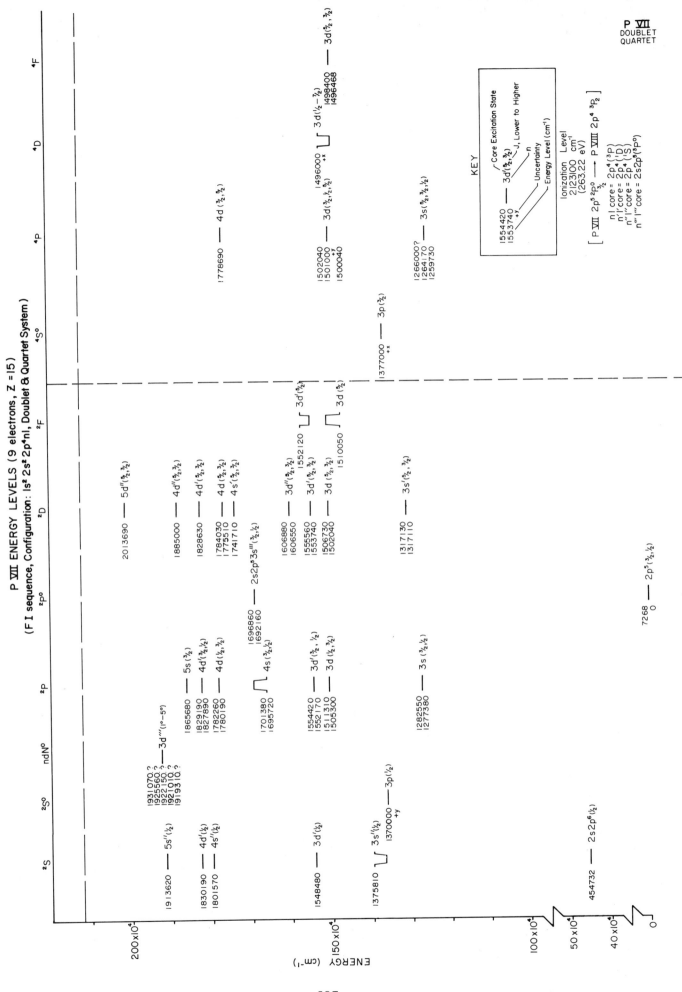

P VII ENERGY LEVELS (9 electrons, Z =15)
(F I sequence, Configuration: 1s² 2s² 2p⁴nl, Doublet & Quartet System)

587

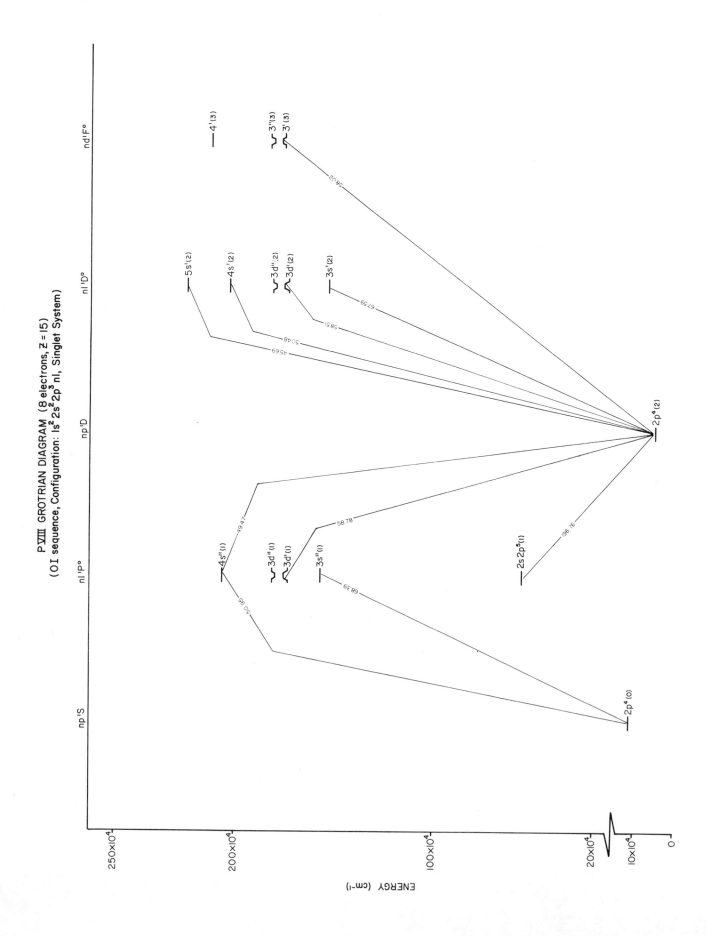

P VIII
SINGLET
GROTRIAN DIAGRAM

P VIII GROTRIAN DIAGRAM (8 electrons, Z = 15)
(OI sequence, Configuration: ls² 2s² 2p³ nl, Singlet System)

588

P VIII ENERGY LEVELS (8 electrons, Z =15)
(O I Sequence, Configuration: 1s²2s²2p³nl, Singlet System)

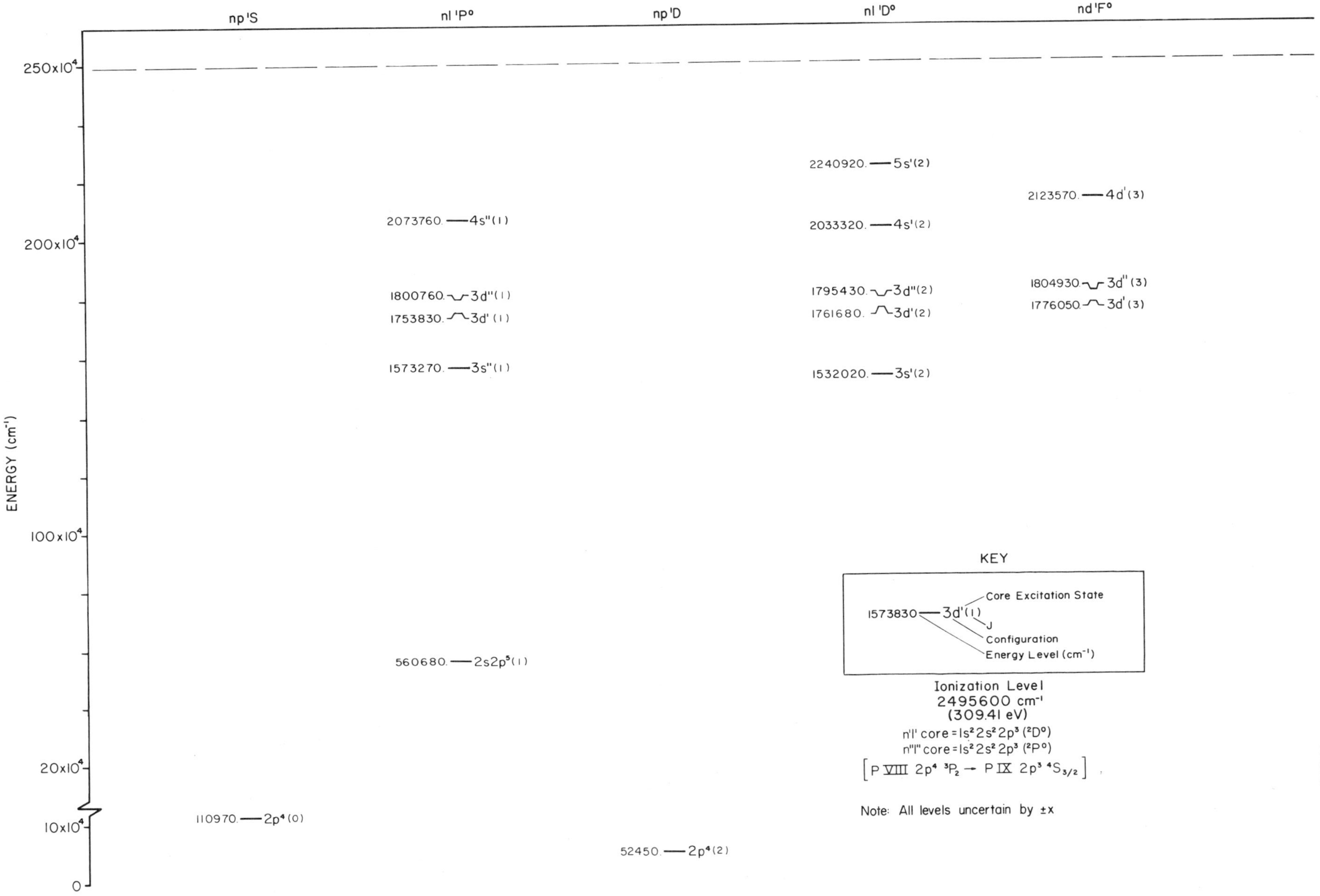

P VIII
SINGLET

589

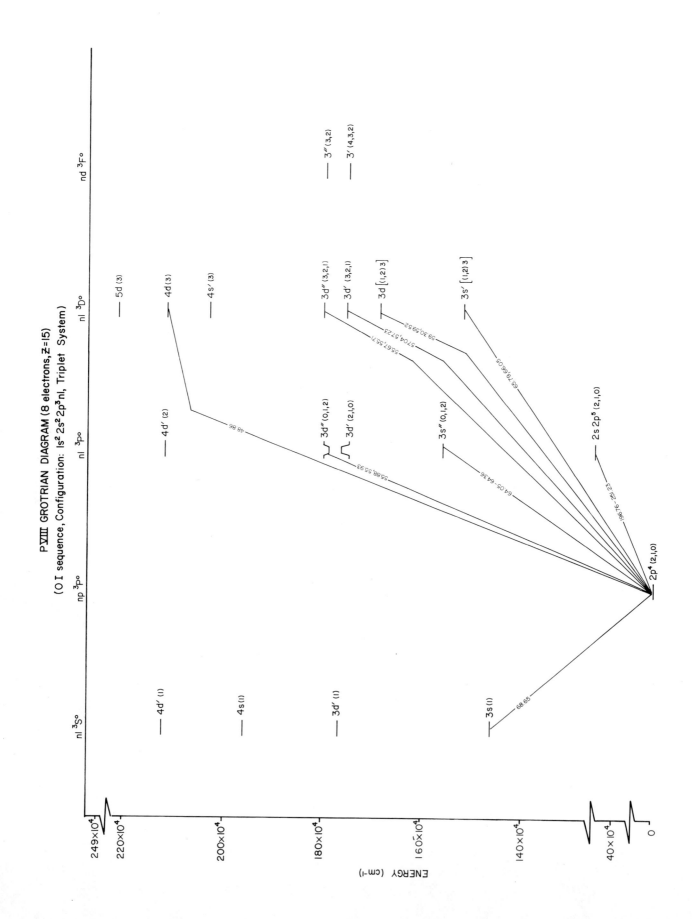

P VIII GROTRIAN DIAGRAM (8 electrons, Z=15)

(O I sequence, Configuration: 1s² 2s² 2p³ nl, Triplet System)

ENERGY (cm⁻¹)

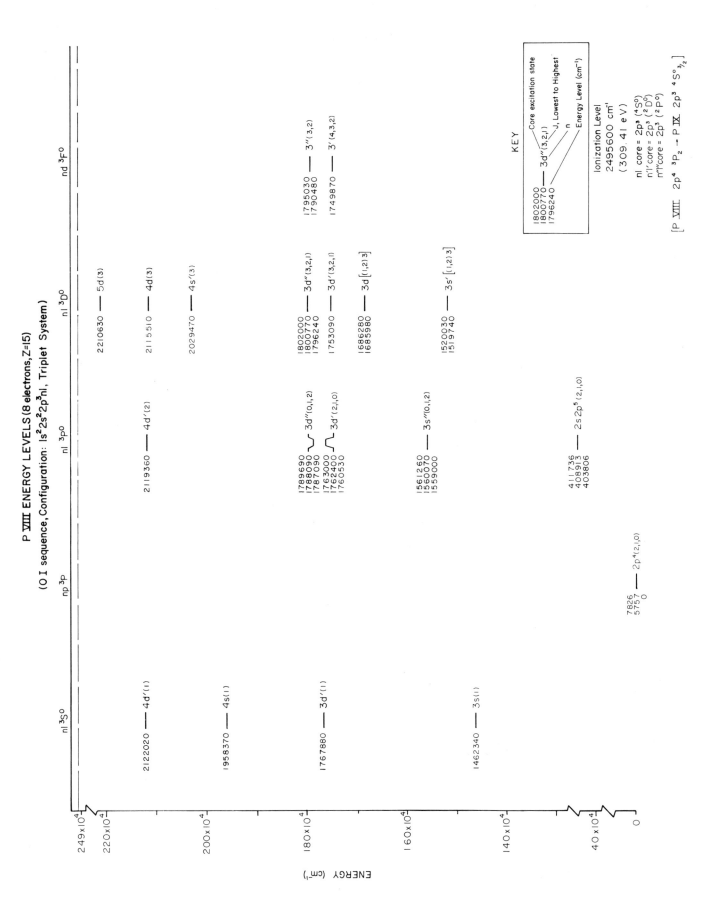

P VIII ENERGY LEVELS (8 electrons, Z=15)

(O I sequence, Configuration: 1s²2s²2p³nl, Triplet System)

P VIII
TRIPLET

591

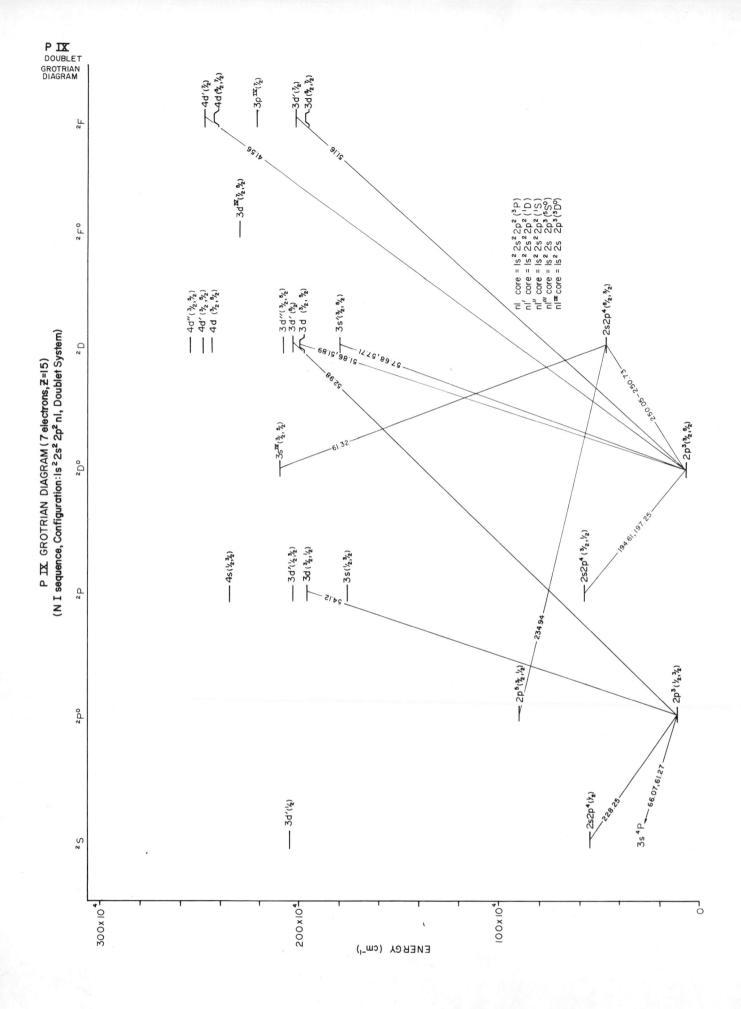

P IX
DOUBLET
GROTRIAN
DIAGRAM

P IX GROTRIAN DIAGRAM (7 electrons, Z=15)
(N I sequence, Configuration: 1s² 2s² 2p² nl, Doublet System)

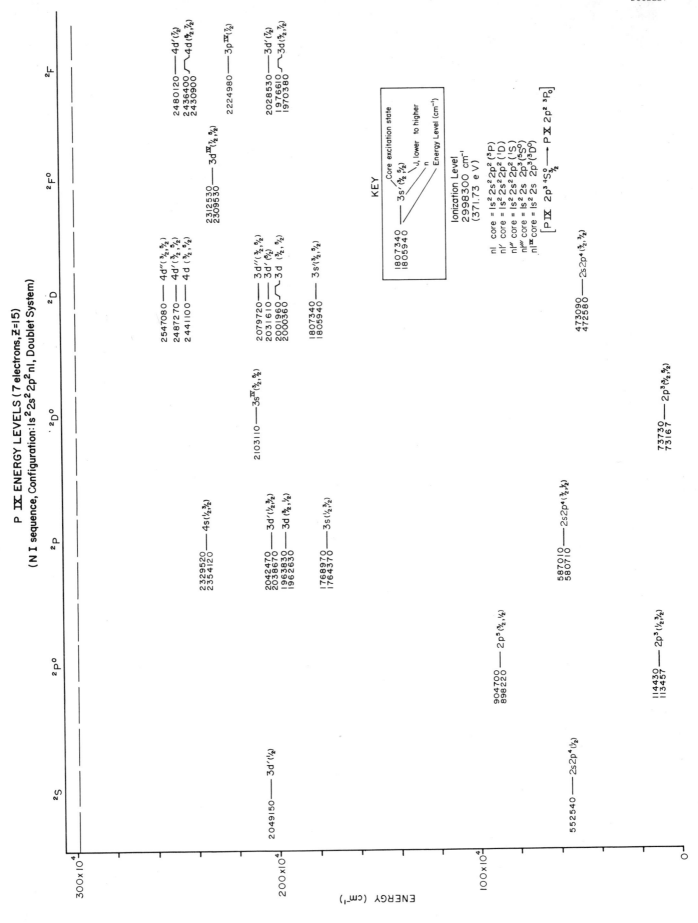

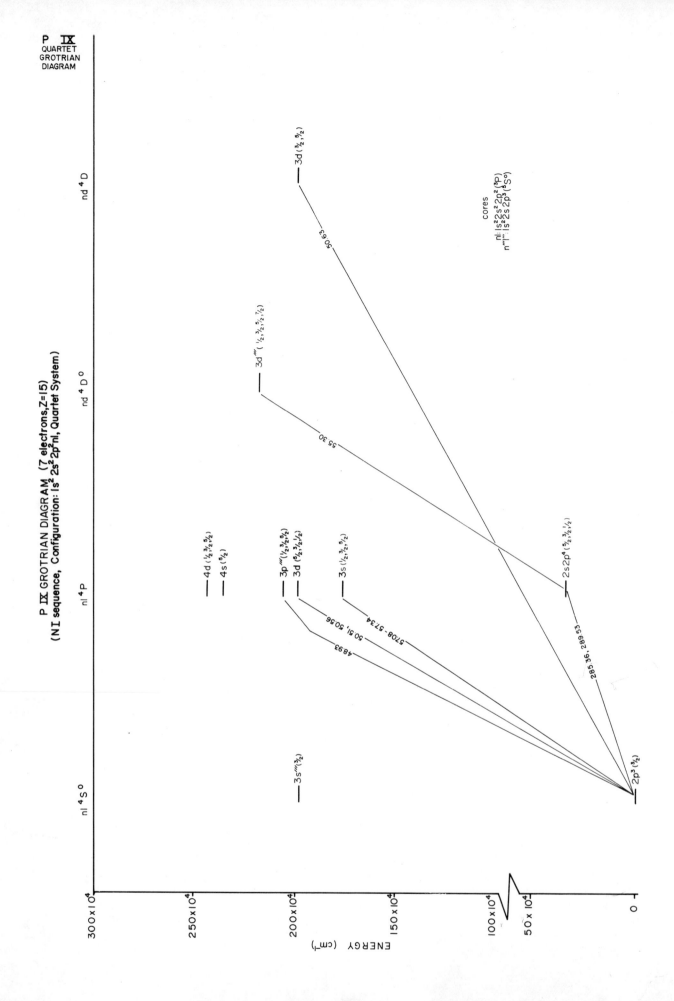

P IX GROTRIAN DIAGRAM (7 electrons, Z=15)
(NI sequence, Configuration: 1s² 2s² 2p² nl, Quartet System)

P IX
QUARTET
GROTRIAN
DIAGRAM

594

P IX ENERGY LEVELS (7 electrons, Z=15)
(NI sequence, Configuration: $1s^2\ 2s^2\ 2p^2 nl$, Quartet System)

KEY

Core exitation state

$3d'''' \left(\tfrac{5}{2}, \tfrac{3}{2}, \tfrac{1}{2}\right)$ — J, lowest to highest
n
Energy Level (cm^{-1})

1980870.
1979750.
1977830.

Ionization Level
2998300 cm^{-1}
(371.73 eV)

cores: nl : $1s^2\ 2s^2\ 2p^2(^3P)$
n'''' : $1s^2\ 2s\ 2p^3(^5S^o)$

$\left[P\,IX\ 2p^3\ {}^4S^o\ {}_{\tfrac{3}{2}} \rightarrow P\,X\ 2p^2\ {}^3P_0 \right]$

nd ^{4}D

1975970. —— 3d $\left(\tfrac{3}{2}, \tfrac{5}{2}\right)$
1973870.

nd ^{4}D^o

2161390. —— 3d''' $\left(\tfrac{1}{2}, \tfrac{3}{2}, \tfrac{5}{2}, \tfrac{7}{2}\right)$

nl ^{4}P

2435220. —— 4d $\left(\tfrac{1}{2}, \tfrac{3}{2}, \tfrac{5}{2}\right)$
2354100. —— 4s $\left(\tfrac{5}{2}\right)$

2043950. —— 3p''' $\left(\tfrac{1}{2}, \tfrac{3}{2}, \tfrac{5}{2}\right)$
1980870. —— 3d $\left(\tfrac{5}{2}, \tfrac{3}{2}, \tfrac{1}{2}\right)$
1977830.

1751850. —— 3s $\left(\tfrac{1}{2}, \tfrac{3}{2}, \tfrac{5}{2}\right)$
1746250.
1744000.

353050. —— 2s2p^4 $\left(\tfrac{5}{2}, \tfrac{3}{2}, \tfrac{1}{2}\right)$
350440.
345390.

nl ^{4}S^o

1965970. —— 3s''' $\left(\tfrac{3}{2}\right)$

0 —— 2p^3 $\left(\tfrac{3}{2}\right)$

ENERGY (cm^{-1})

300 x 10^4
250 x 10^4
200 x 10^4
150 x 10^4
100 x 10^4
50 x 10^4
0

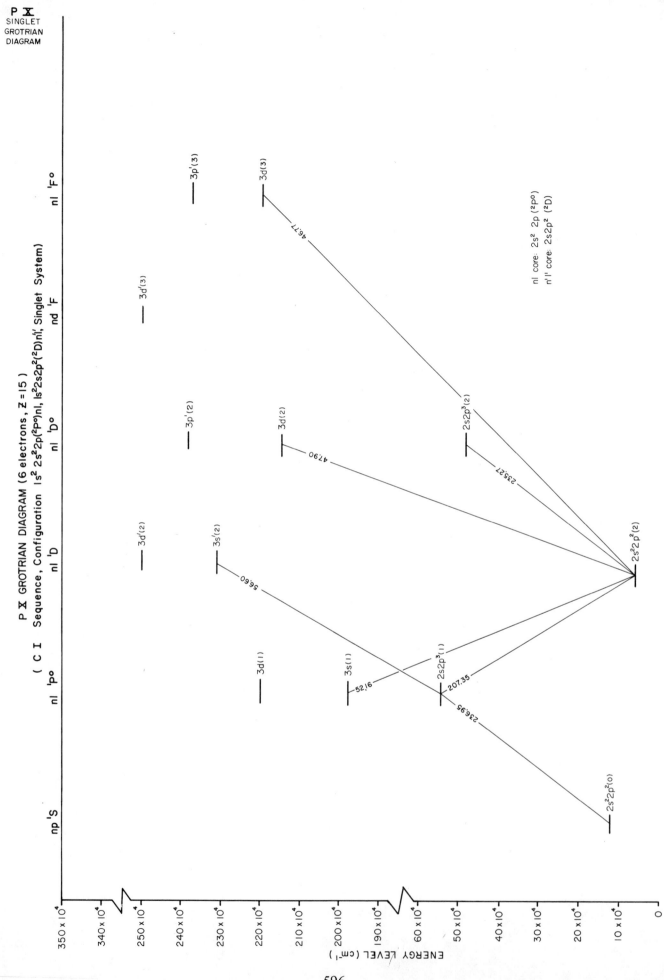

P Ⅹ SINGLET GROTRIAN DIAGRAM

P Ⅹ GROTRIAN DIAGRAM (6 electrons, Z=15)

(C I Sequence, Configuration 1s² 2s²2p(²P°)nl, 1s²2s2p²(²D)nl', Singlet System)

nl core: 2s² 2p (²P°)
nl' core: 2s2p² (²D)

ENERGY LEVEL (cm⁻¹)

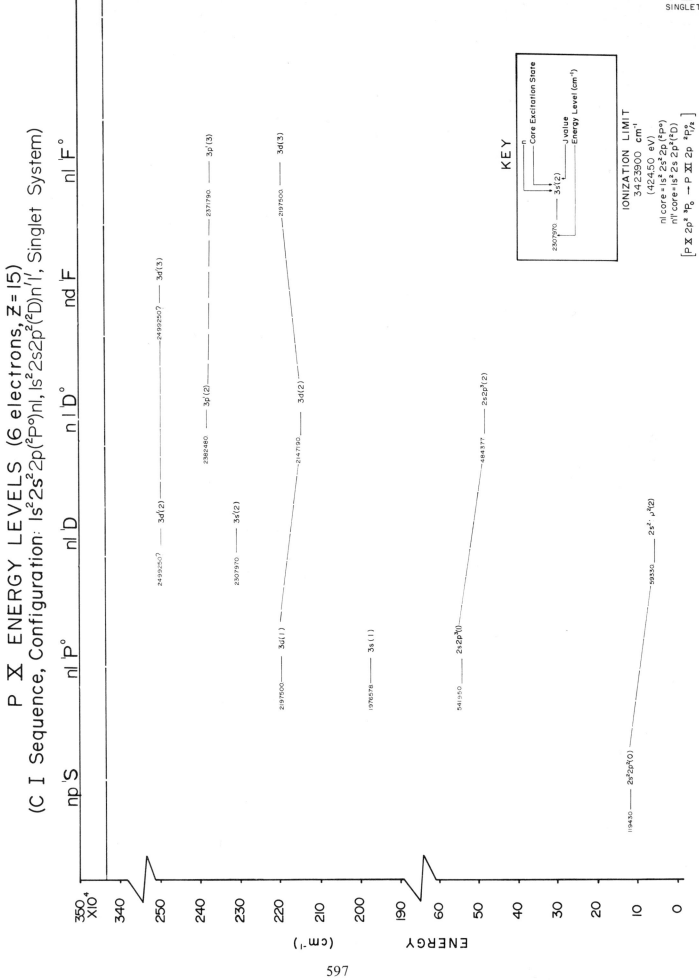

P X ENERGY LEVELS (6 electrons, $\bar{z}$=15)
(C I Sequence, Configuration: $1s^2 2s^2 2p(^2P°)nl$, $1s^2 2s 2p^2(^2D)n'l'$, Singlet System)

P X
SINGLET

KEY

n
Core Excitation State
J value
Energy Level (cm⁻¹)

nl core = $1s^2 2s^2 2p(^2P°)$
n'l' core = $1s^2 2s 2p^2(^2D)$

IONIZATION LIMIT
3423900 cm⁻¹
(424.50 eV)

$[P X 2p^2 \, ^3P_0 \rightarrow P XI 2p \, ^2P°_{1/2}]$

597

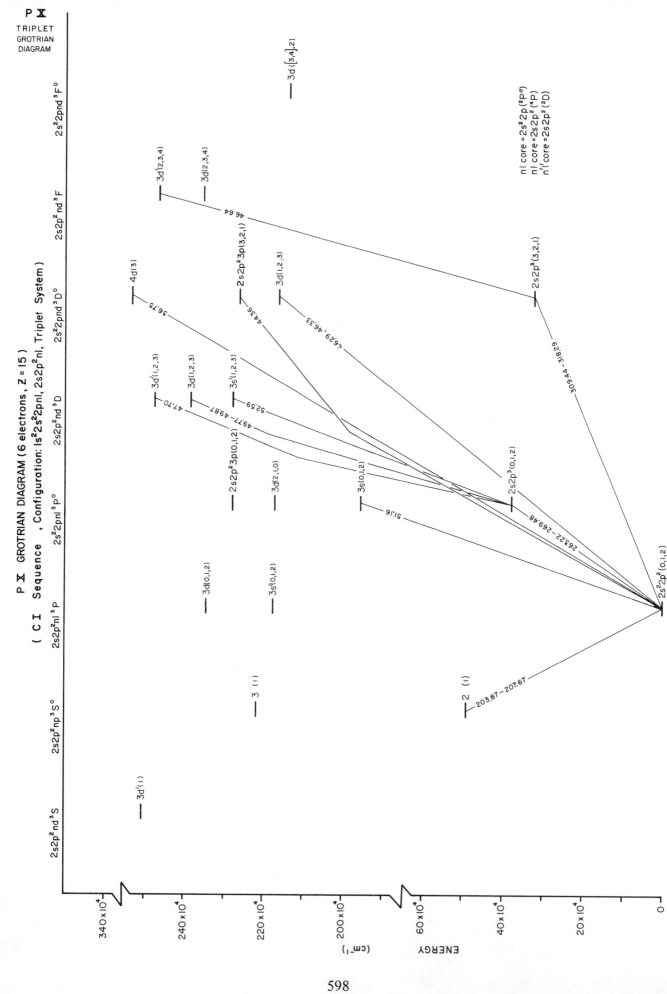

P XⅤ ENERGY LEVELS (6 electrons, Z=15)

(CI Sequence, Configuration: ls²2s²2pnl, 2s2p²nl, Triplet System)

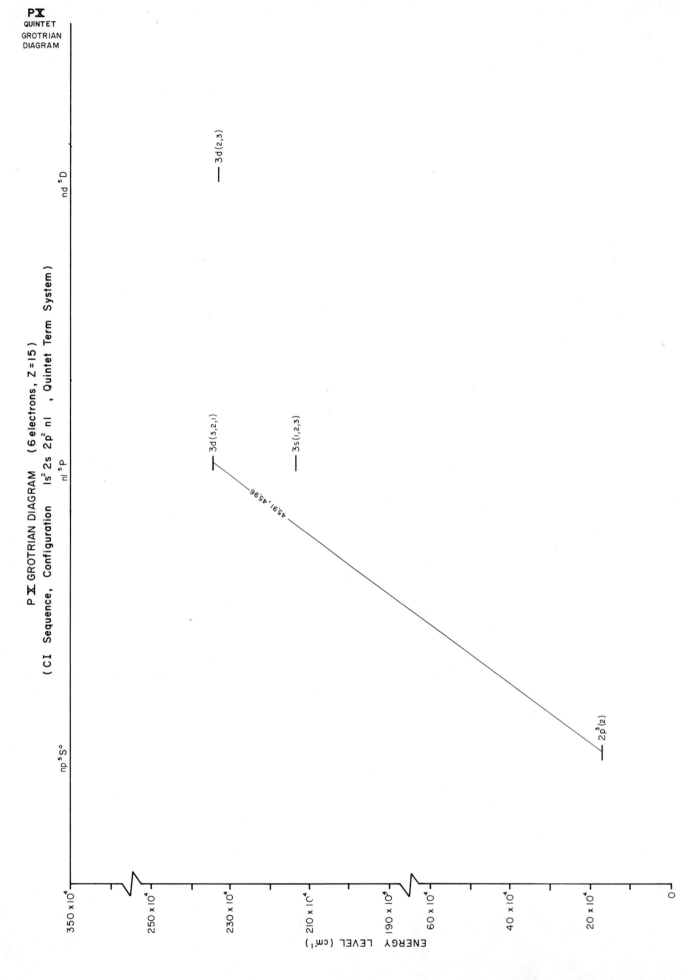

P X QUINTET GROTRIAN DIAGRAM

P X GROTRIAN DIAGRAM (6 electrons, Z = 15)
(CI Sequence, Configuration 1s² 2s 2p² nl , Quintet Term System)

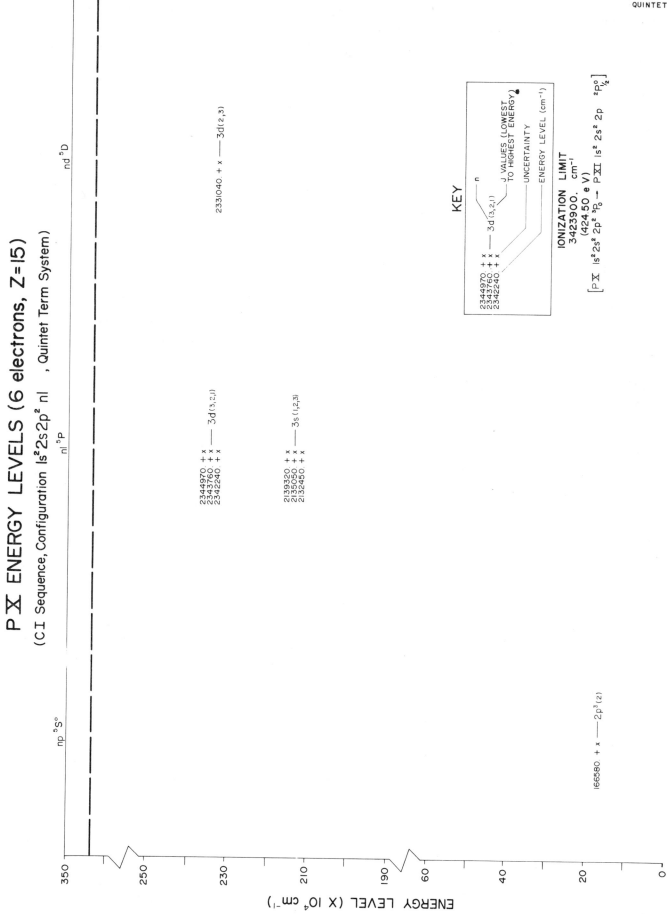

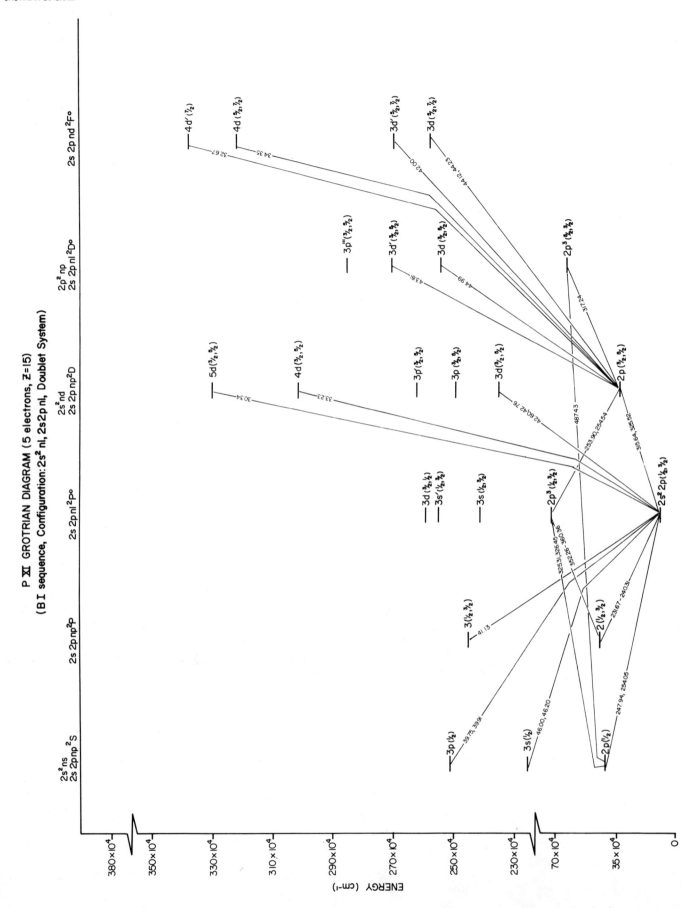

P XI
DOUBLET
GROTRIAN DIAGRAM

P XI GROTRIAN DIAGRAM (5 electrons, Z=15)
(B I sequence, Configuration: 2s² nl, 2s2pnl, Doublet System)

ENERGY (cm⁻¹)

P XI ENERGY LEVELS (5 electrons, Z = 15)
(B I sequence, Configuration: 2s² nl, 2s2pnl, Doublet System)

P XI
DOUBLET

P XI GROTRIAN DIAGRAM (5 electrons, Z=15)
(B I sequence, Configuration: 1s²2s2pnl, 2p²nl, Quartet System)

P XI ENERGY LEVELS (5 electrons, Z = 15)

(B I sequence, Configuration: $1s^2\,2s\,2p\,nl$, Quartet System)

Ionization Level
3868100 cm⁻¹
(479.57 eV)

n l core = 2s 2p ($^3P^o$)

n″ l″ core = $2p^2$ (^3P)

[P XI $2s^2\,2p\,^2P_{1/2}$ → P XII $2s^2\,^1S_0$]

* See Kelly

P XI
QUARTET

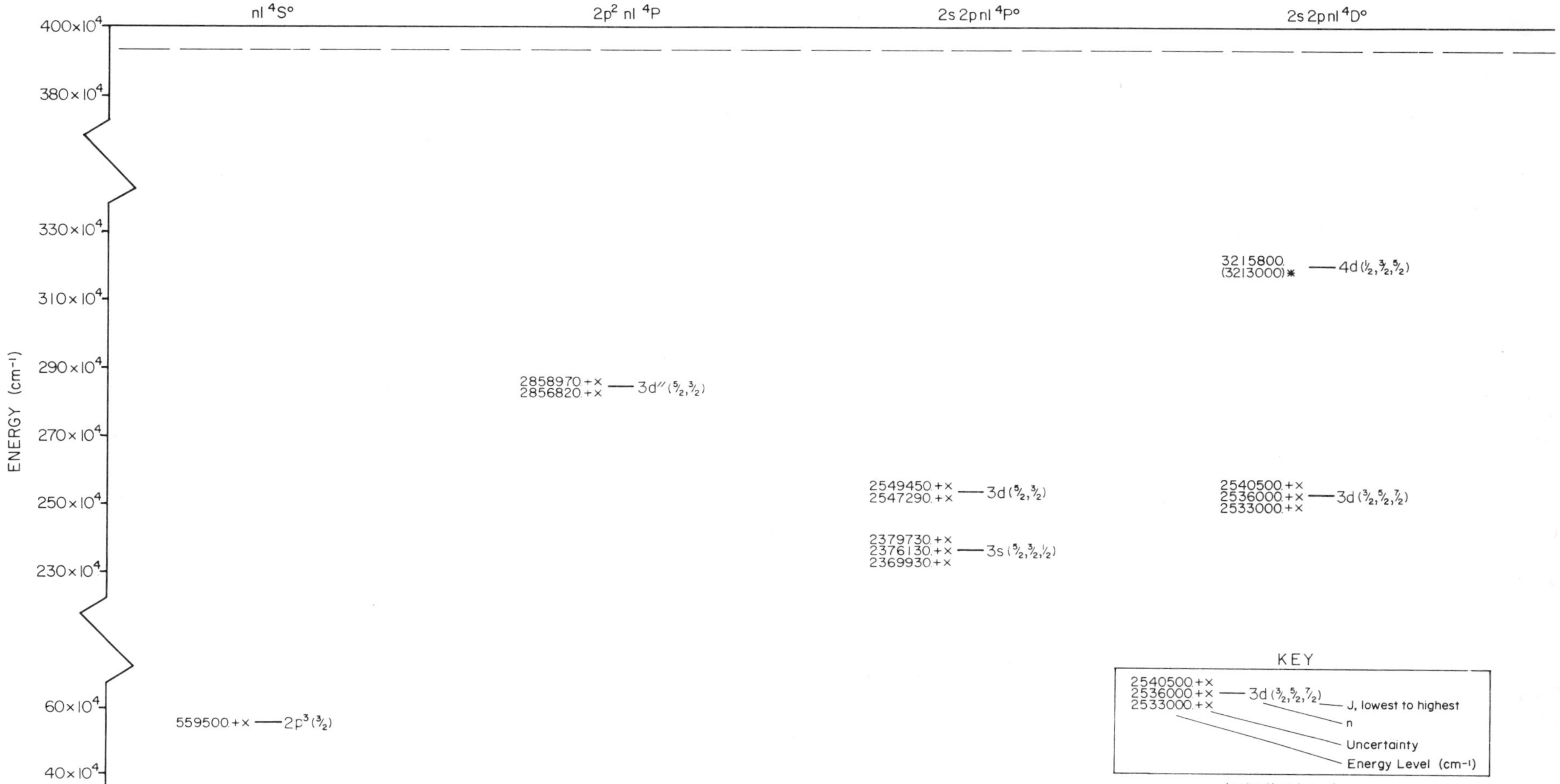

P XII GROTRIAN DIAGRAM (4 electrons, Z = 15)
(Be I sequence, Configuration: $1s^2$ nl n'l', Singlet & Triplet System)

P XII
SINGLET
TRIPLET
GROTRIAN DIAGRAM

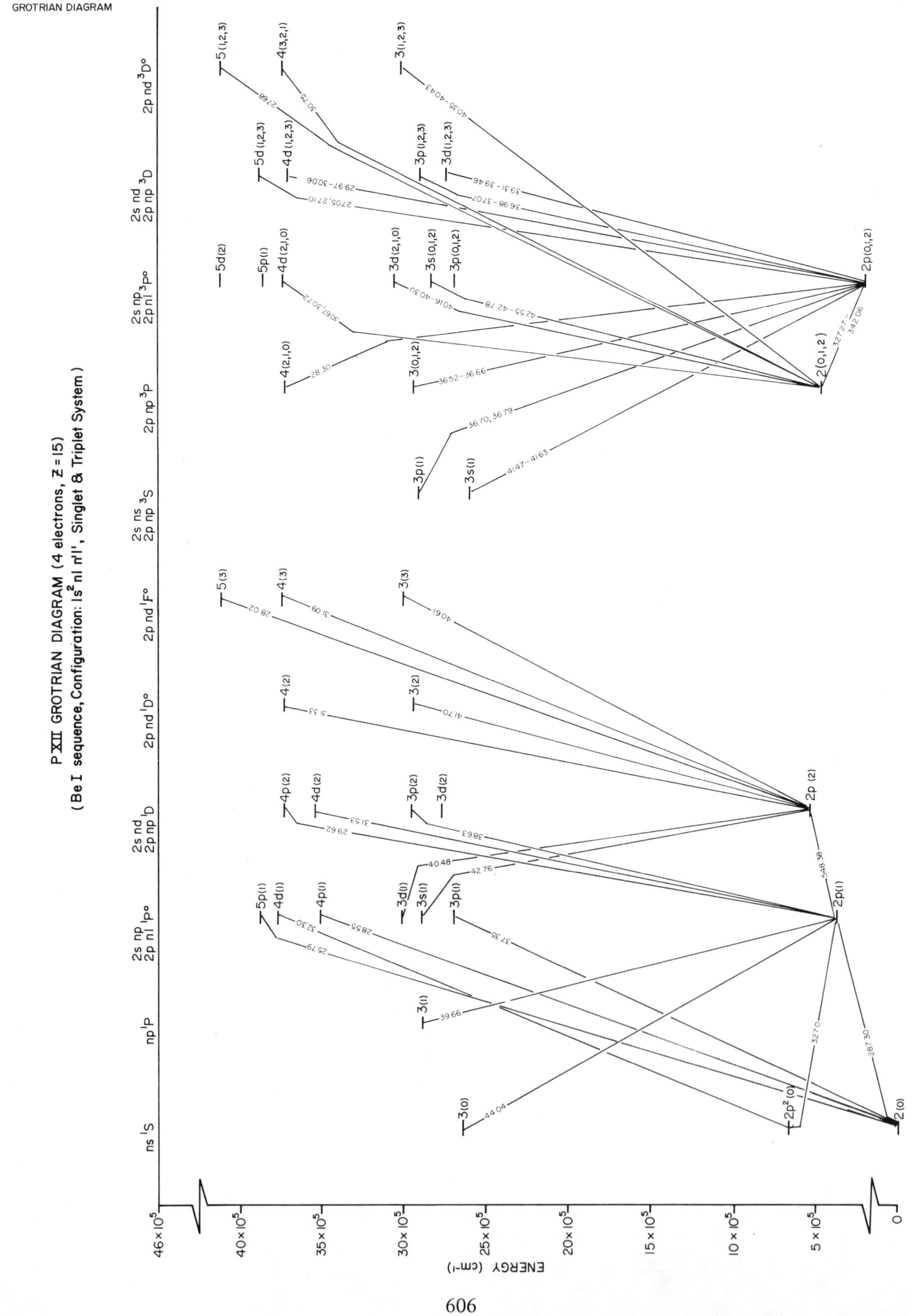

ENERGY (cm⁻¹)

606

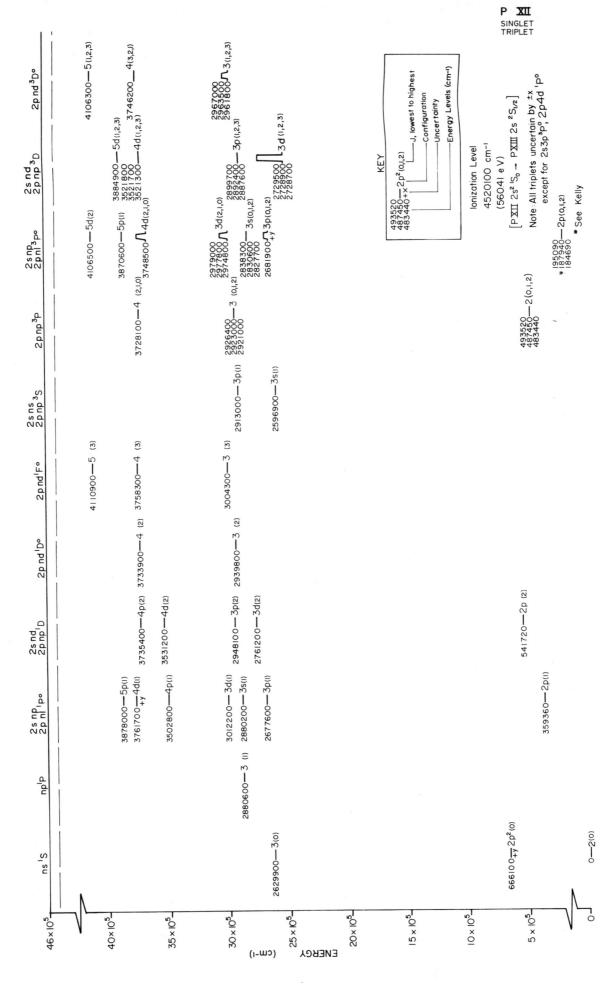

P XII ENERGY LEVELS (4 electrons, Z=15)

(Be I sequence, Configuration: 1s² nl nl'', Singlet & Triplet System)

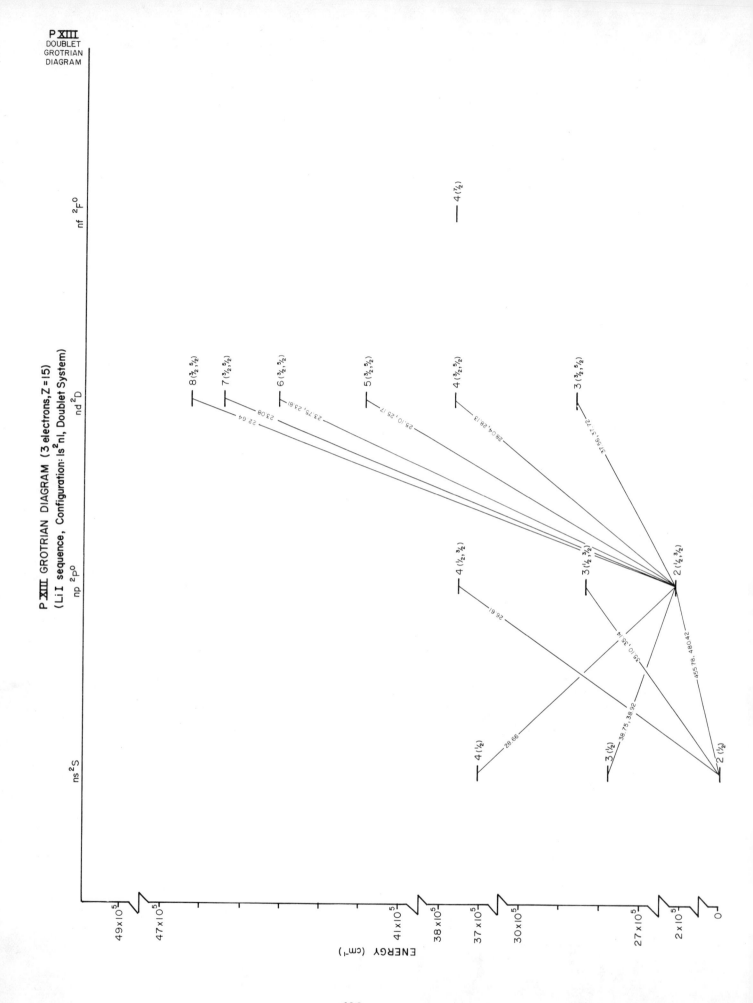

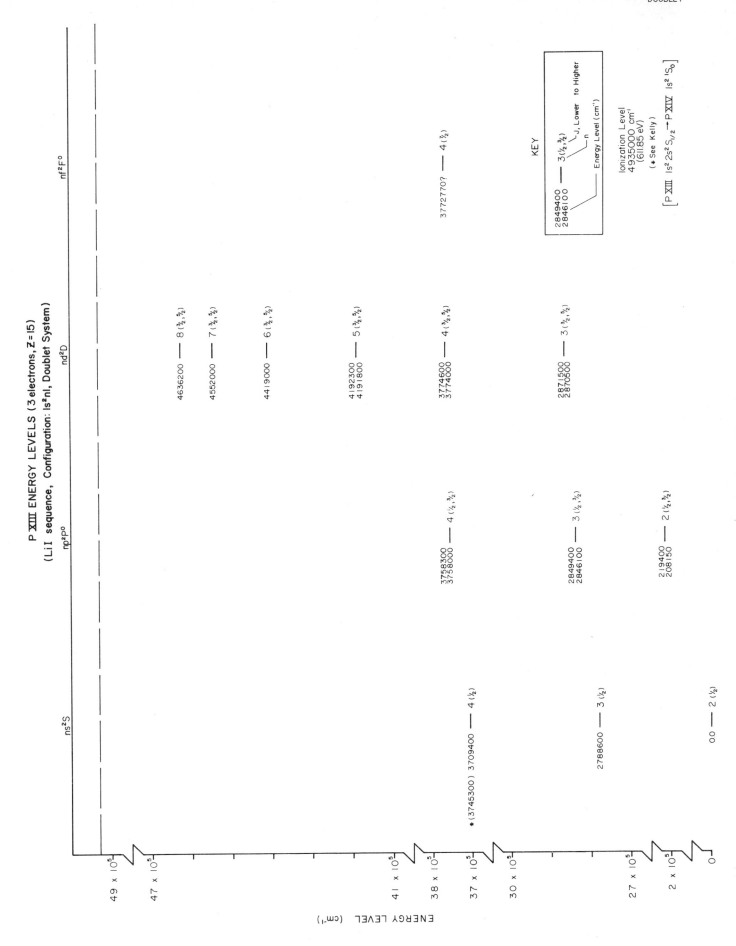

P XIII ENERGY LEVELS (3 electrons, Z=15)
(LiI sequence, Configuration: 1s²nl, Doublet System)

P XIII
DOUBLET

609

P **XIV** GROTRIAN DIAGRAM (2 electrons, Z=15)
(He I sequence, Configuration: ls nl, Singlet & Triplet Systems)

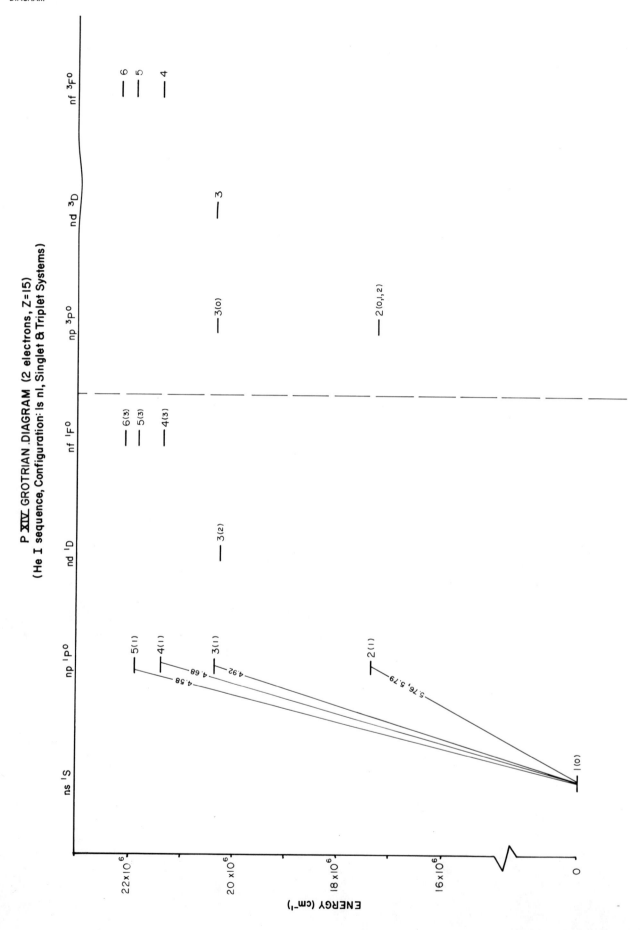

019

P XIV ENERGY LEVELS (2 electrons, Z=15)
(He I sequence, Configuration: 1s nl, Singlet & Triplet Systems)

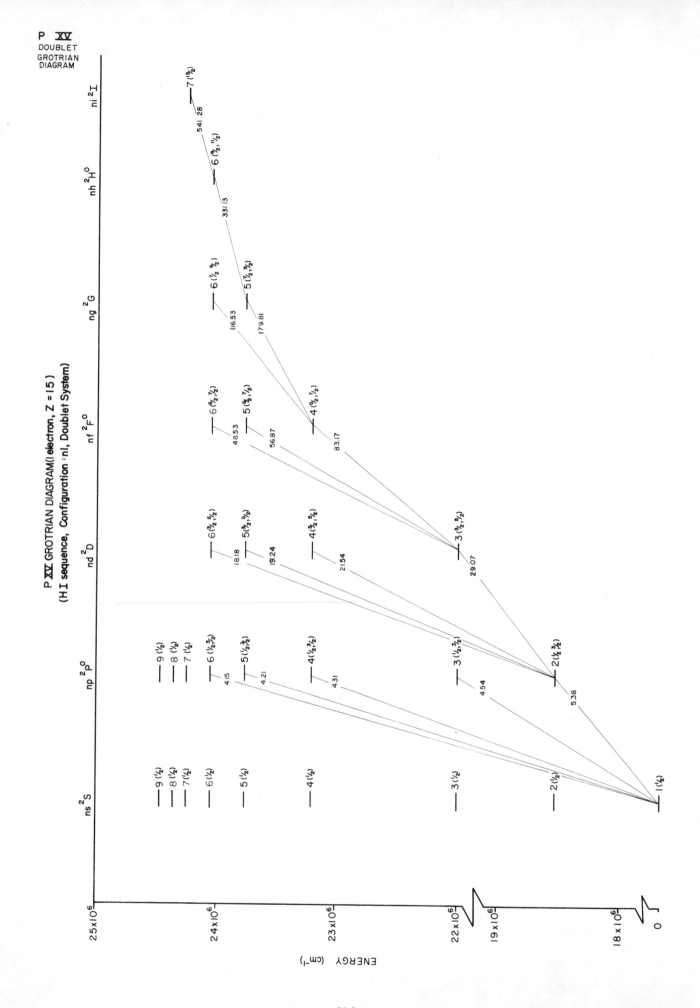

P XV GROTRIAN DIAGRAM(I electron, Z =15)
(H I sequence, Configuration : nl, Doublet System)

P **XV** ENERGY LEVELS (1 electron, Z = 15)
(H I sequence, Configuration : nl, Doublet System)

ENERGY (cm⁻¹)

ns ²S | np ²P° | nd ²D | nf ²F° | ng ²G | nh ²H° | ni ²I

ni ²I
24256025 —— 7 $(^{13}/_2)$

nh ²H°
24074039. —— 6 $(^9/_2, ^{11}/_2)$
24073994.

ng ²G
24073994 —— 6 $(^7/_2, ^9/_2)$
24073925.
23772206. —— 5 $(^7/_2, ^9/_2)$
23772087.

nf ²F°
24073925. —— 6 $(^5/_2, ^7/_2)$
24073811.
23772088. —— 5 $(^5/_2, ^7/_2)$
23771890.
23216500. —— 4 $(^5/_2, ^7/_2)$
23216114.

nd ²D
24073811. —— 6 $(^3/_2, ^5/_2)$
24073582.
23771890. —— 5 $(^3/_2, ^5/_2)$
23771494.
23216115. —— 4 $(^3/_2, ^5/_2)$
23215341.
22015647. —— 3 $(^3/_2, ^5/_2)$
22013813.

np ²P°
24454748. —— 9 $(^1/_2)$
24373628. —— 8 $(^1/_2)$
24255282. —— 7 $(^1/_2)$
24073583. —— 6 $(^1/_2, ^3/_2)$
24072892.
23771496. —— 5 $(^1/_2, ^3/_2)$
23770302.
23215346. —— 4 $(^1/_2, ^3/_2)$
23213012.
22013823. —— 3 $(^1/_2, ^3/_2)$
22008290.
18582723. —— 2 $(^1/_2, ^3/_2)$
18564054.

ns ²S
24454756. —— 9 $(^1/_2)$
24373639. —— 8 $(^1/_2)$
24255298. —— 7 $(^1/_2)$
24072917. —— 6 $(^1/_2)$
23770346. —— 5 $(^1/_2)$
23213098. —— 4 $(^1/_2)$
22008492. —— 3 $(^1/_2)$
18564729. —— 2 $(^1/_2)$
0 —— 1 $(^1/_2)$

KEY

22015647. —— 3 $(^3/_2, ^5/_2)$
22013813. ← J, lower to higher
← n
← Energy Level (cm⁻¹)

Ionization Level
24759943. cm⁻¹
(3069.76 eV)
[P **XV** 1s ²S$_{1/2}$ —→ nucleus]

ENERGY AXIS:
25×10⁶
24×10⁶
23×10⁶
22×10⁶
19×10⁶
18×10⁶
0

ENERGY (cm⁻¹)

613

P I $Z = 15$ 15 electrons

I.S. Bowen, Ap. J. **132**, 1 (1960).

    Author gives line tables of observed forbidden transitions.

P II $Z = 15$ 14 electrons

I.S. Bowen, Ap. J. **132**, 1 (1960).

    Author gives line tables of observed forbidden transitions.

H. Li, J. Opt. Soc. Amer. **62**, 1483 (1972).

    Line and energy-level tables for lines from 2285 to 7840 Å.

P III $Z = 15$ 13 electrons

    Please see the general references.

P IV $Z = 15$ 12 electrons

    Please see the general references.

P V $Z = 15$ 11 electrons

    Please see the general references.

P VI $Z = 15$ 10 electrons

    Please see the general references.

P VII $Z = 15$ 9 electrons

    Please see the general references.

P VIII $Z = 15$ 8 electrons

    Please see the general references.

P IX $Z = 15$ 7 electrons

    Please see the general references.